Industrial Applications of Nanoparticles
A Prospective Overview

T0313080

Editors

Marta I. Litter

School of Habitat and Sustainability
3IA (CONICET-UNSAM)
Universidad Nacional de General San Martín
Campus Miguelete, Av. 25 de Mayo y Francia
1650 San Martín, Provincia de Buenos Aires, Argentina

Arslan Ahmad

Director Technology and Innovation
Sibelco Ankerpoort NV
Op de Bos 300, NL-6223 EP Maastricht
The Netherlands

Affiliate Faculty Professor
Department of Sustainable Development
Environmental Science and Engineering (SEED)
KTH Royal Institute of Techology
Teknikingen 10B, Stockholm, SE-100 44
Sweden

CRC Press
Taylor & Francis Group
Boca Raton London New York

CRC Press is an imprint of the
Taylor & Francis Group, an **informa** business

A SCIENCE PUBLISHERS BOOK

Cover credit: Photo courtesy of KWR, photographer Ivar Pel.

First edition published 2023
by CRC Press
6000 Broken Sound Parkway NW, Suite 300, Boca Raton, FL 33487-2742

and by CRC Press
4 Park Square, Milton Park, Abingdon, Oxon, OX14 4RN

Library of Congress Cataloging-in-Publication Data (applied for)

ISBN: 978-1-032-02476-9 (hbk)
ISBN: 978-1-032-02477-6 (pbk)
ISBN: 978-1-003-18352-5 (ebk)

DOI: 10.1201/9781003183525

Typeset in Times New Roman
by Radiant Productions

Foreword

Nanoparticles are unique - due to their nano-specific properties such as small size and specific surface area which differentiate them from their corresponding bulk materials. Today, the use of nanoparticles is prevalent in many industries and new applications are continuously being explored by academics and industrial researchers at an ever-increasing rate. Despite significant developments in nanoparticle science and application technology, the environmental fate and the interaction of nanoparticles with their biological surroundings should be better understood. It is a collective responsibility of the scientific community, industries and government authorities to develop new best practices, guidelines and binding regulations to ensure the safe production and use of nanomaterials in different industrial applications. Prof. Dr. Marta I. Litter and Dr. Ir. Arslan Ahmad have delivered a comprehensive compilation of the recent developments in nanomaterials science in their new book, *Industrial Applications of Nanoparticles: A Prospective Overview*. This book is certainly a key enabler for future nanotechnology development and related growth in the industrial sector.

Petri Mast

Vice President Technology & Innovation
SCR-Sibelco

Preface

Nanotechnology is one of the most rapidly developing areas of science, offering great potential to solve the current and future developmental challenges in a wide range of industries such as aerospace, agriculture, bioengineering, cosmetics, chemicals, electronics, energy, renewables, surface coatings, textiles, medicine, materials manufacturing, military equipment and beyond. To compile this book, **Industrial Applications of Nanoparticles: A Prospective Overview**, we reached out to distinguished scientists, engineers, and industrial professionals in different parts of the world and invited contributions on topics related to nanoparticle synthesis, characterization, and industrial applications. After an extensive editorial screening and peer-review process, we have been able to present an array of 17 high-quality science-based chapters which cover the recent advancements, challenges and future trends in industrial applications of Nanotechnology.

In Chapter 1, we provide a broad overview of the world of nanotechnology, covering the major historical developments, preparation methods and applications of nanomaterials along with our perspective on the future of Nanotechnology. In Chapter 2, Mukherjee and Pradeep discuss nanomaterial-based technologies for clean water production and their sustainability aspects. In Chapter 3, Britto et al. discuss the application of nanomaterials from bench to industrial-scale along with the critical challenges in translating nanomaterials synthesis into actual products in the context of Argentina. Chapter 4, by Ocampo et al., covers an in-depth discussion on the application of Zerovalent Iron Nanoparticles (nZVI) in industrial wastewater treatment. Chapter 5, by Meichtry et al., is focused on the synthesis of green nanoparticles from plant extracts and their application in a wide range of fields, such as medicine, biosensors, pollutants treatment, agriculture, catalysis and energy storage.

Chapters 6 (Hernández-Ramírez et al.), 7 (Singhal et al.) and 8 (Antonopoulou and Hiskia) discuss photocatalytic nanomaterials, e.g., titanium dioxide (TiO_2), and technologies for environmental applications such as contaminants removal from air, e.g., the NO_x and CO_2 removal, and from (waste)water, e.g., the removal of pathogenic microorganisms. The issues related to the production and emission of nano- and micro-plastics (NPs and MPs) in water, as well as analytical techniques, are covered by Hofman-Caris and Bäuerlein in Chapter 9. The different strategies for the modification of bentonites to increase their versatility and capacity as pollutant adsorbents in water are discussed by Bracco et al. in Chapter 10.

In Chapter 11 the potential of Nanoscale Silica to produce transparent open cellular structures for industrial application is discussed by Bucharsky et al. In Chapter 12, van Genuchten provides a review of the key structural properties of ferrihydrite and birnessite nanoparticles and their application for arsenic (As), antimony (Sb), cadmium (Cd) and lead (Pb) removal from water. Chapter 13 by Martin et al. discusses the different aspects of titanium dioxide photocatalysts doped with cerium ions, such as synthesis, characterization, optical properties, stability, recovery, and performance assessment. In Chapter 14, Vera et al. discuss the advances in nanostructured TiO_2 coatings for application in energy, construction, water treatment, biomedical devices, food additives and food packaging industries. Chapter 15 by Kleiman et al. discusses the nanostructured films by physical vapor deposition for photocatalytic applications. In Chapter 16, Wang et al. provide a view of the recent advances in nanomaterial-based fluorescent development of latent fingerprints from the aspect of developing materials and developing the methods.

Finally, in Chapter 17, Ramanayaka et al. present recent developments in the synthesis of Nanoparticles Assisted Personal Respiratory Protection Equipment and cover issues concerning protection from bioaerosols, different types of nanoparticles used, the real scale production and the challenges to overcome in the current and future production of the face masks.

We are convinced that the knowledge presented in this book is of great value for students and professionals in academia, industry, governments, social services and NGOs. We are always open to receiving feedback and improving further.

29 May 2022   Marta I. Litter
                                                                                                         Arslan Ahmad

Contents

Chapter 1
The World of Nanotechnology

Marta I. Litter[1],* and *Arslan Ahmad*[2]

1. Introduction

Nanotechnology is one of the most rapidly evolving fields in the 21st century. There is not a single strictly established definition of the term nanotechnology, as it covers a wide spectrum of technologies, based on different types of physical, chemical and biological processes (Tolochko 2009), and because of continuously increasing developments. In a simple way, it can be defined as the ability to convert the nanoscience theory to useful applications by observing, measuring, manipulating, assembling, controlling and manufacturing matter at the nanometer scale with atomic precision (Mansoori 2005, Mansoori and Soelaiman 2005). The United States National Nanotechnology Initiative (NNI, Roco 2011) defines nanotechnology as "a science, engineering, and technology conducted at the nanoscale (1 to 100 nm), where unique phenomena enable novel applications in a wide range of fields, from chemistry, physics, and biology, to medicine, engineering, and electronics". Sometimes sizes in the 1000 nm range are considered nanomaterials (Leon et al. 2020). This 1–100 nm size range can correspond to individual molecules for polymers or other macromolecules but can also include higher-order assemblies into nanoparticles. For smaller, angstrom-sized atoms and molecules, this size range consists of small clusters or nanoparticles (Leon et al. 2020, Roco 2011).

The Greek prefix 'nano' means 'something very small' and represents one thousand millionths of a meter (10^{-9} m = 1 nm) (Bayda et al. 2020). The US National Science and Technology Council states that nanoscience and nanotechnology deal with very small sized objects and systems having the following key properties:

1) Dimension: at least one dimension from 1 to 100 nanometers (nm)
2) Process: designed with methodologies showing fundamental control over the physical and chemical attributes of molecular-scale structures
3) Building block property: they can be combined to form larger structures

Nanotechnology allows to create nanomaterials and apply them for special purposes, finding high-tech solutions to a wide range of problems in different fields. For example, it is envisioned that it will radically change the practice of medicine, improving medical devices and drug delivery

[1] School of Habitat and Sustainability, 3IA (CONICET-UNSAM), Universidad Nacional de General San Martín, Campus Miguelete, Av. 25 de Mayo y Francia, 1650 San Martín, Provincia de Buenos Aires, Argentina.
[2] Director Technology and Innovation, Sibelco Ankerpoort NV, Op de Bos 300, NL-6223 EP Maastricht, The Netherlands; Affiliate Faculty Professor, Department of Sustainable Development, Environmental Science and Engineering (SEED), KTH Royal Institute of Technology, Teknikingen 10B, Stockholm, SE-100 44, Sweden.
Email: arslan.ahmad@sibelco.com
* Corresponding author: martalitter24@gmail.com; mlitter@unsam.edu.ar

allowing fully personalized diagnosis and treatment. It will give rise to green and sustainable energy generation and storage, to ensure access to safe water supply through nanotech filtration systems, and to dramatically improve homeland security.

It is important to distinguish between nanoscience and nanotechnology (Mulvaney 2015). Nanoscience is the study of structures and molecules on the nanometer scale (1–100 nm), while nanotechnology utilizes these concepts in practical applications (Mansoori and Soelaiman 2005). Nanoscience is quite natural to microbiological sciences because the sizes of many bioentities (DNA, RNA, proteins, enzymes, viruses, etc.) fall within the nanometer range.

Materials at the nanoscale have chemical, electrical, magnetic, mechanical, optical and other properties considerably different from bulk materials. Some of these properties are intermediate between those of the smallest elements (atoms and molecules) of which they can be composed and those of the macroscopic materials. The properties of nanoparticles are enhanced compared to bulk materials. Nanoscale has become very important due to many specific reasons, including the following (Mansoori and Soelaiman 2005, Zhang 2003):

a) Nanoscale influences the quantum mechanical (wavelike) properties of electrons composing the matter. By nanoscale design of materials, it is possible to vary their micro- and macroscopic properties, such as charge capacity, magnetization and melting temperature, without changing the chemical composition of the materials.

b) Systematic organization of matter on the nanoscale is a key feature of biological entities, allowing to place man-made nanoscale entities inside living cells and to make new materials using the self-assembly features of nature.

c) Nanoscale components have a very high surface to volume ratio. For this reason, they are ideal for use in composite materials, reacting systems, drug delivery, chemical energy storage (e.g., hydrogen and natural gas) and significantly improve reactivity to destroy contaminants.

d) Macroscopic systems made up of nanostructures can have a much higher density than those made up of microstructures and therefore can be better conductors of electricity.

2. History

Here the major developments of the history of nanotechnology are presented, focusing on only the main discoveries without a thorough description.

2.1 The early times

For thousands of years BC, people knew and used natural fabrics such as cotton, wool, silk, flax, etc., which possessed a network of pores of 1–20 nm, i.e., they were typical nanoporous materials. These natural fabrics present high useful properties: they absorb sweat, quickly swell and dry. The ancient manufacturing of bread, wine, beer, cheese and other foodstuffs included fermentation processes at the nanolevel. The dying of hair in black in ancient Egypt was made with a paste composed of lime, lead oxide and water, which reacted with sulfur of keratin, producing nanoparticles of galenite (lead sulfide) (Lewis 2017).

The development of nanoscience started in the time of Democritus (5th century BC), when scientists described if matter was continuous, and thus infinitely divisible into smaller pieces or composed of small, indivisible and indestructible particles (atoms).

In 4th century AD, the Lycurgus cup (now in the British Museum, Fig. 1.1) was made by Romans, being one of the most outstanding achievements in glass manufacturing (Barber and Freestone 1990, Bayda et al. 2020, Lewis 2017, The British Museum).

This cup is the oldest example of a dichroic glass, composed of two different types of glass that change color depending on the lighting conditions: under natural light, the bowl is green, if illuminated from within, it turns red. In 1959, General Electric experts analyzed fragments of

Figure 1.1. The Lycurgus cup. Reproduced with permission from British Museum (The British Museum. Available online: www.britishmuseum.org/research/collection_online/collection).

the bowl, and showed that it consisted of the usual soda-lime-quartz glass with about 1% of gold (Au) and silver (Ag) and 0.5% of manganese (Mn). The phenomenon of dichroism of the cup was explained in 1990 using Transmission Electron Microscopy (TEM) and roentgenograms (Freestone et al. 2007), and it was concluded that it was due to the presence of Au and Ag nanoparticles of 50–100 nm in diameter. X-ray analysis showed that these nanoparticles were an Ag-Au alloy, containing about 10% copper dispersed in a glass matrix. The Au nanoparticles produce a red color as the result of light absorption ($\lambda \sim 520$ nm); the red-purple color is due to the absorption by the bigger particles, while the green color is attributed to the light scattering by colloidal dispersions of Ag nanoparticles (size > 40 nm). Later, Atwater (2007) explained this phenomenon by the effects of plasmon excitation of electrons with metal nanoparticles.

Glowing, glittering "luster" ceramic glazes used in the Islamic world, and later in Europe, developed during the 9th–17th centuries AD, widely contained Ag, Cu and other metallic nanoparticles.

The multi-colored church stained glass windows made during the Middle Age in Europe (6th–15th centuries) contain additives of Au and nanoparticles of other metals and metal oxides. A similar effect as that of the Lycurgus cup can be seen, shining red and yellow colors due to the fusion of Au and Ag nanoparticles into the glass (Bayda et al. 2020, Sciau 2012).

During the 13th–18th centuries AD, Ottomans produced "Damascus" saber blades made of Damask steel from an ore of a special structure, with extraordinary strength; this steel could never be replicated. Paufler et al. (2006) analyzed fragments from the steel using an electronic microscope, finding a nanofibrous structure, formed probably by a special thermomechanical processing. The material contains cementite nanowires and carbon nanotubes (CNTs, see below), which provided strength, resilience, a keen edge and a visible moiré pattern in the steel (Reibold et al. 2006). In the Renaissance (16th century), the Italians made pottery using nanoparticles influenced by those Ottoman techniques.

In 1857, Michael Faraday studied the preparation and properties of colloidal suspensions of "ruby" gold nanoparticles of unique optical and electronic properties, demonstrating that nanostructured Au under certain lighting conditions produced colored materials (Faraday 1857). In 1908, Mie (1908) suggested that the optical properties of Au at small length scales were significantly different from its bulk properties.

2.2 The 20th century

The concept of "nanometer" was first proposed in the early 20th century by the 1925 Nobel Laureate Zsigmondy (1914), who made a detailed study of Au sols and other nanomaterials with sizes down to 10 nm using an ultramicroscope capable of visualizing particles much smaller than the light wavelength.

The first mention of nanotechnology is usually connected with the well-known lecture of Prof. Feynman (Californian Institute of Technology, 1965 Nobel Prize), titled "There's plenty of Room at the bottom (Feynman 1960)", delivered in 1959 at the American Physical Society. Feynman introduced the possibility of the manipulation of matter at the atomic level with microscopic machines able to build complex materials by placing individual atoms in designed arrangements. Thus, nanosized products could be created using atoms as building particles (Tolochko 2009). Feynman stated in his lecture that, e.g., the entire Encyclopedia Britannica could be put on the tip of a needle. He pointed out that, at that time, there was a lack of the appropriate equipment and techniques to work at the molecular level. However, the discussion was considered speculative and lacked importance until it was republished in the 1990s. After Feynman's great contribution, it has been proposed to name the nanometer scale as "the Feynman (φnman) scale", and the notation "φ," for nanometers, i.e., one Feynman (φ) $\equiv$ 1 nanometer (nm) = 10 angstroms (Å) = 10^{-3} microns (μm) = 10^{-9} meter (m).

In the early 1960s, a study on small systems with a different scope was addressed, with the publication of the book "Thermodynamics of Small Systems", followed by a paper on nanothermodynamics (Hill 1963, 2001).

Taniguchi et al., from the Tokyo Science University (1974) were the first to use and define the term "nanotechnology" at the 1974 International Conference on Industrial Production in Japan, describing semiconductor processes that occurred at nanometric scale.

In 1981, Ekimov discovered nanocrystalline, semiconducting quantum dots in a glass matrix and conducted pioneering studies of their electronic and optical properties (Ekimov and Onushchenko 1982). In 1985, Brus (from Bell Laboratories) discovered colloidal semiconductor nanocrystals (quantum dots) (Douglass and Brus 1988), sharing the 2008 Kavli Prize in Nanotechnology with Iijima (see below). In 1993, Bawendi (from the Massachusetts Institute of Technology, MIT) invented a method for the controlled synthesis of quantum dots (Murray and Bawendi 1993).

In 1985, Binnig and Rohrer (IBM Research Laboratory in Zürich, 1986 Nobel Prize) discovered the Scanning Tunneling Microscope (STM) (Binnig and Rohrer 1985, 1987), an equipment able to create direct spatial images of individual atoms. This was followed by the development of the Atomic Force Microscope (AFM) and related techniques that may be classified in the general category of Scanning Probe Microscopy (SPM) techniques. The most important is AFM, developed by Binnig, Quate and Gerber (1986), having the ability to view, measure and manipulate materials to nanometric size fractions, together with the measurement of forces intrinsic to nanomaterials. Feynman's 1959 challenge for miniaturization and his precisely accurate forecast was proved by Eigler and Schweizer (1990) by positioning single atoms using STM, and they manipulated 35 individual xenon atoms spelling the IBM logo (Fig. 1.2).

Drexler (from MIT) published the first book on nanotechnology, titled "Engines of Creation: The Coming Era of Nanotechnology" (Drexler 1986). Based on Feynman's ideas and Taniguchi's term, he proposed the idea of a nanoscale "assembler" that would be able to build a copy of itself and of other more complex items. Drexler's vision of nanotechnology is often called "molecular nanotechnology" or "molecular engineering", by describing the build-up of complex machines from individual atoms, which can independently manipulate molecules and atoms producing self-assembled nanostructures. His vision focused profoundly on biological mechanisms such as protein design. Later, Drexler et al. (1991) published another book titled "Unbounding the Future: the Nanotechnology Revolution", in which they use the terms "nanobots" or "assemblers" for nanoprocesses in medicine, using the word "nanomedicine" for the first time.

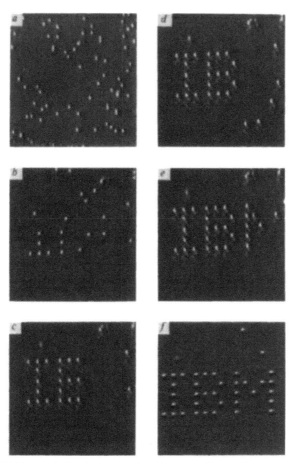

Figure 1.2. Sequence of STM images during the construction of a patterned array of xenon atoms on a nickel surface. The IBM logo is shown. From Eigler and Schweizer 1990 with permission.

Figure 1.3. Scheme of a C_{60} buckyball (fullerene) (extracted from https://bcachemistry.wordpress.com/2014/06/01/balls-from-fury-buckminsterfullerene-c60/).

In 1985, Kroto et al. (from Rice University) (Kroto et al. 1985) discovered the third allotrope of carbon, buckminsterfullerene (fullerene or buckyballs), which formed very stable spheres (Fig. 1.3).

For this discovery, Curl, Kroto, Smalley and Curl received the 1996 Nobel Prize. Later, fullerenes with a larger number of carbon atoms (C_{76}, C_{80}, C_{240}, etc.) were discovered. Researchers from the Rice University built a nanoscale car (https://www.testandmeasurementtips.com/the-instrumentation-of-nanotechnology-faq/) made of oligo(phenylene ethynylene) with alkynyl axles and four spherical C_{60} buckyball wheels. Increasing the temperature, the buckyball wheels moved the nanocar on a gold surface, as in a conventional car.

In 1991, (NEC Corporation) used TEM to observe the hollow graphitic tubes or CNTs belonging to the fullerene family (Iijima and Ichihashi 1993). For these advances, Iijima shared with Brus the Kavli Prize in Nanoscience in 2008. CNTs, like buckyballs, are entirely composed of carbon, but in a tubular shape, exhibiting excellent strength, electrical and thermal conductivity properties. Single-walled carbon nanotubes (SWCNTs) behave like metallic, semimetallic or semiconductive one-dimensional materials, with a longitudinal thermal conductivity that exceeds the thermal conductivity of graphite (Xu et al. 2004), a similar electrical conductivity as copper and thermal conductivity as diamond (Mansoori and Soelaiman 2005). Extremely high tensile strength (~ 100 times that of steel) of ropes made of SWCNTs have been obtained.

2.3 The present times: nanomaterials, carbon dots, nanotubes

In 2000, commercial nanomaterials such as titanium dioxide (TiO_2) and zinc oxide (ZnO) nanoparticles began to be used in sunscreens, photocatalysis, cosmetics and some food products, silver nanoparticles in food packaging, clothing, disinfectants and household appliances, CNTs for stain-resistant textiles and cerium oxide as a fuel catalyst (Nanotechnology Information Center, accessed 2020).

Kresge et al. 1992 (from Mobil Oil) discovered the nanostructured catalytic materials MCM-41 and MCM-48, used largely in refining crude oil, drug delivery, water treatment and other varied applications.

In 2004, carbon dots (C-dots) with size below 10 nm were discovered by Xu et al. (2004) during the purification of SWCNTs. C-dots have benign, an abundant and inexpensive nature, low toxicity and good biocompatibility, together with excellent thermal, optical, mechanical and electronic properties.

Graphene was isolated and characterized in 2004 at the University of Manchester (Novoselov et al. 2004) by pulling graphene layers from graphite with a common adhesive tape (Scotch tape technique). Novoselov and Geim received the 2010 Nobel Prize for these findings. The material exhibits different exceptional quantum mechanical, electrical, chemical, mechanical, optical, magnetic and other properties.

2.4 Governmental, academic and commercial initiatives

Between 2001 and 2004, more than 60 countries initiated their nanotechnology research and development (R&D) programs. Government funding increased from corporations of the USA, Japan and Germany. The most important intellectual patents on R&D nanotechnology were registered between 1970 and 2011 by five private companies: Samsung Electronics (2,578 patents), Nippon Steel (1,490 patents), IBM (1,360 patents), Toshiba (1,298 patents) and Canon (1,162 patents).

The majority of scientific research papers on nanotechnology were published between 1970 and 2012 by the Chinese Academy of Sciences, the Russian Academy of Sciences, the *Centre National de la Récherche Scientifique*, the University of Tokyo and the Osaka University (World Intellectual Property Report 2015). In 1991, the first nanotechnological program of National Scientific Fund started to operate in the USA. During 1996–98, a special committee of the American Center for Global Technology Assessment analyzed the development of nanotechnology in all countries (Tolochko 2009). From 1999 until now, various agencies have been formed to support developments in nanotechnology and nanoscience in the USA, Canada, Japan, Germany, England, France, China, South Korea and the Commonwealth of Independent States (Armenia, Azerbaijan, Belarus, Kazakhstan, Kyrgyzstan, Moldova, Russia, Tajikistan and Uzbekistan) (Tolochko 2009).

3. Classification and Preparation of Nanoparticles

3.1 Classification

Nanomaterials can be classified into several major groups based on their physicochemical composition, including organic-based, carbon-based, metal-based, semiconductor-based, organic-inorganic hybrid, silica based, ceramic, polymeric, biological and self-assembly/biologically directed nanomaterials (Sitharaman 2011).

Organic nanoparticles comprise dendrimers, micelles, liposomes, ferritin, etc., and are biodegradable and non-toxic materials (Fig. 1.4). They are most widely used in the biomedical field, as it will be described later (Ealia and Saravanakumar 2017).

The most common inorganic nanoparticles are metal and metal oxide based nanoparticles. Metal-based nanoparticles are synthesized from metals to nanometric sizes either by destructive or constructive methods. The commonly used metals are Al, Cd, Co, Cu, Au, Fe, Pb, Ag and Zn, while the most common metal oxides are Al_2O_3, CeO_2, Fe_2O_3, Fe_3O_4, SiO_2, TiO_2, ZnO.

The carbon-based nanoparticles are completely composed of carbon: fullerenes, graphene, CNTs, carbon nanofibers and carbon black (Fig. 1.5).

Fullerenes (C_{60}) are spherical carbon molecules formed from about 28 to 1500 carbon atoms, maintained together by sp^2 hybridization. They have diameters up to 8.2 nm for a single layer and

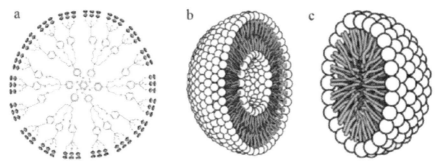

Figure 1.4. Organic nanoparticles: a – Dendrimers, b – Liposomes and c – micelles. Taken from Ealia and Saravanakumar 2017 with permission.

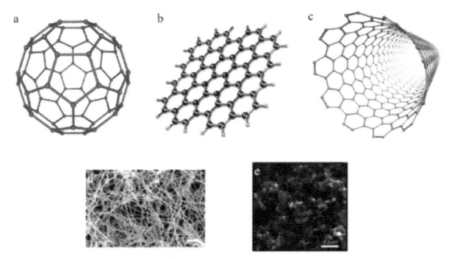

Figure 1.5. Carbon based nanoparticles: a – fullerenes, b – graphene, c – carbon nanotubes, d – carbon nanofibers and e – carbon black. Taken from Ealia and Saravanakumar 2017 with permission.

from 4 to 36 nm for multi-layered materials. The carbon balls with the chemical formula C_{60} or C_{70} are formed when graphite is evaporated by a laser in an inert atmosphere.

Graphene is an allotrope of carbon: it is a hexagonal network of honeycomb lattice made up of carbon atoms in a two dimensional planar surface. The thickness of the graphene sheet is around one nm.

CNTs are graphene nanofoils with a honeycomb lattice of carbon atoms wound into hollow cylinders to form nanotubes; diameters are as low as 0.7 nm for single layered and 100 nm for multi-layered CNTs, while the length varies from a few μm to several mm. The same graphene nanofoils are used to produce carbon nanofibers.

Carbon black is an amorphous material made up of carbon, generally spherical in shape with diameters from 20 to 70 nm. They bound in aggregates of around 500 nm.

3.2 Preparation

Numerous methods of synthesis are either being developed or improved to prepare nanoparticles (Ealia and Saravanakumar 2017, Hasan 2015). The methods of synthesis are categorized into bottom-up or top-down. A simplified representation of the process is presented in Fig. 1.6.

The bottom-up or constructive method is the build-up of a material from atoms to clusters and to nanoparticles. Sol-gel (Ramesh et al. 2016), spinning (Mohammadi et al. 2014, Tai et al. 2007), Chemical Vapor Deposition (CVD, Bhaviripudi et al. 2007), pyrolysis (Kammler et al. 2001) and biosynthesis (Kuppusamy et al. 2016) are the most commonly used bottom-up methods for nanoparticle production.

The top-down or destructive method is the reduction of a bulk material to nanometric scale particles by, e.g., mechanical milling (Yadav et al. 2012), nanolithography (Pimpin and Srituravanich 2012), laser ablation (Amendola and Meneghetti 2009), sputtering (Shah and Gavrin 2006) or thermal decomposition (Salavati-Niasari et al. 2008).

Green synthesis of nanoparticles from plant extracts will be described in Chapter 5.

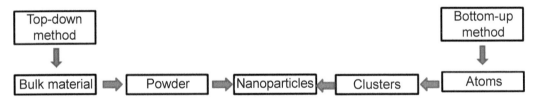

Figure 1.6. Scheme of top-down and bottom-up methods of synthesis of nanoparticles (modified from Ealia and Saravanakumar 2017).

4. Applications of Nanotechnology

4.1 Different areas of application

Nanomaterials have found application in multiple industries such as aerospace, agriculture, bioengineering, biology, energy, environment, materials, food, manufacturing, medicine, military equipment, etc. (Kolahalam et al. 2019). Nanotechnology is widely used in various fields, principally in wear and corrosive environments in the form of nanocomposite materials, nanostructured coatings (in hydroturbine blades, coal fired boilers, etc.), cutting tools, nanofluids (to increase heat transfer rates of fluid), building materials and automotive industries (Sharma et al. 2018).

The recent advancements in analysis tools such as TEM, X-Ray Diffraction (XRD), Fourier transmission infrared (FTIR), etc. is enabling researchers from various disciplines to develop new nanomaterials of different functionality. Presently, some of the active nanotechnology research areas include nanolithography, nanodevices, nanorobotics, nanocomputers, nanopowders, nanostructured catalysts and nanoporous materials, molecular manufacturing, diamondoids, CNTs, fullerene products, nanolayers, molecular nanotechnology, nanomedicine and nanobiology. Some of the advances in nanotechnology and nanoscience are the following (Mansoori and Soelaiman 2005):

a) wear-resistant tires made by combining nanoscale particles of inorganic clays with polymers or other nanoparticle reinforced materials;

b) improved printing by introducing nanoscale particles with good properties in dyes and pigments, as well as advanced inkjet systems;

c) a new generation of lasers, magnetic disk heads, nanolayers with selective optical barriers and chip systems made by controlling layer thickness to less than one nanometer;

d) design of advanced chemical and bio-detectors;

e) advanced drug delivery and drug targeting medicine;

f) chemical-mechanical polishing with nanoparticle slurries, hard coatings and high hardness cutting tools.

As described earlier, in 2000, commercial nanomaterials such as titanium dioxide and zinc oxide nanoparticles began to be used in sunscreens, photocatalysis, cosmetics and some food products, silver nanoparticles in food packaging, clothing, disinfectants and household appliances, CNTs in stain-resistant textiles and cerium oxide as a fuel catalyst (Nanotechnology Information Center).

It has been also been discussed that MCM-41 and MCM-48 (Kresge et al. 1992) are used in refining crude oil as well as for drug delivery, water treatment and other applications.

Carbon-based materials made great progress for computer science and bio-engineering (Bayda et al. 2020). In 2013, the first carbon nanotube computer was developed (Shulaker et al. 2013). C-dots are excellent materials for applications in bioimaging, biosensors and drug delivery, catalysis, energy conversion, photovoltaic devices and nanoprobes for sensitive ion detection (Xu et al. 2004).

Nanotechnology is an important tool for removal of several pollutants from wastewaters and soils, even with the possibility to obtain safe drinking water (Santhosh et al. 2016, Sharma et al. 2009, Tiwari et al. 2008, Tratnyek and Johnson 2006). Nanotechnology has been used to improve the environment and to produce more efficient and cost-effective energy, such as generating less pollution during the manufacture of materials, producing solar cells that generate electricity at a competitive cost, cleaning up organic chemicals polluting groundwater and cleaning Volatile Organic Compounds (VOCs) from air (Bayda et al. 2020).

Nanomaterials have been also used in photosciences, e.g., for constructing light emitting diodes, photovoltaic cells, organic solar cells, etc., in biomedicine and in photocatalysis (Mandal and Ganguly 2011). They are used as antimicrobials (Kim et al. 2007), and as antioxidants (Deligiannakis et al. 2012).

Nanoscience impacts several areas of microbiology as well. Nanomembranes are porous thin-layered materials that allow water molecules to pass through while restricting the passage of bacteria, viruses, salts and metals. Electrospun nanofibers and nanobiocides have shown potential in the improvement of water filtration membranes. Nanobiocides such as metal nanoparticles and engineered nanomaterials have been successfully incorporated into nanofibers showing high antimicrobial activity and stability in water (Botes and Cloete 2010).

Nanomaterials such as zinc oxide quantum dot nanoparticles are also important in the industries of food and cosmetics, improving production, packaging, shelf life, bioavailability, antimicrobial activity and food sensors to test food quality and safety (Mohammad et al. 2022).

Sensors for detection of pesticides, chemicals, heavy metals, food borne pathogens and toxins have also been developed (Sharma et al. 2021).

Use of nanoparticles is convenient for *in situ* applications. The use of nanometals for subsurface remediation of contaminated sites has received significant attention. Nanometals used for these purposes include nanoiron (nZVI) and nanozinc, but nZVI is the most commonly used and is highly effective for the transformation and detoxification of a wide variety of contaminants such as chlorinated organic solvents, pesticides, PCBs, arsenic and heavy metals, among others. In addition, other metals such as palladium or nickel are usually added to increase the reduction rate, forming a bimetallic nanoparticle.

The strength and flexibility of CNTs make them potentially useful in many nanotechnological applications. Currently, CNTs are used as composite fibers in polymers and beton to improve the mechanical, thermal and electrical properties of the bulk product. They also have potential applications as field emitters, energy storage materials, catalysis and molecular electronic components (Kumar et al. 2018).

4.2 Biological and medical advances

Today, nanotechnology has a large impact on human life, e.g., on pharmaceutical applications, medical devices and applications such as diagnostic biosensors, drug delivery systems, imaging probes, tissue engineering, biomaterials, detection of bacteria in bloodstreams, biosensors, bioimaging, antibacterial coatings of medical devices, reduced inflammation, better surgical tissue healing and detection of circulating cancer cells (Lee and Wong 2011, Kinnear et al. 2017).

The amalgamation of nanotechnology with biological fields leads to innovative solutions. Adamantane and diamondoids, which are saturated, polycyclic, cage-like hydrocarbons (Mansoori and Soelaiman 2005, Ramezani and Mansoori 2007) are good examples of molecular building blocks for nanotechnology, drug targeting and gene delivery. The smallest diamondoid molecule (adamantine) was first discovered and isolated from petroleum in 1933. It is generally accompanied by small amounts of alkylated adamantanes: 2-methyl-, 1-ethyl-, and probably 1-methyl-, 1,3-dimethyl, etc. The simplest of these polycyclic diamondoids is adamantane, followed by its homologous diamantine and tri-, tetra-, penta- and hexamantane. Adamantane presents highly unusual physical and chemical properties, having the same structure as the diamond lattice, i.e., highly symmetrical and strain free. The strength, toughness, stiffness rigidity, strength and an assortment of their 3-D shapes make them valuable molecular building blocks.

Bio-nanotechnology is extremely useful for applications in many biology related areas such as diagnosis, drug delivery, molecular imaging, regenerative medicine, antibacterial activities, biomarker detection (e.g., nanobiochips, nanoelectrodes or nanobiosensors) containing nanomaterials are currently on the market (Barreto et al. 2011, Lara and Perez-Potti 2018).

Organic nanoparticles are widely used in the biomedical field, for example as drug delivery systems that can be injected in specific parts of the body (targeted drug delivery) (Ealia and Saravanakumar 2017).

Biologists made advances with other nanoscale structures like cyclodextrins, liposomes and monoclonal antibodies, in many applications, including drug delivery and drug targeting. For biological systems, the nanometer size range is ideal for circulating in the bloodstream, traversing tissues and entering cells. Examples of nanomaterials are abundant in nature, ranging from magnetotactic bacteria that use iron oxide nanoparticles to sense magnetic fields to the unique colors of butterfly wings, which originate from the interactions of light with complex nanoarchitectures and not absorption due to dyes or pigments (Mansoori and Soelaiman 2005).

Cyclodextrins (cyclic oligosaccharides) form inclusion complexes with a wide range of substrates in aqueous solutions and can be applied for encapsulation of drugs in drug delivery. Liposomes are spherical synthetic lipid nanoscale vesicles, produced by dispersion of a phospholipid in aqueous salt solutions. Liposomes self-assemble, encapsulate materials and are used as carriers for small drug molecules, proteins, nucleotides and plasmids to tissues and into cells. A monoclonal antibody protein molecule consists of four protein chains, in a Y-shaped structure of about 10 nm in diameter. The small size of these particles allows intravenous administration, and they can penetrate small capillaries and reach cells in tissues for treatment. Nanostructures smaller than 20 nm can go out of blood vessels. Halas et al. (from Rice University, 2003) developed Au nanoshells, applied in discovery, diagnosis and treatment of breast cancer without chemotherapy, invasive or radiative procedures.

One of the most important applications of nanotechnology in molecular biology is related to nucleic acids. The conceptual foundation for DNA nanotechnology was first established by Seeman (2010), based on the "scaffolded DNA origami", developed in 2006 by Rothemund by enhancing the complexity and size of self-assembled DNA nanostructures in a "one-pot" reaction (Rothemund 2006). DNA nanotechnology has become an interdisciplinary area, gathering concepts from physics, chemistry, materials science, computer science and medicine. Notably, years of extensive studied have made possible to use DNA and other biopolymers directly in array technologies for sensing and diagnostic applications. Remarkable progress has also been made in nano-oncology by improving the efficacy of traditional chemotherapy drugs for large variety of aggressive human cancers (Lee and Wong 2011, Yuan et al. 2019). For this, the tumor site was targeted with several molecules including nanoparticles, antibodies and cytotoxic agents, delivering therapeutic molecules to modulate essential biological processes, like autophagy, metabolism or oxidative stress, exerting anticancer activity (Cordani and Somoza 2019), in addition to result in a significant reduction of the systemic toxicity associated with current chemotherapy treatments (Bayda 2020).

5. Future Outlook

Nanoscience and nanotechnology will have a major impact on several key scientific and technological activities in a not too distant future.

According to Hornigold (2018), a futuristic age of nanorobots that can manipulate matter on the molecular level, leading to 'nanofabrication' of many different products is envisaged. Other nanotech concepts include foglets, i.e., nanorobots that could reassemble themselves into any desired shape, including home furniture.

However, some time will be needed to conclude if nanotechnology is actually the great promise that many consider it to have, and how long this will take. On the other hand, there is a significant concern about the potential health and environmental risks, which are now faced by nanomedicine and nanotoxicology. Due to lack of reliable toxicity data, the potential to affect human health continues to be a major concern.

The differing visions of what nanotechnology is and can become, and the resulting conflicts over funding and over the public image of nanotechnology, are part of the story of the first 30 years of nanotechnology (Lewis 2017).

As an evolving field, the scientific advancements in the field are constantly growing up, including synthesis or discovery of different types of nanoparticles. Nanoparticle research has grown tremendously throughout the years, especially the proportion of research dedicated to the medical field.

The emerging fields of nanoscience and nanotechnology also create the necessary experimental and computational tools for the design and fabrication of nanodimensional electronic, photonic, biological and energy transfer components, such as quantum dots, atomic wires, operating on nanoscopic length scales, etc. Also, the "nano" nomenclature is being constantly updated.

Technological advances in instruments and tools of fabrication and manipulation in nano scale are presently expensive, and as a result, not available to many investigators.

Any development in molecular-based techniques for the study of matter in the fields of nanoscience and nanotechnology will be entirely related to advances in these techniques and will help to understand, simulate, predict and formulate new materials using concepts of quantum and statistical mechanics, intermolecular interaction, molecular simulation and molecular modeling. In this way, the design of new molecular building blocks, allowing self-assembly or self-replication to advance the bottom-up approach to produce the necessary materials for the advancement of nanotechnology will be possible.

The innovative biomedical applications that are presently used in a variety of clinical trials and, in the near future, may support major development in the therapy of cancer (Bayda et al. 2020).

Some selected observations regarding the expected future advances are the following:

a) Nanotechnology could provide a new dimension to the control and improvement of living organisms.

b) Nanotechnology will improve microelectronics and materials design.

c) Biotechnology will be greatly improved by nanotechnology.

d) A modification or change of properties of matter at macro- and microscales, never before identified in nature, will be possible.

e) Nanoscale traps able to remove pollutants from the environment and deactivate chemical warfare agents will be constructed.

f) Inks and dyes, protective coatings, dispersions with optoelectronic properties, nanostructured catalysts, high reactivity reagents, medicine, electronics, structural materials, diamondoids, carbon nanotube and other materials will be also improved.

Undoubtedly, in the near future, nanotechnology will have a direct impact on our lives, independent of our location, career or social position.

6. Conclusions

In only a half century, nanotechnology has become the basis for important industrial applications and is growing exponentially. Presently, nanotechnology and nanoscience are applied in activities concerning physical, chemical, biological and mathematical sciences.

In only a few decades, nanotechnology and nanoscience have become of fundamental importance to industrial applications and medical devices, such as diagnostic biosensors, drug delivery systems and imaging probes. In the food industry, nanomaterials have been exploited to increase drastically the production, packaging, shelf life and bioavailability of nutrients. Some nanostructures display antimicrobial activity against food-borne bacteria, and many different nanomaterials are used as food sensors to detect food quality and safety these days. Nanomaterials are being used to build a new generation of solar cells, hydrogen fuel cells and hydrogen storage systems capable of delivering clean energy.

The most significant advances in nanotechnology are in the broad field of biomedicine and especially in cancer therapeutics with great advantages compared with traditional chemotherapy and radiotherapy.

Finally, it should be highlighted that although nanotechnology is proving beneficial in many fields, there are certain critical environmental issues that need to be addressed. There is still a lack of clear understanding of the (long-term) fate of nanoparticles in the environment. Long-term risk assessment is also crucial when applying nanotechnology to natural environmental systems. In addition, the costs associated with the large-scale production of nanoparticles may hinder their widespread application. For achieving cost efficiency, reuse and recycling of nanomaterials should be considered. However, more studies should be carried out for safety evaluation, sustainable reuse and eventual disposal practices.

References

Amendola, V. and M. Meneghettia. 2009. Laser ablation synthesis in solution and size manipulation of noble metal nanoparticles. Phys. Chem. Chem. Phys. 11: 3805–3821.

Atwater, H.A. 2007. The Promise of Plasmonics. SA Special Editions. 17: 56–63.

Barber, D.J. and I.C. Freestone. 1990. An investigation of the origin of the colour of the Lycurgus Cup by analytical transmission electron microscopy. Archaeometry 32: 33–45.

Barreto, J.A., W. O'Malley, M. Kubeil, B. Graham, H. Stephan and L. Spiccia. 2011. Nanomaterials: applications in cancer imaging and therapy. Adv. Mat. 23: 18–40.

Bayda, S., M. Adeel, T. Tuccinardi, M. Cordani and F. Rizzolio. 2020. The history of nanoscience and nanotechnology: from chemical–physical applications to nanomedicine. Molecules 25: 112.

Bhaviripudi, S., E. Mile, S.A. Steiner, A.T. Zare, M.S. Dresselhaus, A.M. Belcher and J. Kong. 2007. CVD synthesis of single-walled carbon nanotubes from gold nanoparticle catalysts. J. Am. Chem. Soc. 129: 1516–1517.

Binnig, G. and H. Rohrer. 1985. The scanning tunneling microscope. Scientific American. 253.

Binnig, G., C.F. Quate and C. Gerber. 1986. Invention of Atomic Force Microscope (AFM). Phys. Rev. Let. 56: 930–933.

Binnig, G. and H. Rohrer. 1987. Scanning tunneling microscopy from birth to adolescence. Rev. Mod. Phys. 59: 615–625.

Botes, M. and T.E. Cloete. 2010. The potential of nanofibers and nanobiocides in water purification. Critical Reviews in Microbiolog. 36: 68–81.

Cordani, M. and A. Somoza. 2019. Targeting autophagy using metallic nanoparticles: A promising strategy for cancer treatment. Cell. Mol. 76: 1215–1242.

Deligiannakis, Y., G.A. Sotiriou and S.P. Pratsinis. 2012. Antioxidant and antiradical SiO_2 nanoparticles covalently functionalized with gallic acid. ACS APPL Mater Interfaces 4: 6609–6617.

Douglass, D.C. and L.E. Brus. 1988. Surface derivatization and isolation of semiconductor cluster molecules. J. Am. Chem. Soc. 110: 3046–3050.

Drexler, E.K., C. Peterson and G. Pergamit. 1991. Unbounding the Future: The Nanotechnology Revolution. William Morrow and Company, New York.

Drexler, K.E. 1986. Engines of Creation: The Coming Era of Nanotechnology. Anchor Books, Doubleday, New York.

Ealia, A.M. and M.P. Saravanakumar. 2017. A review on the classification, characterisation, synthesis of nanoparticles and their application. 14th ICSET-2017 IOP Publishing IOP Conf. Series: Mater. Sci. Eng. 263.

Eigler, D.M. and E. Schweizer. 1990. Positioning single atoms with a scanning tunnelling microscope. Nature 344: 524–6.

Ekimov, A. and A. Onushchenko. 1982. Quantum size effect in the optical-spectra of semiconductor micro-crystals. Sov. Phys. Semicond. 16: 775–778.

Faraday, M. 1857. The Bakerian Lecture: On the relations of gold and other metals to light. Proceed. Royal Society of London 8: 356–361.

Feynman, R.P. 1960. There's plenty of Room at the bottom—An invitation to enter a new field of physics. Eng. Sci. 23: 22–36.

Freestone, I., N. Meeks, M. Sax and C. Higgitt. 2007. The Lycurgus Cup—A Roman nanotechnology. Gold Bull. 40: 270–277.

Halas, N., J. West and R. Drezek. 2004. Nanoshell-enabled photonics-based imaging and therapy of cancer. Technol. Cancer Res. Treat. 3: 33–40.

Hasan, S. 2015. A review on nanoparticles: their synthesis and types. Res. J. Recent. Sci. 4: 1–3.

Hill, T.L. 1963. Thermodynamics of Small Systems. W.A. Benjamin, New York.

Hill, T.L. 2001. A different approach to nanothermodynamics. Nano Letters 1: 273–275.

Hornigold, T. 2018. What Are the Origins of Nanotechnology? https://www.azonano.com/article.aspx?ArticleID=4858. Accessed on 18 March 2021.

Iijima, S. and T. Ichihashi. 1993. Single-shell carbon nanotubes of 1-nm diameter. Nature 363: 603–605.

Kammler, B.H.K., L. Mädle and S.E. Pratsinis. 2001. Flame synthesis of nanoparticles. Chemical Engineering Technology 24: 583–96.

Kim, J.S., E. Kuk, K.Y. Yu, J.-H. Kim, S.J. Park, H.J. Lee, S.H. Kim, Y.K. Park, Y.H. Park, C-.Y. Hwang, Y-.K. Kim, Y-.S. Lee, D.H. Jeong and M-.H. Cho. 2007. Antimicrobial effects of silver nanoparticles. Nanomedicine 3: 95–101.

Kinnear, C., T.L. Moore, L. Rodriguez-Lorenzo, B. Rothen-Rutishauser and A. Petri-Fink. 2017. Form follows function: nanoparticle shape and its implications for nanomedicine. Chem. Rev. 117: 11476–11521.

Kolahalam, L.A., I.V.K. Viswanath, B.S. Diwakar, B. Govindh, V. Reddy and Y. Murthy. 2019. Review on nanomaterials: Synthesis and applications. Materials Today: Proceedings 18: 2182–2190.

Kresge, C.T., M.E. Leonowicz, W.J. Roth, J.C. Vartuli and J.S. Beck. 1992. Ordered mesoporous molecular sieves synthesized by a liquid-crystal template mechanism. Nature 359: 710–712.

Kroto, H.W., J.R. Heath, S.C. O'Brien, R.F. Curl and R.E. Smalley. 1985. C60: buckminster fullerene. Nature 318: 162–163.

Kumar, S., M. Nehra, D. Kedia, N. Dilbaghi, K. Tankeshwar and K.-H. Kim. 2018. Carbon nanotubes: A potential material for energy conversion and storage. Progress in Energy and Combustion Science 64: 219–253.

Kuppusamy, P., M.M. Yusoff and N. Govindan. 2016. Biosynthesis of metallic nanoparticles using plant derivatives and their new avenues in pharmacological applications—An updated report SAUDI Pharm. J. Saudi Pharmaceutical Journal 2: 473–484.

Lara, S. and A. Perez-Potti. 2018. Applications of nanomaterials for immunosensing. Biosensors (Basel) 8: 104.

Lee, P.Y. and K.K.Y. Wong. 2011. Nanomedicine: A new frontier in cancer therapeutics. Curr. Drug Deliv. 8: 245–253.

Leon, L., E.J. Chung and C. Rinaldi. 2020. A brief history of nanotechnology and introduction to nanoparticles for biomedical applications. pp. 1–4. *In*: Chung, E.J., L. Leon and C. Rinalodi (eds.). Nanoparticles for Biomedical Applications. Elsevier.

Lewis, J. 2017. A brief history of nanotechnology. https://foresight.org/a-brief-history-of-nanotechnology/, accessed on 21 March 2021.

Mandal, G. and T. Ganguly. 2011. Applications of nanomaterials in the different fields of photosciences. Indian Journal of Physics 8: 1229–1245.

Mansoori, G.A. 2005. Principles of Nanotechnology: Molecular Based Study of Condensed Matter in Small Systems. World Scientific Pub. New York.

Mansoori, G.A. and T.A.F. Soelaiman. 2005. Nanotechnology—An introduction for the standards community. J. ASTM Intern. 2: 1–22.

Mie, G. 1908. Beiträge zur Optik trüber Medien, speziell kolloidaler Metallösungen. Ann. Phys. 330: 377–445.

Mohammad, Z.H., F. Ahmad, S.A. Inrahim and S. Zaidi. 2022. Application of nanotechnology in different aspects of the food industry. Discover Food. 2.

Mohammadi, S., A. Harvey and A.K.V.K. Boodhoo. 2014. Synthesis of TiO_2 nanoparticles in a spinning disc reactor. Chemical Engineering Journal 258: 171–184.

Morin, J.-F., Y. Shirai and J.M. Tour. 2006. En route to a motorised nanocar. Org. Lett. 8: 1713–16.

Mulvaney, P. 2015. Nanoscience vs nanotechnology—Defining the field. ACS Nano 9: 2215–7.

Murray, C.B. and M.G. Bawendi. 1993. Synthesis and characterization of nearly monodisperse CdE (E = S, Se, Te) semiconductor nanocrystallites TS nanocrystallites. J. Am. Chem. Soc. 115: 8706–8715.

Nanotechnology Information Center: Properties, Applications, Research, and Safety Guidelines, American Elements. Archived from the original on 26 December 2014, Accessed 07 Oct 2020.

Novoselov, K.S., A.K. Geim, S.V. Morozov, D. Jiang, Y. Zhang, S.V. Dubonos, I.V. Grigorieva and A.A. Firsov. 2004. Electric field effect in atomically thin carbon films. Science 306: 666–669.

Paufler, P., M. Reibold, M.W. Kochmann, A.A. Levin, N. Pätzke and D.C. Meyer. 2006. Towards an understanding of Damascene blades. 23rd European Crystallographic Meeting, ECM23, Leuven, Acta Cryst. A62, p. 55.

Pimpin, A. and W. Srituravanich. 2012. Review on micro- and nanolithography techniques and their applications. Engineering Journal. 16.

Ramesh, S., J.V. Ramaclus, E. Mosquesra and B.B. Das. 2016. Sol-gel synthesis, structural, optical and magnetic characterization of Ag3(2+x)PrxNb4-xO11+ δ (0.0 ≤ x ≤ 1.0) nanoparticles. RSC Advances. 8.

Ramezani, H.H. and G.A. Mansoori. 2007. Diamondoids as molecular building blocks for nanotechnology. pp 44–71. *In*: Mansoori, G.A., F.T. George, L. Assoufid and G. Zhang (eds.). Molecular Building Blocks of Nanotechnology—From Diamondoids to Nanoscale Materials and Applications. Springer. New York.

Reibold, M., P. Paufler, A.A. Levin, W. Kochmann, N. Pätzke and D.C. Meyer. 2006. Materials: Carbon nanotubes in an ancient Damascus sabre. Nature 444: 286.

Roco, M.C. 2011. The long view of nanotechnology development: the National Nanotechnology Initiative at 10 years. J. Nanoparticle Res. 13: 427–45.

Rothemund, P.W.K. 2006. Folding DNA to create nanoscale shapes and patterns. Nature 440: 297–302.

Salavati-Niasari, M., F. Davar and N. Mir. 2008. Synthesis and characterization of metallic copper nanoparticles via thermal decomposition. Polyhedron. 27: 3514–3518.

Santhosh, C., V. Velmurugan, G. Jacob, S.K. Jeong, S.A.N. Grace and A. Bhatnagar. 2016. Role of nanomaterials in water treatment applications: A review. Chemical Engineering Journal 306: 1116–1137.

Sciau, P. 2012. Nanoparticles in Ancient materials: the metallic lustre decorations of medieval ceramics. *In*: Hashim, A.A. (ed.). The Delivery of Nanoparticles. IntechOpen. Rijeka, Croatia.

Seeman, N.D. 2010. Nanomaterials based on DNA. Annu Rev. Biochem. 79: 65–87.

Shah, P. and A. Gavrin. 2006. Synthesis of nanoparticles using high-pressure sputtering for magnetic domain imaging. Journal of Magnetism and Magnetic Materials 301: 118–123.

Sharma, P., V. Pandey, M.M.M. Sharma, A. Patra, B. Singh, S. Mehta and A. Husen. 2021. A review on biosensors and nanosensors application in agroecosystems. Nanoscale Research Letters. 16.

Sharma, V.P., U. Sharma, M. Chattopadhyay and V.N. Shukla. 2018. Advance applications of nanomaterials: a review. Materials Today: Proceedings 5: 6376–6380.

Sharma, Y.C., V. Srivastava, V.K. Singh, S.N. Kaul and C.H. Weng. 2009. Nano-adsorbents for the removal of metallic pollutants from water and wastewater. Environ. Technol. 30: 583–609

Shulaker, M.M., G. Hills, N. Patil, H. Wei, H.-Y. Chen, H.S-. Wong and S. Mitra. 2013. Carbon nanotube computer. Nature 501: 526–530.

Sitharaman, B. 2011. Nanobiomaterials Handbook. Boca Raton, FL. CRC Press.

Tai, C.Y., C. Tai, M. Chang and H. Liu. 2007. Synthesis of magnesium hydroxide and oxide nanoparticles using a spinning disk reactor. Ind. Eng. Chem. Res. 46: 5536–5541.

Taniguchi, N., C. Arakawa and T. Kobayashi. 1974. On the basic concept of nano-technology. *In*: Proceed International Conference on Production Engineering, Tokyo, Japan.

The British Museum. www.britishmuseum.org/research/collection_online/collection_object_details. aspx?objobjec=61219&partId=1, accessed on 17 March 2021.

Tiwari, D.K., J. Behari and P. Sen. 2008. Application of nanoparticles in waste water treatment. World Applied Sciences Journal 3: 417–433.

Tolochko, N.K. 2009. UNESCO – EOLSS sample chapters nanoscience and nanotechnologies - History Of Nanotechnology - ©Encyclopedia of Life Support Systems (EOLSS). https://www.eolss.net/sample-chapters/ C05/E6-152-01.pdf. Accessed 17 March 2021.

Tratnyek, P.G. and R.L. Johnson. 2006. Nanotechnologies for environmental cleanup. Nano Today 1: 44–48.

World Intellectual Property Report: Breakthrough Innovation and Economic Growth. 2015 World Intellectual Property Organization. pp. 112–4. https://www.wipo.int/publications/en/details.jsp?id=3995. Accessed on 22 March 2021.

Xu, X., R. Ray, Y. Gu, H.J. Ploehn, L. Gearheart, K. Raker and W. Scrivens. 2004. Electrophoretic analysis and purification of fluorescent single-walled carbon nanotube fragments. J. Am. Chem. Soc. 126: 12736–12737.

Yadav, T.P., R.M. Yadav and D.P. Singh. 2012. Mechanical milling: a top down approach for the synthesis of nanomaterials and nanocomposites. Nanoscience and Nanotechnology 2: 22–48.

Yuan, Y., Z. Gu, C. Yao, D. Luo and D. Yang. 2019. Nucleic acid–based functional nanomaterials as advanced cancer therapeutics. Small. 15: 1900172.

Zhang, W.X. 2003. Nanoscale iron particles for environmental remediation: an overview. J. Nanoparticle Res. 5: 323–332.

Zsigmondy, R. 1914. Colloids and the Ultramicroscope. J. Wiley and Sons. New York.

Chapter 2

Nanomaterials-enabled Technologies for Clean Water and their Sustainability Aspects

Sritama Mukherjee[#] and T. Pradeep*

1. Sustainability in Materials

Materials are used for sustenance of life and for improving living standards, thus becoming indispensable for human life. They are used across industries and all sectors of the economy. The 21st century saw the rise of several advanced materials that are highly functional and are therefore implemented in areas such as energy, environment, healthcare, construction and many more. Rapid advances in science and technology involving polymers and ceramics in the 20th century were made in the field of structural material science. The path was paved by rigorous research in thermochemistry in the 17th and 18th centuries and in electrochemistry in the 18th and 19th centuries, leading to the entire periodic table being accessed for engineering purposes by the mid-20th century. Subsequently, inventions on materials with greater functionalities such as semiconducting, magnetic, optoelectronic, piezoelectric and thermoelectric materials followed later in the 20th century. Many such products may demand heavy utilization of metals from restricted geographical and industrial sources like highly localized ores or by-product extracts, making their supply limited, costly and non-eco-friendly (Koltun 2010). In addition, governments may categorize such materials as "critical" due to economic, political or environmental reasons. Reckless production of materials without considering their long-term environmental, economic and societal impacts would lead to undesired consequences such as degradation of air, water and land, loss of biodiversity, resource depletion and increasing disparity (Qu et al. 2013). Due to growing concerns of the magnitude of the impact on the planet by global population, there is an emergence of the concept of sustainable development, if that ideates "development that meets the needs of the present without compromising the ability of future generations to meet their own needs", according to the Brundtland Report (1987) (Mensah 2019). Since then, a wide range of interpretations of the concept can be seen. Under the umbrella of the United Nations (UN), key agencies like ECOSOC (Economic and Social Council) have prioritized their work towards sustainable development encompassing all

DST Unit of Nanoscience (DST UNS) and Thematic Unit of Excellence (TUE), Department of Chemistry, Indian Institute of Technology Madras, Chennai 600036, India.
[#] Present address: Department of Bioproducts and Biosystems, School of Chemical Engineering, Aalto University, Finland. Email: sritama.pu13@gmail.com
* Corresponding author: pradeep@iitm.ac.in

its three pillars, namely environmental, economic and social sustainability. Sustainable Development Goals (SDGs) containing 17 action plans were adopted by the UN as "a universal call to action to end poverty, protect the planet and improve the lives and prospects of everyone, everywhere" as a part of the 2030 Agenda for Sustainable Development. Environmental sustainability emerged to be most important when it was realized that the earth has limited natural resources like water, land, etc., and there exist boundaries within which equilibrium between natural capital for economic input and treating it as a sink for waste has to be maintained. It is about conserving the quality of the natural environment and climate, and its ability to support healthy generations. Clean water forms the basis of SDG 6, which promises to 'ensure availability and sustainable management of water and sanitation for all' ('GOAL 6: Clean water and sanitation | UNEP - UN Environment Programme'). Under this goal, there are clean water-related targets like (a) access to safe and affordable drinking water for all; (b) improvement in water quality by decrease in pollution, eliminating or minimizing release of untreated hazardous chemicals and materials, halving the proportion of untreated wastewater and substantially increasing recycling and safe reuse globally; (c) amplify international cooperation and capacity building for developing countries in water and sanitation activities and programs, including water reclamation, desalination, water efficiency, wastewater treatment, recycling and reuse technologies, etc. Materials Science is central to achieving the aforementioned targets under SDG6 (Green et al. 2012).

An interesting fact that the physical and chemical properties of matter change with size as the nanometer length scale is reached makes this dimension appealing for every subject area. Nanoscience offers the unique ability to understand physical and chemical processes at the most fundamental level and manipulate and control nano-objects to produce a desired effect or a property that cannot be achieved in the bulk ('A matter of scale' 2016).

2. Nanomaterials Production at Industrial Level

Rapid advancement in technological innovations has generated immense interest across the globe to explore and utilize novel aspects of nanotechnology. The application of engineered nanomaterials (ENMs) is an emerging field and has gained huge attention in recent years. Engineered nanomaterials are those where any one of the dimensions of materials falls in the nanoscale range (1–100 nm). Particles generated at the nanoscale regime show unique structural and functional (optical, magnetic, electrical and others) properties because of their very large specific surface area, high surface energy and quantum confinement (Colvin 2003). Due to these exclusive physicochemical properties, nanoparticles find endless applications in medicine, electronics, cosmetics, textiles, renewable energy, environmental protection, surface coatings, chemicals and other industries. For example, TiO_2, ZnO and Ag nanoparticles (AgNP) are used in many consumables ranging from personal care products to surgical instruments due to their UV-blocking and antibacterial properties. It was claimed that nanoproducts make up just 0.01% of the 400 million tons per annum of chemicals manufactured globally in 2004, while it was predicted that, with growth in each sector of the nanotech industry, production will grow by at least 1000 times by 2020 ('Nanotechnology: small science on a big scale | Feature | Chemistry World', https://www.chemistryworld.com/features/nanotechnology-small-science-on-a-big-scale/3004705.article). The synthesis of metal nanoparticles falls into two main categories: top-down approach and bottom-up approach. The top-down approach involves the process by which nanoparticles are produced by size reduction, while the bottom-up approach involves nanoparticles being formed from tiny entities such as atoms and molecules. Both approaches can be implemented in different experimental conditions; however, the key attributes that need to be controlled are particle size, particle shape, size distribution, particle composition, degree of particle agglomeration, etc., as per requirement (Gottschalk and Nowack 2011). Key parameters that are to be considered for technology selection on a large scale are scalability, cost-effectiveness, stability of the final product, reproducibility, contamination levels and abrasion of tools (Rastogi et al. 2020). The types of materials that undergo large scale production are: (a) carbon-based materials like fullerenes, carbon nanotubes (CNT) and carbon nanofibers; (b) inorganic nanoparticles of metals (Au, Ag, Pt and Pd), metal oxides (ZnO, TiO_2, CeO_2, ZrO_2, CuO, Al_2O_3, NiO, Fe_2O_3, Fe_3O_4, etc.), metal/metal oxide-CNT composites, Quantum Dots (QD), and so on (Shegokar and Nakach 2020). The common physical synthetic methods of nanoparticles are laser ablation, electric arc discharge, RF plasma method, mechanical attrition and lithography. These physical methods are very expensive as well as time and energy consuming. The conventional chemical synthesis involves the use of reducing agents, organic solvents, etc., creating toxicity along with environmental issues. The scientific community is trying to develop novel and industrially viable production methods. Until today, the fabrication of bulk nanostructures is often carried out using top-down conventional technologies, and the facilities needed for large-scale production have a huge environmental impact. Parameters like energy consumption both for the production of nanomaterials and the construction of facilities and equipment, water/solvent utilization, greenhouse gas emissions, waste generation, etc., should be critically assessed (Charitidis et al. 2014).

2.1 History of materials used for clean water application and the need for the use of nanomaterials

Boiling is a simple water purification method, but it is effective when the water temperature is 160 °F (70°C) or higher for 30 minutes, above 185 °F (85°C) for 3 minutes, and at a vigorous boil at 212 °F (100°C) for 1 minute to kill thermally sensitive microbes. However, it does not help in removing other toxins like metal ions (Smith 2017). Bone charcoal is one of the oldest and cheapest materials for removing fluoride and heavy metals. Bone charcoal is a natural porous black granular carbon material produced by carbonizing animal bones (usually cow bones) to mineralize residual organic carbon into graphite and leaving it in the porous structure of apatite. Due to vegetarian concerns and religious sentiment, its use has diminished. Bone charcoal alternatives were activated carbon and ion exchange resins. The three main categories of industrial adsorbents were carbon-based compounds (graphite, activated carbon), oxygen-based compounds (zeolites, silica) and polymer-based compounds. Other adsorbents such as zero-valent iron, iron oxides, alumina and various experimental media are considered. Materials used for coagulation-flocculation often included aluminum and iron salts to separate the suspended solids from water. Household gravity-based sand filtration has been explored for more than 100 years to remove turbidity, reduce Fe and Mn and kill pathogens. The first ceramic filters were made in the 19th century, and the porous ceramic candle-style drip filter became popular worldwide, where water flows through the candle (0.2 mm size pores), eliminating 99.9% of waterborne bacteria and suspended solids. Impregnation of silver and activating carbon centers enhanced the antimicrobial and organics removal activity of the filters. The hollow fiber membrane filter unit uses light hollow fiber membrane tubes to separate microorganisms from water in a size dependent manner. A 0.1 mm tube is expected to remove 99.9% of bacteria from the water. These filters are easy to assemble, require a simple backwash depending on the turbidity of the water, and have a long service life of over 10 years. One of the major challenges was to provide drinking water of high purity using efficient, cost-effective, safe and robust materials, which became a major interest for water scientists. The technology to remove some contaminants must reach molecular limits to ensure that purified water meets the newest drinking water standards. In the recent past, several materials like clay/mud, alum, chlorine-based chemicals, potassium permanganate, activated carbon and other low-cost adsorbents developed from various resources have been explored for decontamination of water. Recently, the application of nanostructured materials for scavenging and degradation of water contaminants is gaining large popularity due to their unique size-dependent properties like large surface area, porous morphology, short intra-particle diffusion distance, the scope of appropriate surface functionalization, etc., and a potential to impart stability, reusability and recyclability to the materials (Nagar and Pradeep 2020). Engineered nanomaterials can facilitate multiple treatment methods like adsorption, disinfection or catalytic degradation of pollutants targeting a single or an array of pollutants that may otherwise require different removal strategies. Other advantages that nanomaterials can offer over bulk materials are effective contaminant removal even at low concentrations, generation of less waste after treatment as less quantity of nanomaterial will be required compared to the corresponding bulk form.

2.2 Role of nanomaterials for clean water

Extensive research continues to address water remediation of industrial effluents or water containing bacteria, viruses, harmful pathogens, dyes, heavy metals (lead, cadmium, zinc), pesticides, insecticides, etc. Several types of nanomaterials such as metal nanoparticles, polymer/biopolymer nanoparticles, zeolites, carbon-based nanomaterials, nanocomposites, metal oxide nanoparticles, nanoscale semiconductor photocatalysts, etc., have been tested for water remediation by researchers (Beyene and Ambaye 2019). Several different water treatment pathways using nanomaterials have been incorporated for the enhancement of their efficiency. Among them, the most commonly used technologies are based on adsorption, membranes, photocatalytic degradation and disinfection by anti-microbial nanomaterials, along with sensing and monitoring technology. Various nanomaterials developed for treating water contamination are presented in Fig. 2.1 and briefly discussed below.

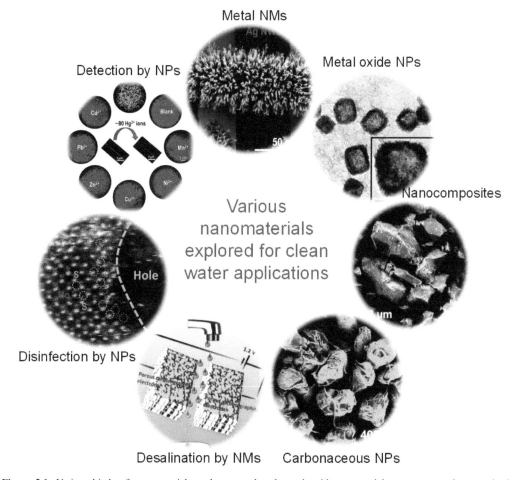

Figure 2.1. Various kinds of nanomaterials such as metal and metal oxide nanoparticles, nanocomposites, graphenic nanoparticles and metal nanoclusters used for clean water applications like contaminant removal, desalination, disinfection and sensing.

2.2.1 Metal nanoparticles

Metal nanoparticles possess a large surface area, thus improving catalytic properties and significantly higher reactivity, allowing small organic and inorganic molecules to be adsorbed on the surface. The most widely studied metal nanoparticles are Ag (AgNP), Au (AuNP) and Fe (nZVI) nanoparticles (Mathew and Pradeep 2014). While AgNP has been well established for its antimicrobial properties, AuNP and nZVI have been explored extensively for various strategies of detection and sequestration of heavy metals, respectively. AuNP and AgNP in suspension or supported over activated alumina can effectively remove chlorpyrifos, endosulfan and malathion, common pesticides found in surface waters of developing countries, from contaminated water (Nair and Pradeep 2007). Magnetic nanoparticles such as Fe, Ni and Co can be removed from the system by an external magnetic field and can be used for many purposes based on biocompatibility, magnetic sensitivity, and so on (Pradeep and Anshup 2009).

2.2.2 Metal oxide nanoparticles

These types of nanoparticles often exhibit ion-exchange sites within their framework for the removal of heavy metals, fluoride, nitrate, pesticides, etc., from contaminated water. Amongst the existing adsorbents, ferric oxides, aluminum oxides, magnesium oxides, manganese oxides, titanium oxides and cerium oxides have been tested for their high uptake capacity and selectivity. Most iron oxide minerals such as goethite (α-FeOOH), hematite (α-Fe$_2$O$_3$) and magnetite (Fe$_3$O$_4$) are abundant and thermodynamically stable in natural systems, while others (e.g., ferrihydrite (Fe$_5$HO$_8$·4H$_2$O), maghemite (γ-Fe$_2$O$_3$)) can be designated as intermediates. All of them are considered as low-cost and efficient alternatives as effective household treatments for arsenic removal from water (Mohan and Pittman 2007). Zeolites minerals primarily constitute aluminosilicates with a 3D backbone structure containing AlO$_4$ and SiO$_4$ tetrahedra. They communicate with each other by sharing oxygen atoms and forming interconnected cages and channels containing moving water molecules, and they are often loaded with MnO$_2$, Ce (III), La (III), mono- or bimetallic Fe-Al oxides (Pradeep and Anshup 2009). These materials are very efficient in reducing harmful cation and anion concentrations from drinking water and wastewater.

2.2.3 Nanocomposites

Nanocomposites are generally comprised of two parts, the matrix support and a surface-functionalized filler reinforcement. Different nanoparticles of metal/metal oxides (FeOOH, Al$_2$O$_3$, TiO$_2$), nanoAg, CNT, activated carbon, zeolites, ion-exchange species, etc. are added to the matrix/functionalized nanofibers of polymers and biopolymers (Li et al. 2015). These nanofillers have high surface area and they assist to improve hydrophilicity, photocatalytic effects and antimicrobial activity of the nanocomposite by their unique physico-chemico-mechanical properties. For example, according to a survey, some filters remove pesticides from drinking water reducing pesticide levels from very high concentration levels to well below the safety standards (0.5 μg L^{-1}, for all pesticides taken together) using nanocomposites, that had reached over 7.5 million people by 2016 (Nair and Pradeep 2007).

2.2.4 Nanomembranes

Membranes composed of nanofibers that can eliminate various contaminants from the aqueous phase with high rejection rates are being exploited as a pretreatment approach before Reverse Osmosis (RO). Numerous reports show multifunctional membranes by incorporation of nanomaterials into polymeric or inorganic membranes known as nanocomposite membranes. RO and nanofiltration (NF) membranes are usually made of polyamides, cellulose acetate/diacetate/triacetate, etc. Apart from polymeric membranes, membranes prepared by Graphene Oxide (GO), CNT and Mixed Matrix Materials (MMMs) have gained great attention due to their desirable properties such as tunable pore structure, excellent chemical, mechanical and thermal properties, good salt rejection and/or high water permeability (Al Aani et al. 2020). An electrospinning technique enables the use of a wide range of polymers, namely polyamide-66, polycaprolactone (PCL), polystyrene (PS) fibers, polymethylmethacrylate (PMMA), cellulose, chitosan fibers, etc., that are spun into nanofiber mats that can be used for NF (Anis et al. 2019). Aquaporins, the transmembrane water channel proteins have been selectively transporting water molecules across the cells of living organisms for millions of years. Several water purification membranes have been synthesized with aquaporin incorporation, following an understanding of aquaporin protein structure by Peter Agre in the early 1990s, for which he was awarded the Nobel prize (Agre 2004).

2.2.5 Carbonaceous nanomaterials

Carbon allotropes can be categorized according to the coordination number of the carbon and its atomic arrangement in the lattice structure into 0D materials like carbon dots and fullerenes, 1D materials like CNT, 2D materials such as graphene, GO, reduced GO, etc. (Zeng et al. 2017) and 3D entities like amorphous or porous carbon materials. These have been studied extensively owing to their low cost, nontoxicity and advantageous physicochemical properties for the removal of a range of organic, inorganic and microbial contaminants (Colvin 2003). Biochar made from feedstocks and activated carbon from phenolic resins has large surface areas of typically 300–2500 m^2 g^{-1} and highly microporous structures are often the standard means of removing pesticides like MCPA (2-methyl-4-chlorophenoxyacetic acid), diuron, metaldehyde and clopyralid from drinking water or wastewater by adsorption (Peng et al. 2020).

2.2.6 Dendrimers

Dendrimers are a new class of environmentally friendly polymer molecules with multiple functions. They can be used as stabilizers for metal nanoparticles and as contaminant-specific functional groups on the surface of nanoparticles. The two most commonly used dendrimers are the polyamide amine (PAMAM) dendrimer and the polypropylene imine dendrimer (Saleem and Zaidi 2020). Their reactivity, shape and size depend on the core formation and chemical composition, internal branching and surface functional groups. The dendrimer-enabled filtration process exploits these unique properties of dendritic nanopolymers to develop a new generation of low-pressure filtration processes for removal of water contaminants such as Cu^{2+}, U^{6+}, ClO_4^-, etc. (Diallo 2014). Due to their large size, dendrimers attach entities such as cations, anions, organic substances and biological compounds. It then retains them in their bifurcated structure and prevents them from passing through the membrane. Another example, the PAMAM-CD adsorbent used an insoluble crosslinked copolymer (PAMAM, 2nd generation) containing β-cyclodextrin (β-CD) moiety and polyamidoamine. It was used to adsorb Cu (II) and Pb (II) ions from water and organic compounds such as 2,4-dichlorophenol, 2,4,6-trichlorophenol, etc. (Li et al. 2011).

2.2.7 Nanotechnology for desalination of water

Membrane-based separation technologies such as RO, Forward Osmosis (FO), electrodialysis (ED) and NF have gained commercial success compared with thermal desalination methods. Polyamide membranes are widely used due to their high selectivity and permeability compared with conventional membranes made of cellulose acetate. Other types of membranes like the nanocomposite polymeric membranes integrated with CNT, graphene and other 2D frameworks like MoS_2 and holey MoS_2 and defect-free aquaporin-based membranes are interesting as they show very high efficiency (Rout et al. 2021, Qu et al. 2013). Ion-exchange processes like ED and capacitive deionization (CDI) are most commonly used for chemical desalination where oxide NPs of silica, titania, carbonaceous nanomaterials, Ag NPs, zeolites, etc., are incorporated into ion-exchange membranes for ED to tune properties such as surface area, ionic conductivity, tensile strength, energy efficiency, thermal stability, etc. (Gupta et al. 2018).

2.2.8 Nanotechnology for disinfection of water

Several nanomaterials having strong antimicrobial properties, including that of Ag, ZnO, TiO_2, Ce_2O_4, CNT, fullerenes and so on, which can enable physical-based, peroxy-based, photocatalytic and electro-assisted disinfection processes (Qu et al. 2013). Nanomaterials with antimicrobial properties interact directly with microorganisms by interrupting transmembrane electron transfer, destroying/penetrating microbial envelopes (bacterial membranes and viral capsids) or oxidizing microbial components (proteins, DNA/RNA, etc.). Graphene, CNT, MXene, Au/Cu nanowires, etc. rely on physical contact with the microbes for disinfection (Hossain et al. 2014). MXene is a class of 2D nanomaterials derived from the family of transition metal carbides, nitrides and carbonitrides, expressed by the general formula $M_{n+1}X_nT_x$ (n = 1–3), where M is an early transition metal group, X is carbon and/or nitrogen and Tx are surface terminal groups (–OH, –O–, and/or –F). For example, $Ti_3C_2T_x$ and its composites show effective removal and separation of heavy metals, dyes and radionuclides, while acting as an antibacterial agent for disinfecting water (Rasool et al. 2019). The Fenton approach involves activation of H_2O_2 to generate $HO^{\bullet}$ through catalytic reaction with the dissolved Fe^{2+}/Fe^{3+} redox pair. Semiconductors like TiO_2 P25 modified by metallic co-catalysts (e.g., Pt, Pd and Au), present improved antimicrobial activity under UV irradiation, reacting with O_2 in water to generate a series of powerful Reactive Oxygen Species (ROS), including O^{2-}/HO_2, $HO^{\bullet}$ and H_2O_2, which act in photocatalytic water disinfection (Huo et al. 2020). For the electrochemistry-based disinfection process, specific nanomaterial modified electrodes including Boron-Doped Diamond (BDD), SnO_2, PbO_2 and doped-TiO_2 electrodes are used as anode materials to *in situ* generate highly reactive ROS (such as $HO^{\bullet}$) and $Cl^{\bullet}$ for indirect oxidation (Pichel et al. 2019).

2.2.9 Nanotechnology for detection and monitoring water quality

A number of techniques utilizing nanomaterial surfaces are being studied for sensing contaminant species; these techniques use changes in physical and chemical properties of the nanomaterials (e.g., optical absorption, photoluminescence, surface enhancement of Raman signals) on interaction with contaminants (Willner and Vikesland 2018). The selectivity is an important aspect in the detection of contaminants and can be ensured by using the right nanomaterials and choosing the appropriate ligand immobilization on the surface of the nanoparticles. An example is the selective detection of cysteine and glutathione at micromolar concentrations using gold nanorods (Sudeep et al. 2005). Researchers are investigating emerging materials and their properties, such as atomically precise noble metal clusters like Au_{25}@BSA, where the Au core is protected by Bovine Serum Albumin (BSA) ligand (Mathew and Pradeep 2014). The emission of the cluster can be enhanced by fixing the cluster to plasmonic particles or electrospun fibers. In both cases, contaminants such as mercury ions can be detected and quantified up to a few ions at the level of a single fiber (Ghosh et al. 2014). Other nanomaterials explored for sensing application include semiconductor QDs (Xavier et al. 2010), carbon QDs (Guo et al. 2015), Metal-Organic Frameworks (MOFs), micellar nanoparticles, CeO_2@ZrO_2 nanocages (Mukherjee et al. 2020b), Au@(SiO_2-FITC)@Ag_{15} mesoflowers (where FITC is fluorescein isothiocyanate) (Mathew et al. 2012), CdSe NP-modified TiO_2 nanotubes arrays, and so on. By creating more sensitive, compact and cost-effective sensor accessories, one can integrate such devices into smartphone and facilitate point-of-use applications (Zhou et al. 2014).

While nanomaterials such as CNT, cellulose-based nanomaterials, MXenes, aquaporins, etc., have shown outstanding performance at the laboratory scale, several of them have also been successfully mass-produced and translated into commercial products. As an example, CNTs are among the widely used nanomaterials in filtration systems that can efficiently remove inorganic, organic and biological contaminants from water. Similarly, commercial TiO_2 nanoparticles have been used extensively in photocatalytic technologies for water treatment.

3. Implementation of Concepts of Sustainability on Nanomaterials

Nanoscale particles occur in nature as a result of natural and anthropogenic processes. For example, a large number of nanoparticles are generated in the atmosphere by natural events such as volcanic eruptions and forest fires, while intentional or unintentional anthropogenic activities may bring nanoparticles to the environment by fumes generated during welding, smelting and burning of fossil fuel and plastic waste. There can be other industrial processes as well (Martínez et al. 2020). The unusual characteristics of the engineered nanomaterials like surface properties, chemical reactivity, physical adsorption, etc., which are dependent on their nanoscale dimensions and surface area may enable them for greater interactions with biological systems that cause toxicity and necessitate novel investigative approaches to assess their nanotoxicological behavior (Qu et al. 2013). It has been reported that the size of particles is inversely proportional to their toxicity, which implies toxic properties of nanomaterials even if they are relatively inert in their bulk form (e.g., carbon black and TiO_2). Nanomaterials exposed to the environment by any of the methods could be deposited on agricultural areas and water resources and hamper the fertility of soil and purity of water. This mobility can even lead to their bioaccumulation in the food chain and may be transferred to multiple generations. Those released into the atmosphere may result in a decrease in the stratospheric temperature. Nano ecotoxicological studies can analyze the release, important pathways, transformation and fate of nanomaterials in the environment. When nanoscale entities are in the aquatic environment, they can undergo various transformations, usually oxidation and reduction reactions. Dissolution, sulfidation, aggregation and adsorption of macromolecules and molecules/ions occur readily in the environment and in biological systems (Louie et al. 2014). These reactions can enhance or reduce their toxicity potential. The rate, extent and type of possible transformations depend on the properties of the core nanoparticle, its electronic properties, protecting groups and its surrounding chemical and biological environment (Fojtů et al. 2017). The reactive sites on the surface of the nanomaterials have extraordinary or altered electronic properties that have the ability to generate ROS and activate Fenton chemistry under the stress of dynamic redox environments catalyzed by chemicals or light. Graphene oxide, single-walled CNT, C_{60}, TiO_2 and carbon black nanoparticles have been studied using animal models and they have shown certain cytotoxicity. AgNPs that have been widely used for antimicrobial treatment of water, have been demonstrated to be most toxic to living cell lines. Bioaccumulation of nanoparticles has been observed in the human liver and kidney due to blood circulation, when these particles enter via the food chain, while those entering through inhalation tend to accumulate in the lungs and become the cause of asthma and other respiratory-system-related diseases (Handy et al. 2008). This deposition of nanoparticles in human organs results in the generation of ROS at the accumulation site, causing inflammatory cytokines and cytotoxic cellular responses.

Facing these challenges, clean water technologies involving nanomaterials that seek the potential of resource exploitation and environmental pollution require strict assessments to evaluate environmental sustainability. Post-assessment of the environmental impact of nanomaterials and related technologies has been one of the significant fields in sustainability research that would provide the foundation for pollution reduction and introduction of cleaner technologies and operations (Livingston et al. 2020). Figure 2.2 depicts the lifecycle of the release of nanoparticles in the environment from industrial processes from raw material extraction to nanomaterial/device manufacturing to the use phase, the end-of-life and nanotechnology-specific releases of nanomaterials across the lifecycle (dotted lines). The nanomaterial/device manufacturing includes synthesis of nanomaterials, post-synthesis processing and integration into nano-enabled devices. The use phase comprises drinking water and wastewater treatment, and the end-of-life includes incineration and landfills/solid waste) (solid lines) and nanotechnology-specific releases of nanomaterials (dotted lines).

3.1 *Tools for assessing sustainability and environmental impact*

Tools and concepts that help in studying the sustainability of materials and water-related technologies such as (a) Life Cycle Assessment (LCA) (Salieri et al. 2018), (b) Risk Assessment (RA) (Guineé et al. 2017), (c) GUIDEnanotool (Fernández-Cruz et al. 2018), (d) LICARA nanoSCAN (van Harmelen et al. 2016), (e) quantitative structure-activity relationships (Lamon

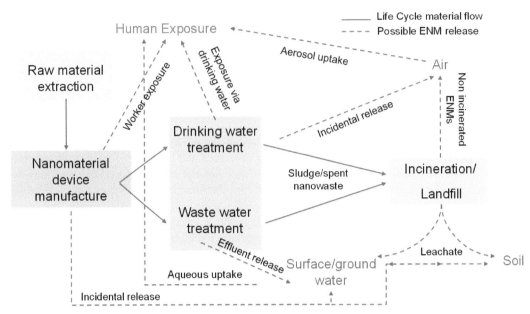

Figure 2.2. Plausible lifecycle of the release of nanoparticles in the environment from industrial processes. Adapted from (Falinski et al. 2020).

et al. 2019), (f) techno-economic assessment (Siegert et al. 2019), (g) Ashby-based nanomaterial selection (Ashby 2012), (h) fuzzy-Delphi method (Kamali et al. 2019), (i) sustainability metrics (Sheldon 2018), and so on will be presented in some detail below and are summarized in Table 2.1.

ISO 14040:2006 briefly covers LCA studies and Life Cycle Inventory (LCI) studies. LCA consists of four iterative stages according to ISO 2006 (a, b): (i) definition of the goal/scope; (ii) inventory assessment; (iii) impact assessment; and (iv) interpretation (Guineé et al. 2017). Goal/scope definition (i) means setting research goals for a particular application to the target audience. The scope includes the system under investigation, its features and boundaries, functional units and the categories and methods used. Then, in step (ii), an inventory is created containing all the data that goes in and out of the system from each process unit from inputs (materials, energy, secondary processes, etc.) to outputs (products, by-products, waste and environmental emissions). Impact assessment step (iii) reveals the environmental importance of the inventory. The results of the class provide information about environmental issues related to system inputs and outputs. Finally, interpretation (iv) describes the results of the previous steps. This allows to make decisions and recommendations so that you can score and weigh the entire process. In the case of nanomaterials, the stage of their life cycle, i.e., from the production of nanomaterials to the products that contain these nanomaterials, the use or application of such products and their ultimate end-of-life management/disposal, all of these are considered together as one lifecycle (Pesqueira et al. 2020).

Risk Assessment (RA) refers to the quantitative and qualitative assessment of risks to human health and/or the environment due to the presence or mixture of specific pollutants. Risk is defined as the probability of exposure to a hazard. In this case, exposure can be estimated by measuring the environmental concentration or by modeling the fate of pollutants in the environment to obtain the Predicted Environmental Concentration (PEC) (Leeuwen 2007). A basic idea related to adverse effects is commonly expressed in terms of laboratory-derived dose-response relationships (Sperber 2001). There are four main stages in assessing environmental risks. (1) Hazard evaluation: Identification of existing hazards. (2) Exposure assessment: Determination of predicted environmental concentration (PEC) or Predicted Daily Intake (PDI). (3) Impact Assessment: Determination of critical levels of exposure based on predicted no impact concentration (PNEC) or Acceptable Daily Intake (ADI). (4) Risk characterization: PEC/PNEC or PDI/ADI quotient calculation (Guineé et al. 2017).

Table 2.1. Various assessment tools that can be used for evaluating the sustainability of water purification technologies.

Assessment tool number	Sustainability assessment tools	Parameters for analyzing nanomaterials-enabled technologies	References
1.	Life cycle assessment	Definition of goal/scope, inventory assessment, impact assessment and interpretation considering the stages of production of the nanomaterials to the products that contain these nanomaterials, use or application of such products, and their ultimate end-of-life management/disposal	Upreti et al. 2015
2.	Risk assessment	Hazard assessment, exposure assessment, effect assessment and risk characterization	Fadeel et al. 2018
3.	Sustainability metrics	Mass intensity, solvent intensity, reaction mass efficiency, energy consumption, E-factor, CO_2 emission and many more	Sheldon 2018
4.	Ashby-based selection	Price, ecotoxicity and associated energy consumption, health risk, life-cycle environmental impact and techno-economic performance	Babbitt and Moore 2018
5.	GUIDEnanotool	Whole product lifecycle of nanomaterial-enabled products, risk management considering safer-by-design, exposure control and waste management	Falinski et al. 2020
6.	LICARA nanoSCAN	Risks to public, workers and consumers; benefits for economic, environmental and societal opportunities by the use of the product	Van Harmelen et al. 2016

GUIDEnanotool is an interactive digital guidance tool that develops innovative methodologies to assess and manage human and environmental health risks of nanomaterial-enabled products considering the whole product lifecycle including nanomaterial production, product manufacturing, its use, recycling and end-of-life (Fadeel et al. 2018). The tool conducts risk assessment related to release/exposure assessment of nanomaterials, their environmental fate, their impact on target organisms and extends to risk management considering safer-by-design, controlling exposure and waste management ('Development of a risk assessment strategy within the GUIDEnano project').

LICARA nanoSCAN is another modular web-based tool that assesses benefits and risks associated with new or existing nanoproducts to enable the development of sustainable and competitive nanoproducts (van Harmelen et al. 2016). Risks to the public, workers and consumers are assessed and benefits are evaluated based on the economic, environmental and social opportunities associated with the use of the product. Tools based on risk assessment and life cycle assessment that do not use quantitative data can identify gaps, missing data and associated uncertainties. Areas are identified where more detailed assessment, improved product design or risk management measures need to be applied.

In Ashby-based nanomaterial selection, Ashby diagrams can be adapted and applied to the streamlined selection of nanomaterials. This tool helps users identify applications and compare a wide range of nanomaterial categories based on relevant functional and sustainability characteristics (price, ecotoxicity and energy expenditure) (Babbitt and Moore 2018). Once some potential nanomaterials have been identified, further assessments are made to assess health risks, lifecycle environmental impacts and technological economic performance. These methods are applied in the design phase prior to the initial screening phase, taking into account the development rate of many known and novel nanomaterials (Falinski et al. 2018).

Sustainability or green metrics is an attempt to quantify the progress toward the broader goal of environmental sustainability by taking a holistic approach. The calculations consider parameters such as mass intensity, solvent intensity, reaction mass efficiency, energy consumption, E-factor and CO_2 emission (Mukherjee et al. 2019). To understand the principles behind these parameters, one needs to consider two main concepts that are responsible for making any material or process 'green': (a) efficient utilization of resources (raw materials, energy resources, chemicals, etc.) in the process of manufacture and application of materials, and (b) the minimization of waste generation and use of hazardous solvents and reagents in the manufacture and application of materials (Jiménez-González et al. 2012). Energy consumption and carbon footprint estimation are crucial sustainability parameters as they relate to climate change and also to the total cost for the utilities, which necessitates identifying effective and efficient ways of reducing energy use and greenhouse gas emissions.

3.2 *Sustainability assessment of various water treatment technologies: case studies*

There are a few reports where researchers have tried to analyze the environmental impacts of various developing and already developed water/wastewater treatment technologies.

a) A study was conducted by Kamali et al. 2019 using a multi-criteria decision approach (assisted by a fuzzy-Delphi method) integrating the technical, economic, environmental and social criteria in order to rank the nine treatment processes among the various sewage treatment technologies for processing industrial effluents. This study includes 17 criteria of technical standards (treatment efficiency, ease of implementation, integration, process stability, health and safety risks, etc.), environmental aspects (biowaste generation, chemical release, CO_2 emissions, water consumption, possibility of reuse, by-product recovery), economical parameters (initial investment, maintenance and operating costs) and social factors (odor effects, noise effects, visual effects and general acceptance) to prioritize the methods investigated. The study considered membrane-based technologies followed by adsorption, oxidation with nanomaterials and Fenton process in the decreasing order of sustainable physicochemical methods for the treatment of industrial effluents.

b) Memon et al. 2007 studied the Life Cycle Impact Assessment (LCIA) of four treatment technologies including reed beds, membrane bioreactors (MBR); Membrane Chemical Reactors (MCR) and an innovative green roof water recycling system (GROW). Materials and energy required for technology building and operation were measured and a lifecycle inventory based on the collected data was created. An inventory was used as an input to the *Simapro* software to run LCIA. The results showed that, among the technologies studied, GROW had the best performance in most impact assessment categories, MCRs appeared to be less environmentally friendly, thus concluding that natural treatment processes have lower environmental impact than membrane-based technologies. Power consumption has been found to have a significant impact on all environmental categories.

c) Ren et al. 2013 evaluated the social, economic and environmental sustainability of ceramic filters impregnated with silver nanoparticles for Point-Of-Use (POU) drinking water treatment in developing countries. The research focused on evaluating the cultural acceptability of this treatment among targeted users, understanding the effect of raw water quality on disinfection efficiency, changing material composition and optimizing disinfection. This conclusion derived from social, economic and environmental parameters clearly showed ceramic filter POU technology for drinking water treatment as a more sustainable option in developing countries, rather than the centralized treatment system widely used in industrialized countries. Ceramic filters outperformed on the levels of environmental sustainability in four of the five life cycle impacts assessed: energy use, water use, global warming potential and emission of Particulate Matter PM10 describes inhalable particles, with diameters that are generally 10 micrometers and smaller. Under the Clean Air Act, EPA sets and reviews national air quality standards for PM.

d) A case study of the synergy between wastewater treatment plants and photovoltaic systems was presented by Andrei et al. 2020 with the target of improving the energetic, environmental and economic impacts. Season-wise energy consumption analysis of wastewater treatment plants was done that showed the highest and lowest demands during April and November, respectively. Photovoltaic modules were used in the plant in order to obtain a power output as high as possible for partially compensating the power need. The impact of the photovoltaic system connectivity on the power grid was assessed by means of the matching-index method and the storage battery significantly improved this parameter, while carbon credit and energy payback time were used to assess the environmental impact. The results proved that the photovoltaic system mitigated 12,118 tons of carbon and the embedded energy got compensated by production in 8½ years. From the economic viewpoint, it showed leveled cost of energy since the price of energy from the photovoltaic source is below the current market price of energy.

e) In another study, Mukherjee et al. 2019a calculated the sustainability metrics of ferrihydrite incorporated microcrystalline cellulose and nanocellulose/polyaniline nanocomposites (MCCFH and CNPFH) for the removal of arsenic and fluoride from water, respectively (Mukherjee et al. 2019a). Na-Zn-Se-based composite for the release of essential minerals to desalinated water was also considered for a similar evaluation (Ravindran et al. 2019). All the three materials were assessed by the parameters including mass intensity, solvent intensity, reaction mass efficiency, energy consumption, E-factor and CO_2 emission. The nanocomposites were products of water-based green synthesis, hence exhibited excellent numbers with respect to waste generation (E-factor) and energy consumption, while they still required to improve their yield, thus the reaction mass efficiency. These studies on sustainability metrics are summarized in Table 2.2.

Table 2.2. Sustainability metrics were calculated for various nanomaterials used in applications for clean water.

Nanomaterials	Sustainability metrics						Reference
	Mass Intensity (kg/kg)	Water Intensity (kg/kg)	Reaction Mass Efficiency (%)	Energy Intensity (kW.h/kg)	E factor (kg/kg)	CO_2 emission (mg/kg)	
MCCFH	1.9	29.2	52	2.3	0.3	-	(Mukherjee et al. 2019)
CNPFH	1.84	38.8	54	1.8	0.6	-	(Mukherjee et al. 2020)
M1(Na-Zn-Se)	1.204	4.7	83	3.2	0.02	40	(Ravindran et al. 2019)
CAlFeC	5.68	36.67	17.6	2.0	0.12	-	(Egor et al. 2021)

4. Summary and Future Directions

Before using any nano-enabled technologies in the field, it is essential to examine factors including efficiency, cost, product lifetime and also environmental and social impacts. These factors are important for both end-users and industries, and understanding each of them can answer fundamental concerns such as the advantages that nanotechnology offer in addressing global water challenges over existing technologies. The importance of using nanomaterials is that they have the potential to purify water at a reduced expense as the nanomaterial-loading requirement is low and the working efficiency in removing pollutants is high. Clarifications on aspects like safety, stability, regeneration, reuse and disposal aspects of nanomaterials can help in the selection and design of nanomaterials to maximize structure-property–function relationships. Reports show that the production phases result in the release of less than 2% of all nanomaterials, while the remaining 98–99.9% of release occurs during the use and end-of-life phases to air, water and soil, which call for improvement in the design of nano-enabled devices, such as the use of nanomaterials impregnated or immobilized on various substrates. For example, AMRIT (Arsenic and Metal Removal by Indian Technology) consists of a nanostructured FeOOH reinforced with a chitosan template that can selectively remove arsenic from drinking water. Technology supplies about one million people with affordable clean water every day, giving hope to others. Eighty million people in India are affected by this problem

(Kumar et al. 2017). This solution does not require electricity, does not allow nanomaterials to seep into the environment, and is affordable for people living in the world's poorest regions. As a result, this technology has been approved for national implementation by the Government of India.

In general, various optimized analytical techniques exist for the detection and characterization of nanoparticles, although numerous inadequacies need to be addressed in applying current toxicity testing methods to evaluate nanomaterials. Scalability is another parameter that needs to be studied for the commercial viability of nano-based technologies. Factors that affect the scalability of nanomaterials include the availability of raw material resources, extraction and manufacturing costs and compatibility with common water treatment infrastructures. The production processes need to be improved in terms of following sustainable practices. Understanding and assessing the impact of various methods of synthesis of nanomaterials and post-synthesis techniques with sustainability metrics can boost the efficiency of processes. However, comprehensive and robust analytical methods have become prime requirements. At present, in the absence of adequate experimental data, most tools assume that all engineered nanomaterials with the same composition have a similar performance and possess the same levels of risk, ignoring their physicochemical properties. This ultimately results in the alteration in the accuracy of the sustainability assessment.

For creating novel materials, considerable effort should be made on research concerning regeneration, recycling, remediation, etc. Effective implementation of any water treatment technology in a community requires a complete analysis of its ability to benefit from technology by evaluating their interest in adapting and maintenance of those services. Better public-private partnerships are needed for establishing guidelines to address the transition from a traditional linear flow of materials in a "take–make–use–dispose of" economy to a greener, circular economy, also to address the inconsistencies in the regulations and raise public awareness to avoid negative approach towards nano-enabled technologies.

Acknowledgments

The authors thank IIT Madras and the Department of Science and Technology, Government of India for constantly supporting the research program on water and nanomaterials.

References

A matter of scale. 2016. Nature Nanotechnology 11: 733–733.

Agre, P. 2004. Aquaporin Water Channels (Nobel Lecture). Angewandte Chemie International Edition 43: 4278–4290.

Al Aani, S., T.N. Mustafa and N. Hilal. 2020. Ultrafiltration membranes for wastewater and water process engineering: A comprehensive statistical review over the past decade. Journal of Water Process Engineering 35: 101241.

Andrei, H., C.A. Badea, P. Andrei and F. Spertino. 2020. Energetic-environmental-economic feasibility and impact assessment of grid-connected photovoltaic system in wastewater treatment plant: case study. Energies 14: 100.

Anis, S.F., R. Hashaikeh and N. Hilal. 2019. Microfiltration membrane processes: A review of research trends over the past decade. Journal of Water Process Engineering 32.

Ashby, M. 2012. Materials and the Environment: Eco-informed Material Choice: Second Edition. Materials and the Environment: Eco-informed Material Choice: Second Edition: 1–616.

Babbitt, C.W. and E.A. Moore. 2018. Sustainable nanomaterials by design. Nature Nanotechnology 13: 621–623.

Beyene, H.D. and T.G. Ambaye. 2019. Application of sustainable nanocomposites for water purification process. Sustainable Polymer Composites and Nanocomposites: 387–412.

Charitidis, C.A., P. Georgiou, M.A. Koklioti, A.F. Trompeta and V. Markakis. 2014. Manufacturing nanomaterials: From research to industry. Manufacturing Review 1.

Colvin, V.L. 2003. The potential environmental impact of engineered nanomaterials. Nature Biotechnology 21: 1166–1170.

Development of a risk assessment strategy within the GUIDEnano project - PDF Free Download.

Diallo, M.S. 2014. Water Treatment by Dendrimer-Enhanced Filtration: Principles and Applications. Nanotechnology Applications for Clean Water: Solutions for Improving Water Quality: Second Edition: 227–239.

Egor, M., A.A. Kumar, T. Ahuja, S. Mukherjee, A. Chakraborty, C. Sudhakar et al. 2021. Cellulosic ternary nanocomposite for affordable and sustainable fluoride removal. ACS Sustainable Chemistry and Engineering 9: 12788–12799.

Fadeel, B., L. Farcal, B. Hardy, S. Vázquez-Campos, D. Hristozov, A. Marcomini et al. 2018. Advanced tools for the safety assessment of nanomaterials. Nature Nanotechnology 13: 537–543.

Falinski, M.M., D.L. Plata, S.S. Chopra, T.L. Theis, L.M. Gilbertson and J.B. Zimmerman. 2018. A framework for sustainable nanomaterial selection and design based on performance, hazard, and economic considerations. Nature Nanotechnology 13: 708–714.

Falinski, M.M., R.S. Turley, J. Kidd, A.W. Lounsbury, M. Lanzarini-Lopes, A. Backhaus et al. 2020. Doing nano-enabled water treatment right: sustainability considerations from design and research through development and implementation. Environmental Science: Nano 7: 3255–3278.

Fernández-Cruz, M.L., D. Hernández-Moreno, J. Catalán, R.K. Cross, H. Stockmann-Juvala et al. 2018. Quality evaluation of human and environmental toxicity studies performed with nanomaterials—the GUIDEnano approach. Environmental Science: Nano 5: 381–397.

Fojtů, M., W.Z. Teo and M. Pumera. 2017. Environmental impact and potential health risks of 2D nanomaterials. Environ. Sci.: Nano 4: 1617–1633.

Ghosh, A., V. Jeseentharani, M.A. Ganayee, R.G. Hemalatha, K. Chaudhari, C. Vijayan and T. Pradeep. 2014. Approaching sensitivity of tens of ions using atomically precise cluster-nanofiber composites. Analytical Chemistry 86: 10996–11001.

GOAL 6: Clean water and sanitation | UNEP - UN Environment Programme.

Gottschalk, F. and B. Nowack. 2011. The release of engineered nanomaterials to the environment. Journal of Environmental Monitoring 13: 1145–1155.

Green, M.L., L. Espinal, E. Traversa and E.J. Amis. 2012. Materials for sustainable development. MRS Bulletin 37: 303–308.

Guineé, J.B., R. Heijungs, M.G. Vijver and W.J.G.M. Peijnenburg. 2017. Setting the stage for debating the roles of risk assessment and life-cycle assessment of engineered nanomaterials. Nature Nanotechnology 12: 727–733.

Guo, Y., L. Zhang, S. Zhang, Y. Yang, X. Chen and M. Zhang. 2015. Fluorescent carbon nanoparticles for the fluorescent detection of metal ions. Biosensors and Bioelectronics 63: 61–71.

Gupta, S.S., M.R. Islam and T. Pradeep. 2018. Chapter 7—Capacitive Deionization (CDI): An Alternative Cost-Efficient Desalination Technique. Elsevier Inc.

Handy, R.D., R. Owen and E. Valsami-Jones. 2008. The ecotoxicology of nanoparticles and nanomaterials: Current status, knowledge gaps, challenges, and future needs. Ecotoxicology 17: 315–325.

Harmelen, T. van, E.K. Zondervan-van den Beuken, D.H. Brouwer, E. Kuijpers, W. Fransman, H.B. Buist et al. 2016. LICARA nanoSCAN—A tool for the self-assessment of benefits and risks of nanoproducts. Environment International 91: 150–160.

Hossain, F., O.J. Perales-Perez, S. Hwang and F. Román. 2014. Antimicrobial nanomaterials as water disinfectant: Applications, limitations and future perspectives. Science of the Total Environment 466–467: 1047–1059.

Huo, Z.Y., Y. Du, Z. Chen, Y.H. Wu and H.Y. Hu. 2020. Evaluation and prospects of nanomaterial-enabled innovative processes and devices for water disinfection: A state-of-the-art review. Water Research 173.

Jiménez-González, C., D.J.C. Constable and C.S. Ponder. 2012. Evaluating the "Greenness" of chemical processes and products in the pharmaceutical industry—a green metrics primer. Chem. Soc. Rev. 41: 1485–1498.

Kamali, M., K.M. Persson, M.E. Costa and I. Capela. 2019. Sustainability criteria for assessing nanotechnology applicability in industrial wastewater treatment: Current status and future outlook. Environment International 125: 261–276.

Koltun, P. 2010. Materials and sustainable development. Progress in Natural Science: Materials International 20: 16–29.

Kumar, A.A., A. Som, P. Longo, C. Sudhakar, R.G. Bhuin, S.S. Gupta et al. 2017. Confined metastable 2-line ferrihydrite for affordable point-of-use arsenic-free drinking water. Advanced Materials 29: 1604260.

Lamon, L., D. Asturiol, A. Vilchez, R. Ruperez-Illescas, J. Cabellos, A. Richarz and A. Worth. 2019. Computational models for the assessment of manufactured nanomaterials: Development of model reporting standards and mapping of the model landscape. Computational Toxicology 9: 143–151.

Leeuwen, C.J. Van. 2007. General Introduction. Risk Assessment of Chemicals: 1–36.

Li, J., Y. Huang and D. Shao. 2015. Conjugated polymer-based composites for water purification. Fundamentals of Conjugated Polymer Blends, Copolymers and Composites: 581–618.

Li, N., X. Wei, Z. Mei, X. Xiong, S. Chen, M. Ye and S. Ding. 2011. Synthesis and characterization of a novel polyamidoamine–cyclodextrin crosslinked copolymer. Carbohydrate Research 346: 1721–1727.

Livingston, A., B.L. Trout, I.T. Horvath, M.D. Johnson, L. Vaccaro, J. Coronas et al. 2020. Challenges and directions for green chemical engineering—role of nanoscale materials. Sustainable Nanoscale Engineering: From Materials Design to Chemical Processing: 1–18.

Louie, S.M., R. Ma and G.V. Lowry. 2014. Transformations of nanomaterials in the environment. Frontiers of Nanoscience 7: 55–87.

Martínez, G., M. Merinero, M. Pérez-Aranda, E.M. Pérez-Soriano, T. Ortiz, B. Begines and A. Alcudia. 2020. Environmental impact of nanoparticles' application as an emerging technology: a review. Materials 14: 166.

Mathew, A., P.R. Sajanlal and T. Pradeep. 2012. Selective visual detection of TNT at the sub-zeptomole level. Angewandte Chemie - International Edition 51: 9596–9600.

Mathew, A. and T. Pradeep. 2014. Noble metal clusters: Applications in energy, environment, and biology. Particle and Particle Systems Characterization 31: 1017–1053.

Memon, F.A., Z. Zheng, D. Butler, C. Shirley-Smith, S. Lui, C. Makropoulos and L. Avery. 2007. Life cycle impact assessment of greywater recycling technologies for new developments. Environmental Monitoring and Assessment 129: 27–35.

Mensah, J. 2019. Sustainable development: Meaning, history, principles, pillars, and implications for human action: Literature review. Cogent Social Sciences 5.

Mohan, D. and C.U. Pittman. 2007. Arsenic removal from water/wastewater using adsorbents—A critical review. Journal of Hazardous Materials 142: 1–53.

Mukherjee, S., A.A. Kumar, C. Sudhakar, R. Kumar, T. Ahuja, B. Mondal et al. 2019. Sustainable and affordable composites built using microstructures performing better than nanostructures for arsenic removal. ACS Sustainable Chemistry & Engineering 7: 3222–3233.

Mukherjee, S., H. Ramireddy, A. Baidya, A.K. Amala, C. Sudhakar, B. Mondal et al. 2020. Nanocellulose-reinforced organo-inorganic nanocomposite for synergistic and affordable defluoridation of water and an evaluation of its sustainability metrics. ACS Sustainable Chemistry and Engineering 8: 139–147.

Mukherjee, S., M. Shah, K. Chaudhari, A. Jana, C. Sudhakar, P. Srikrishnarka et al. 2020. Smartphone-based fluoride-specific sensor for rapid and affordable colorimetric detection and precise quantification at sub-ppm levels for field applications. ACS Omega 5: 25253–25263.

Nagar, A. and T. Pradeep. 2020. Clean water through nanotechnology: needs, gaps, and fulfillment. ACS Nano 14: 6420–6435.

Nair, A.S. and T. Pradeep. 2007. Extraction of chlorpyrifos and malathion from water by metal nanoparticles. Journal of Nanoscience and Nanotechnology 7: 1871–1877.

Nanotechnology: small science on a big scale | Feature | Chemistry World. https://www.chemistryworld.com/features/nanotechnology-small-science-on-a-big-scale/3004705.article Accessed: 2022.01.13.

Peng, Z., X. Liu, W. Zhang, Z. Zeng, Z. Liu, C. Zhang et al. 2020. Advances in the application, toxicity and degradation of carbon nanomaterials in environment: A review. Environment International 134: 105298.

Pesqueira, J.F.J.R., M.F.R. Pereira and A.M.T. Silva. 2020. Environmental impact assessment of advanced urban wastewater treatment technologies for the removal of priority substances and contaminants of emerging concern: A review. Journal of Cleaner Production 261: 121078.

Pichel, N., M. Vivar and M. Fuentes. 2019. The problem of drinking water access: A review of disinfection technologies with an emphasis on solar treatment methods. Chemosphere 218: 1014–1030.

Pradeep, T. and Anshup. 2009. Noble metal nanoparticles for water purification: A critical review. Thin Solid Films 517: 6441–6478.

Qu, X., J. Brame, Q. Li and P.J.J. Alvarez. 2013. Nanotechnology for a safe and sustainable water supply: Enabling integrated water treatment and reuse. Accounts of Chemical Research 46: 834–843.

Rastogi, S., G. Sharma and B. Kandasubramanian. 2020. Nanomaterials and the environment. The ELSI Handbook of Nanotechnology: Risk, Safety, ELSI and Commercialization: 1–23.

Rasool, K., R.P. Pandey, P.A. Rasheed, S. Buczek, Y. Gogotsi and K.A. Mahmoud. 2019. Water treatment and environmental remediation applications of two-dimensional metal carbides (MXenes). Materials Today 30: 80–102.

Ravindran, S.J., A. Mahendranath, P. Srikrishnarka, A.A. Kumar, M.R. Islam, S. Mukherjee et al. 2019. Geologically inspired monoliths for sustainable release of essential minerals into drinking water. ACS Sustainable Chemistry & Engineering 7: 11735–11744.

Ren, D., L.M. Colosi and J.A. Smith. 2013. Evaluating the sustainability of ceramic filters for point-of-use drinking water treatment. Environmental Science and Technology 47: 11206–11213.

Rout, P.R., T.C. Zhang, P. Bhunia and R.Y. Surampalli. 2021. Treatment technologies for emerging contaminants in wastewater treatment plants: A review. Science of The Total Environment 753: 141990.

Saleem, H. and S.J. Zaidi. 2020. Developments in the application of nanomaterials for water treatment and their impact on the environment. Nanomaterials 10: 1764.

Salieri, B., D.A. Turner, B. Nowack and R. Hischier. 2018. Life cycle assessment of manufactured nanomaterials: Where are we? NanoImpact 10: 108–120.

Shegokar, R. and M. Nakach. 2020. Large-scale Manufacturing of Nanoparticles—An Industrial Outlook. Elsevier Inc.

Sheldon, RA. 2018. Metrics of green chemistry and sustainability: past, present, and future. ACS Sustainable Chemistry & Engineering 6: 32–48.

Siegert, M., J.M. Sonawane, C.I. Ezugwu and R. Prasad. 2019. Economic assessment of nanomaterials in bio-electrical water treatment. Nanotechnology in the Life Sciences: 1–23.

Smith, L. 2017. Historical Perspectives on Water Purification. Chemistry and Water: The Science Behind Sustaining the World's Most Crucial Resource: 421–468.

Sperber, W.H. 2001. Hazard identification: from a quantitative to a qualitative approach. Food Control 12: 223–228.

Sudeep, P.K., S.T.S. Joseph and K.G. Thomas. 2005. Selective detection of cysteine and glutathione using gold nanorods. Journal of the American Chemical Society 127: 6516–6517.

Upreti, G., R. Dhingra, S. Naidu, I. Atuahene and R. Sawhney. 2015. Life cycle assessment of nanomaterials. *In*: Basiuk, V. and E. Basiuk (eds.). Green Processes for Nanotechnology. Springer, Cham.

Willner, M.R. and P.J. Vikesland. 2018. Nanomaterial enabled sensors for environmental contaminants Prof Ueli Aebi, Prof Peter Gehr. Journal of Nanobiotechnology 16: 1–16.

Xavier, P.L., K. Chaudhari, P.K. Verma, S.K. Pal and T. Pradeep. 2010. Luminescent quantum clusters of gold in transferrin family protein, lactoferrin exhibiting FRET. Nanoscale 2: 2769–2776.

Zeng, X., G. Wang, Y. Liu and X. Zhang. 2017. Graphene-based antimicrobial nanomaterials: rational design and applications for water disinfection and microbial control. Environmental Science: Nano 4: 2248–2266.

Zhou, Y., J.F. Zhang and J. Yoon. 2014. Fluorescence and colorimetric chemosensors for fluoride-ion detection. Chemical Reviews 114: 5511–5571.

Chapter 3

Nanomaterials from the Bench to Industry

An Experience in an Emerging Country

Fiona M. Britto,[1] *M. Jazmín Penelas,*[1] *Verónica Lombardo,*[2] *Mara Alderete*[1] and
G.J.A.A. Soler-Illia[1,*]

1. Introduction

The last 20 years have witnessed an explosive advance in the design, synthesis and processing of nanomaterials with highly controlled dimensions. The possibilities of harnessing the chemical composition and morphology of nanomaterials allow the tuning of magnetic, optical, electrical, biological or mechanical properties. The first generation of the "nanotechnology revolution" led to amazing changes in the electronic structure of nanomaterials due to electron confinement, as well as the drastic increase in the surface to volume ratio of the nanostructures for small size systems, which enhances the role of interfacial energy or surface sites (Baig et al. 2021). During the last decades, nanomaterials have evolved from passive nanostructures (i.e., based on the control of nanoscale properties) to active nanostructures (i.e., responsive to external stimuli). The next generations of nanotechnologies aim at creating nanosystems with different functional domains that will evolve to molecular nanosystems with precisely crafted functions located in space, akin to biological systems (Roco 2011, Soler-Illia and Azzaroni 2011). Nanotechnology is nowadays a key enabling technology that constitutes the base of a transdisciplinary set of foundational technological fields integrated with biotechnology, information technology and cognition (Roco 2020) emerging as a cornerstone for knowledge-based industries originated in partnerships between the university, industry, and government (Nayfeh 2018).

Currently, nanomaterials are produced through two main ways, known as the top-down and the bottom-up approaches. The top-down approach consists of the breakdown of a bulk material into nanoscale objects by mechanical milling, chemical etching, thermal or laser ablation, while bottom-up methods rely on physicochemical processes from molecular precursors, which includes sol-gel synthesis, chemical precipitation and spray pyrolysis, among others.

[1] Instituto de Nanosistemas, Universidad Nacional de San Martín, Av. 25 de Mayo 2021, San Martín, 1650, Buenos Aires, Argentina.
[2] Gerencia Química and Instituto de Nanociencia y Nanotecnología-CONICET, CNEA, Av. Gral. Paz 1455, San Martín, 1650, Buenos Aires, Argentina.
Emails: fbritto@unsam.edu.ar; mpenelas@unsam.edu.ar; marialombardo@cnea.gov.ar; malderete@unsam.edu.ar
* Corresponding author: gsoler-illia@unsam.edu.ar

Materials present nanoscale effects due to confinement in one dimension (e.g., thin films, graphene), in two dimensions (e.g., nanowires and nanotubes) or three dimensions (e.g., quantum dots). The term *nanoparticle* encompasses a wide range of nanostructures of different chemical composition, shape and size in the range of 1–100 nm. Metal oxide nanoparticles such as SiO_2, TiO_2, Fe_3O_4, nanometals such as Au, Ag or Pt, semiconductor nanocrystals like CdS, CdSe, InAs or carbon-based nanotubes are among the most extended examples in industrial applications. Nanoparticles are of great interest as additives to improve properties such as hardness, UV radiation absorption, magnetization, color, etc. of industrial chemicals. The small dimensions greatly facilitate the incorporation of inorganic components into fluid products, in terms of costs and process design. Furthermore, for catalytic, antimicrobial and electronic properties, the dramatic increase in the interface given by the large surface area of the nanoparticles adds significant value to the final product (Stark et al. 2015).

It is important to emphasize that, unlike atoms, which are naturally identical, nanoparticles are not strictly monodisperse, rather presenting size and shape distribution. Since their physicochemical properties strictly depend on the shape and size, synthesis methods must be optimized to achieve the highest possible precision, control and repeatability, in order to generate reliable industrial products.

At the laboratory, nanoparticles can be produced with high purity and customized properties but, in general, synthesis methods allow from a few mg to the gram scale per batch, which is very far from a commercial scale, which was estimated in 2020 between 10 and 10^5 ton/year, depending on the type of nanomaterial and the application. For this reason, commercially available nanomaterials are mainly based on relatively inexpensive inorganic nanomaterials (metal oxides, clays, carbon nanotubes, phosphates, carbonates) and synthetic polymers, as well as well-established scaling-up procedures such as spray-pyrolysis (Bang and Suslick 2010) or fluidized bed reactors (Jia and Wei 2017). Its commercial uses are primarily associated with the automotive, packaging, construction, textile and recreational applications (Charitidis et al. 2014, Sanchez et al. 2011).

The path towards real-world applications of sophisticated nanomaterials implies, on the one hand, technical challenges associated with the scaling-up of synthesis processes to generate nanostructures with a high production rate, preserving the original physicochemical properties (Butler 2016, Sebastian et al. 2014, Stark et al. 2015). On the other hand, practical, safety and regulatory barriers must be taken into account for a successful translation to the industry.

In the particular case of the production of nanoparticles, strict requirements are imposed on their physical and chemical characteristics, including size, shape, chemical composition, type of functional group on their surface and dispersibility (Sebastian et al. 2014). The upscaling may significantly alter heat/mass transfer rates, which leads to different regimes for the nucleation and growth of nanoparticles that affect, in turn, their size, crystallinity or agglomeration.

Different strategies have been explored to produce nanomaterials from bottom-up methods on a larger scale by controlling the properties. Usual synthesis can be in the gas phase (aerosol, plasma, CVD) or the liquid phase, the latter taking place in milder conditions while allowing good control of size, monodispersion and agglomeration. The use of different configurations of batch reactors, continuous reactors, microreactors and continuous flow microfluidic devices has been studied for a variety of up-scaled synthesis in the liquid phase (Duanmu et al. 2018, Khan et al. 2004, Miroshnychenko and Beznosyk 2015, Paliwal et al. 2014, Shiba and Ogawa 2018). Less conventional methods, such as hydrothermal synthesis using supercritical water, have also been evaluated (Tighe et al. 2013). On the other hand, several works have emphasized sustainability through the selection and evaluation of non-toxic reagents agents and solvents and the development of energy-efficient synthetic methods (Duan et al. 2015).

The development of new large-scale processes for nanomaterial fabrication is expensive and needs large investment; therefore, traditional industries are reluctant to invest, unless they are guaranteed profitable returns. On the other hand, toxicology and environmental issues related to nanoscale materials are relatively unexplored in a systematic way and lead to caution in the market towards nanoenabled products. This factor adds to the costs. Therefore, techno-economic analysis and life cycle analysis are mandatory for improving the impact of new nanomaterials and creating a sustainable ecosystem for the development of nanotechnologies.

The situation is especially relevant in developing countries, in which a critical mass of researchers in the field is beginning to emerge, although a lack of infrastructure is evident. However, there is still a disconnection between laboratory- and industrial-scale investigation, as well as a lack of knowledge in areas critical to translation such as intellectual property and regulations. Solving these aspects is crucial to generate economic growth through high added-value nanotechnologies, necessary for improving the economies of that particular region.

This chapter discusses some critical aspects in the process of translating nanomaterials synthesis into actual products, in the context of an emergent country. Argentina has a rich research system with high standards and traditions that include three Nobel Prizes in Chemistry and Medicine. The nanotechnology community in the country was established around 2000 and received particular support as a vacancy area between 2004 and 2010. This allowed to establish a sound scientific community with high-quality researchers, and the advent of a small yet interesting and relatively productive science-technology ecosystem (Berger et al. 2021, Foladori and Carrozza 2017). This ecosystem enabled a fast response to the Covid-19 pandemic with commercial products such as the antibacterial facemasks manufactured by Kovi, fast nanoparticle-based Co-SARS-2 sensing kits made by Chemtest or antimicrobial coatings developed by Hybridon.

Next the possibilities and limitations of the direction that spans from materials design, laboratory production, scaling-up, registration of nanostructured materials to a final tradeable product will be illustrated using case studies carried out in our laboratories. We will analyze the technological and economic motivations of the project, and present a brief overview of the technical solution proposed, to finally show the outcomes and limitations. We believe that the accumulated experience illustrates the motivations, the technical and economical complexity of the nanotechnology market in an emergent country with limited infrastructure but a well-established scientific community.

2. Development of a Nanostructured Inorganic Antifungal: From Lab Design to Plant Scaling

One of the modern challenges in agriculture is the sustainable production and use of antimicrobial agents, pesticides and fertilizers, in order to optimize their performance and minimize the leakage of potentially damaging chemical species. For example, copper-based pesticides that include metallic copper, hydroxides, oxides or basic salts as active ingredients are well known and have demonstrated high activity against a variety of bacterial and fungal diseases. However, copper accumulation might lead to soil contamination and affect biomass (Keiblinger et al. 2018). It is expected that the use of copper-based nanoparticles might lead to enhanced antimicrobial action, together with a reduced metal charge that minimizes contamination.

In this framework, Tort Valls S. A. (Pilar, Argentina), a chemical company that produces and markets antifungal compounds for agriculture had a dual motive for developing a nanostructured cuprous oxide to meet the international market demands. First, their technology based on Cu (II) basic chlorides had become less competitive with respect to cuprous species that deliver copper ions more efficiently than cupric species. In addition, smaller particles were sought for, which can have an enhanced effect due to a size-dependent change in reactivity explained by its surface area. Thus, decreasing particle size would lead to materials that are more effective as antimicrobials.

The first contact came through the Argentine National Research Council (CONICET) Liaison Office, which put our group in contact with the company. Preliminary meetings were focused on understanding the company's needs and expectations, as well as in defining a product that could outperform the antimicrobial performance of its competitors. The objective of the project was to develop a protocol to produce nano-cuprous oxide in the laboratory (200 mL and 2 L reactors), to further develop a pre-pilot scale batch test (20 L). These meetings were followed by visits to the production plant, in order to get familiar with Tort Valls' ongoing technologies, processes and equipment, which was useful to design a process compatible with the resources and know-how of the company.

A second important issue was the search for regulatory requirements. We followed the FAO (Food and Agriculture Organization of the United Nations) requirements to evaluate the quality of the product throughout the project. These requirements particularly specify the percentage of cupric species, metallic copper and the minimum content of cuprous species that the product must reach.

The first approach proposed sugar as a reducing agent, a well-known way to produce cuprous species from cupric precursors (Fehling's reaction, Fig. 3.1a). The first protocols were aimed at optimizing the quality and concentrations of the reagents to obtain a product that meets the regulatory requirements. The process was performed at room temperature and in a batch process that involved a reaction in a stirred tank followed by a separation step of Cu_2O particles, as depicted in Fig. 3.1b.

Despite obtaining a high-quality nanostructured product that met regulatory requirements, the company decided not to follow the developed protocol, because this synthesis strategy produced a significant amount of effluent. On the other hand, the product was slightly contaminated with unreacted sugar molecules, which could function as a carbon source for microorganisms. Last, there were some mass transfer issues that hindered translation to the plant scale from laboratory-size reactors.

With the lessons learned, we set our goal to use cleaner reducing agents both industrially applicable and that minimize residues in the sample. In order to gain insight into the scaling-up process, we carried out intermediate-scale tests (20 L) of the protocols developed, which helped to reduce the gap between reactions developed on a laboratory scale and plant scale. After evaluating several options and careful optimization, a reducing agent that met all the technical and economic requirements was selected, which is industrially protected.

After carrying out a post-laboratory synthesis of the developed protocol, the company took the task to scale up the production in the plant, in pilot-scale (4000 L) shown in Fig. 3.2a. The XRD patterns of the final product were in good agreement with the Cu_2O pattern (Fig. 3.2b) and SEM characterization of the product obtained showed nanostructured particles (Fig. 3.2c). The obtained products were re-tested with satisfactory results regarding antifungal capacity. This test was carried out in citrus plantations in the province of Tucumán and it was found that the product performance was equivalent to a premium product, used as a control sample.

(a)

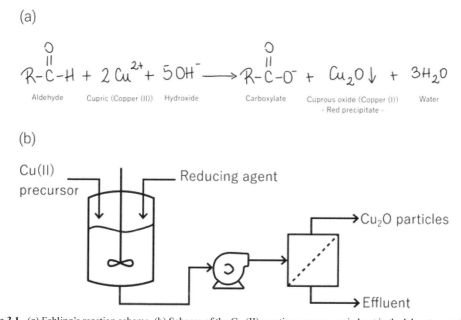

(b)

Figure 3.1. (a) Fehling's reaction scheme. (b) Scheme of the Cu (II) reaction process carried out in the laboratory and plant.

300 ml   1000 ml   20 L   4000 L

Figure 3.2. (a) Photographs of the different setups used in the successive scaling-up processes, from 200 mL to 4000 L. (b) XRD patterns (Source CuKα: 1.541 Å) for the cuprous oxide obtained in the laboratory (blue curve) and the 4000 L reactor (red curve). Diffractograms present peaks corresponding to Cu$_2$O (cuprite) with a cubic lattice of cell parameter $a = 4.26$ Å. (c) Representative SEM micrograph and (d) Obtained powder of the product obtained in the plant scale production using the final developed protocol.

In conclusion, this project proved the importance of the efficient communication between the company and the research group. This resulted in the development of a sustainable and robust synthetic method towards cuprous oxide nanoparticles that was easy to scale up with the already existing competencies in the company. Another important issue was the establishment of clear goals and well-defined roles for both teams at the company and the university. Critical to the success was to establish partial stages and verifiable milestones, in order to optimize the project flow.

3. Synthesis Scalable to Pilot Plant of Bifunctional Silica Colloids with Application in Nanostructured Coatings and Nanocomposites

Surface-modified silica nanoparticles constitute an essential building block in a wide variety of platforms for modifying functional properties of polymers or surfaces (Sanchez et al. 2011). Nanosilica offers high stability and surface area, optical transparency and low toxicity (Iler 1979). In addition, the well-established silica chemistry allows modifying the nanoparticle surface with active agents such as molecular ligands, biomolecules or polymers (Beltrán-Osuna and Perilla 2016, De Crozals et al. 2016). The synergistic combination of these characteristics allows to tailor special properties with great specificity for novel materials, catalysis, environmental remediation and biomedical applications (Banerjee et al. 2014, Jawed et al. 2020, Sharma et al. 2015, Walmsley et al. 2015). For instance, organic modifiers allow to increase chemical compatibility at the silica-polymer interface, improving the dispersibility of silica nanoparticles in any organic matrix (Lee and Yoo 2016, Rahman and Padavettan 2012) or improving the adhesion and durability properties of non-stick or superhydrophobic coatings (de Francisco et al. 2014).

Here the development of bifunctionalized silica nanoparticles from the laboratory-scale to the design of a pilot-plant scaled process, carried out in collaboration with U.B. Rhein

Chemie, Lanxess S.A. (Burzaco, Argentina) are presented; this company unit was dedicated mainly to the tire industry. The goal of the project was to develop a robust coating with non-stick and lubricating properties to pretreat rubber surfaces. To this end, it was proposed to use surface-modified silica nanoparticles (SiO_2-NPs) with two complementary functional groups: Vinyl (V) and polydimethylsiloxane (PDMS). Vinyl is a polymerizable group that can constitute an anchoring group of the nanoparticle to a reactive polymeric surface such as an uncured rubber (Ketelson et al. 1995) while polydimethylsiloxane (PDMS) is a silicone capable of imparting hydrophobicity to the surface, which can be used to control lubricity or adhesion properties (de Francisco et al. 2014).

Hence, the project inferred two aspects to be developed:

- Obtaining a prototype nanomaterial on a laboratory scale and characterizing its properties.
- Design of a pilot plant to carry out production, according to the requirements of the synthesis, purification and separation of nanomaterials.

The development of the prototype nanomaterial began with the synthesis of SiO_2 NPs by the well-known Stöber method that suggests the hydrolysis and condensation of an alkoxide precursor (tetraethoxysilane, TEOS) in an ethanolic solution, in the presence of ammonia (Stöber et al. 1968). While this synthetic path is not the best choice for commercial materials due to the cost, it provides a reproducible route to monodisperse particles, constituting a model system. Spherical monodisperse SiO_2-NPs of 100 nm diameter were synthesized. Subsequently, surface functionalization was carried out by post-synthesis grafting, where the silanol ($\equiv SiOH$) or silanolate surface groups ($\equiv Si\text{-}O^-$) can react with functional organosilanes through controlled condensation.

To meet the needs of the project, an alkoxysilane, vinitrimethoxysilane (VTMS) and a hydroxyl terminal reactive silicone (PDMS-OH) were selected as surface modifying agents. The surface functionalization process schematized in Fig. 3.3a was designed to take place by silanization in two stages. First, the VTMS was added in an adequate quantity to achieve a partial coverage of the surface and then, the PDMS-OH was incorporated through condensation of the silicone chains on the available sites. Finally, the separation and purification of the bifunctional nanoparticles, SiO_2-(V+PDMS) NPs, was performed by successive cycles of washing, centrifugation and resuspension of the particles, until total elimination of the components of the synthesis medium and possible excess of functionalizing agents. Lastly, nanoparticles were oven-dried and ground into a fine powder ready for further characterization and use.

The prototype nanomaterial was successfully obtained at the laboratory scale with high control of morphology and the incorporation of both functions. The synthesis of SiO_2-(V+PDMS) NPs turned out to be robust and reproducible. The SEM micrograph of Fig. 3.3b shows the spherical and monodisperse morphology, while the IR spectrum of the NPs indicates the presence of both functional groups, vinyl and polydimethylsiloxane, in the final nanomaterial (Fig. 3.3c). Contact angle experiments demonstrated the increase in the surface hydrophobicity of the NPs throughout the process (Fig. 3.3d).

The challenge consisted of designing a pilot plant that would be especially built to produce this nanomaterial. When considering scaling up this nanomaterial, several concerns arose. It was necessary to review various synthetic aspects and to adapt them in order to develop a viable production of this material. In the first place, the large-scale production of colloidal silica by the Stöber method was considered impractical, since the synthesis has a relatively low yield (approximately 1% m/V), and also requires the use of relatively expensive precursors (i.e., TEOS) and flammable organic solvents (ethanol) (Hyde et al. 2016). Furthermore, special care must be taken with the handling and storage of TEOS to preserve its purity since it was observed that the presence of prehydrolyzed by-products in the reagent affects the final monodispersion of the silica NPs. In this context, adapting the surface functionalization process to commercial fumed silica (Aerosil®) emerged as an interesting alternative, with the aim of obtaining a high value-added nanomaterial from a commercial silica precursor.

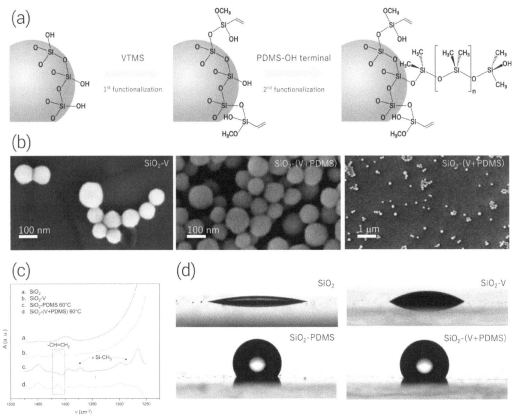

Figure 3.3. (a) Scheme of the synthesis process of bifunctional SiO$_2$-(V+PDMS) nanoparticles; (b) SEM micrographs of SiO$_2$-V (left) and SiO$_2$-(V+PDMS) nanoparticles at different magnifications (center, right); (c) detailed FTIR spectrum of functionalized NPs indicating the presence of vinyl and polydimethylsiloxane functional groups in the nanoparticles; (d) contact angle characterization of the NPs along the functionalization process.

The most convenient process for the silica functionalization is a one-pot reaction process. In a one-pot synthesis, the entire process takes place in the same vessel, on which the reagents are successively added. This approach allowed us to save time and energy and minimize waste. For this route to be viable, the design of orthogonal surface modification strategies is essential, as this process enables the preservation of integrity and reactivity of the different functional groups present during the reaction (Ho Wong and Zimmerman 2013). These requirements are fulfilled in the case of the functionalizing agents involved in this development, VTMS and PDMS-OH.

After functionalization, the next steps are the purification and separation of the nanoparticles. As mentioned above, centrifugal wash and resuspension cycles are not easily accessible on a larger scale for nanocolloids. In the case of surface modification by silanization, distillation is an interesting alternative. The by-products of the reaction using alkoxysilanes are volatile: alcohol, water and ammonia or another amine that acts as a basic catalyst. It is also possible that high-boiling oligomers generated by the condensation of the silane in solution are present. However, if the functionalization conditions are optimized, its concentration is expected to be low.

Distillation would not only fulfill the function of purification, but also that of separation and transfer to a suitable medium. Once the nanoparticles have been functionalized and the volatile by-products removed, they can be concentrated as much as desired by reducing the solvent. Likewise, a second liquid can be incorporated into the suspension and make a solvent exchange by distillation in order to obtain the suspended NPs in a more convenient solvent. In our case, dimethicone, a low toxicity and low cost high-boiling cyclic silicone, was chosen. As a result, a suspension of the bifunctionalized nanosilica in dimethicone was obtained, which was suitable to be applied directly as a spray coating.

The bifunctionalized silica NPs obtained can be processed and incorporated into various base materials, in order to obtain surfaces with specific functional properties as schematized in Fig. 3.4a. The compatibility tests of SiO_2-(V+PDMS) NPs with the rubber compound indicated a good dispersibility of the nanoparticles while conserving the curing kinetics and the mechanical behavior of the polymeric material. A considerable increase in the tear resistance of the compound was also found, which is related to the ability of NPs to act as reinforcement. A partial transfer of the nanoparticles was observed by SEM after a series of curing cycles, which is due to the low curing of the tires in the process (Fig. 3.4b), making this formulation impractical for the intended use.

However, a dispersion of the functionalized nanoparticles could be used to impart super hydrophobicity to surfaces. Commercial nanosilica functionalized with PDMS in alcoholic suspension was deposited on filter paper using a manual atomizer to obtain a superhydrophobic coating. Figure 3.4c shows the typical "pearl" shape of the water droplets deposited on superhydrophobic surfaces obtained by dispersion of the bifunctional nanoparticles, showing a contact angle greater than 160°. Furthermore, by using a mask, superhydrophobic and other highly hydrophilic areas can be obtained on the same surface (Fig. 3.4d). The low surface energy of the modified nanoparticles, combined with the hierarchical micro and nano-scale topography of the aggregates formed during synthesis, contribute to achieving the architecture that gives rise to the phenomenon (Liu and Jiang 2012).

In summary, this example illustrates the limitations that arise when transferring the conventional laboratory functionalization scheme directly to a larger scale. In this case, it was necessary to design an alternative scheme, in which four main changes were introduced in the pre-pilot scale with respect to the laboratory-scale synthesis:

- High surface area commercial silica was used
- The bi-functionalization was performed in one pot two-step strategy,

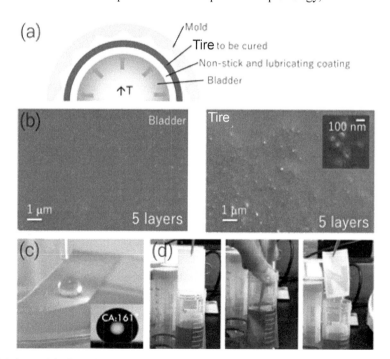

Figure 3.4. (a) Scheme of the tire mold production, indicating the location of the non-sticking surface location on the bladder surface. (b) SEM pictures of the bladder and the nanocomposite surfaces after five curing cycles. (c) Superhydrophobic paper surface after spray coating of modified silica dispersion. (d) Hydrophilic pattern on a hydrophobic surface.

- Purification and separation were performed through distillation.
- Functional NPs were obtained in a concentrated suspension in a suitable solvent, instead of a dry powder, being ready to apply.

It is important to mention that the guidelines that emerge from the analysis for this particular case of nanoparticles and functional groups can be extrapolated to the production of other hybrid particulate nanomaterials by the sol-gel approach.

4. Scaling-Up, Functionalization and Reusability of Mesoporous Adsorbents

The emergence of ordered mesoporous materials in the mid-90s paved the way to high surface area matrices with a great potential in fields such as environment, energy, catalysis and biomedicine. In spite of their straightforward and reproducible synthesis, upscaling of mesoporous matrices and the associated costs are still significant hurdles to industrial applications.

In this context, a joint project to explore the use of mesoporous adsorbents was performed between the Argentine National Atomic Energy Commission (CNEA) and U.B. Rhein Chemie, Lanxess S.A. (Burzaco, Argentina), who were interested in developing adsorbents for their clients in the mining industry. The removal of heavy metals from wastewater is a relevant issue, as most of them are toxic even at very low concentrations. In particular, this project proposed developing a functionalized mesoporous silica matrix with high selectivity and effectivity for Cu (II) adsorption at low metal ion concentrations; due to the cost and waste issues, the material should be reusable (Aguado et al. 2009).

There were thus two development challenges here: the synthesis of an organic-inorganic mesoporous hybrid matrix with superior performance and cyclability for metal recovery, and the possibility of scaling up the material in an economically convenient manner. In particular, Cu (II) was selected as a model of heavy metal ion that constitutes an important residue from mine tailing dams.

Due to its open porous structure and high surface area, mesoporous SBA-15 particles were chosen as the adsorbent. Mesoporous SBA-15 particles were synthesized through precipitation and then functionalized with amino groups ($-NH_2$) to impart selectivity towards Cu (II) ions through metal complexation.

Although mesoporous silica can be reproducibly obtained in the laboratory, its large-scale production is a challenge. In particular, SBA-15 is synthesized under strong acidic conditions, using PEO-PPO-PEO triblock copolymer (P123 Pluronic) as a structure-directing agent and a silica source, usually TEOS or sodium silicate. SBA-15 synthesis has five main steps: (1) surfactant dissolution in an acidic medium; (2) addition of TEOS to the mixture under stirring at 35°C; (3) aging at 90°C under static conditions; (4) separation; and (5) calcination. A process flow diagram is presented in Fig. 3.5.

Several efforts have been made over the past two decades to fully understand the formation mechanism of this material. However, detailed information on large-scale production strategies for SBA-15 is still lacking. Typically, SBA-15 synthesis is carried out in a batch process at low pH; the long synthesis and aging times at elevated temperatures lead to time and energy consumption and make it difficult to scale up. Under these conditions, the maximum reported as up-scaled SBA-15 synthesis was for a volume of 1.5 L with a yield of 24–27 g per batch (Thielemann et al. 2011).

The strategy implemented by our group to adapt a laboratory synthesis to a scalable and cost-effective protocol consisted of the evaluation of risk factors that appear when increasing the reaction volume at different synthesis stages, as depicted in Fig. 3.5. We were mainly interested in the effect that different physical parameters have on the final structural and morphological properties of the synthesized material, which are the most sensitive to scale change, such as temperature, stirring rate and time reaction. One of the major drawbacks to achieve a large-scale production of SBA-15 is the requirement for long synthesis times (i.e., 20 hours at 35°C and 24 hours at 90°C). We were able to reduce the reaction times to 2 and 15 hours for the first and second reaction steps,

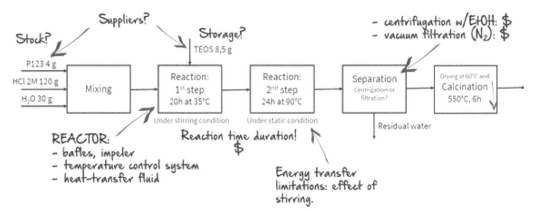

Figure 3.5. Scheme of the different stages of SBA-15 production, indicating challenges in the process of scaling up.

respectively, with only minor differences in structural and adsorption properties. This result was in good agreement with works carried out by several groups who studied the influence of reaction time during mesophase formation and aging (Flodström et al. 2004a, Flodström et al. 2004b, Flodström et al. 2004c, Ruthstein et al. 2006) aiming at optimizing the synthesis times (Fulvio et al. 2005a, 2005b, Jaroniec and Fulvio 2005).

Generally, SBA-15 synthesis protocols tend to differentiate only between experimental conditions under 'vigorous stirring' or a 'static condition' and no information is given about the stirring rate and the geometry and size of the reactor or the structure of the impellers/magnetic stirring bars. These are key factors that affect the fluid dynamic conditions of the reaction and therefore are especially important at scalingup conditions. We have recently found that SBA-15 synthesized at a low stirring rate (200 rpm) showed poor mesoscopic order but at a higher stirring rate (400 rpm), a highly ordered mesoporous structure was obtained. *Ex situ* SAXS characterization (Fig. 3.6) showed the broadening of scattering peaks and the disappearance of higher-order peaks for SBA-15 synthesized at the low stirring rate. This result was further confirmed by nitrogen adsorption, where the obtained hysteresis loops are different from those typical of highly ordered mesoporous materials, indicating an extra disorder in the channels (Fig. 3.6).

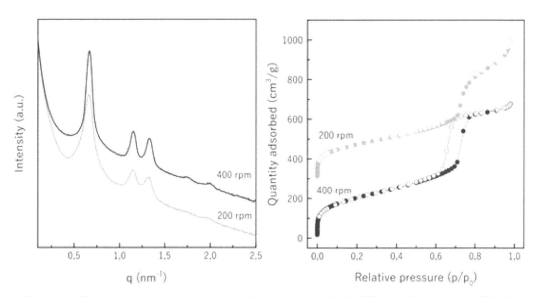

Figure 3.6. Effect of stirring in the mesostructure (left) and pore texture (right) of SBA-15 (Britto et al. unpublished).

An alternative route towards industrial production of templated mesoporous materials is through the aerosol method. A solution or suspension is atomized into very small droplets that can be considered as individual reactors. These droplets dispersed in a carrier gas come into contact with a hot gas stream. Particle production takes place on drying or pyrolysis of these droplets, the resulting powders are recovered in a collector at the end of the equipment (Lombardo et al. 2021). The synthesis of mesoporous materials by combining sol-gel and aerosol, using a pore templating agent, is known as aerosol-based Evaporation-Induced Self-Assembly (EISA) (Debecker 2018). As with the typical synthesis of mesoporous materials by EISA, the gradual drying of droplets allow the formation of a liquid crystal mesophase, assisting the formation of the inorganic (or hybrid) network around the soft template. A scheme of the process can be seen in Fig. 3.7. This type of synthesis often involves volatile solvents that are generally flammable, as well as strong inorganic acids, such as HCl, which is either the catalyst for the silica sol-gel reaction or an agent to control the condensation reaction of transition metal centers (Araujo et al. 2010, Guo et al. 2016).

The presence of flammable organic solvents as well as inorganic acids that can corrode parts of the aerosol equipment at high temperatures is a limitation for the industrial synthesis of mesoporous aerosols. In this context, recent works aim at obtaining synthesis methods with non-toxic or less flammable solvents and safer chemical precursors and reaction media (Xiong et al. 2016). Zelcer et al. introduced the use of a mixture of acetylacetonate and acetic acid for the synthesis of mesoporous oxides in an aqueous medium. This is an interesting approach since acetic acid catalyzes sol-gel hydrolysis and acetate can form complexes with hydrolyzed metallic species, modulating the condensation kinetics of the system. This original approach allows the continuous production of mesoporous oxide spherical particles with high control of the surface area and porosity, in an aqueous and less aggressive medium (Soler-Illia et al. 2016, Zelcer et al. 2020).

Concerning hybrid functional materials, SBA-15 is a suitable large-pore matrix to be readily modified. A post-grafting route allows to obtain accessible amino functions, which improves the charge of Cu (II) ions. From an industrial perspective, amine functionalization presents limitations to large-scale implementation, as high temperatures and aromatic solvents are involved. In order to evaluate the feasibility of working under more sustainable conditions, a study of the degree of functionalization achieved at different temperatures and reaction solvent was carried out (Lombardo et al. 2012, Lombardo 2013). Table 3.1 shows the N/Si ratio obtained by Elemental Analysis (EA)

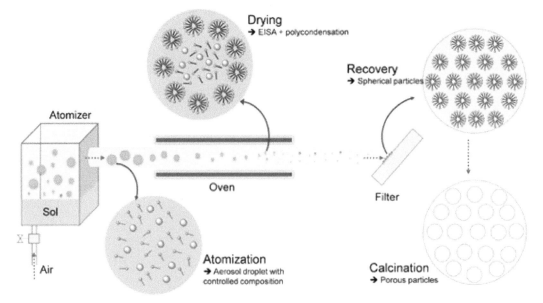

Figure 3.7. (a) Aerosol-assisted sol-gel process for mesoporous materials combined with the EISA process. From Reference (Debecker 2018), with permission from John Wiley and Sons.

Table 3.1. N/Si ratio by elemental analysis and XPS for different samples of SBA-15 functionalized with APTES by post-functionalization in different solvents and maximum adsorption capacity (Q_0) of Cu (II) per gram of material.

N/Si	EA	XPS	Sup/Bulk	Q_0 (mg/g)
SBA-N-C T-R	0.16 ± 0.02	0.49 ± 0.05	3.1	89 ± 1
SBA-N-E T-R	0.19 ± 0.02	-	-	72 ± 3
SBA-N-C T-RT	0.14 ± 0.02	0.42 ± 0.04	3.0	77 ± 2
SBA-N-E T-RT	0.18 ± 0.02	-	-	64 ± 3
SBA-N-C THF	0.17 ± 0.02	0.49 ± 0.05	2.9	
SBA-N-C EtOH	0.05 ± 0.01	0.16 ± 0.02	3.2	23 ± 4
SBA-N-C EtOH 40	0.05 ± 0.01	0.15 ± 0.02	3.0	20 ± 4
SBA-N-C iPrOH	0.05 ± 0.01	0.17 ± 0.02	3.4	22 ± 5

and by XPS for different samples of SBA-15 post-grafted with 3-(aminopropyl)-triethoxysilane (APTES). In some cases, the porogen agent was removed by calcination (SBA-N-C) and in others by extraction with an EtOH/HCl solution (SBA-N-E). The solvents in which the functionalization was carried out were toluene, at room temperature (T-RT) and at reflux (TR), ethanol at room temperature (EtOH), and at 40°C (EtOH 40) and tetrahydrofuran (THF) and isopropanol (iPrOH) at room temperature. A comparison of the N/Si ratio obtained by EA (bulk) with that obtained by XPS (surface) permits to establish that in all cases, surface functionalization is preferential.

The materials in which the surfactant was extracted (SBA-N-E T-R and SBA-N-E T-RT) present a higher degree of functionalization with respect to the calcined samples (SBA-N-C T-R and SBA-N-E T-RT). This is due to the fact that extracted samples present a higher amount of surface Si-OH groups, which are ideal anchoring sites for APTES, similar to that observed by Aguado et al. (2009). The functionalization in THF was similar to that obtained in toluene; the use of this solvent must be evaluated in each particular case, as THF has the disadvantage of being more hydrophilic than toluene. The lower degree of functionalization exhibited by the functionalized materials in alcohols may be due to the fact that the protic polar solvents (EtOH and iPrOH) effectively solvate the nucleophile, reducing its reactivity, which leads to a lower degree of functionalization. For this, it must be assumed that there are no significant differences in solvent diffusion within the porous structure and that the functionalization reaction has an SN_2 mechanism (Bento and Bickelhaupt 2007).

These materials were evaluated as Cu (II) adsorbents, and the maximum adsorption capacity (Q_0) is also shown in Table 3.1. In the case of samples functionalized in alcohols, Q_0 was three times lower than for the other materials, in agreement with the lower degree of functionalization observed.

The materials functionalized in toluene at reflux have greater adsorption capacity than the materials functionalized in toluene at room temperature. However, an analysis of the N/Si ratio for these materials (Table 3.1) does not show a significant difference in the degree of functionalization. This can be explained by the lability of the functional groups since the APTES on the surface is susceptible to hydrolysis. Using elevated temperature during functionalization or a subsequent heat treatment leads to a more effective condensation of the APTES on the silica surface (Lombardo et al. 2012).

Finally, the performance as a Cu (II) absorbent of the material with the highest degree of functionalization was compared with that of a commercial exchange resin, Lewatit® MonoPlus TP207, finding similar metal ion uptake capacities. Despite good Cu (II) adsorption capacity was shown by SBA-15-N matrices and faster adsorption/desorption, the use of these sophisticated materials in real situations was still limited by their cost. Therefore, efforts were made to maximize the reusability of these adsorbents in a process that could recover the Cu (II). Repeated adsorption-desorption and regeneration cycles were tested with an especially designed bicarbonate regeneration step. This procedure showed excellent reuse properties in repeated adsorption-desorption cycles. Moreover, the SBA-15 modified with amino groups could be incorporated into a porous polymer matrix such as polyacrylonitrile to obtain a robust and reusable adsorbent, easy to process in a column under flow conditions (Lombardo and Soler-Illia 2020). In spite of its promising technical features, the preparation and cost issues made SBA-15 based adsorbents not competitive with regards to the already used exchange resins. However, the adsorbent might be of use for recovering higher added value molecules from complex media.

5. Antibacterial Coatings with Residual Activity: Adapting the Technology to the Product

According to the World Health Organization (WHO), approximately 700 thousand deaths are caused worldwide every year by healthcare-related diseases (HCRD). On the other hand, these kinds of diseases produced more than US$ 6.5 billion of economic losses only in the USA.[1] In addition, the emergence of bacteria resistant to antibiotics has raised the interest in nanostructured inorganic materials such as silver or transition metal oxides (Cu, Zn, Mn…). These nanospecies provide metallic ions in their surroundings, which present affinity to sulfur proteins. Metal ions adhere to the cell wall and cytoplasmic membrane, leading to the disruption of the bacterial envelope. Interestingly, bacteria cannot adapt to these toxic agents. Therefore, nanostructured particles, the source of metallic ions are promising antimicrobial agents.

In this framework, a medium-sized company (ADOX SA, Ituzaingó, Argentina) which produces and markets disinfectant products for health institutions, was looking for a technology with high disinfection power, broad spectrum and residual effect to prevent infections and contaminations in a hospital environment. Although the products of the company were efficient, they lacked residuality, so they could not protect the surface against new contamination events after disinfection.

Our groups had developed a technology with long-lasting antimicrobial effects on treated surfaces, based on Mesoporous Thin Film (MTF) ceramic coatings capable of adsorbing and releasing silver ions over time (Catalano et al. 2016). This development received the main national prize at the INNOVAR 2016 Fair. The mesoporous film technology allowed to modify glass, tiles or other ceramic surfaces with a permanent porous layer that was *per-se* antibiofilm (Pezzoni et al. 2017) and could be further loaded with antibiotics. However, this process was not compatible with materials that undergo thermal decomposition, such as plastic surfaces.

ADOX saw the potential of this nano-enabled platform and decided to establish a joint public-private project with our laboratories, with the idea of incorporating long-lasting antimicrobial technology into its portfolio of products marketed to health institutions. The company needed a powerful yet flexible product to treat a variety of surfaces, easy to apply and eventually remove. This led to rethinking of the material and the technology, based on market needs and aimed at non-expert users.

The first product design meetings were focused on defining the problem, and the product desired characteristics. The product, like any disinfectant used in health institutions, had to comply with all the requirements requested by the Argentine regulatory health institution (Argentine National Administration of Drugs, Food and Medical Technology, ANMAT), in terms of effectiveness and safety. At this stage, the company's guidance was the key to define the problems that this technology had to solve, such as user behavior, regulatory issues, costs, necessary technical and user validations.

A work team with complementary skills was formed, composed of the ADOX CEO (commercial and regulatory), senior research scientists with interdisciplinary backgrounds (chemists, biochemists, material engineers, pharmacists), and junior researchers (biotechnologists, project and management specialists and pharmacists). The team was challenged to adapt the technology to develop a product that matched the ADOX business model in less than three months. A specific spin-off technology company, Hybridon, was created to develop a new line of products and technologies to prevent

---

[1] https://www.who.int/gpsc/country_work/gpsc_ccisc_fact_sheet_en.pdf.

infections and contaminations. This company was born after an agreement between the public institutions and ADOX.

The first approach involved a mesoporous particle as the active principle, within a formulation that contains a film-forming agent (polymer) and a solvent that facilitated the application of the product, making it possible to deposit it by spraying. Several options of mesoporous particles were explored to retrieve the functionality of the coatings and their ability to adsorb and desorb over the time metallic ions that act as antimicrobials. The first candidate was SBA-15, which was well-known by the team, and had shown excellent performance to adsorb and desorb metallic ions, such as copper and silver. SBA-15 synthetic protocols were improved with regards to time and temperature, as discussed earlier. A second relevant issue was the separation and purification process, which represents one of the most important factors in the final cost.

Microbiological tests demonstrated a fast disinfection, as well as prevention of microbial growth on the treated surfaces for the required period of 24 hours; actual disinfection persisted for much longer times. The short-term efficacy is shown in Table 3.2 as the number of colonies at the beginning of the experiment and after 20 minutes of exposure (N_0 and N_{20}, respectively).

The reduction percentage of bacteria after 24-hour exposure ($D\%$) and the effectiveness (an efficient antibacterial performance is considered for $R > 2$) are shown in Table 3.3. Both parameters clearly indicated that the product was compatible with cleaning procedures established by health institutions, which involve cleaning procedures at the most every 24 hours.

In order to demonstrate that the product does not produce any undesirable effect in contact with skin, its potential skin-irritating effect was tested using the OECD Guidelines for the testing of

Table 3.2. Antimicrobial effect assessment by long-lasting antimicrobial coating in a short time. Test performed using International Standard Organization (ISO) 18593:2004.

Microorganism	Tiles		Ceramic		Steel	
	N_0	N_{20}	N_0	N_{20}	N_0	N_{20}
E. coli	334	ND	303	ND	298	14
P. aeruginosa	575	ND	383	ND	410	ND
S. cholerasuis	363	ND	419	ND	543	ND
S. aureus	213	ND	263	ND	250	ND
B. cereus	383	9	297	5	388	ND

N_0: initial number of colonies on the surface; N_{20}: number of colonies on the surface after 20 minutes exposure to formulation; *ND*: No detection of colonies.

Table 3.3. Residual effect evaluation in different surfaces by Japanese Industrial Standard- Antibacterial Products-Test for antibacterial activity and efficacy-JIS Z 2801: 2010 (E)-normative.

Ceramic				
Cycle	**Control (CFU)***	**I-MSN** (CFU)**	**D% *****	**R ******
1 (24 hours)	1.45×10^5	< 10	> 99.999	5.16
2 (48 hours)	1.13×10^5	< 20	> 99.999	5.84
3 (72 hours)	3.28×10^7	1.92×10^3	99.994	4.86
Tiles				
Cycle	Control (CFU)*	I-MSN (CFU)	D% **	R ***
1 (24 hours)	5.5×10^4	< 10	> 99.999	4.74
2 (48 hours)	2.51×10^5	< 20	> 99.999	5.39
3 (72 hours)	2.88×10^7	4.7×10^2	99.998	4.77

* CFU: Colony Forming Units; ** I-MSN: ion-adsorbed Mesoporous Silica Nanoparticles; *** *D%*: reduction percentage, **** *R*: effectiveness parameter; R > 2 indicates biocidal effect. **Note:** In this evaluation, the same surface was re-contaminated after 24, 48 and 72 hours.

chemicals 404 (adopted 24th–April 2002). Both the liquid product in direct contact with skin, and a treated surface in contact with skin were evaluated. No irritation effects were observed. An evaluation of the corrosion effect produced by the coating was performed by electrochemical techniques, in order to validate the application on metal surfaces. The results showed that the product did not produce corrosion in carbon steel SAE-1020 and stainless steel AISI-304.

The following step was to evaluate the product performance in real use situations, which was carried out in three different health institutions. The tests were performed by the technical team, using swabs to collect contamination in surfaces, and then assessing the number of colonies in treated and untreated surfaces, after and before applying the product. These field experiments led to R > 4 for more than 24 hours in different surfaces of hospital rooms and operating rooms, validating the problem and allowing to verify the quality of the product under real-world situations. Interestingly, the coated surface also revealed an effective agent against *Coronaviridae*, which brought much interest to be applied in public spaces or public transport.

After submission of the technical reports to the regulatory authority (ANMAT) and further refinement, the commercialization of this product was approved, four years after its first design. This is the first nanotechnology-enabled disinfectant product developed in Argentina approved by this institution, which is an interesting milestone.

In summary, the development process of the Hybridon technology gives a good example of some critical steps required to develop a nano-enabled product from a scientific advance:

1) Build a work team with complementary skills.

2) Define the problem and empathize with those who are currently facing it. In this case, the 25-year market experience of ADOX in developing solutions for health institutions was crucial.

3) *Ideation*: With the problem and needs defined, all knowledge and background were used in order to propose different technological alternatives.

4) *Prototyping*: Build a prototype with selected technological alternatives.

5) *Prototype Assessment*: Evaluate the performance of different developed prototypes, both in the laboratory and real situations.

6) *Implementation stage*: This stage implies critical issues such as process optimization, cost reduction opportunity, safety assessment of materials in the medium term and long term.

7) *Regulations*: It is essential to count with laboratories compliant with Good Manufacturing Practice (GMP), in order to optimize the development stage of the products and the requirements of the regulatory offices.

8) The project funding was varied, and came from different institutions, mostly public (universities, CONICET, ministries) and private (ADOX). The funding dynamics was very slow, and a critical issue for the success of the project. In particular, public funding in Argentina is relatively scarce and the actual rate of funding delivery is slow and bureaucratic. The local conditions of high inflation and devaluation had a negative impact on the product development, especially in the first stages. Although these limitations were eventually surpassed, this took precious energy and time, which delayed the expected milestones.

Figure 3.8 summarizes the road from an initial concept to the applied product, which had to cover several practical issues, including a technical and economically viable formulation based on market needs (an efficient antimicrobial with no impact in aspect).

The positive outcome resulted in the creation of a spin-off company and led to everyone learning new competencies and ways of collaborating. The ADOX team developed advanced antimicrobial nanotechnologies, and an experienced group of researchers learned that advanced scientific concepts need flexibility.

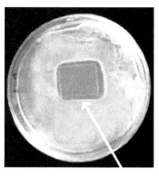  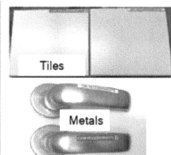

Figure 3.8. Left: Antibacterial effect of a concept thin film. Center: the different powder formulations of Hybridon, and the product presentation. Right: examples of applications of the product on tiles and door handles, showing no apparent modification of the surfaces.

6. Conclusions and Outlook

The examples discussed above present some common patterns that help to understand the way to develop nano-enabled products in a developing country with a traditionally rich scientific community in the nanoscience area, yet lacking an effective nanotechnology transfer ecosystem.

We found that the successful interactions between our laboratories and the companies were a combination of "science-push" (mesoporous adsorbent for Lanxess, mesoporous thin films for antibacterials for ADOX) with "market pull" (antifungal particles for Tort-Valls, additives for the tire industry for Lanxess). This approach is based on a negotiation between the public and private actors and enriches both of them, and some lessons can be derived.

The incorporation of silica additives to rubber bladders and the mesoporous adsorbents were not industrially implemented, partially due to higher costs and also because they required several changes in the usual technologies, which imposed an extra restriction. These examples demonstrate that nanomaterial-based technologies are in competition with well-established technologies. The development of novel nanotechnologies, therefore, relies in great part on overcoming these economic/market barriers by finding adequate precursors or processes that lower costs. However, the know-how developed had a deep impact both in the research group, leading to opening a composite materials development line with highly trained personnel and novel strategies of interaction with companies.

On the other hand, the two other products described above are making their way to the market. In the case of cuprous oxide, while the inventive height of the product itself is relatively low, the "market pull" made the company invest in a fast development that ended up in a breakthrough for the company. Indeed, some key innovations emerged along with the processing of the material, but patenting was relegated to a second stage. This development illustrates the important role of regulations regarding nanostructured materials; in this particular case, the Argentine National Agri-Food Health and Quality Service (SENASA) is a last but necessary bottleneck in the way to a successful product. The antibacterial platform co-developed by Hybridon-ADOX is also an example of a well-conducted development after a "science push" situation. The key to success here was the adequate assessment of the niche market for a product from the established company, as well as the plasticity demonstrated by the research team, who adapted their initial advanced materials to the actual needs for a product. Another important point for success was the continuous feedback of our laboratory and the company with the regulatory agency in order to solve unexpected technical questions that often limit the fast approval of products.

Concerning technical aspects, while the fundamental issues were mostly solved, in these particular cases some critical issues were detected in engineering nanomaterials, particularly in scaling up, process optimization, cost reduction opportunity, safety assessment of materials in the medium term and long term. A lesson learned is to evaluate these aspects in the ideation stage and to use these criteria to choose the technology alternatives to be built and tested in the prototyping and prototype assessment stage. If this feasibility assessment is left to the end, there is a risk of building something impossible to implement, thereby wasting an enormous amount of time and resources.

Other more general aspects to be taken into account when developing a technology to solve a real problem emerged, such as regulatory issues (related to validation and GMP, added to the already mentioned approval by the agencies) and a sound analysis of the environmental and health-related hazards. These aspects involve design, production and understanding the life cycle analysis of nanomaterials, which are crucial and still need investigation.

In addition, there are economic and societal factors that limit the translation of basic nanoscience to real-world technological applications. In the particular case of Argentina, an adequate management of IP status, and techno-economic analysis that includes market research, competitive technologies, strategic precursor availability and supplier selection is still lacking. These constitute critical barriers towards nanomaterials commercialization, and they should be solved in order to bring up a clear governance for nanotechnologies and the creation of high added-value products.

Finally, it is worth emphasizing the importance to perform technology validation according to specific standards. One is tempted to design a proprietary validation protocol instead of following an international standard. This might work for laboratory use, but the results of such a procedure are not useful as a valid result when registering the product.

The technologies and developments discussed above show that even though in a small size ecosystem with considerable infrastructure and culture barriers, a combination of a sound scientific community and a close collaboration and feedback between academic and industrial R&D teams leads to the successful development of real-world nanotechnology products.

References

Aguado, J., J.M. Arsuaga, A. Arencibia, M. Lindo and V. Gascón. 2009. Aqueous heavy metals removal by adsorption on amine-functionalized mesoporous silica. J. Hazard. Mater. 163: 213–21.

Araujo, Paula Z., Vittorio Luca, Patricia B. Bozzano, Hugo L. Bianchi, Galo Juan De Ávila Arturo Soler-Illia and Miguel A. Blesa. 2010. Aerosol-assisted production of mesoporous titania microspheres with enhanced photocatalytic activity: The basis of an improved process. ACS Applied Materials and Interfaces 2(6): 1663–73. https://doi.org/10.1021/am100188q.

Baig, N., I. Kammakakam and W. Falath. 2021. Nanomaterials: A review of synthesis methods, properties, recent progress, and challenges. Mater. Adv. 2: 1821–1871.

Banerjee, D.A., A.J. Kessman, D.R. Cairns and K.A. Sierros. 2014. Tribology of silica nanoparticle-reinforced, hydrophobic sol-gel composite coatings. Surf. Coatings Technol. 260: 214–19.

Bang Jin Ho and Kenneth S. Suslick. 2010. Applications of ultrasound to the synthesis of nanostructured materials. Advanced Materials 22(10): 1039–59. https://doi.org/10.1002/adma.200904093.

Beltrán-Osuna, A.A. and J.E. Perilla. 2016. Colloidal and spherical mesoporous silica particles: synthesis and new technologies for delivery applications. J. Sol-Gel Sci. Technol. 77: 480–96.

Bento, A. Patrícia and F. Matthias Bickelhaupt. 2007. Nucleophilic substitution at silicon (SN@Si) via a central reaction barrier. Journal of Organic Chemistry 72(6): 2201–7. https://doi.org/10.1021/jo070076e.

Berger Mauricio, Tomás Carrozza and Gonzalo Bailo (eds.). 2021. Nanotecnología y Sociedad En Argentina Para Una Agenda Inter y Transdisciplinaria (Vol. I). Universidad Nacional de Córdoba.

Butler, Richard. 2016. A matter of scale. Nat. Nanotechnol. 11: 733.

Catalano, P.N., M. Pezzoni, C. Costa, G.J.A.A. Soler-Illia, M.G. Bellino and M.F. Desimone. 2016. Optically transparent silver-loaded mesoporous thin film coating with long-lasting antibacterial activity. Microporous Mesoporous Mater. 236: 158–66.

Charitidis, C.A., P. Georgiou, M.A. Koklioti, A.-F. Trompeta and V. Markakis. 2014. Manufacturing nanomaterials: from research to industry. Manufacturing Rev. 1: 11.

Crozals, Gabriel De, R. Bonnet, C. Farre and C. Chaix. 2016. Nanoparticles with multiple properties for biomedical applications: a strategic guide. Nano Today 11: 435–63.

Debecker, Damien P. 2018. Innovative sol-gel routes for the bottom-up preparation of heterogeneous catalysts. Chemical Record 18: 662–75. https://doi.org/10.1002/tcr.201700068.

Duan, H., D. Wang and Y. Li. 2015. Green chemistry for nanoparticle synthesis. Chem. Soc. Rev. 44: 5778–92.

Duanmu Yanqing, Carson T. Riche, Malancha Gupta, Noah Malmstadt and Qiang Huang. 2018. Scale-up modeling for manufacturing nanoparticles using microfluidic T-junction. IISE Transactions 50(10): 892–99. https://doi.org/10.1080/24725854.2018.1443529.

Flodström, K., C.V. Teixeira, H. Amenitsch, V. Alfredsson and M. Lindén. 2004. *In Situ* synchrotron small-angle X-ray scattering/X-ray diffraction study of the formation of SBA-15 mesoporous silica. Langmuir 20: 4885–91.

Flodström, K., H. Wennerström and V. Alfredsson. 2004. Mechanism of mesoporous silica formation, a time-resolved NMR and TEM study of silica-block copolymer aggregation. Langmuir 20: 680–88.

Flodström, K., H. Wennerström, C.V. Teixeira, H. Amenitsch, M. Lindén and V. Alfredsson. 2004. Time-resolved *in situ* studies of the formation of cubic mesoporous silica formed with triblock copolymers. Langmuir 20: 10311–16.

Foladori Guillermo and Tomás Carrozza. 2017. Políticas de Nanotecnología En Argentina a La Luz de Criterios de La OCDE. Ciencia, Docencia y Tecnología 28(55): 115–40.

Francisco, R.D., P. Tiemblo, M. Hoyos, C. González-Arellano, N. García, L. Berglund and A. Synytska. 2014. Multipurpose ultra and superhydrophobic surfaces based on oligodimethylsiloxane-modified nanosilica. ACS Applied Mater. Interfaces 6: 18998–10. https://doi.org/10.1021/am504886y.

Fulvio, P.F., S. Pikus and Mietek Jaroniec. 2005a. Short-time synthesis of SBA-15 using various silica sources. J. Colloid Interface Sci. 287: 717–20.

Fulvio, P.F., S. Pikus and M. Jaroniec. 2005b. Tailoring properties of SBA-15 materials by controlling conditions of hydrothermal synthesis. J. Mater. Chem. 15: 5049–53.

Guo Zhendong, Guang Xiong, Liping Liu, Peng Li, Liang Hao, Yuanyuan Cao and Fuping Tian. 2016. Aerosol-assisted synthesis of hierarchical porous titanosilicate molecular sieve as catalysts for cyclohexene epoxidation. Journal of Porous Materials 23(2): 407–13. https://doi.org/10.1007/s10934-015-0094-7.

Ho Wong, C. and S.C. Zimmerman. 2013. Orthogonality in organic, polymer, and supramolecular chemistry: from merrifield to click chemistry. Chem. Commun. 49: 1679–95.

Hyde, E.D.E.R., A. Seyfaee, F. Neville and R. Moreno-Atanasio. 2016. Colloidal silica particle synthesis and future industrial manufacturing pathways: a review. Ind. Eng. Chem. Res. 55: 8891–8913.

Iler, R.K. 1979. The Chemistry of Silica: Solubility, Polymerization, Colloid and Surface Properties, and Biochemistry. John Wiley & Sons, New York.

Jaroniec, M. and P.F. Fulvio. 2005. Optimization of synthesis time for SBA-15 materials. Stud. Surf. Sci. Catal. 156: 75–82.

Jawed, A., V. Saxena and L.M. Pandey. 2020. Engineered nanomaterials and their surface functionalization for the removal of heavy metals: a review. J. Water Proc. Eng. 33: 101009.

Jia Xilai and Fei Wei. 2017. Advances in production and applications of carbon nanotubes. Topics in Current Chemistry 375(1). https://doi.org/10.1007/s41061-017-0102-2.

Keiblinger, Katharina M., Martin Schneider, Markus Gorfer, Melanie Paumann, Evi Deltedesco, Harald Berger, Lisa Jöchlinger et al. 2018. Assessment of Cu applications in two contrasting soils—Effects on soil microbial activity and the fungal community structure. Ecotoxicology 27(2): 217–33. https://doi.org/10.1007/s10646-017-1888-y.

Ketelson, H.A., M.A. Brook and R.H. Pelton. 1995. Sterically stabilized silica colloids: radical grafting of poly(methyl methacrylate) and hydrosilylative grafting of silicones to functionalized silica. Polym. Adv. Technol. 6: 335–44.

Khan, Saif A., Axel Günther, Martin A. Schmidt and Klavs F. Jensen. 2004. Microfluidic synthesis of colloidal silica. Langmuir 20(20): 8604–11. https://doi.org/10.1021/la0499012.

Lee, D.W. and B.R. Yoo. 2016. Advanced silica/polymer composites: materials and applications. J. Ind. Eng. Chem. 38: 1–12.

Liu, K. and L. Jiang. 2012. Bio-inspired self-cleaning surfaces. Ann. Rev. Mater. Res. 42: 231–63.

Lombardo, M.V., M. Videla, A. Calvo, F.G. Requejo and G.J.A.A. Soler-Illia. 2012. Aminopropyl-modified mesoporous silica SBA-15 as recovery agents of Cu (II)-sulfate solutions: adsorption efficiency, functional stability and reusability aspects. J. Hazard. Mater. 223–224: 53–62.

Lombardo, M.V. 2013. Síntesis y Propiedades de Sílice Mesoporosa Híbrida y Su Uso En Recuperación Secundaria de Iones Divalentes. Ph. D. Thesis, Universidad Nacional de San Martín.

Lombardo, M.V. and G.J.A.A. Soler-Illia. 2020. Polyacrylonitrile and hybrid SBA-15: A robust composite material for use as copper (II) adsorbent in flow conditions. J. Inorg. Organomet. Polym. Mater. 30: 1206–17.

Lombardo María Verónica, Andrea Verónica Bordoni and Alejandro Wolosiuk. 2021. Mesoporous particles by combination of aerosol route and sol-gel process. pp. 24–49. *In*: Franceschini, E.A. (ed.). Nanostructured Multifunctional Materials Synthesis, Characterization, Applications and Computational Simulation, 1st edition. CRC Press. https://doi.org/10.1201/9780367822194-2.

McConville, Francis X. 2002. The Pilot Plant Real Book: A Unique Handbook for the Chemical Process Industry. FXM Engineering and Design.

Miroshnychenko Yuliia and Yuriy Beznosyk. 2015. Simulation of the process of silica functionalization in the microreactor. Eastern-European Journal of Enterprise Technologies 2(5): 46–53. https://doi.org/10.15587/1729-4061.2015.39417.

Nayfeh, M. 2018. Nanotechnology and Society: From Lab to Consumer. In Fundamentals and Applications of Nano Silicon in Plasmonics and Fullerines. Current and future Trends. Nayfeh, M., pp. 519–69.

Paliwal, Rishi R., Jayachandra Babu and Srinath Palakurthi. 2014. Nanomedicine scale-up technologies: Feasibilies and challenges. AAPS PharmSciTech 15: 1527–34. https://doi.org/10.1208/s12249-014-0177-9.

Pezzoni Magdalena, Paolo N. Catalano, Ramón A. Pizarro, Martín F. Desimone, Galo J.A.A. Soler-Illia, Martín G. Bellino and Cristina S. Costa. 2017. Antibiofilm effect of supramolecularly templated mesoporous silica coatings. Materials Science and Engineering C 77: 1044–49. https://doi.org/10.1016/j.msec.2017.04.022.

Rahman, I.A. and V. Padavettan. 2012. Synthesis of silica nanoparticles by sol-gel: size-dependent properties, surface modification, and applications in silica-polymer nanocomposites: a review. J. Nanomaterials Article ID 132424.

Raquel de Francisco, Pilar Tiemblo, Mario Hoyos, Camino González-Arellano, Nuria García, Lars Berglund and Alla Synytska. 2014. Multipurpose ultra and superhydrophobic surfaces based on oligodimethylsiloxane-modified nanosilica. ACS Applied Materials and Interfaces 6(21): 18998–10.

Roco, M.C. 2011. The long view of nanotechnology development: the national nanotechnology initiative at 10 years. J. Nanopart. Res. 13: 427–45.

Roco, M.C. 2020. Principles of convergence in nature and society and their application: from nanoscale, digits, and logic steps to global progress. J. Nanopart. Res. 22: 321.

Ruthstein, S., J. Schmidt, E. Kesselman, Y. Talmon and D. Goldfarb. 2006. Resolving intermediate solution structures during the formation of mesoporous SBA-15. J. Am. Chem. Soc. 128: 3366–74.

Sanchez, C., P. Belleville, M. Popall and L. Nicole. 2011. Applications of advanced hybrid organic–inorganic nanomaterials: from laboratory to market. Chem. Soc. Rev. 40: 453–1152.

Sebastian, V., M. Arruebo and J. Santamaria. 2014. Reaction engineering strategies for the production of inorganic nanomaterials. Small 10: 835–53.

Sharma, R.K., S. Sharma, S. Dutta, R. Zboril and M.B. Gawande. 2015. Silica-nanosphere-based organic-inorganic hybrid nanomaterials: synthesis, functionalization and applications in catalysis. Green Chem. 17: 3207–30.

Shiba Kota and Makoto Ogawa. 2018. Precise synthesis of well-defined inorganic-organic hybrid particles. Chemical Record 18(7): 950–68. https://doi.org/10.1002/tcr.201700077.

Soler-Illia, G.J.A.A. and O. Azzaroni. 2011. Multifunctional hybrids by combining ordered mesoporous materials and macromolecular building blocks. Chem. Soc. Rev. 40: 1107–50.

Soler-Illia, Galo J.A.A., M.V. Lombardo, A. Zelcer and E.A. Franceschini. 2016. Method for producing spherical particles of mesoporous metal oxides having a controlled composition, surface area, porosity and size. WO2016IB58022, issued 2016.

Stark, W.J., P.R. Stoessel, W. Wohlleben and A. Hafner. 2015. Industrial applications of nanoparticles. Chem. Soc. Rev. 44: 5793–5805.

Stöber, W., A. Fink and E. Bohn. 1968. Controlled growth of monodisperse silica spheres in the micron size range. J. Colloid Interface Sci. 26: 62–69.

Thielemann, J.P., F. Girgsdies, R. Schlögl and Christian, Hess. 2011. Pore structure and surface area of silica SBA-15: Influence of washing and scale-up. Beilstein J. Nanotechnol. 19: 110–18.

Tighe, C.J., R. Quesada Cabrera, R.I. Gruar and J.A. Darr. 2013. Scale up production of nanoparticles: continuous supercritical water synthesis of Ce-Zn oxides. Ind. Eng. Chem. Res. 52: 5522–28.

Walmsley, G.G., A. McArdle, R. Tevlin, A. Momeni, D. Atashroo, M.S. Hu, A.H. Feroze et al. 2015. Nanotechnology in bone tissue engineering. Nanomed. Nanotechnol. Biol. Med. 11: 1253–63.

Xiong Guang, Jinpeng Yin, Jiaxu Liu, Xiyan Liu, Zhendong Guo and Liping Liu. 2016. Aerosol-assisted synthesis of nano-sized ZSM-5 aggregates. RSC Advances 6(103): 101365–71. https://doi.org/10.1039/c6ra22564k.

Zelcer, A., E.A. Franceschini, M.V. Lombardo, A.E. Lanterna, G.J.A.A. Soler-Illia. 2020. A general method to produce mesoporous oxide spherical particles through an aerosol method from aqueous solutions. Journal of Sol-Gel Science and Technology 94(1): 195–204. https://doi.org/10.1007/s10971-019-05175-0.

Chapter 4

Application of Zerovalent Iron Nanoparticles (nZVI) in Industrial Wastewater Treatments

Santiago Ocampo,[1] *Luciano Carlos*[1,]* and *Fernando García Einschlag*[2]

1. Introduction

In the past decades, nanotechnology has provided promising alternative approaches to water treatment. Due to their excellent performance against contaminants and their wide applications, a variety of nanomaterials is the subject of active research and development. Among them, nanozerovalent iron (nZVI) is one of the most widespread nanomaterials due to its enhanced reactivity compared to conventional bulk ZVI materials, and it has now become one of the most important engineered nanomaterials (Pasinszki and Krebsz 2020). Owing to its nanometer length scale (1–100 nm) and hence large specific surface area, nZVI has higher reducing ability and activity, stronger adsorption properties and better mobility than microscale iron particles. These characteristics allow nZVI to work as an effective remediating agent for a wide range of contaminants, including halogenated hydrocarbons, nitroaromatic compounds, organic dyes, antibiotics, pesticides, nitrates, heavy metal ions, etc. (Raychoudhury and Scheytt 2013, Fu et al. 2014, Xu et al. 2017, Hassan et al. 2019, Foltynowicz et al. 2020). Furthermore, it has shown a strong bactericidal and toxic effect against microorganisms. However, nZVI has a high tendency to agglomerate and reacts quickly with oxygen and water giving rise to the formation of various iron oxides, which results in surface passivation that decreases the treatment efficiency. To overcome these issues, nZVI particles are usually capped with different inorganic and organic materials or supported on various substrates (Adusei-Gyamfi and Acha 2016, Lu et al. 2016).

The high effectiveness and versatility of nZVI have demonstrated that it may be used as a suitable technology for practical applications in water treatment. nZVI-based technologies have been used extensively for the remediation of polluted groundwater, but limited information is available on their use in industrial wastewater treatment. The potential of nZVI for water remediation has triggered laboratory-scale, pilot-scale and industrial-scale studies dedicated to evaluate the effectiveness of nZVI-based technologies in different types of industrial effluents (Kamali et al. 2019). However, most studies focus on the removal of single pollutants, while relatively few articles deal with real

[1] Instituto de Investigación y Desarrollo en Ingeniería de Procesos, Biotecnología y Energías Alternativas, PROBIEN (CONICET-UNCo), Buenos Aires 1400, Neuquén, Argentina.
[2] Instituto de Investigaciones Fisicoquímicas Teóricas y Aplicadas (INIFTA), CCT-La Plata-CONICET, Universidad Nacional de La Plata, Diag. 113 y 64, La Plata, Argentina.
Emails: santiago.ocampo@probien.gob.ar; fgarciae@quimica.unlp.edu.ar
* Corresponding author: luciano.carlos@probien.gob.ar

wastewater remediation. Moreover, the feasibility and efficiency of nZVI in practical industrial applications still require closer examination, since long-term field studies for the environmental impact of nZVI are very scarce.

In this chapter, we present the process and mechanisms involved in the removal of pollutants and recent developments on the use of nZVI-based technologies aimed at the remediation of different real industrial effluents at different scales (laboratory, pilot and industrial).

2. Processes and Mechanisms of Contaminant Removal by nZVI

ZVI is a reactive metal that easily yields ferrous species due to its standard redox potential ($E^0 = -0.44$ V). Thus, even under anoxic conditions and at near-neutral pH values, water can act as an oxidant when ZVI is introduced into an aqueous solution. Water-mediated ZVI corrosion leads to the production of molecular hydrogen (Eq. (1)):

$$Fe^0 + 2\ H_2O \rightarrow Fe^{2+} + H_2 + 2\ HO^- \tag{1}$$

On the other hand, in the presence of dissolved oxygen, the production of ferrous species is driven by the cathodic reduction of O_2 (Eq. (2)):

$$2\ Fe^0 + O_2 + 2\ H_2O \rightarrow 2\ Fe^{2+} + 4\ HO^- \tag{2}$$

Under weak acidic to near neutral and oxygenated conditions, Fe^{2+} ions in an aqueous phase may be further oxidized to yield ferric hydroxide (Eq. (3)).

$$4\ Fe^{2+} + O_2 + 10\ H_2O \rightarrow 4\ Fe\ (OH)_3\downarrow + 8\ H^+ \tag{3}$$

Furthermore, in the presence of Fe^{2+} and O_2, H_2O_2 may also be produced, through the intermediate formation of superoxide radical anion ($O_2^{\cdot-}$) (Eqs. (4) and (5)).

$$Fe^{2+} + O_2 \rightarrow Fe^{3+} + O_2^{\cdot-} \tag{4}$$

$$Fe^{2+} + O_2^{\cdot-} + 2\ H^+ \rightarrow Fe^{3+} + H_2O_2 \tag{5}$$

The *in situ* produced H_2O_2 may react with Fe^0 to generate Fe^{2+} (Eq. (6)) or with Fe^{2+} to yield $HO^\cdot$ radicals through the Fenton reaction (Eq. (7)).

$$Fe^0 + H_2O_2 \rightarrow Fe^{2+} + HO^- \tag{6}$$

$$Fe^{2+} + H_2O_2 \rightarrow Fe^{3+} + HO^- + HO^\cdot \tag{7}$$

Ferric species are rather insoluble above pH 4 and their precipitation usually leads to the formation of an iron oxide layer or Corrosion Layer (CL) onto the surface of ZVI particles. The type and characteristics of the CL formed depend on water chemistry, solution pH, hydrodynamic conditions and the contribution of aging processes. Reported corrosion products include several oxides such as magnetite, lepidocrocite, goethite, maghemite, akaganeite, wustite and ferrihydrite (Huang and Zhang 2005) as well as a set of unstable Fe(II)-Fe(III) hydroxy salts (containing chloride, sulfate or carbonate anions) that are usually termed as green rusts. The properties of the CL, such as the electronic conductivity and porosity, can influence the reactivity of iron in natural waters since the formation of some oxides can enhance or decrease the reactivity of the iron surface.

Several studies have shown that many factors affect nZVI reactivity, including: (1) degree of Fe core crystallinity, (2) surface area, (3) aging time, (4) aqueous phase pH, (5) nanoparticle loading, (6) concentrations of contaminants and other reactive water constituents, and (7) presence of nZVI stabilizers (Mpouras et al. 2014). These factors collectively influence the degree to which nZVI reacts with dissolved oxygen and H_2O to form a CL on its surface, as well as the tendency of the nZVI particles to aggregate. Specifically, high pH, aging, small particle size and high suspension density all favor faster aggregation. Aggregation causes a decrease in surface area, which consequently affects nZVI reactivity. Additionally, aging and competitive oxidation cause a loss of Fe content that

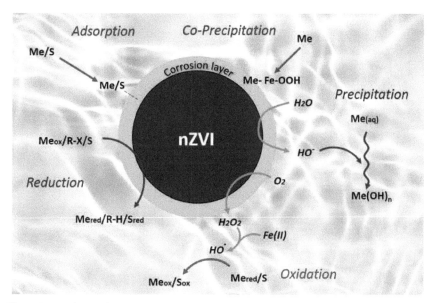

Figure 4.1. Various action mechanisms of nZVI. Me = metal; S = organic specie; RX = halocarbons.

reduces nZVI redox activity and efficiency. The formation of CL passivates nZVI particles and, in some cases, limits their application.

nZVI has a complex chemistry that allows the elimination of contaminants through different mechanisms depending on the nature of contaminants and the experimental conditions. The removal of contaminants in nZVI systems occurs mainly by five mechanisms: reduction, sorption, precipitation, co-precipitation and oxidation, which are summarized in Fig. 4.1.

A brief description of each mechanism is given below:

1) **Reduction:** The chemical reduction of organic compounds by nZVI systems is a surface-mediated reaction, which results in the direct reduction of the contaminant. Consequently, the removal of an organic contaminant by nZVI requires direct contact between the contaminant and reactive sites on the nZVI surface. The contaminant molecules first diffuse through the solution to the nZVI surface, where they adsorb onto the reactive sites. Then, electrons are transferred from nZVI to the contaminant molecules yielding reduced compounds and various oxides on the surface. The anodic reaction for the dissolution of ZVI is illustrated in Eq. (8), whereas the reductive reactions for halogenated compounds, azo dyes, and nitro compounds are represented by Eqs. (9), (10), and (11), respectively.

$$Fe^0 \rightarrow Fe^{2+} + 2\ e^- \tag{8}$$

$$RX + 2\ e^- + H^+ \rightarrow RH + X^- \tag{9}$$

$$R\text{-}N{=}N\text{-}R' + 4\ e^- + 4\ H^+ \rightarrow R\text{-}NH_2 + R'\text{-}NH_2 \tag{10}$$

$$R\text{-}NO_2 + 6\ e^- + 6\ H^+ \rightarrow R\text{-}NH_2 + 2\ H_2O \tag{11}$$

2) **Sorption:** Pollutants in solution can be removed via sorption onto the ZVI surface, which is one of the major uptake pathways for inorganic contaminants such as Pb(II), Cd(II), U(VI), As(V) and could be also a significant removal mechanism for organic pollutants (Gunawardana et al. 2011). Organic contaminants with high hydrophobicity are more susceptible to sorption, regardless of their reactivity towards nZVI. Besides, the CL of the ZVI nanoparticles may provide active sites for the chemical adsorption of pollutants on its surface achieving the removal of pollutants from water (Foltynowicz et al. 2020).

3) **Precipitation:** nZVI could induce an increase in the solution pH, which, in turn, leads to the precipitation of insoluble compounds. This mechanism has been reported for the removal of some heavy metals including Cu, Pb, Cd, Co and Zn ions (Lu et al. 2016, Ullah et al. 2020).

4) **Co-precipitation:** The unspecific co-precipitation of contaminants during iron corrosion is an important mechanism of contaminant removal by nZVI (Noubactep 2008). At pH values ranging from 4 to 9, ZVI corrosion results in the formation of a CL on its surface. At lower pH values, the CL is relatively porous, unstable and detachable, while at higher pH values, the CL is more stable. At neutral pH values, various iron oxides (green rusts, hydroxides and oxyhydroxides) can precipitate and several contaminants both organic and inorganic may be entrapped in the matrix of iron hydroxides on the ZVI surface. The important role of co-precipitation was reported for a number of pollutants including Cr, As, Ni and chlorinated phenols (Gunawardana et al. 2011, Ullah et al. 2020).

5) **Oxidation:** The oxidation of pollutants can also take place in nZVI applications (Litter and Slodowicz 2017, Litter 2020). Considering the reactions that are triggered by the presence of ZVI and O_2 in water, the combination of H_2O_2 and Fe^{2+} ions produce $HO^\bullet$ radicals (i.e., Fenton reaction, Eq. (7)), which are capable of oxidizing organic compounds. Fe^{2+} is generated *in situ* by the corrosion of the ZVI mediated by H_2O, O_2, H_2O_2 or reducible pollutants (Eqs. (1), (2), (6), and (9) to (11)), while H_2O_2 can be generated *in situ* via Eq. (5), or added externally in a process called nZVI-assisted Fenton. In addition to H_2O_2, other types of oxidants have been used in nZVI-assisted Fenton systems to produce reactive radicals able to react with the organic pollutants, the most important ones being salts of peroxydisulfate ($S_2O_8^{2-}$) and peroxymonosulfate ($S_2O_5H^-$). These species lead to the formation of $SO_4^{\bullet-}$ (Eqs. (12)–(15)), a strong oxidant with a redox potential of 2.6 V similar to that of $HO^\bullet$ (2.8 V) (Devi et al. 2016).

$$Fe^0 + HSO_5^- + H^+ \rightarrow Fe^{2+} + SO_4^{2-} + H_2O \tag{12}$$

$$Fe^{2+} + HSO_5^- \rightarrow SO_4^{\bullet-} + Fe^{3+} + HO^- \tag{13}$$

$$Fe^0 + S_2O_8^{2-} \rightarrow Fe^{2+} + 2\ SO_4^{2-} \tag{14}$$

$$Fe^{2+} + S_2O_8^{2-} \rightarrow SO_4^{\bullet-} + SO_4^{2-} + Fe^{3+} \tag{15}$$

The pH significantly affects the process performance in nZVI-assisted Fenton systems for the removal of contaminants (Rezaei and Vione 2018). High degradation efficiencies are usually obtained in the pH range 3–4, which is mainly attributed to acceleration of iron corrosion, dissolution of the passive oxide layers on the ZVI surface and efficient generation of hydroxyl radicals. In contrast, neutral and alkaline pH values decrease the efficiency of nZVI-assisted Fenton systems due to Fe^{3+} precipitation and less effective $HO^\bullet$ production.

3. Application of nZVI-Based Technologies

Here some advances that have been made in the application of nZVI-based technologies for the remediation of different real industrial effluents at different scales will be presented. It is worth mentioning that many industrial applications of zerovalent iron in powder, granular and fibrous forms have been reported during the past two decades. These large-scale applications are well established and usually involve mainly macro- or micro-ZVI particles (Foltynowicz et al. 2020). In contrast, most of the reported work concerning nanoscale zerovalent iron was performed at the laboratory scale, while there are few studies reported on pilot and field applications. Despite nZVI particles being much more reactive than other iron sources, the relatively high production costs associated with nZVI-based materials have limited their large-scale application.

3.1 Laboratory-scale experiments on real industrial samples

Hamdy et al. (Hamdy et al. 2018) reported that nZVI can improve the quality of textile industry effluents at a laboratory scale. From real samples of wastewater collected from a textile industry located in the city of Sadat, Egypt, they found that the nZVI-based treatment produced removal efficiencies of COD, BOD, TN (Total Nitrogen), TP (Total Phosphorous), Cu^{2+}, Zn^{2+} and Pb^{2+} with values close to 91, 87, 65, 78, 100, 29 and 99%, respectively via batch experiments. In this study, nZVI was prepared via the chemical reduction method with $NaBH_4$, which resulted in nZVI particles with diameters ranging from 10 to 100 nm. In another study performed by the authors (Hamdy et al. 2019), nZVI was used to treat electroplating wastewater; an effluent containing severe levels of heavy metals. In this case, continuous-feed tests were performed attaining removal efficiencies for TSS, COD, nitrogen, phosphorus, Cr^{6+}, Pb^{2+}, Ag^+, Cu^{2+}, Ni^{2+}, Mn^{2+}, Zn^{2+}, Fe^{3+}, Al^{3+} and Co^{2+} of 91, 68, 94, 98, 66, 91, 83, 80, 17, 47, 54, 94, 100 and 42%, respectively. Homhuol et al. (Homhoul et al. 2011) investigated the treatment of wastewater from a distillery industry in Thailand by using nZVI and chitosan-supported nZVI. The synthesis method employed for the production of the reactive nanomaterials was the chemical reduction with $NaBH_4$, in the absence and the presence of chitosan, previously dissolved in an acidic water solution. The wastewater used by the authors was collected from the washing unit of the distillery without any pre-treatment process. In particular, this type of effluent includes melanoidins (a dark brown pigment), together with high levels of COD and BOD. The optimal condition for the treatment was achieved at pH 3, obtaining removal efficiencies higher than 95, 94 and 64% for color, COD and BOD, respectively. Compared with free nZVI, lower amounts of chitosan-supported nZVI are needed to obtain the same removal efficiency values. Rahmani et al. (Rahmani et al. 2020) applied a heat-activated persulfate system in combination with reduced graphene oxide-supported nZVI (nZVI-rGO) for oil refinery wastewater treatment. The nZVI-rGO was prepared via the chemical reduction of Fe^{2+} salt with $NaBH_4$ in the presence of graphene oxide, obtaining nZVI aggregates of spherical particles with a diameter range of 86–147 nm distributed on the rGO surface. The authors used the optimal conditions, obtained for the degradation of furfural in preliminary laboratory tests, in the treatment of the real sample of petrochemical wastewater. The reported removal efficiencies were 72.4 and 43.8% for COD and BOD, respectively. In addition, after the treatment, the biodegradability index (BOD/COD) increased to more than 0.4, resulting in a biodegradable effluent that can be easily treated by microorganisms. In another study involving petroleum refinery wastewater, Rasheed et al. (Rasheed et al. 2011) studied the application of nZVI particles in the presence of ultrasonication in a probe-type sonochemical reactor (24 kHz, 400 W ultrasound). In this work, the authors used nZVI produced via the liquid-phase reduction method, with an average crystalline particle size of 31.1 nm. They found an optimal condition for the effective degradation of effluents at $0.15 \, g \, L^{-1}$ nZVI and an initial pH of 5. On the other hand, Haneef et al. (Haneef et al. 2020) applied an nZVI-based heterogeneous Fenton-like process to treat real oil and gas effluents at a laboratory scale. In this study, a commercial nZVI with 50 nm particle size and $20–25 \, m^2 \, g^{-1}$ surface area was used as the catalyst, while the Polycyclic Aromatic Hydrocarbon (PAH) and COD removals were monitored to evaluate the efficiency of the process. At the optimal working conditions, nZVI $(4.35 \, g \, L^{-1})$, H_2O_2 $(1.60 \, g \, L^{-1})$ and pH (2.94), the authors found removal efficiencies of both PAHs and COD near to 89.3 and 75.7%, respectively.

3.2 Pilot-scale experiments on real industrial samples

Li et al. (Li et al. 2014a) investigated the treatment of smelting wastewater using zerovalent iron nanoparticles (nZVI) at a pilot scale. The process setup developed by the authors consisted of an nZVI reactor, a clarifier and an nZVI recirculation pump. The system was operated at a flowrate of $400 \, L \, h^{-1}$ and 35000 L of wastewater were treated using 75 kg of nZVI. In this case, nZVI particles were produced via the reduction of ferric ions with $NaBH_4$. The treatment plant showed a good performance, achieving an As removal efficiency higher than 99.9% with an average removal capacity of 239 mg As/g Fe. Furthermore, others metal ions, such as Ni, Cu and Zn, were reduced to concentrations less than $0.1 \, mg \, L^{-1}$. Mahmoud et al. (Mahmoud et al. 2021) treated wastewater collected from the effluent of a textile factory located in the city of El-Obour, Egypt, in a pilot prototype system that included tanks for coagulation/flocculation, sedimentation and filtration before the degradation/adsorption unit based on the use of bimetallic Fe/Cu nanoparticles. The bimetallic nanoparticles were synthesized from fresh nZVI and cupric sulfate solution. At the optimum conditions, 0.5 g L^{-1} ferric chloride as a coagulant, pH 6.0, 1.4 g L^{-1} nanoparticles dosage, and 80 minutes contact time at room temperature, the removals of COD, BOD, color, TN, TP and TSS were 96, 98, 82, 69, 88, and 97%, respectively. Viraldi et al. (Vilardi et al. 2018) studied the heterogeneous Fenton processes at a large laboratory-scale plant for the treatment of tannery wastewater using nZVI particles. They evaluated the efficiency of the treatment based on the decrease in COD, total polyphenols and Cr (VI) at the end of each experiment performed in the batch mode. They found that the heterogeneous Fenton process showed better performance than the homogeneous one towards both the removal of COD and TP, achieving removal efficiencies of 75 and 85%, respectively, at the pH of the tannery wastewater (pH 4.8).

3.3 Large scale applications of nZVI-based technologies

Li et al. (Li et al. 2014b) reported a field application of nZVI-based technology to treat concentrated Cu^{2+} wastewater from a printed circuit board manufacturing plant. The pilot plant consisted of an nZVI reactor that was operated as a Continuous Flow Stirred Tank Reactor (CFSTR), a clarifier and an integrated coagulation sedimentation unit. In addition, a recovery unit was incorporated after the CFSTR to recycle the nZVI particles. The nZVI reactor (1600 L) was operated continuously with flow rates ranging from 1000–2500 L h^{-1}. With this pilot system, an average Cu^{2+} removal efficiency higher than 96% was achieved using 0.20 g L^{-1} nZVI and a hydraulic retention time of 100 minutes. Likewise, the nZVI reactor reached a high volumetric loading rate of 1876 g Cu per m^3 per day for Cu^{2+} removal. In the global process, 250000 L of wastewater containing 70 mg L^{-1} Cu (II) were treated by using 55 kg nZVI. The same research group built an nZVI treatment facility in Jiangxi, China, which is operated by the Jiangxi Copper Company (Li et al. 2017). The nZVI treatment plant was used as a pretreatment stage to remove As and metals from wastewater. The plant consisted of a two-stage nZVI reactor and a polishing treatment, and it was operated at an average flow rate of 30 m^3 h^{-1} and with an average nZVI dose of 0.4–0.5 g L^{-1}. Nearly 10000 m^3 of wastewater were treated with 4400 kg of nZVI during a 28-day operating period.

In another work, Li et al. (Li et al. 2019) reported a large-scale application of the nZVI-based technology for the enrichment and recovery of gold from industrial wastewater. The plant-treated wastewater came from a smelting factory that produced 20 tons of gold and 1.5 million tons of copper every year. The nZVI used in this study was prepared via high-energy precision milling and resulted in nanoparticles with a zerovalent iron content of ~ 70% (w/w) and a specific surface area of 16–19 m^2 g^{-1}. In particular, the plant treated 350 m^3 d^{-1} of wastewater using nZVI (150 kg d^{-1}) and more than 5 kg of gold was recovered from almost 120000 m^3 of wastewater in 12 months.

Figure 4.2 summarizes the most common configurations for large-scale application of nZVI-based water treatment processes.

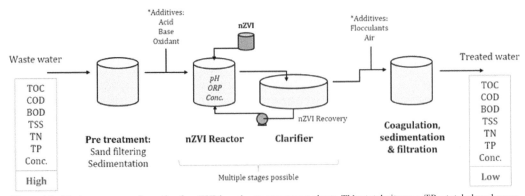

Figure 4.2. Most common configuration for nZVI-based water treatment plants. TN = total nitrogen. TP = total phosphorous. Conc. = concentration of relevant species. * = optional additive.

4. Conclusions

The application of nZVI for the development of industrial wastewater treatment technologies is gaining interest in the research community, and field application studies with different types of effluents are being developed. Among the main advantages of these technologies, it is notable to cite the ability to remove a wide range of pollutants by different types of processes, the possibility of reusing the nanomaterials and the low toxicity of the materials.

nZVI particles have proven to possess several advantages over bulk or microscale ZVI materials, such as higher surface area and reactivity, enhanced adsorption capacity and better mobility. However, owing to their high initial oxidation rates, nZVI particles may require, in some cases, some surface modification to improve their stability and long-term activity. Moreover, given the variety of removal mechanisms involved, nZVI synthetic routes should be carefully selected considering the nature of the wastewater to be treated. Finally, studies concerning the fate and long-term environmental impact on nZVI-based materials are scarce, and hazardous effects of NPs on ecosystems should be carefully assessed.

The development of simple and low-cost synthesis methods is necessary to increase the production of nZVI on a large scale since the widely used borohydride synthetic route is limited due to the relatively high costs and toxicity of the reagent. Further research should be devoted to designing cost-effective and highly efficient nanomaterials. The use of green synthesis routes for nanomaterials, such as the use of natural extracts, reutilization of waste materials as precursors, and the application of these eco-friendly nanomaterials will reduce the overall production cost and help reduce the environmental impact, thus allowing the widespread use and development of these technologies at large scale.

Acknowledgements

The authors would like to thank Prof. Dr. Marta I. Litter for her careful revision, valuable comments and suggestions that helped to improve the chapter draft.

References

Adusei-Gyamfi, J. and V. Acha. 2016. Carriers for nano zerovalent iron (nZVI): Synthesis, application and efficiency. RSC Adv. 6: 91025–91044.

Devi, L.G., M. Srinivas and M.L. Arunakumari. 2016. Heterogeneous advanced photo-Fenton process using peroxymonosulfate and peroxydisulfate in presence of zero valent metallic iron: A comparative study with hydrogen peroxide photo-Fenton process. J. Water Process Eng. 13: 117–126.

Foltynowicz, Z., A. Maranda, B. Czajka, L. Wachowski and T. Sałaciński. 2020. The effective removal of organic and inorganic contaminants using compositions based on zero-valent iron nanoparticles (n-ZVI). Materially Wysokoenergetyczne/High Energy Mater. 37–74.

Fu, F., D.D. Dionysiou and H. Liu. 2014. The use of zero-valent iron for groundwater remediation and wastewater treatment: A review. J. Hazard. Mater. 267: 194–205.

Gunawardana, B., N. Singhal and P. Swedlund. 2011. Degradation of chlorinated phenols by zero valent iron and bimetals of iron: A review. Environ. Eng. Res. 16: 187–203.

Hamdy, A., M.K. Mostafa and M. Nasr. 2018. Zero-valent iron nanoparticles for methylene blue removal from aqueous solutions and textile wastewater treatment, with cost estimation. Water Sci. Technol. 78: 367–378.

Hamdy, A., M.K. Mostafa and M. Nasr. 2019. Techno-economic estimation of electroplating wastewater treatment using zero-valent iron nanoparticles: batch optimization, continuous feed, and scaling up studies. Environ. Sci. Pollut. Res. 26: 25372–25385.

Haneef, T., M.R. Ul Mustafa, K. Rasool, Y.C. Ho and S.R. Mohamed Kutty. 2020. Removal of polycyclic aromatic hydrocarbons in a heterogeneous fenton like oxidation system using nanoscale zero-valent iron as a catalyst. Water (Switzerland) 12: 1–19.

Hassan, M.M.A., A.K. Hassan, A.H. Hamas and D.E. Shther. 2019. Treatment of contaminated water with industrial dyes by using nano zero valent iron (NZVI) and supported on pillared clay. Adv. Anal. Chem. 1: 1–7.

Homhoul, P., S. Pengpanich and M. Hunsom. 2011. Treatment of distillery wastewater by the nano-scale zero-valent iron and the supported nano-scale zero-valent iron. Water Environ. Res. 83: 65–74.

Huang, Y.H. and T.C. Zhang. 2005. Effects of dissolved oxygen on formation of corrosion products and concomitant oxygen and nitrate reduction in zero-valent iron systems with or without aqueous Fe^{2+}. Water Res. 39: 1751–1760.

Kamali, M., K.M. Persson, M.E. Costa and I. Capela. 2019. Sustainability criteria for assessing nanotechnology applicability in industrial wastewater treatment: Current status and future outlook. Environ. Int. 125: 261–276.

Li, S., W. Wang, Y. Liu and Zhang, W. xian. 2014a. Zero-valent iron nanoparticles (nZVI) for the treatment of smelting wastewater: A pilot-scale demonstration. Chem. Eng. J. 254: 115–123.

Li, S., W. Wang, W. Yan and Zhang, W.X. 2014b. Nanoscale zero-valent iron (nZVI) for the treatment of concentrated Cu(II) wastewater: A field demonstration. Environ. Sci. Process. Impacts 16: 524–533.

Li, S., W. Wang, F. Liang and W.X. Zhang. 2017. Heavy metal removal using nanoscale zero-valent iron (nZVI): Theory and application. J. Hazard. Mater. 322: 163–171.

Li, S., J. Li, W. Wang and Zhang, W. xian. 2019. Recovery of gold from wastewater using nanoscale zero-valent iron. Environ. Sci. Nano 6: 519–527.

Litter, M.I. and M. Slodowicz. 2017. An overview on heterogeneous Fenton and photoFenton reactions using zerovalent iron materials. J. Adv. Oxid. Technol. 20: 2371–1175.

Litter, M.I. 2020. Introduction to oxidative technologies for water treatment. pp. 119–175. *In*: Filip, J., T. Cajthaml, P. Najmanová, M. Černík and R. Zbovril (eds.). Advanced Nano-Bio Technologies for Water and Soil Treatment. Springer, Cham, Switzerland.

Lu, H.J., J.K. Wang, S. Ferguson, T. Wang, Y. Bao and H.X. Hao. 2016. Mechanism, synthesis and modification of nano zerovalent iron in water treatment. Nanoscale 8: 9962–9975.

Mahmoud, A.S., M.K. Mostafa and R.W. Peters. 2021. A prototype of textile wastewater treatment using coagulation and adsorption by Fe/Cu nanoparticles: Techno-economic and scaling-up studies. Nanomater. Nanotechnol. 11: 1–21.

Mpouras, T., I. Panagiotakis, D. Dermatas and M. Chrysochoou. 2014. Nano-zero valent iron: An emerging technology for contaminated site remediation. Geo-Congress 2014: Geo-characterization and Modeling for Sustainability, 2206–2215.

Noubactep, C. 2008. Processes of contaminant removal in "Fe^0-H_2O" systems revisited: the importance of co-precipitation. Open Environ. Sci. 1: 9–13.

Pasinszki, T. and M. Krebsz. 2020. Synthesis and application of zero-valent iron nanoparticles in water treatment, environmental remediation, catalysis, and their biological effects. Nanomaterials 10: 917.

Rahmani, A., M. Salari, K. Tari, A. Shabanloo, N. Shabanloo and S. Bajalan. 2020. Enhanced degradation of furfural by heat-activated persulfate/nZVI-rGO oxidation system: Degradation pathway and improving the biodegradability of oil refinery wastewater. J. Environ. Chem. Eng. 8: 104468.

Rasheed, Q.J., K. Pandian and K. Muthukumar. 2011. Treatment of petroleum refinery wastewater by ultrasound-dispersed nanoscale zero-valent iron particles. Ultrason. Sonochem. 18: 1138–1142.

Raychoudhury, T. and T. Scheytt. 2013. Potential of zerovalent iron nanoparticles for remediation of environmental organic contaminants in water: A review. Water Sci. Technol. 68: 1425–1439.

Rezaei, F. and D. Vione. 2018. Effect of pH on zero valent iron performance in heterogeneous Fenton and Fenton-like processes: A review. Molecules 23: 3127.

Ullah, S., X. Guo, X. Luo, X. Zhang, S. Leng, N. Ma and P. Faiz. 2020. Rapid and long-effective removal of broad-spectrum pollutants from aqueous system by ZVI/oxidants. Front. Environ. Sci. Eng. 14: 1–10.

Vilardi, G., J. Rodríguez-Rodríguez, J.M. Ochando-Pulido, N. Verdone, A. Martinez-Ferez and L. Di Palma. 2018. Large laboratory-plant application for the treatment of a tannery wastewater by Fenton oxidation: Fe(II) and nZVI catalysts comparison and kinetic modelling. Process Saf. Environ. Prot. 117: 629–638.

Xu, Y., C. Wang, J. Hou, P. Wang, G. You, L. Miao, B. Lv, Y. Yang and F. Zhang. 2017. Application of zero valent iron coupling with biological process for wastewater treatment: a review. Rev. Environ. Sci. Biotechnol. 16: 667–693.

Chapter 5

Green Synthesized Nanoparticles via Plant Extracts
Present and Potential Applications

Jorge Martín Meichtry,[1,2,*] *Loredana Vitola,*[1] *Abril Ferrero,*[1]
Alejandro Marcelo Senn[2] and *Marta I. Litter*[3]

1. Introduction

There is an increasing interest in the present and potential applications of metal and metal oxide nanoparticles synthesized with plant extracts. A search in Scopus including the words "plant extract", "metal" and "nanoparticles" indicates that more than 2800 articles, including 212 reviews, have been published in the last two decades, most of them since 2010, as it can be seen in Fig. 5.1.

Biological systems such as bacteria, fungi, actinomycetes, yeasts, viruses and plants have been reported to be useful for the synthesis of various metal and metal oxide nanoparticles. Among these, the biosynthesis of nanoparticles from plants seems to be a very effective method in developing a rapid, clean, non-toxic and eco-friendly technology. The use of plant biomass or extracts for the biosynthesis of novel metal nanoparticles (silver, gold, platinum and palladium) would be more significant if the nanoparticles are synthesized extracellularly and in a controlled manner according to their dispersity of shape and size (Akhtar et al. 2013).

In this chapter, the focus will be centered on the green synthesis of nanoparticles (gNPs) using plant extracts, described as an emerging technology with several advantages compared with the conventional chemical synthetic procedures: it is safer, ecofriendly, simpler, faster, energy-efficient, of low-cost and the compounds obtained are more stable and biocompatible, presenting low toxicity (Ovais et al. 2017, Paiva-Santos et al. 2021). Besides, monodisperse gNPs can be obtained with a precise control of the synthesis conditions (Singh et al. 2016), and the materials present an improved interaction with living cells due to the presence of the capping compounds (Kralova and Jampilek 2021). In this review, emphasis will be given to silver, gold and iron gNPs, the most studied in the literature.

[1] Centro de Tecnologías Químicas, Facultad Regional Buenos Aires, Universidad Tecnológica Nacional, Medrano 951, 1425 Buenos Aires, Argentina.
[2] Comisión Nacional de Energía Atómica, CONICET, Gerencia Química, Av. Gral. Paz 1499, San Martín, 1650 Prov. de Buenos Aires, Argentina.
[3] Habitat and Sustainability School, IIIA (CONICET-UNSAM), Universidad Nacional de General San Martín, Campus Miguelete, Av. 25 de Mayo y Francia, 1650 San Martín, Prov. de Buenos Aires, Argentina.
Emails: abrillferrero@gmail.com; lore.vitola12@gmail.com; amsenn@cnea.gov.ar; martalitter24@gmail.com
* Corresponding author: jmeichtry@frba.utn.edu.ar

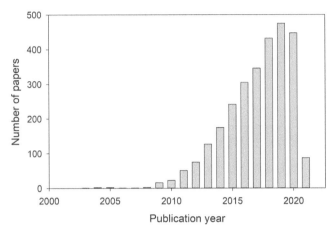

Figure 5.1. Annual publications related to plant extract synthesis of metal and metal oxide nanoparticles. Search performed in Scopus (29 March 2021) using "plant", "extract", "metal" and "nanoparticle" in the title, abstract and keywords.

2. Synthesis of Nanoparticles from Plant Extracts

In Chapter 1, a brief section on the preparation of nanomaterials can be consulted. However, conventional methods present some drawbacks, especially regarding the high costs and toxicity of raw materials used, e.g., sodium borohydride ($NaBH_4$). The green syntheses are eco-friendly alternatives to chemical and physical methods for preparing NPs. They comprise biological ways, i.e., the use of bacteria, enzymes, fungi, plants or plant extracts (Abdelghany et al. 2018, Hasan 2015, Tarannum et al. 2019, Vaseghi et al. 2018). Among them, the use of plant extracts seems to be the best choice, because it is faster, cheaper and less dangerous than other processes (e.g., those using microorganisms) (Paiva-Santos et al. 2021). Plant extracts have been considered suitable due to their nontoxicity for flora-fauna, reliability and environmentally-ecologically acceptability (Kharissova et al. 2013). The polyphenols and caffeine in the plant extracts act as stabilizing agents of the nanoparticles (Weng et al. 2013), and the produced NPs present a variety of size, shape and characteristics with different abilities for practical applications, e.g., in medicine, removal of pollutants, etc. (Khashij et al. 2020). Green methods are also cost-effective and can be easily scaled up, not requiring high-temperature, energy, pressure or harmful chemicals (Rafique et al. 2017).

Regarding the sources, extraction procedures and composition of plant extracts, carbohydrates, flavonoids, fatty acids, polyphenols, among others, have been reported to be the compounds responsible for the synthesis and stabilization of gNPs (Singh et al. 2020, Paiva-Santos et al. 2021, Wu et al. 2021). However, in a study that analyzed the contents of four main active biomolecules in 15 different plant extracts, it was found that polyphenols are the leading ingredients (Xiao et al. 2016). Leaves are the most common parts used of the plants for the synthesis of gNPs, due to their high content of polyphenols. Among others, the following leaves have been tested: *S. Jambos* (L.) *Alston* (Xiao et al. 2016), *yerba mate* tea (*Ilex paraguariensis* Saint Hilaire) (García et al. 2019), green tea (*Camellia sinensis*) (García et al. 2019, Lin et al. 2020), *Eucalyptus* (Liu et al. 2018, Wu et al. 2019), *Cinnamomum tamala* (bay leaf, Ealias et al. 2016), *Solanum nigrum* (Anjana et al. 2019), *S. Canadensis* (Botha et al. 2019), *Indigofera tinctoria* (Vijayan et al. 2017), *Piper longum* (Nakkala et al. 2016), *Coriandrum sativum*, and *Phyllanthus amarus* (Singh et al. 2020). Peel, considered a waste of fruits, has also been used to obtain extracts (Kokila et al. 2015, Ramar et al. 2015), using, e.g., korla fragrant pear (*Pyrus sinkiangensis*) (Rong et al. 2020), red peanut skin (Pan et al. 2020), mango peel (Desalegn et al. 2019), among others. Flowers (Baharara et al. 2015, Manikandan et al. 2015), like *Gardenia jasminoides* (Espinoza-Gómez et al. 2019), barks (Yugandhar et al. 2015), gums (Velusamy et al. 2015), seeds (Haghighi Pak et al. 2016, Sadeghi and Gholamhoseinpoor

2015), roots (Pugazhendhi et al. 2015, Rathi Sre et al. 2015) and green alga genus of *Ulva* spp. (Ergüt et al. 2019) have also been used as the source of extracts.

Although different organic solvents like methanol (Rahimi-Nasrabadi et al. 2014, Sadeghi and Gholamhoseinpoor 2015) and ethyl acetate (Rajesh et al. 2012) have been used to prepare the extracts, the use of hot water is by far the most studied and cheapest strategy, being a greener and a lower cost process (Paiva-Santos et al. 2021); optimal extract temperatures were determined to be around 80°C (García et al. 2019, Liu et al. 2018, Machado et al. 2013). Sometimes, the extracts were concentrated or purified, but as stated, they are usually used as obtained (Paiva-Santos et al. 2021).

Using Fourier Transform Infrared (FTIR) spectroscopy, peaks attributed to O-H, C-H, C=C, C-N, C-C, C-O, C-N and C-H (aromatic) groups were revealed in the extracts (Anjana et al. 2019, Vijayan et al. 2017). In general, -OH groups are related to the reduction of the NPs, while –COOH, –CO and –NH$_2$ are responsible for reducing and capping the nanoparticles (Paiva-Santos et al. 2021).

3. Silver-based Nanoparticles

Silver NPs (AgNPs) exhibit attractive properties, such as high electrical and thermal conductivity, surface-enhanced Raman scattering, chemical stability, high catalytic activity and antimicrobial activities. Physical and chemical methods for synthesizing AgNPs include thermal decomposition, laser ablation, microwave irradiation, sonochemical procedures, use of reverse micelles, salt reduction, radiolysis, etc. NPs with the desired size and shape as well as antimicrobial activities have been reported (Kheybari et al. 2010, Yadi et al. 2018).

3.1 Synthesis

AgNPs can be synthesized by the bottom-up approach, and chemical reduction is the most general method. Various organic and inorganic reducing agents in aqueous or non-aqueous solutions are used for the reduction of silver ions, e.g., polyethylene glycol block copolymers, sodium citrate, Tollens' reagent, ascorbate, hydrogen, N,N-dimethyl formamide and NaBH$_4$. This method allows obtaining a substantial amount of NPs in a short time. However, the chemicals used are toxic and lead to non-ecofriendly byproducts. On the other hand, the development of the green synthesis of AgNPs is advancing as a key branch of nanotechnology, using biological entities like living plants, plant extracts, plant biomass, bacteria, fungi, algae, yeasts and biopolymers (Abdelghany et al. 2018, Tarannum et al. 2019, Vaseghi et al. 2018, Vijayan et al. 2017). The green synthesis of AgNPs using plant extracts (gAgNPs) is a very simple method, with low energy requirements and environmental impact. Besides, the high specificity of biomolecules involved in the biosynthesis process enables an efficient control of the AgNPs size and shape, critical to optimizing applications. The use of plants in the synthesis of AgNPs is more advantageous over the use of microorganisms, which need expensive methodologies for maintaining microbial cultures, and more time for synthesis; additionally, the plant extracts have a higher bioreduction potential compared with microbial cultures.

The green synthesis is normally performed using silver nitrate (AgNO$_3$) as the precursor, together with the natural reducing agents. The synthesis proceeds in three steps: (a) reduction of Ag+ to Ag0, (b) agglomeration and (c) stabilization of the suspension (Srikar et al. 2016). The active components responsible for the reduction are those containing carboxyl and hydroxyl groups like antioxidants, vitamins, alkaloids, alcohols and polyphenols; it has been reported that Nicotinamide Adenine Dinucleotide (NAD$^+$) also takes part in the synthesis process (Kumar et al. 2020). The stabilization is carried out by flavonoids and terpenoids (Tarannum et al. 2019). The reduction potential of the polyphenols is in the range of 0.30–0.80 V, enough for the reduction of Ag+ to Ag0 ($E^0_{Ag+/Ag0} = 0.80$ V).

Besides AgNO$_3$ and concentrations of polyphenols, other parameters such as temperature and pH can alter the characteristics of the products, like the size and shape. For example, an increase in temperature leads to smaller gAgNPs, with shapes changing from triangular at 20°C, to octahedral at 30–40°C and spherical at higher temperatures. An increase in pH increases the nucleation centers, leading to NPs of smaller size (Singh et al. 2020). Longer reaction times give larger and more crystalline gAgNPs (Akthar et al. 2013).

There is vast literature devoted to the synthesis of gAgNPs using a large number of plants (Akthar et al. 2013, Vanlalveni et al. 2021, Vaseghi et al. 2018, Vijayan et al. 2017). As mentioned earlier, leaves are the most frequently used extract source, although other parts of the plants have also been tested (Abdelghany et al. 2018, Akthar et al. 2013, Mystrioti et al. 2016, Vanlalveni et al. 2021, Vaseghi et al. 2018).

For example, extracts of leaves of *Abutilon indicum, Azadirachta indica, Crotolaria retusa, Terminalia arjuna, Ocimum tenuiflorum, Solanum trilobatum, Syzygium cumini, Centella asiatica, Citrus sinensis, Justicia glauca, Hemidesmus indicus, Cassia tora, Ziziphus jujube, Andrographis echioides, Eucalyptus camaldulensis, Lawsonia inermis* (Henna), *Aloe vera, Convolvulus pluricaulis, Casuarina equisetifolia* L., *Prunus amygdalus, Aloysia triphylla, Eichhornia crassipes, Lantana camara, Vitis vinifera, Indoneesiella echioides* (L.) and *Rauvolfia serpentina* Benth were used. Extracts of *Abelmoschus esculentus* (L.) pulp and banana peel were also employed. The root extracts of *Potentilla fulgens* and *Erythrina indica* were also reported. Mixed gum extract of neem (*Azadirachta indica*), *Alpinia calcarata* roots, ethanolic extracts of rose (*Rosa indica*) petals were also reported. Aqueous extracts of *Pistacia atlantica* and *Dracocephalum moldavica* seeds, unripe fruits of *S. trilobatum, Emblica officinalis, Prosopis farcta, Syzygium alternifolium*, bark extract of *S. alternifolium, Vigna radiata*, kenaf (*Hibiscus cannabinus*) were also tested (Abdelghany et al. 2018, Rafique et al. 2017).

Extracts of micro- and macroalgae were also employed for the biosynthesis of gAgNPs, e.g., *Pithophora oedogonia, Botryococcus braunii, Coelastrum* sp., *Spirulina* sp., *Limnothix* sp., *Chlorella, Anabaena oryzae, Nostoc muscorum*, and *Calothrix marchic, Amphiroa fragilissima, Caulerpa racemosa, Laurencia aldingensis, Laurenciella* sp., *Spirogyra varians, B. braunii, Coelastrum* sp., *Spirulina* sp., *Limnothix* sp., *Chlorella vulgaris, Chlamydomonas reinhardtii, Spirulina platensis, Chlorella pyrenoidosa, Sargassum polycystum* C. Agardh, among others (Abdelghany et al. 2018).

3.2 Characterization

Ag0 nanoparticles show a characteristic absorption peak in the UV-Vis region due to the Surface Plasmon Resonance (SPR) effect, a resonant oscillation of conduction electrons induced by the electromagnetic field of the incident light. It is absent in the individual atoms or the bulk substance. The SPR peak of gAgNPs has been reported to appear in the 400–420 nm region (around 3 keV, Vijayaraghavan et al. 2012), and it has been used to monitor the size and stability of the particles (Balavigneswaran et al. 2014, Rajeshkumar and Bharath 2017, Vijayaraghavan et al. 2012). Other absorption bands appear along the UV-Vis spectrum (200–800 nm) and depend on the morphology, shape, size and chemical surroundings of the gAgNPs (Rajeshkumar and Bharath 2017).

The SEM analysis of gAgNPs showed spherical particles in most cases (Kumar et al. 2020, Vanlalveni et al. 2021). However, other shapes, as triangular and flake-like, were also obtained depending on the characteristics of the natural extract used, the reaction time, pH, etc. (Akhtar et al. 2013). EDS analysis of many gAgNPs indicated a silver purity ranging from 45 to 80% w/w (Srikar et al. 2016).

The Face Center Cubic (FCC) was the most commonly reported crystalline structure (Ali et al. 2016). Accordingly, the gAgNPs prepared from apple extracts showed diffraction peaks at 38.15°, 44.35°, 64.59°, 77.47°, 81.60°, among others (Ali et al. 2016). Cubic and hexagonal structures have also been reported.

Several functional groups on the surface of gAgNPs have been reported by the FTIR analysis. Amine, aromatic, alkynes and ethylene groups were detected (Ali et al. 2016, Tho et al. 2013). Bands corresponding to alkyl halides in the range 600–680 cm^{-1} were also detected (Bagherzade et al. 2017).

3.3 Applications

AgNPs have antibacterial, antifungal and antiparasitic action, and have found application in health, food storage, textile coatings, agriculture and environmental fields (Paul and Roychoudhury 2021, Verma and Maheshwari 2019), playing a major part in the commercial applications of pharmaceutical and other medical sciences (Abdelghany et al. 2018). Other uses, as in analytical chemistry and photocatalysis (Kumar et al. 2013b, Vaseghi et al. 2018), have also been reported. Figure 5.2 summarizes the main applications of gAgNPs.

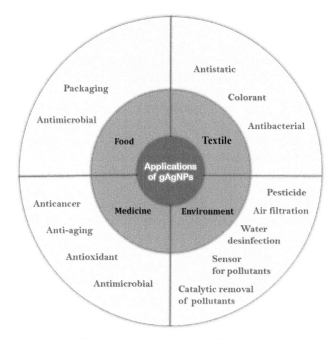

Figure 5.2. Main applications of gAgNPs.

3.3.1 Textile

The incorporation of gAgNPs into textile materials has been extensively studied, and it can be found in a wide range of commercially available products (Radetić 2013). The principal reason for using gAgNPs in textiles is their antibacterial properties but they could also be used as a colorant and antistatic agent. For example, gAgNPs can be directly synthesized onto chitosan-coated cotton, the fabric showing beautiful yellow hues with antioxidant and antibacterial properties (Shahid-ul-Islam et al. 2020). gAgNPs synthesized with Curcuma longa tuber powder and extract have also been immobilized onto cotton cloth, showing good bactericidal activity; this activity diminished with successive washings depending on the immobilization method (Sathishkumar et al. 2010).

3.3.2 Medical

AgNPs have been broadly used in the healthcare industry as antibacterial and anticancer agents (Kumar et al. 2020, Lee and Jun 2019). Biosynthesized AgNPs present superior biocompatibility than the chemically synthesized ones, and they lead to an enhanced chemotherapeutic efficacy along with minimal systemic toxicity. Some AgNPs based chemotherapy drugs have recently been approved by the U.S. Food and Drug Administration (FDA, Ratan et al. 2020). gAgNPs present lower toxicity compared with that of chemically synthesized AgNPs and their selectivity toward cancerous cells, being able to be used as drug vehicles (Karmous et al. 2020, Paiva-Santos et al. 2021). In particular, gAgNPs displayed a cytotoxic effect against several cancer cell lines, such as epithelial (HeLa), breast (MCF-7), lung (A549) and colon (HT 29) cell lines (Gengan et al. 2013, Jeyaraj et al. 2013,

Rasheed et al. 2017, Sanpui et al. 2011, Vijayan et al. 2017). Nanoparticle-based combinatorial therapies, utilizing NPs with anticancer activity in combination with a chemotherapeutic agent have been reported (Yadi et al. 2018).

The cytotoxic activity of gAgNPs on cancer cells is dose-related and linked to a decrease in cell viability, fragmentation in nucleic acids, inhibition of the proliferation and induction of apoptosis (Baharara et al. 2015). However, gAgNPs have shown toxicity toward mammal cells and organs, and more toxicological studies should be carried out (Ferdous and Nemmar 2020).

The antibacterial activity of gAgNPs is related to the presence of both the extracts and the Ag nanoparticles, which release Ag ions; in general, the bactericidal activity increases with the decrease in the size of the gAgNPs (Paiva-Santos et al. 2021, Singh et al. 2020). gAgNPs interact with cellular organelles and biomolecules like DNA, enzymes, ribosomes, among others, and this can have an effect on the cell membrane permeability, oxidative stress, gene expression, protein activation, etc.; these multiple attacks prevent the generation of resistance in microorganisms (Ferdous and Nemmar 2020, Singh et al. 2020).

Several synthesized gAgNPs demonstrated high catalytic activity as well as excellent antimicrobial properties against both gram-negative and gram-positive bacteria (Rafique et al. 2017). gAgNPs also inhibit biofilm formation (Burlacu et al. 2019). Regarding the effect of the shape on the efficiency of the disinfection, it has been reported that triangular gAgNPs are more efficient than spherical ones (Ferdous and Nemmar 2020, Singh et al. 2020).

gAgNPs have also shown interesting insecticidal activity, killing different mosquito larvae while being less toxic to non-targeted insects (Kumar et al. 2020). In the case of malaria, gAgNPs have shown the potential to treat affected patients and to act as a biosensor (Mohammadi et al. 2021).

gAgNPs have shown wound healing activity, by increasing the generation of fibroblast and collagen and decreasing the amount of inflammatory cells (Paiva-Santos et al. 2021).

3.3.3 Environmental

The antimicrobial properties of gAgNPs have been tested in air and water disinfection with optimal results. gAgNPs deposited onto carbon nanotubes and tested in an air filtration medium reached a relative viability lower than 1% in the filtered air (Jung et al. 2011). Other gAgNPs, anchored onto macroporous methacrylic acid copolymer beads, were highly effective to remove gram-positive and gram-negative bacteria from water (Gangadharan et al. 2010). Suspended gAgNPs were found efficient for sewage disinfection (Raja et al. 2012).

gAgNPs synthesized from Ginkgo biloba leaf aqueous extract enhanced the $NaBH_4$ reduction removal of azo dyes (Huang et al. 2020). Carmoisine, tartrazine and brilliant blue FCF were efficiently removed by gAgNPs obtained using the Viburnum opulus fruit extract (David and Moldovan 2020). gAgNPs with and without H_2O_2 showed a fast, efficient and stable catalytic activity for the degradation of cationic organic dyes, but they were not efficient against anionic ones (Qing et al. 2017). gAgNPs also showed photocatalytic activity for the degradation of methyl orange (Kumar et al. 2013b).

gAgNPs have also demonstrated to be an efficient probe for hexavalent chromium detection due to the red shift of the SPR peak from 400 to 510 nm (Balavigneswaran et al. 2014).

3.3.4 Other applications

In agriculture, the bactericidal, fungicidal and nematicidal properties of gAgNPs have attracted attention as substitutes of agrochemicals; gAgNPs damage the membrane of pathogens and cause oxidative stress. They also modify the defense mechanisms of the plant, increasing its resistance to plagues (Paul and Roychoudhury 2021). gAgNPs favored the regeneration of shoot cultures of S. chirata by generating Reactive Oxygen Species (ROS) that induced antioxidant defense (Saha and Gupta 2018).

4. Gold Nanoparticles

Likewise gAgNPs, gAuNPs have also been one of the first gNPs studied due to their numerous potential applications as a bactericidal agent, in cancer treatment, drug delivery and as a catalyst; the NPs also show good electronic, optical and thermal properties (Anjana et al. 2019, Singh et al. 2020, Zhang et al. 2020). gAuNPs are also easy to synthesize, as the extracts can readily reduce Au (III) into Au (0) ($E^0_{Au3+/Au0}$ = +1.50 V) (Nadeem et al. 2017).

4.1 Synthesis

As other gNPs, gAuNPs can be prepared by biological synthesis using bacteria, fungi, yeast, cyanobacteria, algae or viruses (Akhtar et al. 2013). However, those prepared by mixing plant extracts with $HAuCl_4$ aqueous solution were found to be much more stable than the others (Singh et al. 2020). During the synthesis, some functional extract groups like polyols and carboxylic acids were responsible for the formation of the gAuNPs, identified by the appearance of a violet coloration in the mixture (Anjana et al. 2019). In some cases, gAuNPs were dried and calcined at 500°C for the elimination of remaining organic compounds (Anjana et al. 2019). Synthesis conditions such as pH, extract and Au concentrations and temperature can affect the characteristics of the obtained gAuNPs: larger NPs were synthesized at acidic pH (Zhang et al. 2020) because fewer nucleation centers were formed; for this reason, Au(0) deposits over the scarce centers formed increasing their size (Singh et al. 2020). Besides, at pH 2, there are fewer functional groups in the extracts to stabilize and avoid the aggregation of the NPs (Zhang et al. 2020). Nanoscale spheres and plates were obtained at higher Au concentrations (Zhang et al. 2020). Higher extract concentrations gave spherical NPs, while lower concentrations produced triangular and hexagonal NPs (Zhang et al. 2020 and references therein). Finally, at 50°C, rod-like and platelike gAuNPs with a smaller size were obtained, while at 25°C, spherical gAuNPs were produced (Zhang et al. 2020 and references therein), with the most homogeneous distribution around 25–30°C (Singh et al. 2020).

4.2 Characterization

Reported sizes of gAuNPs were in the 7–58 nm range (Anjana et al. 2019, Zhang et al. 2020 and references therein); XRD analysis indicated a face-centered cubic structure (Vijayan et al. 2017). UV-Vis spectroscopy confirmed the formation of gAuNPs by the detection of an SPR absorption in the 530–545 nm range (Anjana et al. 2019, Nakkala et al. 2016, Vijayan et al. 2017); the narrow SPR peaks indicated that the gAuNPs were spherical and monodisperse (Anjana et al. 2019). FTIR indicated the presence of phytochemicals around the gAuNPs (Zhang et al. 2020 and references therein), responsible for the reduction and stabilization of the NPs (Nadeem et al. 2017). TEM images confirmed the spherical shape of the gAuNPs, and their mild agglomeration, without gold nanoflakes or nanorods (Anjana et al. 2019, Nakkala et al. 2016). gAuNPs exhibited superior antioxidant behavior, with good 2,2-diphenyl-1-picrylhydrazyl (DPPH), superoxide radical, good nitric oxide radical and hydrogen peroxide scavenging activities (Nakkala et al. 2016). Although the plant extracts themselves have antioxidant properties, gAuNPs present a higher antioxidant activity due to their spheric shape that exposed more freely the -OH groups of the extracts (Vijayan et al. 2017).

4.3 Applications

Due to their good electronic, optical and thermal properties and chemical stability, gAuNPs were commonly used in electrochemical energy storage applications, i.e., as supercapacitors (Anjana et al. 2019). As other AuNPs, these green materials can have a very remarkable use in the biomedical field, as antimicrobial agents, biosensors, in hyperthermia therapy, delivery systems for therapeutic drugs and genetic materials, as well as antibacterial drugs, detection of metabolites, tumors, endocytosis and receptors in cells for the treatment of different cancers (Nadeem et al. 2017, Vijayan et al. 2017, Zhang et al. 2020 and references therein). They have low toxicity and high biostability (Paiva-Santos et al. 2021).

gAuNPs also absorb visible light and have applications in light-mediated clinical treatments (Botha et al. 2019); they can be used as components of sunscreens for skin protection (Paiva-Santos et al. 2021).

In the case of anticancer applications, lung carcinoma and melanoma cells could be treated with gAuNPs (Paiva-Santos et al. 2021), although their efficiency for treating lung cells was lower than that of gAgNPs (Vijayan et al. 2017). A change in the morphology and a shrinking of cancer cells could be appreciated for both types of NPs, with ROS being the most probable candidates to induce the damage in the cells (Vijayan et al. 2017).

Anti-diabetic properties of gAuNPs have also gained attention, as an increase in pancreatic cell survival has been reported in zebrafish (Burlacu et al. 2019).

gAuNPs have very high antimicrobial activity due to their morphology, chemistry surface, polyvalent and photothermic nature; the antimicrobial activity decreased with the increase of the size of the NPs (Nadeem et al. 2017). Some studies found that gAuNPs were less efficient than gAgNPs, although this was probably related to the higher aggregation of gAuNPs in the agar medium used in the test (Vijayan et al. 2017). The main antimicrobial mechanism was the link between the gAuNPs to functional groups of diverse enzymes, interrupting the respiratory chains and generating the production of free radicals that attack DNA and other molecules (Nadeem et al. 2017). Other possible mechanisms can be the decrease in ATPase activities. Similar to gAgNPs, gAuNPs have the potential to be used for the treatment of malaria (Mohammadi et al. 2021).

gAuNPs are useful for the catalytic degradation of dyes like methylene blue, methyl red, crystal violet and acridine orange, although the degradation rates were rather slow (Nakkala et al. 2016). Nitrophenol reduction to aminophenol could also be achieved (Burlacu et al. 2019). When the surface area increased, the catalytic efficiency also increased (Nakkala et al. 2016 and references therein). gAuNPs have a lower antioxidant activity than gAgNPs, due to their lower tendency to lose electrons (Burlacu et al. 2019).

5. Iron-Based Nanoparticles

Iron-based nanoparticles (FeNPs) have gained increasing interest in recent years for their use in the removal of pollutants in waters and soil such as halogenated organic compounds, heavy metals, arsenic and nitrate (Litter et al. 2018 and references therein). Zerovalent iron nanoparticles (nZVI) were the most studied, although iron-oxide NPs like nanomagnetite and nanohematite have also shown interesting environmental and medical applications.

5.1 Synthesis

nZVI were prepared usually by reduction of ferric or ferrous salts with $NaBH_4$ (Wang and Zhang 1997) or by H_2 mediated reduction (Pasinszki and Krebsz 2020). However, as the reagents are expensive, hazardous and toxic, the search for more environmentally friendly synthetic ways such as the use of plant extracts was proposed (Crane and Scott 2012, García et al. 2019, Lin et al. 2020, Paiva-Santos et al. 2021, Rong et al. 2020). The reaction of iron salts with extracts of leaves or grains of natural plants, such as GT, coffee and wine (Oakes 2013) was reported. Hoag et al. (2009) reported the first synthesis of iron nanoparticles prepared from GT extracts and Fe $(NO_3)_3$ solutions. Later, similar products using a variety of natural materials and other Fe common salts ($FeCl_3$, $FeSO_4$, Fe-EDTA and mixtures of Fe (III) and Fe (II)) have been reported, and this topic has been reviewed (Espinoza-Gómez et al. 2019, Genuino et al. 2013, Herlekar et al. 2014, Kharissova et al. 2013). When the Fe (III) solutions were mixed with the extracts, a black suspension instantaneously appeared, indicating the formation of the iron-based nanoparticles. The degree of the color of this black suspension changed according to the type of extract used, and this can be explained by the different Fe^{3+} reducing the capacity of each plant extract (Xiao et al. 2016). The active compounds participating in the synthesis of gFeNPs starting from plant extracts are polyphenols, caffeine or L-theanine, among others (Anesini et al. 2008, Wu et al. 2021). The main reaction involves the Fe complexation by the -OH groups of polyphenols, followed by their oxidation to carbonyl groups and the corresponding metal reduction (Anesini et al. 2008, García et al. 2019, Paiva-Santos et al. 2021). The extracts act as capping agents, avoiding gFeNPs sintering by electrostatic and steric interactions (Paiva-Santos et al. 2021). As with all gNPs, the key parameters that affect the synthesis are the extract source and storage condition, ratio of metal to extract (higher metal to extract ratio gives bigger NPs), pH, temperature and O_2 concentration (Paiva-Santos et al. 2021).

However, a discrepancy exists in the composition and chemical structure of these green materials, and there is still a dispute in literature regarding the possibility that polyphenols can reduce Fe (II) to give Fe (0) (Lin et al. 2020, Rong et al. 2020, Smuleac et al. 2011). Among the different structures proposed for gFeNPs, low solubility mono- and bis-type complexes between Fe (III) and the extracts are the expected products within pH 1–7, presenting a variable Fe coordination ratio, depending on the structure of the polyphenols and the pH (Gust and Suwalski 1994, Oakes 2013, Perron and Brumaghim 2009, Ryan and Hynes 2007, Wang 2013, 2015). This inhibits the formation of oxides and hydroxides. Although tea polyphenols are strong reducing chelators, their reduction potentials range between 0.3 and 0.8 V, a value not enough for reduction of Fe (II) to the zerovalent state ($E^0_{Fe2+/Fe0} = -0.44$ V) and only reduction of the initial Fe^{3+} to Fe^{2+} is possible ($E^0_{Fe3+/Fe2+} = 0.77$ V) (Markova et al. 2014, Wang 2013, 2015). However, although Fe (II) structures can be formed, this process is slow at neutral pH values, and the polyphenol ligands strongly stabilize Fe^{3+} over Fe^{2+}. As Fe (III) complexes are more stabilized by polyphenols than Fe (II), O_2 would easily oxidize Fe (II) (García et al. 2019). Consequently, if formed, the Fe^{2+} complexes would rapidly auto oxidize in the presence of O_2, reverting to Fe^{3+}-polyphenol complexes (Perron and Brumaghim 2009). The characterization confirms this hypothesis. Therefore, a structure similar to that of Fig. 5.3 can be suggested for the iron complexes, schematized for catechins, taking into account that these organic species are abundant in the extracts.

For samples prepared from ferrous sulfate or under reducing N_2 atmosphere, the presence of Fe (0) in the product was indicated (Hoag et al. 2009, Huang et al. 2015, Madhavi et al. 2013, Ravikumar et al. 2015, Smuleac et al. 2011, Sunardi et al. 2017, Wang et al. 2014a,b, Weng et al. 2017). However, iron nanoparticles prepared by a very rapid procedure starting from Fe (II) or Fe (III) salts, generally at Room Temperature (RT) and acid pH, with no further treatment, were amorphous, did not present peaks of Fe (0) or iron oxides in the XRD patterns, or show ferromagnetic behavior (Carvalho and Carvalho 2017, Luo et al. 2014, 2016, Machado et al. 2015, Martínez-Cabanas et al. 2016, Nadagouda et al. 2010, Njagi et al. 2010, Oakes 2013). Some authors indicated that the interaction of natural extracts with iron produces organic complexes, the formation of oxides and hydroxides being generally inhibited (Markova et al. 2014, Oakes 2013, Wang 2013, 2015). In other

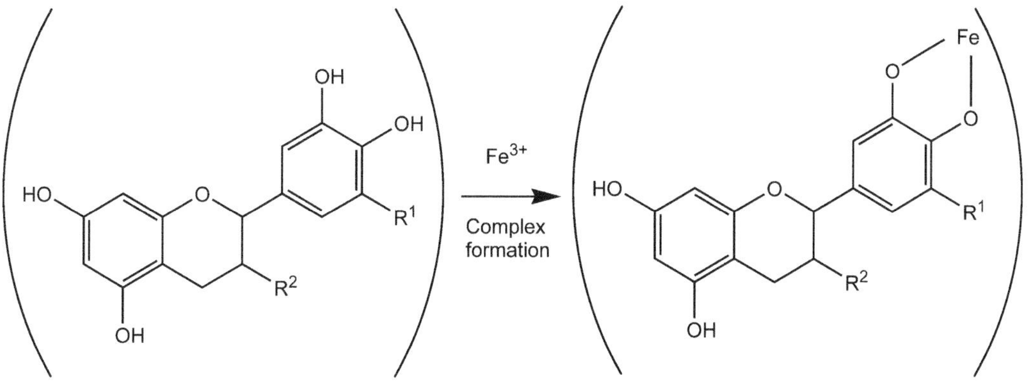

Figure 5.3. Iron complexes formed by catechins and Fe (III) (García et al. 2019, with permission).

cases, the nature of the iron nanoparticles was not clear, with some of the samples presenting XRD peaks of iron crystalline phases (Kiruba Daniel et al. 2012, Kumar et al. 2013a, Luo et al. 2015, Machado et al. 2013, Makarov et al. 2014, Mystrioti et al. 2016, Poguberović et al. 2016, Wei et al. 2017). Pan et al. (2020) reported XRD results showing that the organic components of the extracts developed a covering on the nanoparticles, revealing a core-shell structure (as indicated in Fig. 5.4).

At the time the synthesis was performed using an alkali (e.g., NaOH), when ammoniacal iron(II) salts were used, or while the preparation was done (or finished) at high temperatures (ca. 70°C), the samples presented well-defined phases of iron oxides (Ealias et al. 2016, Ergüt et al. 2019, Herrera-Becerra et al. 2010, Huang et al. 2015, Kuang et al. 2013, Murgueitio et al. 2016, 2018, Önal et al. 2017, Rao et al. 2013, Rong et al. 2020, Shahwan et al. 2011, Srivastava et al. 2011, Venkateswarlu et al. 2013). Senthil and Ramesh (2012), Weng et al. (2013), Thakur and Karak (2014) and Groiss et al. (2017) prepared samples in the absence of NaOH that contained magnetite and other iron oxides, but no explanation was given about the formation of oxides in those samples.

5.2 Characterization

SEM-EDS, TEM and HR-TEM analyses indicated that gFeNPs are formed by dispersed spherical NPs as well as aggregates of irregular shape or in the form of chains (Ealias et al. 2016, García et al. 2019, Lin et al. 2020, Liu et al. 2018, Rong et al. 2020), with diameters in the range of 15–100 nm, being C, O and Fe the main components, on a percentage proportional to the Fe: extract Molar Ratio (MR) used. By changing the synthesis conditions, diameters of up to 1 nm could be obtained (Pan et al. 2020). Rong et al. (2020) indicated that the extracts form a layer on the nanoparticle surfaces that inhibits gFeNPs aggregation, although Ealias et al. (2016) reported that gFeNPs agglomerates with a cluster size varying from 10 to more than 100 μm. It has been reported that the exterior layer of the FeNPs was thick, with a uniform distribution between the NPs. The Fe-based NPs appeared encapsulated by biomolecules from the extract (Lin et al. 2020). Irregular agglomerates of amorphous gFeNPs of different size together with some free, spherical particles were observed (Ealias et al. 2016, García et al. 2019, Lin et al. 2020, Liu et al. 2018).

Numerous publications indicate that gFeNPs prepared from Fe (II) or Fe (III) salts at RT and acid pH, with no further treatment, were amorphous, without peaks of Fe (0) or iron oxides in the XRD patterns and without ferromagnetic behavior (Carvalho and Carvalho 2017, Desalegn et al. 2019, García et al. 2019, Kumar et al. 2013a, Lin et al. 2020, Luo et al. 2014, 2015, 2016, Machado et al. 2015, Oakes 2013, Wei et al. 2017, Wu et al. 2019). UV-Vis spectra confirmed that gFeNPs are mainly composed of Fe (III)-extract complexes, while amorphous carbon or carbon partially bound to Fe was detected by Raman spectroscopy (García et al. 2019). Chemical generation of H_2 after reaction of gFeNPs with H_2SO_4, which only takes place if Fe (0) is present, was not formed in the gFeNPs obtained from *yerba mate* extracts (García et al. 2019). FTIR measurements confirmed the presence of Fe-O bonds (Carvalho and Carvalho 2017, García et al. 2019, Luo et al. 2014).

Polyphenols, flavonoids, caffeine, caffeic acid, catechols and aminoacids were detected by FTIR in gFeNPs synthesized using GT extracts (Lin et al. 2020, Rong et al. 2020). O-H stretching vibration of alcoholic or carboxylic groups were reported (Desalegn et al. 2019); phenolic hydroxyl group signals initially present in the pure extracts weakened or disappeared, suggesting that they participated in the complexation with Fe (III)/Fe (II) (Wu et al. 2021), or in the stabilization of Fe oxides (Espinoza-Gómez et al. 2019).

When the Fe-based nanomaterials were composed of iron oxides, Fe-O bonds were detected by FTIR (Ealias et al. 2016, Ergüt et al. 2019, Espinoza-Gómez et al. 2019, Rong et al. 2020). XPS results showed that these NPs contained hydroxyl and aliphatic/aromatic groups, together with Fe_3O_4 and minor amounts of Fe (0) (Ealias et al. 2016, Espinoza-Gómez et al. 2019, Lin et al. 2020, Rong et al. 2020), although Wu et al. (2021) reported that Fe was mostly as Fe (III). Fe_3O_4 was detected by Raman (Espinoza-Gómez et al. 2019, Rong et al. 2020).

gFeO$_x$NPs: Characterization and aqueous pollutants removal mechanisms

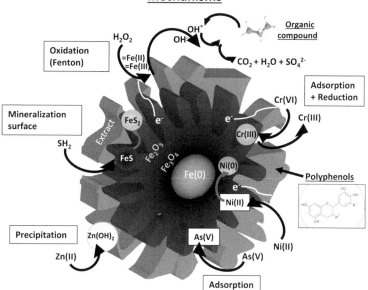

Figure 5.4. Characterization of gFeNPs and removal mechanisms of different aqueous pollutants.

Surface areas reported for gFeNPs are smaller (17.5–37.2 m^2 g^{-1}) (García et al. 2019, Lin et al. 2020) or similar (51.14 m^2 g^{-1}) (Wu et al. 2019) to those reported for nZVI obtained from NaBH$_4$ (56.5 m^2 g^{-1}) (García et al. 2019), probably due to the coverage of the gFeNPs by the extracts.

5.3 *Applications*

5.3.1 *Treatment of inorganic pollutants*

The most studied applications for gFeNPs are the removal of dyes and inorganic pollutants from water. In particular, the treatment of Cr (VI) and As (III)/(V) that are considered priority pollutants is not efficient by conventional technologies, especially when compared to advanced processes, such as heterogeneous photocatalysis and use of iron-based nanomaterials (Litter et al. 2016, 2018). In Table 5.1, some works regarding the removal efficiency of these pollutants by gFeNPs are presented and compared with nZVI.

For these inorganic pollutants, the removal mechanism involves the adsorption on the active sites of the gFeNPs, followed by reduction by Fe (II) and/or the extracts as in the case of Cr (VI) or formation of Fe complexes for As. Langmuir and Freundlich were the most used isotherms to evaluate the gFeNPs adsorption capacity toward these pollutants (Rong et al. 2020, Wu et al. 2019, 2021); from the reviewed literature, no model proved to be the more suitable one, with the best fitting depending on the nature of gFeNPs and the treated pollutant. Removal kinetics also depends on the gFeNPs evaluated and the treated pollutant, being pseudo-first and pseudo-second-order regimes the most used models; in particular, pseudo-second-order kinetics have been associated with an adsorption process via chemisorption (Rong et al. 2020, Wu et al. 2019). In the case of Cr (VI), faster and more efficient removal was observed at low pH (García et al. 2019, Liu et al. 2018); higher Cr (VI) removal rates were also valued at higher [Poll.]/[Fe] MR (Pan et al. 2020). Cr (III) was the only reported product of Cr (VI) reduction, which remained partially in solution; direct Cr (VI) reduction by the extracts alone was slower and less efficient than with gFeNPs (García et al. 2019). The NPs could be reused up to four times, keeping a removal efficiency of 55.8% after the fourth reuse (Liu et al. 2018).

Table 5.1. Applications of gFeNPs for the treatment of inorganic pollutants in aqueous phase.

Pollutant	Exp. conditions	Optimal [Poll.]/[Fe] MR	Reference
Cr (VI)	[Cr (VI)] = 0.3 mM, pH 3	1:0.6	Garcia et al. 2019
Cr (VI)	[Cr (VI)] = 0.19 mM, pH 4	1:130	Liu et al. 2018
Cr (VI)	[Cr (VI)] = 0.19 mM, pH 5	1:20	Rong et al. 2020
Cr (VI)	[Cr (VI)] = 0.19 mM, pH 4.7	1:186	Pan et al. 2020
Cr (VI)	[Cr (VI)] = 0.48 mM, pH 7	1:2.1	Poguberović et al. 2016
Cr (VI)	[Cr (VI)] = 0.19 mM*	1:1.3	Xiao et al. 2016
Cr (VI)**	[Cr (VI)] = 0.3 mM, pH 3	1:1.25	García et al. 2019
Cr (VI)***	[Cr (VI)] = 0.96 mM, pH 4.58	1:1.89	Mystrioti et al. 2016
As (III)	[As (III)] = 0.33 mM, pH 4–8	1:4.5	Poguberović et al. 2016
As(V)	[As(V)] = 0.3 mM, pH 4–6	1:92	Wu et al. 2019
As(V)	[As(V)] = 0.003–1.3 mM, pH 6	1:67	Wu et al. 2021

*: pH not reported.
**: experiment done with commercial nZVI.
***: pilot-plant scale experiment.

In the case of As (V), no changes in the redox state were reported; the adsorption took place by bidentate complexation with Fe (III), reflecting a chemisorption process in agreement with equilibrium (Langmuir isotherm) and kinetic (pseudo-second-order) results (Wu et al. 2019). Free hydroxyl groups of the organic moiety of the gFeNPs also participate in As (V) complexation (Wu et al. 2021). The optimal reaction pH for As (V) removal was at pH 5, while for As (III) was at pH 10 (Poguberović et al. 2016).

After use to remove pollutants, the aggregation of gFeNPs increased and the morphology became coarser, due to the filling of the pores by the trapped pollutants and by changes in the surface charge distribution, causing a decrease in BET surface area (Lin et al. 2020, Rong et al. 2020, Wu et al. 2019).

Despite many laboratory studies regarding the treatment of pollutants, only Mystrioti et al. (2018) reported a pilot-scale application of gFeNPs. The authors studied Cr (VI) removal in a tank containing 24 tons of soil, where 3 m^3 of a gFeNPs suspension (prepared from GT) was injected; 100 L day^{-1} of a 5 mg L^{-1} Cr (VI) contaminated water flowed through the tank. The system was tested for over a year, reaching a total Cr content below 10 μg L^{-1}. This example proved that the *in situ* injection technique, with the generation of a reactive zone perpendicular to the flow of groundwater, is an efficient remediation alternative to the conventional pump-and-treat technology. This aspect of the transport characteristics of gFeNPs in sand for *in situ* soil and aquifer treatments (see Fig. 5.5) have been studied (Chrysochoou et al. 2012). *In situ* treatments offer a cheaper alternative to on-site treatment of groundwater (Fig. 5.5), which implies the extraction of the polluted water or soil and their mix with gFeNPs in a chemical reactor.

5.3.2 *Treatment of organic pollutants*

gFeNPs have been used to remove organic pollutants mainly by three mechanisms: adsorption, photocatalytic degradation and the Fenton process (with the addition of H$_2$O$_2$). In Fenton reactions, hydroxyl radicals and/or high Fe oxidation states are formed after the reaction of Fe (II) with H$_2$O$_2$ (Bataineh et al. 2012, Litter and Slodowicz 2017); these powerful species have the capacity of oxidizing the organic pollutant. Organic pollutants such as dyes and antibiotics (Espinoza-Gómez et al. 2019, Lin et al. 2020) were removed by adsorption on gFeNPs.

The use of gFeNPs enables the use of the Fenton reaction at circumneutral pH conditions, with Fe acting as a catalyst, avoiding Fe (III) precipitation. Dyes such as Putnam sky blue 39 (Espinoza-Gómez et al. 2019), methyl orange (Desalegn et al. 2019), malachite green (Ergüt et al.

Different alternatives for pollutants treatment using gFeO$_x$NPs

1. *In situ*

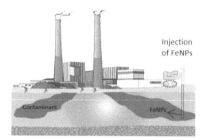

2. *On site* or *ex situ*

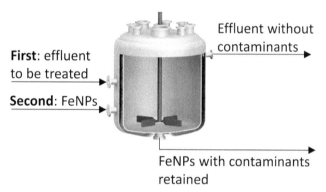

Figure 5.5. Different applications of gFeNPs.

2019) and others (Carvalho and Carvalho 2017) were the most studied compounds. For example, iron oxide nanoparticles synthesized from *Ulva* spp. extract and loaded into carbonated hydroxyapatite were reusable in the malachite green removal by a heterogeneous Fenton-like process for as much as six cycles without significant activity loss. Moreover, a real textile wastewater treated with this material by the same process showed an almost total discoloration (94%) and a significant COD reduction (42%) (Ergüt et al. 2019). However, the removal of Putnam sky blue 39 by dark adsorption using magnetite particles prepared from *Gardenia jasminoides* flower extract, was not very efficient, requiring more than 1 g L^{-1} of the NPs to remove 0.05 mg L^{-1} of the dye; in contrast, when the material was used as a photocatalyst under sunlight, a complete removal could be obtained, with the dye not only adsorbed over the gFeNPs but also degraded (Espinoza-Gómez et al. 2019).

5.3.3 *Other applications*

Heterogeneous Fenton using gFeNPs was also employed for increasing the dewatering of a sludge, due to the breakdown of Extracellular Polymeric Substances (EPS) that lead to a significant reduction in proteins, polysaccharides, water content and heavy metals in the sludge (Ealias et al. 2016).

Several gFeNPs also showed antibacterial properties, as the positively charged NPs get attached to the negatively charged cells, causing the rupture of the cell walls; however, their efficiency was lower than that of gAgNPs (Singh et al. 2020).

6. Conclusions

Metal nanoparticles synthesized from plant extracts have been prepared and characterized as a green and low-cost materials, presenting many potential applications in fields such as environment, medicine, agriculture, etc. Despite these numerous potential applications, the reports regarding its full-scale or pilot-plant scale applications of gNPs are still scarce. For example, regarding medical applications, even clinical trials have not been published (Paiva-Santos et al. 2021). Therefore, many more studies are still required to improve this very attractive field in order to reach commercial applications.

Acknowledgements

L. Vitola and A. Ferrero thank the Facultad Regional Buenos Aires (Universidad Tecnológica Nacional, UTN, Argentina) for their scholarships. This work was performed within the PID-UTN MSUTIBA0004700TC project, financed by the UTN.

References

Abdelghany, T.M., A.M.H. Al-Rajhi, M.A. Al Abboud, M.M. Alawlaqi, A. Ganash Magdah, E.A.M. Helmy and A.S. Mabrouk. 2018. Recent advances in green synthesis of silver nanoparticles and their applications: about future directions: a review. BioNanoSci. 8: 5–16.

Akhtar, M.S., J. Panwar and Y.-S. Yun. 2013. Biogenic synthesis of metallic nanoparticles by plant extracts. ACS Sust. Chem. Eng. 1: 591–602.

Ali, Z.A., R. Yahya, S.D. Sekaran and R. Puteh. 2016. Green synthesis of silver nanoparticles using apple extract and its antibacterial properties. Adv. Mat. Sci. Eng. e4102196.

Anesini, C., G.E. Ferraro and R. Filip. 2008. Total polyphenol content and antioxidant capacity of commercially available tea (*Camellia sinensis*) in Argentina. J. Agric. Food Chem. 56: 9225–9229.

Anjana, P.M., M.R. Bindhu and R.B Rakhi. 2019. Green synthesized gold nanoparticle dispersed porous carbon composites for electrochemical energy storage. KeAi Chinese Roots Global Impact 2: 389–395.

Bagherzade, G., M.M. Tavakoli and M.H. Namaei. 2017. Green synthesis of silver nanoparticles using aqueous extract of saffron (*Crocus sativus* L.) wastages and its antibacterial activity against six bacteria. Asian Pacific J. Tropical Biomed. 7: 227–233.

Baharara, J., F. Namvar, T. Ramezani, M. Mousavi and R. Mohamad. 2015. Silver nanoparticles biosynthesized using *Achillea biebersteinii* flower extract: apoptosis induction in MCF-7 cells via caspase activation and regulation of Bax and Bcl-2 gene expression. Molecules 20: 2693–2706.

Balavigneswaran, C.K., T.S.J. Kumar, R. Packiaraj and S. Prakash. 2014. Rapid detection of Cr(VI) by AgNPs probe produced by Anacardium occidentale fresh leaf extracts. Appl. Nanosci. 4: 367–378.

Bataineh, H., O. Pestovsky and A. Bakac. 2012. pH-induced mechanistic changeover from hydroxyl radicals to iron(IV) in the Fenton reaction. Chem. Sci. 3: 1594–1599.

Botha, T.L.E., E. Elemike, S. Horn, D.C. Onwudiwe, J.P. Giesy and V. Wepener. 2019. Cytotoxicity of Ag, Au and Ag-Au bimetallic nanoparticles prepared using golden rod (*Solidago canadensis*) plant extract. Nature 9: 4169.

Burlacu, E., C. Tanase, N.-A. Coman and L. Berta. 2019. A review of bark-extract-mediated green synthesis of metallic nanoparticles and their applications. Molecules 24: 4354.

Carvalho, S.S.F. and N.M.F. Carvalho. 2017. Dye degradation by green heterogeneous Fenton catalysts prepared in presence of *Camellia sinensis*. J. Environ. Manag. 187: 82–88.

Chrysochoou, M., C.P. Johnston and G. Dahal. 2012. A comparative evaluation of hexavalent chromium treatment in contaminated soil by calcium polysulfide and greentea nanoscale zero-valent iron. J. Hazard. Mater. 201–202: 33–42.

Crane, R.A. and T.B. Scott. 2012. Nanoscale zero-valent iron: Future prospects for an emerging water treatment technology. J. Hazard. Mater. 211-212: 112–125.

David, L. and B. Moldovan. 2020. Green synthesis of biogenic silver nanoparticles for efficient catalytic removal of harmful organic dyes. Nanomaterials 10: 202.

Desalegn, B., M. Megharaj, Z. Chen and R. Naidu. 2019. Green synthesis of zero valent iron nanoparticle using mango peel extract and surface characterization using XPS and GC-MS. Heliyon 5: e01750.

Ealias, A.M., J.V. Jose and M.P. Saravanakumar. 2016. Biosynthesised magnetic iron nanoparticles for sludge dewatering via Fenton process. Environ. Sci. Pollut. Res., doi 10.1007/s11356-016-7351-4.

Ergüt, M., D. Uzunoğlu and A. Özer. 2019. Efficient decolourization of malachite green with biosynthesized iron oxide nanoparticles loaded carbonated hydroxyapatite as a reusable heterogeneous Fenton-like catalyst. J. Environ. Sci. Health A, doi: 10.1080/10934529.2019.1596698.

Espinoza-Gómez, H., L.Z. Flores-López, K.A. Espinoza and G. Alonso-Nuñez. 2019. Microstrain analyses of Fe_3O_4NPs greenly synthesized using *Gardenia jasminoides* flower extract, during the photocatalytic removal of a commercial dye. App. Nanosci., doi: 10.1007/s13204-019-01070-w.

Ferdous, Z. and A. Nemmar. 2020. Health impact of silver nanoparticles: a review of the biodistribution and toxicity following various routes of exposure. Int. J. Mol. Sci. 21: 2375.

Gangadharan, D., K. Harshvardan, G. Gnanasekar, D. Dixit, K.M. Popat and P.S. Anand. 2010. Polymeric microspheres containing silver nanoparticles as a bactericidal agent for water disinfection. Water Res. 44: 5481–5487.

García, F.E., A.M. Senn, J.M. Meichtry, T.B. Scott, H. Pullin, A.G. Leyva et al. 2019. Iron-based nanoparticles prepared from *yerba mate* extract. Synthesis, characterization and use on chromium removal. J. Environ. Manag. 235: 1–9.

Gengan, R.M., K. Anand, A. Phulukdaree and A. Chuturgoon. 2013. A549 lung cell line activity of biosynthesized silver nanoparticles using *Albizia adianthifolia* leaf. Coll. Surf. B 105: 87–91.

Genuino, H.C., Z. Luo, N. Mazrui, M.I. Seraji and G.E. Hoag. 2013. Green synthesized iron nanomaterials for oxidative catalysis of organic environmental pollutants. pp. 41–61. *In*: Suib, S.L. (ed.). New and Future Developments in Catalysis. Catalysis for Remediation and Environmental Concerns. Elsevier. Amsterdam, The Netherlands.

Groiss, S., S. Raja, V. Thivaharan and V. Ramesh. 2017. Structural characterization, antibacterial and catalytic effect of iron oxide nanoparticles synthesised using the leaf extract of Cynometra ramiflora. J. Mol. Struct. 1128: 572–578.

Gust, J. and J. Suwalski. 1994. Use of Mössbauer spectroscopy to study reaction products of polyphenols and iron compounds. Corrosion 50: 355–365.

Haghighi Pak, Z., H. Abbaspour, N. Karimi and A. Fattahi. 2016. Eco-friendly synthesis and antimicrobial activity of silver nanoparticles using *Dracocephalum moldavica* seed extract. App. Sci. 6: 69.

Hasan, S. 2015. A review on nanoparticles: their synthesis and types. Res. J. Recent Sci. 4: 9–11.

Herlekar, M., S. Barve and R. Kumar. 2014. Plant-mediated green synthesis of iron nanoparticles. J. Nanopart. 9: 140614.

Herrera-Becerra, R., J.L. Rius and C. Zorrilla. 2010. Tannin biosynthesis of iron oxide nanoparticles. Appl. Phys. A 100: 453–459.

Hoag, G.E., J.B. Collins, J.L. Holcomb, J.R. Hoag, M.N. Nadagouda and R.S. Varma. 2009. Degradation of bromothymol blue by 'greener' nano-scale zero-valent iron synthesized using tea polyphenols. J. Mater. Chem. 19: 8671–8677.

Huang, L., F. Luo, Z. Chen, M. Megharaj and R. Naidu. 2015. Green synthesized conditions impacting on the reactivity of Fe NPs for the degradation of malachite green. Spectrochim. Acta A 137: 154–159.

Huang, L., Y. Sun, S. Mahmud and H. Liu. 2020. Biological and environmental applications of silver nanoparticles synthesized using the aqueous extract of *Ginkgo biloba* leaf. J. Inorg. Organomet. Polym. 30: 1653–1668.

Jeyaraj, M., G. Sathishkumar, G. Sivanandhan, D. Mubarak Ali, M. Rajesh, R. Arun et al. 2013. Biogenic silver nanoparticles for cancer treatment: An experimental report. Coll. Surf. B 106: 86–92.

Jung, J.H., G.B. Hwang, J.E. Lee and G.N. Bae. 2011. Preparation of airborne Ag/CNT hybrid nanoparticles using an aerosol process and their application to antimicrobial air filtration. Langmuir 27: 10256–10264.

Karmous, I., A. Pandey, K.B. Haj and A. Chaoui. 2020. Efficiency of the green synthesized nanoparticles as new tools in cancer therapy: insights on plant-based bioengineered nanoparticles, biophysical properties, and anticancer roles. Biol. Trace Elem. Res. 196: 330–342.

Kharissova, O.V., H.V. Dias, B.I. Kharisov, B.O. Pérez and V.M. Pérez. 2013. The greener synthesis of nanoparticles. Trends Biotechnol. 31: 240–248.

Kheybari, S., N. Samadi, S.V. Hosseini, A. Fazeli and M.R. Fazeli. 2010. Synthesis and antimicrobial effects of silver nanoparticles produced by chemical reduction method. DARU J. Pharm. Sci. 18: 168–172.

Khashij, M., S. Sadoughi, A. Dalvand, M. Mehralian, A.A. Ebrahimi and R. Khosravi. 2020. Removal of reactive black 5 dye using zerovalent iron nanoparticles produced by a novel green synthesis method. Pigment Resin Technol. 49: 215–221.

Kiruba Daniel, S.C.G., G. Vinothini, N. Subramanian, K. Nehru and M. Sivakumar. 2012. Biosynthesis of Cu, ZVI, and Ag nanoparticles using Dodonaea viscosa extract for antibacterial activity against human pathogens. J. Nanopart. Res. 15: 1319.

Kokila, T., P.S. Ramesh and D. Geetha. 2015. Biosynthesis of silver nanoparticles from Cavendish banana peel extract and its antibacterial and free radical scavenging assay: a novel biological approach. Appl. Nanosci. 5: 911–920.

Kralova, K. and J. Jampilek. 2021. Responses of medicinal and aromatic plants to engineered nanoparticles. Appl. Sci. 11: 1813.

Kuang, Y., Q. Wang, Z. Chen, M. Megharaj and R. Naidu. 2013. Heterogeneous fenton-like oxidation of monochlorobenzene using green synthesis of iron nanoparticles. J. Colloid Interface Sci. 15: 67–73.

Kumar, M.K., B.K. Mandal, S.K. Kumar, P.S. Reddy and B. Sreedhar. 2013a. Biobased green method to synthesise palladium and iron nanoparticles using *Terminalia chebula* aqueous extract. Spectrochim. Acta Part A Mol. Biomol. 102: 128–133.

Kumar, P., M. Govindaraju, S. Senthamilselvi and K. Premkumar. 2013b. Photocatalytic degradation of methyl orange dye using silver (Ag) nanoparticles synthesized from *Ulva lactuca*. Coll. Surf. B 103: 658–661.

Kumar, D., P. Kumar, H. Singh and V. Agrawal. 2020. Biocontrol of mosquito vectors through herbal-derived silver nanoparticles: prospects and challenges. Env. Sci. Poll. Res. 27: 25987–26024.

Lee, S.H. and B.-H. Jun. 2019. Silver nanoparticles: synthesis and application for nanomedicine. Int. J. Mol. Sci. 20: 865.

Lin, Z., X. Weng, G. Owens and Z. Chen. 2020. Simultaneous removal of Pb(II) and rifampicin from wastewater by iron nanoparticles synthesized by a tea extract. J. Cleaner Prod. 242: 118476.

Litter, M.I., N. Quici, J.M. Meichtry and A.M. Senn. 2016. Photocatalytic removal of metallic and other inorganic pollutants. pp. 35–71. *In*: Dionysiou, D.D., G. Li Puma, J. Ye, J. Schneider and D. Bahnemann (eds.). Photocatalysis: Applications. Royal Society, doi: 10.1039/9781782627104-00035.

Litter, M.I. and M. Slodowicz. 2017. An overview on heterogeneous fenton and photofenton reactions using zerovalent iron materials. J. Adv. Oxid. Technol. ISSN (Online) 2371–1175, doi: 10.1515/jaots-2016-0164.

Litter, M.I., N. Quici and J.M. Meichtry. 2018. Iron Nanomaterials for Water and Soil Treatment. Pan Stanford Publishers, New York.

Liu, Y., X. Jin and Z. Chen. 2018. The formation of iron nanoparticles by *Eucalyptus* leaf extract and used to remove Cr(VI). Sci. Tot. Environ. 627: 470–479.

Luo, F., Z. Chen, M. Megharaj and R. Naidu. 2014. Biomolecules in grape leaf extract involved in one-step synthesis of iron-based nanoparticles. RSC Adv. 4: 53467–53474.

Luo, F., D. Yang, Z. Chen, M. Megharaj and R. Naidu. 2015. The mechanism for degrading Orange II based on adsorption and reduction by ion-based nanoparticles synthesized by grape leaf extract. J. Hazard. Mat. 296: 37–45.

Luo, F., Z. Chen, M. Megharaj and R. Naidu. 2016. Simultaneous removal of trichloroethylene and hexavalent chromium by green synthesized agarose-Fe nanoparticles hydrogel. Chem. Eng. J. 294: 290–297.

Machado, S., W. Stawinski, P. Slonina, A. Pinto, J. Grosso, H. Nouws et al. 2013. Application of green zero-valent iron nanoparticles to the remediation of soils contaminated with ibuprofen. Sci. Total Environ. 461: 323–329.

Machado, S., J.G. Pacheco, H.P.A. Nouws, J.T. Albergaria and C. Delerue-Matos. 2015. Characterization of green zero-valent iron nanoparticles produced with tree leaf extracts. Sci. Total Environ. 533: 76–81.

Madhavi, V., T.N.V.K.V. Prasad, A.V.B. Reddy, B.R. Reddy and G. Madhav. 2013. Application of phytogenic zerovalent iron nanoparticles in the adsorption of hexavalent chromium. Spectroch. Acta Part A 116: 17–25.

Manikandan, R., B. Manikandan, T. Raman, K. Arunagirinathan, N.M. Prabhu, M. Jothi Basu et al. 2015. Biosynthesis of silver nanoparticles using ethanolic petals extract of Rosa indica and characterization of its antibacterial, anticancer and anti-inflammatory activities. Spectroch. Acta Part A 138: 120–129.

Markova, Z., P. Novak, J. Kaslik, P. Plachtova, M. Brazdova, D. Jancula, K.M. Siskova, B. Machala, B. Marsalek, R. Zboril and R. Varma. 2014. Iron (II,III) polyphenol complex nanoparticles derived from green tea with remarkable ecotoxicological impact. ACS Sustainable Chem. Eng. 2(7): 1674–1680.

Martnez-Cabanas, M., M. López-García, J. Barriada, R. Herrero and M.E. Sastre de Vicente. 2016. Green synthesis of iron oxide nanoparticles. Development of magnetic hybrid materials for efficient As(V) removal. Chem. Eng. J. 301: 83–91.

Mohammadi, L., K. Pal, M. Bilal, A. Rahdar, G. Fytianos and G.Z. Kyza. 2021. Green nanoparticles to treat patients with Malaria disease: An overview. J. Mol. Struct. 1229: 129857.

Murgueitio, E., L. Cumbal, A. Debut and J. Landvar. 2016. Synthesis of iron nanoparticles through extracts of natives fruits of Ecuador as Capul (*Prunus serotina*) and Mortio (*Vaccinium floribundum*). Biol. Med. (Aligarh) 8. https://doi.org/10.4172/0974-8369.1000282.

Murgueitio, E., L. Cumbal, M. Abril, A. Izquierdo, A. Debut and O. Tinoco. 2018. Green synthesis of iron nanoparticles: Application on the removal of petroleum oil from contaminated water and soils. J. Nanotechnol. 4184769. https://doi.org/10.1155/2018/4184769.

Mystrioti, C., T.D. Xanthopoulou, N. Papassiopi and A. Xenidis. 2016. Comparative evaluation of five plant extracts and juices for nanoiron synthesis and application for hexavalent chromium reduction. Sci. Total Environ. 539: 105–113.

Mystrioti, C., A. Toli, N. Papassiopi, D. Dermatas and S. Thimi. 2018. Chromium removal with environmentally friendly iron nanoparticles in a pilot scale study. Bull. Environ. Contam. Toxicol. 101: 705–710.

Nadagouda, M., A. Castle, R. Murdock, S. Hussain and R. Varma. 2010. *In vitro* biocompatibility of nanoscale zerovalent iron particles (NZVI) synthesized using tea polyphenols. Green Chem. 12: 114–122.

Nadeem, M., B.H. Abbasi, M. Younas, W. Ahmad and T. Khan. 2017. A review of the green syntheses and antimicrobial applications of gold nanoparticles. Green Chem. Let. Rev. 10: 216–227.

Nakkala, J.R., R. Mata and S.R. Sadras. 2016. The antioxidant and catalytic activities of green synthesized gold nanoparticles from *Piper longum* fruit extract. Process Safety Environ. Protec. 100: 288–294.

Njagi, E.C., H. Huang, L. Stafford, H. Genuino, H. Galindo, J. Collins, G. Hoag and S. Suib. 2010. Biosynthesis of iron and silver nanoparticles at room temperature using aqueous sorghum bran extracts. Langmuir 27: 264–271.

Oakes, J.S. 2013. Investigation of Iron Reduction by Green Tea Polyphenols for Application in Soil Remediation. University of Connecticut Graduate School, Storrs, Connecticut, USA.

Önal, E.S., T. Yatkin, M. Ergüt and A. Özer. 2017. Green synthesis of iron nanoparticles by aqueous extract of *Eriobotrya japonica* leaves as a heterogeneous Fenton-like catalyst: degradation of basic red 46. Int. J. Chem. Eng. Appl. 8: 327–333.

Ovais, M., A. Raza, S. Naz, N.U. Islam, A.T. Khalil, S. Ali et al. 2017. Current state and prospects of the phytosynthesized colloidal gold nanoparticles and their applications in cancer theranostics. Appl. Microbiol. Biotechnol. 101: 3551–3565.

Paiva-Santos, A.C., A.M. Herdade, C. Guerra, D. Peixoto, M. Pereira-Silva, M. Zeinali et al. 2021. Plant-mediated green synthesis of metal-based nanoparticles for dermopharmaceutical and cosmetic applications. Int. J. Pharm. 597: 120311.

Pan, Z., Y. Lin, B. Sarkar, G. Owens and Z. Chen. 2020. Green synthesis of iron nanoparticles using red peanut skin extract: Synthesis mechanism, characterization and effect of conditions on chromium removal. J. Coll. Interf. Sci. 558: 106–114.

Paul, A. and A. Roychoudhury. 2021. Go green to protect plants: repurposing the antimicrobial activity of biosynthesized silver nanoparticles to combat phytopathogens. Nanotech. Environ. Eng. 6: 10.

Pasinszki, T. and M. Krebsz. 2020. Synthesis and application of zero-valent iron nanoparticles in water treatment, environmental remediation, catalysis, and their biological effects. Nanomaterials 10: 917; doi:10.3390/nano10050917.

Perron, N. and J. Brumaghim. 2009. A review of the antioxidant mechanisms of polyphenol compounds related to iron binding. Cell Biochem. Biophys. 53: 75–100.

Poguberović, S.S., D.M. Krcmar, S.P. Maletić, Z. Kónya, D.D. Tomasević Pilipović, D.V. Kerkez et al. 2016. Removal of As(III) and Cr(VI) from aqueous solutions using "green" zero-valent iron nanoparticles produced by oak, mulberry and cherry leaf extracts. Ecol. Eng. 90: 42–49.

Pugazhendhi, S., E. Kirubha, P.K. Palanisamy and R. Gopalakrishnan. 2015. Synthesis and characterization of silver nanoparticles from *Alpinia calcarata* by green approach and its applications in bactericidal and nonlinear optics. App. Surf. Sci. 357: 1801–1808.

Qing, W., K. Chen, Y. Wang, X. Liu and M. Lu. 2017. Green synthesis of silver nanoparticles by waste tea extract and degradation of organic dye in the absence and presence of H_2O_2. App. Surf. Sci. 423: 1019–1024.

Radetić, M. 2013. Functionalization of textile materials with silver nanoparticles. J. Mater. Sci. 48: 95–107.

Rafique, M., M.I. Sadaf, S. Rafique and M.B. Tahir. 2017. A review on green synthesis of silver nanoparticles and their applications. Artif. Cells, Nanomed. Biotechnol. 45: 1272–1291.

Rahimi-Nasrabadi, M., S.M. Pourmortazavi, S.A.S. Shandiz, F. Ahmadi and H. Batooli. 2014. Green synthesis of silver nanoparticles using *Eucalyptus leucoxylon* leaves extract and evaluating the antioxidant activities of extract. Nat. Product Res. 28: 1964–1969.

Raja, K., A. Saravanakumar and R. Vijayakumar. 2012. Efficient synthesis of silver nanoparticles from *Prosopis juliflora* leaf extract and its antimicrobial activity using sewage. Spectrochim. Acta Part A 97: 490–494.

Rajesh, S., D.P. Raja, J.M. Rathi and K. Sahayaraj. 2012. Biosynthesis of silver nanoparticles using *Ulva fasciata* (Delile) ethyl acetate extract and its activity against *Xanthomonas campestris pv. malvacearum*. J. Biopest. 5: 119–128.

Rajeshkumar, S. and L.V. Bharath. 2017. Mechanism of plant-mediated synthesis of silver nanoparticles—A review on biomolecules involved, characterisation and antibacterial activity. Chem.-Biol. Interac. 273: 219–227.

Ramar, M., B. Manikandan, P.N. Marimuthu, T. Raman, A. Mahalingam, P. Subramanian et al. 2015. Synthesis of silver nanoparticles using *Solanum trilobatum* fruits extract and its antibacterial, cytotoxic activity against human breast cancer cell line MCF 7. Spectrochim. Acta Part A 140: 223–228.

Rao, A., A. Bankar, A.R. Kumar, S. Gosavi and S. Zinjarde. 2013. Removal of hexavalent chromium ions by *Yarrowia lipolytica* cells modified with phyto-inspired Fe^0/Fe_3O_4 nanoparticles. J. Contam. Hydrol. 146: 63–73.

Rasheed, T., M. Bilal, H.M.N. Iqbal and C. Li. 2017. Green biosynthesis of silver nanoparticles using leaves extract of *Artemisia vulgaris* and their potential biomedical applications. Coll. Surf. B 158: 408–415.

Ratan, Z.A., M.F. Haidere, M. Nurunnabi, S.M. Shahriar, A.J.S. Ahammad, Y.Y. Shim et al. 2020. Green chemistry synthesis of silver nanoparticles and their potential anticancer effects. Cancer 12: 855.

Rathi Sre, P.R., M. Reka, R. Poovazhagi, M.A. Kumar and K. Murugesan. 2015. Antibacterial and cytotoxic effect of biologically synthesized silver nanoparticles using aqueous root extract of *Erythrina indica* lam. Spectrochim. Acta Part A 135: 1137–1144.

Ravikumar, K.V.G., D. Kumar, A. Rajeshwari, G.M. Madhu, P. Mrudula, N. Chandrasekaran et al. 2015. A comparative study with biologically and chemically synthesized nZVI: applications in Cr(VI) removal and ecotoxicity assessment using indigenous microorganisms from chromium-contaminated site. Environ. Sci. Pollut. Res. 23: 2613–2627.

Rong, K., J. Wang, Z. Zhang and J. Zhang. 2020. Green synthesis of iron nanoparticles using *Korla fragrant* pear peel extracts for the removal of aqueous Cr(VI). Ecol. Eng. 149: 105793.

Ryan, P. and M.J. Hynes. 2007. The kinetics and mechanisms of the complex formation and antioxidant behaviour of the polyphenols EGCg and ECG with iron(III). J. Inorg. Biochem. 101: 585–593.

Sadeghi, B. and F. Gholamhoseinpoor. 2015. A study on the stability and green synthesis of silver nanoparticles using *Ziziphora tenuior* (Zt) extract at room temperature. Spectrochim. Acta Part A 134: 310–315.

Sadeghi, B., A. Rostami and S.S. Momeni. 2015. Facile green synthesis of silver nanoparticles using seed aqueous extract of *Pistacia atlantica* and its antibacterial activity. Spectrochim. Acta Part A 134: 326–332.

Saha, N. and S.D. Gupta. 2018. Promotion of shoot regeneration of *Swertia chirata* by biosynthesized silver nanoparticles and their involvement in ethylene interceptions and activation of antioxidant activity. Plant. Cell Tissue Organ. Cult. 134: 289–300.

Sanpui, P., A. Chattopadhyay and S.S. Ghosh. 2011. Induction of apoptosis in cancer cells at low silver nanoparticle concentrations using chitosan nanocarrier. ACS Appl. Mater. Interf. 3: 218–228.

Sathishkumar, M., K. Sneha and Y.-S. Yun. 2010. Immobilization of silver nanoparticles synthesized using Curcuma longa tuber powder and extract on cotton cloth for bactericidal activity. Biores. Technol. 101: 7958–7965.

Senthil, M. and C. Ramesh. 2012. Biogenic synthesis of Fe_3O_4 nanoparticles using *Tridax procumbens* leaf extract and its antibacterial activity on *Pseudomonas aeruginosa*. Dig. J. Nanomater. Biostruct. 7: 1655–1661.

Shahid-ul-Islam, B., S. Butola and A. Kumar. 2020. Green chemistry based in-situ synthesis of silver nanoparticles for multifunctional finishing of chitosan polysaccharide modified cellulosic textile substrate. Int. J. Biol. Macromol. 152: 1135–1145.

Shahwan, T., S.A. Sirriah, M. Nairat, E. Boyacı, A.E. Eroğlu, T.B. Scott et al. 2011. Green synthesis of iron nanoparticles and their application as a Fenton-like catalyst for the degradation of aqueous cationic and anionic dyes. Chem. Eng. J. 172: 258–266.

Singh, A., P.K. Gautam, A. Verma, V. Singh, P.M. Shivapriya, S. Shivalkar et al. 2020. Green synthesis of metallic nanoparticles as effective alternatives to treat antibiotics resistant bacterial infections: A review. Biotechnol. Rep. 25: 1–11.

Singh, P., Y.-J. Kim, D. Zhan and D.-C. Yang. 2016. Biological synthesis of nanoparticles from plants and microorganisms. Trends Biotech. 34: 588–599.

Smuleac, V., R. Varma, S. Sikdar and D. Bhattacharyya. 2011. Green synthesis of Fe and Fe/Pd bimetallic nanoparticles in membranes for reductive degradation of chlorinated organics. J. Membr. Sci. 379: 131–137.

Srikar, S.K., D.D. Giri, D.B. Pal, P.K. Mishra and S.N. Upadhyay. 2016. Green synthesis of silver nanoparticles: a review. GSC 06: 34–56.

Srivastava, S., R. Awasthi, N.S. Gajbhiye, V. Agarwal, A. Singh, A. Yadav and R.K. Gupta. 2011. Innovative synthesis of citrate-coated superparamagnetic Fe_3O_4 nanoparticles and its preliminary applications. J. Colloid Interface Sci. 359: 1041–11.

Sunardi Ashadi, S.B. Rahardjo and Inayati. 2017. Ecofriendly synthesis of nano zero valent iron from banana peel extract. J. Phys. Conf. Ser. 795: 012063.

Tarannum, N., D. Divya and Y.K. Gautam. 2019. Facile green synthesis and applications of silver nanoparticles: a state-of-the-art review. RSC Adv. 9: 34926–34948.

Thakur, S. and N. Karak. 2014. One-step approach to prepare magnetic iron oxide/reduced graphene oxide nanohybrid for efficient organic and inorganic pollutants removal. Mater. Chem. Phys. 144: 425–432.

Tho, N.T.M., T.N.M. An, M.D. Tri, T.V.M. Sreekanth, J.-S. Lee, P.C. Nagajyothi et al. 2013. Green synthesis of silver nanoparticles using *Nelumbo nucifera* seed extract and its antibacterial activity. Acta Chim. Slov. 60: 673–678.

Vanlalveni, C., S. Lallianrawn, A. Biswas, M. Selvaraj, B. Changmai and S.L. Rokhum. 2021. Green synthesis of silver nanoparticles using plant extracts and their antimicrobial activities: a review of recent literature. RSC Adv. 11: 2804–2837.

Vaseghi, Z., A. Nematollahzadeh and O. Tavakoli. 2018. Green methods for the synthesis of metal nanoparticles using biogenic reducing agents: a review. Rev. Chem. Eng. 34: 529–559.

Velusamy, P., J. Das, R. Pachaiappan, B. Vaseeharan and K. Pandian. 2015. Greener approach for synthesis of antibacterial silver nanoparticles using aqueous solution of neem gum (*Azadirachta indica* L.). Ind. Crops Prod. 66: 103–109.

Venkateswarlu, S., Y.S. Rao, T. Balaji, B. Prathima and N.V.V. Jyothi. 2013. Biogenic synthesis of Fe_3O_4 magnetic nanoparticles using plantain peel extract. Mater. Lett. 100: 241–244.

Verma, P. and S.K. Maheshwari. 2019. Applications of silver nanoparticles in diverse sectors. Int. J. Nano Dim. 10: 18–36.

Vijayan, R., S. Joseph and B. Mathew. 2017. *Indigofera tinctoria* leaf extract mediated green synthesis of silver and gold nanoparticles and assessment of their anticancer, antimicrobial, antioxidant and catalytic properties. Artif. Cells Nanomed. Biotechnol. 46: 861–871

Vijayaraghavan, K., S.P.K. Nalini, N.U. Prakash and D. Madhankumar. 2012. One step green synthesis of silver nano/microparticles using extracts of *Trachyspermum ammi* and *Papaver somniferum*. Coll. Surf. B 94: 114–117.

Wang, C.B. and W.X. Zhang. 1997. Synthesizing nanoscale iron particles for rapid and complete dechlorination of TCE and PCBs. Environ. Sci. Technol. 31: 2154–2156.

Wang, Z. 2013. Iron complex nanoparticles synthesized by eucalyptus leaves. ACS Sustain. Chem. Eng. 1: 1551–1554.

Wang, T., X. Jin, Z. Chen, M. Megharaj and R. Naidu. 2014a. Green synthesis of Fe nanoparticles using eucalyptus leaf extracts for treatment of eutrophic wastewater. Sci. Total Environ. 466–467: 210–213.

Wang, T., J. Lin, Z. Chen, M. Megharaj and R. Naidu. 2014b. Green synthesized iron nanoparticles by green tea and eucalyptus leaves extracts used for removal of nitrate in aqueous solution. J. Clean. Prod. 83: 413–419.

Wang, Z. 2015. Iron–Polyphenol Complex Nanoparticles Synthesized by Plant Leaves and Their Application in Environmental Remediation. PhD Thesis. University of South Australia.

Wei, Y., Z. Fang, L. Zheng and E.P. Tsang. 2017. Biosynthesized iron nanoparticles in aqueous extracts of *Eichhornia crassipes* and its mechanism in the hexavalent chromium removal. Appl. Surf. Sci. 399: 322–329.

Weng, X., L. Huang, Z. Chen, M. Megharaj and R. Naidu. 2013. Synthesis of iron-based nanoparticles by green tea extract and their degradation of malachite. Ind. Crops Prod. 51: 342–347.

Weng, X., M. Guo, F. Luo and Z. Chen. 2017. One-step green synthesis of bimetallic Fe/Ni nanoparticles by eucalyptus leaf extract: biomolecules identification, characterization and catalytic activity. Chem. Eng. J. 308: 904–911.

Wu, Z., X. Su, Z. Lin, G. Owens and Z. Chen. 2019. Mechanism of As(V) removal by green synthesized iron nanoparticles. J. Hazard. Mat. 764: 120811.

Wu, Z., X. Su, Z. Lin, N.I. Khan, G. Owens and Z. Chen. 2021. Removal of As(V) by iron-based nanoparticles synthesized via the complexation of biomolecules in green tea extracts and an iron salt. Sci. Tot. Environ. 379: 142883.

Xiao, Z., M. Yuan, B. Yang, Z. Liu, J. Huang and D. Sun. 2016. Plant-mediated synthesis of highly active iron nanoparticles for Cr (VI) removal: Investigation of the leading biomolecules. Chemosphere 150: 357–364.

Yadi, M., E. Mostafavi, B. Saleh, S. Davarana, I. Aliyevac, R. Khalilov et al. 2018. Current developments in green synthesis of metallic nanoparticles using plant extracts: a review. Art. Cells Nanomed. Biotechnol. 46: 336–343.

Yugandhar, P., R. Haribabu and N. Savithramma. 2015. Synthesis, characterization and antimicrobial properties of green-synthesised silver nanoparticles from stem bark extract of *Syzygium alternifolium* (Wt.) Walp. 3 Biotech. 5: 1031–1039.

Zhang, D., X-l. Ma, Y. Gu, H. Huang and G-w. Zhang. 2020. Green synthesis of metallic nanoparticles and their potential applications to treat cancer. Front. Chem. 8: 1–18.

Chapter 6

Developing Photocatalytic Nanomaterials for Environmental Applications

Aracely Hernández-Ramírez, * *Laura Hinojosa-Reyes, Jorge Luis Guzmán-Mar,*
María de Lourdes Maya-Treviño and *Minerva Villanueva-Rodríguez*

1. Introduction

The exceptional development of materials has involved the design of nanomaterials with special characteristics for many applications related to the environment. Among environmental concerns, air and water pollution have been recognized as a threat to humans, flora and fauna. Many efforts have been made worldwide to implement technologies that allow eliminating different kinds of pollutants present in air and water.

A promising technology as a viable alternative to the conventional methods of pollutants treatment is heterogeneous photocatalysis. Heterogeneous photocatalysis belongs to the advanced oxidation processes and implies the use of a semiconductor material that is activated with light. When the energy of the photons exceeds the energy of the semiconductor bandgap (E_g), electrons (e^-) are promoted from the Valence Band (VB) to the Conduction Band (CB), generating holes (h^+). In semiconductors, these photoexcited e^-/h^+ pairs participate in redox reactions with the adsorbed donor or acceptor molecules. The electrons can also react with oxygen to form $O_2^{\cdot-}$ radicals and the holes react with surface hydroxyl groups to form HO$^\cdot$, which can eliminate many toxic and recalcitrant pollutants.

The crucial step in this process is the nature and the properties of the nanoparticles used as the photocatalyst. The most used material is the TiO_2 semiconductor; nevertheless, it has some disadvantages such as high e^-/h^+ recombination and the need for ultraviolet irradiation due to its wide bandgap of 3.2 eV, limiting its industrial applications.

Therefore, the development of new types of photocatalysts includes the change in the size of the particles; the nanometric size of photocatalytic materials substantially improves their optical properties and specific surface area, compared with bulk materials (Radhika et al. 2016). New nanophotocatalysts are being developed to take advantage of visible light or solar energy to implement green sustainable technology for the decrease of contamination.

Universidad Autónoma de Nuevo León, Facultad de Ciencias Químicas, Ave. Universidad S/N, Cd. Universitaria, San Nicolás de los Garza N.L. México, C.P. 66455.
Emails: laura.hinojosary@uanl.edu.mx; jorge.guzmanmr@uanl.edu.mx; maria.mayatv@uanl.edu.mx; minerva.villanuevardr@uanl.edu.mx
* Corresponding author: aracely.hernandezrm@uanl.edu.mx

In this chapter, the synthesis and applications of novel nanostructured photocatalysts for the efficient removal of various pollutants will be described, and some relevant applications will be presented.

2. Active Photocatalytic Nanomaterials for Degradation of Pharmaceuticals

In recent years, Pharmaceutical Compounds (PhCs) have become of interest in the environment since they have been classified as emerging compounds due to their continuous release to the environment, persistence and adverse effects on aquatic organisms.

The environmental effects related to PhCs have grown due to population increase, the elevated demand for pharmaceutical drugs with and without prescription, the inadequate disposal of expired antibiotics and the overuse in animal farming (Calvete et al. 2019). The primary detected PhCs are antibiotics, β-blockers, anti-inflammatories, psychiatric medicines, antivirals, anticancer drugs and anesthetics. Most of these molecules are excreted by urine or feces as the original compound or as their metabolites. Wastewater treatment plants (WWTP) and hospital discharges are the main factors where PhCs are released into the environment, hospital discharges being a critical point since the concentration of drugs in effluents can be higher than municipal discharges. The Non-Steroidal Anti-Inflammatories Drugs (NSAID) (acetaminophen, naproxen and ibuprofen), β-blockers (losartan) antibiotics (sulfamethoxazole) (Hernández-Tenorio et al. 2021), antidiabetics (glibenclamide) and hormones (17-βestradiol) (Luja-Mondragon et al. 2019) are the most detected PhCs in hospital discharges at concentrations from 0.16 to 51.22 µg L^{-1}.

Antibiotics are one of the most relevant compounds in the environment since they are recalcitrant molecules and can provoke the formation of antibiotic-resistant pathogenic microorganisms. Penicillins, cephalosporins, fluoroquinolones, aminoglycosides and macrolides are the most detected compounds in wastewater (2.2–6200 ng L^{-1}), WWTP influents (2.05–14,000,000 ng L^{-1}), and WWTP effluents (0.5–6,500,000 ng L^{-1}) (Lofrano et al. 2017). Moreover, different adverse effects have been reported due to PhCs exposure, such as alterations to embryonic development and teratogenic effects on aquatic organisms exposed to hospital effluents (Luja-Mondragon et al. 2019), even at very low concentrations (ng L^{-1}). Therefore, several studies related to PhCs treatment for degradation and mineralization by heterogeneous photocatalysis have been reported. Many of these studies mainly used TiO$_2$ or modified TiO$_2$ as the photocatalyst. For instance, commercial TiO$_2$ as Aeroxide or Degussa P25 demonstrated high efficiency under UV light for PhCs removal at the laboratory scale to degrade an NSAIDs mixture (Villanueva-Rodríguez et al. 2019), and for a single anti-inflammatory degradation at pilot-scale under natural sunlight (Diaz-Angulo et al. 2020). TiO$_2$ P25 was also evaluated to degrade isoniazid and pyrazinamide antibiotics in a mixture, leading to > 92% of degradation and 63% of mineralization in 300 minutes, with evident superiority over ZnO under UV light (Guevara-Almaraz et al. 2015). The pollutant mineralization is determined by the Total Organic Carbon (TOC) abatement, which means that the original molecule is transformed into inorganic ions, CO$_2$ and H$_2$O.

As TiO$_2$ is mainly activated under UV light, some strategies have been described to enhance the photocatalytic activity using visible light. One of these strategies is the sensitization of the semiconductor with dyes or metallic complexes, where the excitation of the colored compound with visible radiation leads to the subsequent electron injection to the semiconductor band (Góngora et al. 2017). Sensitization with dyes can also induce oxidant species generation such as singlet oxygen and superoxide anion radical, O$_2$$^{\cdot-}$ (Diaz-Angulo et al. 2019) as shown in Fig. 6.1.

In this sense, TiO$_2$ and ZnO have been modified to degrade some anti-inflammatory drugs. Góngora et al. (2017) incorporated Ru(II) polyaza complexes on the TiO$_2$ surface, achieving close to 80% of degradation and 65% of mineralization of ibuprofen in 300 minutes, using TiO$_2$-RuL2 (1.5 wt.%, L2 = N1-(2-aminobenzyliden)-N2,N2-bis(2-(2-aminobenzyliden)aminoethyl) ethane-1,2-diaminoruthenium(II)). Similar results were obtained using the RuL3 complex:

Photocatalysis with TiO₂   **Dye-sensitized TiO₂**

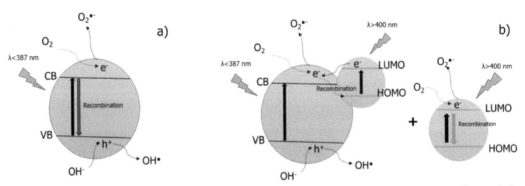

Figure 6.1. Involved mechanisms of (a) bare TiO₂ and (b) dye-sensitized TiO₂ during heterogeneous photocatalytic processes (Reprinted with permission from Diaz-Angulo et al. 2020).

N1,N2-bis(2-aminobenzyliden)ethane-1,2-diaminoruthenium(II), while using TiO₂ P25, a lower degradation was observed (45%).

TiO₂ Degussa and ZnO have also been sensitized with the perinaphthenone dye (Diaz-Angulo et al. 2019). It was found that diclofenac (DFC, 30 mg L^{-1}) degradation under UV-Vis radiation (290–690 nm, maxima at 385 and 540 nm) was higher by using the coupled process compared with only photocatalysis and photosensitized oxidation. The photosensitized oxidation (no photocatalyst) achieved 71% of DCF removal. However, for the coupled process the diclofenac degradation was 98 and 99% using TiO₂ and ZnO, respectively.

TiO₂ sensitized with the methyl red dye also has demonstrated elevated activity, achieving almost complete mineralization of DCF under sunlight at a total accumulated energy of 270 kJ m^{-2}. The mechanism of degradation depends on the amounts of the photocatalyst, dye and contaminant. Moreover, heterogeneous photocatalysis degradation can be controlled by the activation of the photocatalysts and/or dye, or by degradation of both contaminant and dye (Diaz-Angulo et al. 2020).

Coupled semiconductors also improve the photocatalytic treatment through a type II heterojunction, leading to a decrease of the E_g value and the e$^-$/h$^+$ recombination by efficient charge transference between the oxides (Hernández-Coronado 2020, Núñez-Salas et al. 2021a). In particular, ZnO has been coupled with FeTiO₃ (Núñez-Salas et al. 2021a), achieving a better photocatalytic activity and lower recombination rate with an increased amount of FeTiO₃ (from 1 to 10 wt.%) in ZnO, reaching almost complete degradation and 17% mineralization of ciprofloxacin after 180 minutes of reaction. In another research, α-Fe₂O₃ incorporation into TiO₂ enhanced the mineralization of cefuroxime (62%) compared with bare TiO₂ (35.8% in 100 kJ m^{-2}). However, it is worth noting that, using materials with supported iron, it is also possible to take advantage of the presence of iron to react with H₂O₂ and obtain a synergetic effect from the combination of heterogeneous photocatalysis and photo-Fenton reaction (72% of mineralization) (Hernández-Coronado 2020). Table 6.1 shows some relevant works of recent years associated with different modified photocatalysts.

On the other hand, the incorporation of non-metal elements into the semiconductor also enhances photocatalytic performance by promoting a decrease in the bandgap, reducing the recombination of the e$^-$/h$^+$ pairs and increasing the specific surface area. Rueda-Salaya et al. (2020) doped ZnO with fluoride to degrade DCF. In this case, fluoride can efficiently substitute oxide ions due to their similar ionic radius (1.31 and 1.38 Å for F$^-$ and O^{2-}, respectively). The complete degradation of DCF and 90% of mineralization was obtained using ZnO-F (20 wt.%) under simulated sunlight. Likewise, phosphorous incorporation into TiO₂ is also of interest because P^{5+} (0.35 Å) can substitute Ti^{4+} (0.68 Å), forming Ti-O-P bindings. Besides, P can be incorporated superficially as PO₄$^{3-}$ reducing the recombination rate and increasing the photocatalytic activity for the degradation of sulfamethazine, an antibiotic for veterinary use (Mendiola-Alvarez et al. 2019a).

Table 6.1. Some strategies to enhance the photocatalytic performance for the degradation of different PhCs.

Photocatalyst strategy	PhCs	Comments	Reference
Sensitization of photocatalysts with dyes or metallic complexes	Ibuprofen (10 mg L^{-1})	80% of degradation and 65% mineralization in 300 minutes with TiO$_2$ sensitized with Ru(II) polyaza complex (1.2 wt. %)	Góngora et al. 2017
	DCF (20 mg L^{-1})	Almost complete mineralization after 270 kJ m^{-2} with TiO$_2$ Degussa sensitized with dye methyl red (4 mg L^{-1})	Diaz-Angulo et al. 2019
Coupled semiconductors	Ciprofloxacin (10 mg L^{-1})	100% degradation and 27% mineralization with FeTiO$_3$/ZnO (10 wt.%)	Núñez-Salas et al. 2021a
	Tetracycline (20 mg L^{-1})	Complete degradation in 90 minutes and 50% mineralization using (5 wt.%) WO$_3$/ZnS	Murillo-Sierra et al. 2021a
Metal and non-metal doping	Sulfamethazine (10 mg L^{-1})	Complete degradation and 32% mineralization in 300 with (1.0 wt.%) P-TiO$_2$	Mendiola-Alvarez et al. 2019a
		Complete degradation and 30% mineralization within 300 minutes using (1.2 wt.%) P-Fe$_2$O$_3$-TiO$_2$	Mendiola-Alvarez et al. 2019b
	Amoxicillin, streptomycin (30 mg L^{-1} each one)	Amoxicillin, 58.61% degradation and 41.51% mineralization after 300 minutes. Streptomicyn solution, 49.67% degradation, and 34.72% mineralization after 300 minutes using Fe^{3+}-TiO$_2$-xNx	Aba-Guevara et al. 2017
	DCF (10 mg L^{-1})	Complete DCF degradation and 90% mineralization at 400 kJ m^{-2} accumulated energy with ZnO-F (20 wt.%)	Rueda-Salaya et al. 2020
		Complete degradation at 250 kJ m^{-2} of accumulated energy and 82.4% mineralization at 400 kJ m^{-2} with WO$_3$/TiO$_2$-C (0.18%)	Cordero-García et al. 2016
		100% DCF transformation at 250 kJ m^{-2} and complete mineralization at 400 kJ m^{-2} with WO$_3$/TiO$_2$-N 0.18 wt.%	Cordero-García et al. 2017
Doping with carbon-based material	Ibuprofen, carbamazepine, and sulfamethoxazole (5 mg L^{-1} each one)	54, 81 and 92% of degradation, and 54, 52 and 59% mineralization for carbamazepine, ibuprofen and sulfamethoxazole, respectively after 180 minutes with TiO$_2$-rGO (2.7%)	Lin et al. 2017

Heterojunctions between coupled semiconductors with non-metals have also been studied to degrade PhCs. In this sense, incorporating (1.2 wt.%) P into Fe_2O_3-TiO_2 enhanced the photocatalytic activity for mineralization of sulfamethazine until 30% in 300 minutes compared with 27% obtained with TiO_2-Fe_2O_3 (Mendiola-Alvarez et al. 2019b).

Nitrogen (anionic radius 1.71 Å) is another non-metal widely studied in doping TiO_2, due to its similar ionic radius and chemical properties with O^{2-}. Aba-Guevara et al. (2017) tested Fe^{3+}-TiO_2-Nx material synthesized by sol-gel and microwave methods to degrade amoxicillin, streptomycin and DCF, obtaining better results with materials with N incorporation under visible light. Cordero-García et al. (2016) reported interstitial incorporation of carbon in WO_3/TiO_2, causing the reduction of the bandgap from 3.10 to 2.89 eV for WO_3/TiO_2 and WO_3/TiO_2-C (0.18%), achieving with the last material complete degradation of DCF at 250 kJ m^{-2} of accumulated energy (simulated solar radiation), and 82.4% of mineralization at 400 kJ m^{-2}. In another study, Cordero-García et al. (2017) described WO_3/TiO_2-N preparation where N was incorporated in a substitutional way, which provoked the displacement of the bandgap to the visible range from 3.10 to 2.78 eV for WO_3/TiO_2 and WO_3/TiO_2-N (0.18%), respectively. DCF was degraded entirely with the co-doped material at 250 kJ m^{-2} and mineralized after 400 kJ m^{-2} under simulated sunlight. These results were attributed to the efficient e$^-$ transference from TiO_2 CB to WO_3 CB, charge transference from the WO_3 VB to TiO_2 VB and introduction of new N2p states over Ti3d level, reducing Ti^{4+} to Ti^{3+} (Fig. 6.2).

The direct Z-scheme heterojunction is another type of heterostructure recently described as an effective strategy, and the most relevant mechanism to promote superior charge separation of e$^-$/h$^+$ pairs since the photogenerated carriers are maintained in a different spatial site (Murillo-Sierra et al. 2021a). Recently, WO_3/ZnS (5/95 wt.%) was prepared by a hydrothermal method and completely degraded tetracycline within 90 minutes under UV-Vis light, exhibiting superior activity than bare WO_3 and ZnS. The nanoheterostructure analyzed by HR-TEM showed a close atomic arrangement in the interface between the two semiconductors, a large surface area (167.45 m^2 g^{-1}) and high stability after four successive cycles of use. The direct Z-scheme charge transfer mechanism was confirmed by electrochemical and theoretical DFT studies determining the formation of the built-in electric field at the interface, explaining the efficient e$^-$/h$^+$ pair separation in the heterojunction and the enhanced photocatalytic activity (Murillo-Sierra et al. 2021a).

The TiO_2 photocatalyst has also been immobilized in carbon-based materials to improve TiO_2 photocatalytic activity under visible light. Among them, graphene has been used because it can act

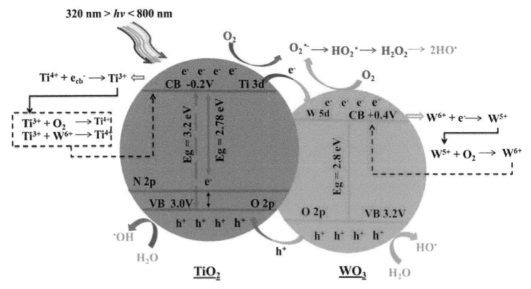

Figure 6.2. Activation mechanism proposed for WO_3/TiO_2-N (Reprinted with permission from Cordero-García et al. 2017).

as an electron acceptor/transporter, reducing the e^-/h^+ recombination rate and increasing the surface area of TiO_2. However, due to its elevated production cost, graphene is commonly used in the form of reduced Graphene Oxide (rGO, obtained from graphene by reduction with hydrazine and sodium borohydride). Lin et al. (2017) reported ibuprofen, carbamazepine and sulfamethoxazole elimination using TiO_2-rGO (2.7% of rGO). This material was immobilized on Side-glowing Optical Fibers (SOFs), which consist of a nude quartz glass fiber as the core and coated with silicone rubber, allowing more uniform light irradiance along the fiber length than other optical fibers. According to their results, all the studied PhCs were degraded as follows: 54, 81, and 92% degradation of carbamazepine, ibuprofen and sulfamethoxazole respectively, after 180 minutes, and mineralization higher than 50% was attained under UV-Vis light.

Bismuth-based materials such as Bi_2O_3, Bi_2WO_6, $BiVO_4$, Bi_2MoO_6, $BiPO_4$, $Bi_2Ti_2O_7$, $(BiO)_2CO_3$, $BiOCOOH$ and $BiOX$ (X = Cl, Br, I) are alternative semiconductors different from TiO_2 or ZnO. BiOXs have attracted attention to degrade PhCs under visible light due to their low E_g value (from 1.76 to 3.19 eV), making them suitable to work under visible light. However, BiOXs present a fast recombination rate. In this sense, these semiconductors have been modified to increase their photocatalytic activity and take advantage of the individual materials. Bismuth-rich photocatalysts, such as Bi_3O_4Cl, $Bi_{12}O_{15}Cl_6$, $Bi_{24}O_{31}Cl_{10}$, Bi_3O_4Br, $Bi_4O_5I_2$ are another strategy to increase their activity (Gao et al. 2021). Most of these materials have demonstrated a higher photocatalytic activity than TiO_2 P25 under visible light. Table 6.2 shows some recent studies using bismuth-based photocatalysts. Bi_2O_3 is a semiconductor considered a non-toxic and non-carcinogenic compound, being the β-phase one of the most efficient polymorphs. The β-Bi_2O_3 synthesized by Coronado-Castañeda et al. (2020) and its performance was compared with the commercial β-Bi_2O_3 on the degradation of isoniazid, observing a similar performance for both materials; however, the synthesized β-Bi_2O_3 reached a higher mineralization degree (24% in 120 minutes) and higher stability than the commercial material.

Other photocatalysts such as $AgBr$-Ag_3PO_4, g-C_3N_4-NiS-MoS_2, CdS-$SrTiO_3$, CeO_2-$NaNbO_3$, g-C_3N_4-Ag_3PO_4, g-C_3N_4-WO_3 have also been studied for PhCs degradation (Calvete et al. 2019). In recent years, ferrous oxalate dihydrate ($FeC_2O_4 \cdot 2H_2O$) has also attracted attention for its photocatalytic potential to degrade PhCs. Conde-Morales et al. (2020) tested $FeC_2O_4 \bullet 2H_2O$ (E_g = 1.9 eV) as an iron source for photo-Fenton degradation of pyrazinamide; this material also showed photocatalyst behavior for oxidation of this antibiotic in the absence of H_2O_2, degrading 35.8% of pyrazinamide in 180 minutes.

Table 6.2. Some recent studies using bismuth-based photocatalysts in the degradation of PhCs.

Photocatalyst	PhCs	Results	Reference
β-Bi_2O_3	Isoniazid (15 mg L^{-1})	93.7% degradation and 24.1% mineralization in 120 minutes	Coronado-Castañeda et al. 2020
BiOI, BiOCl	17 β-ethinyl estradiol, estriol (10 mg L^{-1})	Almost complete removal of both hormones with BiOI under 350 nm irradiation BiOI > BiOCl > TiO_2.	Ahern et al. 2015
BiOCl/BiOI ($BiOCl_{0.75}I_{0.25}$)	Acetaminophen and hydroxyphenyl acetic acid (5-80 mg L^{-1})	100% removal of hydroxyphenyl acetic acid and 80% removal of acetaminophen under solar light for 3 hours	Wang et al. 2016
(30%) SnO_2-BiOI	Oxytetracycline hydrochloride (10 mg L^{-1})	94% degradation in 90 minutes	Wen et al. 2017
$BiVO_4$/BiOI deposited on FTO (fluorine-doped tin oxide)	Acetaminophen and ciprofloxacin (10 mg L^{-1})	68 and 62% degradation values for acetaminophen and ciprofloxacin, in 2 hours Photoelectrocatalysis > Electrooxidation > Heterogeneous photocatalysis	Orimolade et al. 2019

3. Photocatalytic Nanomaterials for Removal of Pesticides

Among the scientific and technological developments, pesticide production is of great concern due to the toxic chemical nature of these compounds, which are designed to control insects, plants, fungi and animals considered as plagues. Due to their chemical properties, pesticides are persistent pollutants in the environment. They resist photochemical, chemical and biochemical degradation. The primary sources of pesticide contamination are agricultural activities, and the chemical or agrochemical industries involved in their synthesis. During manufacturing, transportation, application and final disposal, pesticides or their metabolites can enter the environment and reach the water bodies. The removal of these pollutants from water can be accomplished through heterogeneous photocatalysis technology using novel nanophotocatalysts.

Various research articles have been published related to the photocatalytic degradation of pesticides using nanoparticles of TiO_2 (Hadei et al. 2021, Islam et al. 2020) or TiO_2 modified with metals and non-metals. Different degradation efficiency percentages can be achieved depending on the characteristics of nanomaterials, and the class of pesticide as well as its stability (e.g., organophosphorus triazine derivatives, chloropyridines or organochlorides).

In this context, TiO_2 doped with Cr^{3+} synthesized by the sol-gel microwave-assisted method was evaluated on the degradation of the (4-chloro-2-methylphenoxy) acetic acid (MCPA) under visible (450 nm) and UV (365 nm) light. MCPA is a herbicide, highly water-soluble and highly mobile, which can leach from the soil, exhibiting a potential risk for groundwater contamination (Mendiola-Alvarez et al. 2017). The Cr^{3+} ions were substitutionally incorporated into the TiO_2 lattice improving the textural and optical properties obtaining materials with smaller crystallite and particle size, higher surface area and lower bandgap compared with undoped TiO_2 prepared by the same microwave-assisted method. The Cr^{3+}-doped TiO_2 nanophotocatalyst exhibited a high photocatalytic activity on the degradation of MCPA under visible light, achieving complete degradation (10 mg L^{-1}) in 240 minutes, being superior to undoped TiO_2 and commercial P25, which only degraded 45.7 and 31.1%, respectively. The improved behavior of doped TiO_2 was attributed to its increased surface area and acidity along with increased surface defects (oxygen vacancies) that reduced the e^-/h^+ recombination process.

In another work, MCPA and 2,4-dichlorophenoxyacetic acid (2,4-D), both phenoxy acid herbicides (4.5 mg L^{-1}), were degraded in a mixture under sunlight irradiation by using nanostructures of an Ag(I) Metal-Organic Framework (Ag-MOF) $[Ag(p-OH-C_6H_4COOH)_2(NO_3)]_n$ $[Ag(PHBA)_2(NO_3)]_n$, (1) (PHBA: $C_8H_6O_4$ {p-hydroxybenzoic acid}). In this study, 95% of mineralization of the herbicide mixture was assessed in 20 minutes of solar radiation. Ag-MOF obtained via sonochemical irradiation exhibited nanorod morphology with an average size of 92 nm, improving the degradation process compared with that obtained with a larger particle size (Hayati et al. 2020).

As it was mentioned earlier, titanium oxide has also coupled with other oxide semiconductors with a reduced bandgap to extend its ability to absorb lower-energy photons to harness sunlight. The performance of WO_3/TiO_2, Fe_2O_3/TiO_2 and TiO_2 were studied in the degradation of the herbicide 2,4-D and its main by-product 2,4-dichlorophenol (2,4-DCP) under different radiation sources: natural solar light, visible and UV light (Macías et al. 2017). Complete degradation of 2,4-D aqueous solution (50 mg L^{-1}) and near 89% of mineralization using WO_3/TiO_2 and Fe_2O_3/TiO_2 was achieved after 150 minutes of reaction under solar light, while TiO_2 prepared by sol-gel allowed a lower degradation rate. The degradation and mineralization rates were lower using artificial light (UV and visible lamp) than those obtained under natural sunlight. The produced 2,4-DCP intermediate was completely degraded after 240 minutes under sunlight only with modified photocatalysts. In terms of 2,4-D mineralization, the modified photocatalysts exhibited improved performance due to the narrow bandgap of Fe_2O_3 (2.2 eV) and WO_3 (2.8 eV), which allowed activation of the coupled photocatalysts under visible and solar light. Furthermore, the alignment of the redox potentials of the VB and CB of both oxides promotes the efficient separation of the charge carriers reducing the recombination rate.

In the same way, malathion (S-1,2-bis(ethoxycarbonyl)ethyl 0,0-dimethyl-phosphorodithioate) is a broad-spectrum pesticide used on a wide range of crops, including citrus fruits, which are essential agricultural products in the northeast region of Mexico. Photocatalytic trials using the coupled WO_3/TiO_2 photocatalyst were conducted outdoors, using natural solar irradiation (Ramos-Delgado et al. 2013a). The 2% WO_3/TiO_2 (2 wt.% of WO_3 content with respect to TiO_2) prepared via a sol-gel, process allowed the incorporation of very reactive WO_3 nanoclusters over the anatase TiO_2 surface, with a good homogeneity in spatial and size distribution. Figure 6.3 depicts the 2% WO_3/TiO_2 nanoparticles image with bipyramidal geometry where the predominant planes are (101) type. The WO_3 clusters were preferentially anchored at the edges of the faces of the nanoparticle, suggesting a high reactivity since these sites have a high amount of highly reactive unbounded or low-coordinated atoms. This behavior reduced the charge recombination and increased the surface area of the photocatalytic material (Ramos-Delgado 2013).

The textural and electronic characteristics of 2% WO_3/TiO_2 were compared with bare TiO_2 obtained by sol-gel and depicted in Table 6.3. The higher surface area compared with bare TiO_2 favored the adsorption of more hydroxyl groups on the photocatalyst, contributing to the increase in the generation of $HO^\cdot$ radicals to attack the organophosphorus molecule. The complete degradation of malathion (12 mg L^{-1}) was reached at 120 minutes of reaction, while only 80% was attained using TiO_2. Furthermore, substantial malathion mineralization (63%) was achieved with the coupled photocatalyst exhibiting good stability against photocorrosion, indicating that it is a promising photocatalyst for treating contaminated water in a solar photocatalytic technology (Ramos-Delgado et al. 2013b).

On the other hand, it is well known that zinc oxide is a semiconductor with an energy band similar to that of TiO_2 (E_g = 3.1–3.2 eV), which has shown good photocatalytic activity, in some cases, higher than TiO_2. However, it has the same drawbacks: faster e^-/h^+ recombination and the necessity of UV light for its activation. Therefore, the strategies to overcome these disadvantages, in an analogous way that TiO_2, are doping with metals and non-metals, and coupling with other mixed oxides; the materials were explored for removing some herbicides.

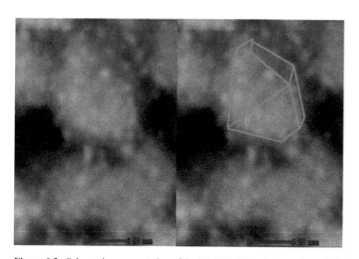 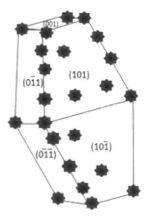

Figure 6.3. Schematic representation of the 2% WO_3/TiO_2 photocatalyst with bipyramidal structure (Ramos-Delgado 2013).

Table 6.3. E_g values, surface area and crystallite size of bare TiO_2 and 2% WO_3/TiO_2.

Photocatalyst	E_g (eV)	S_{BET} ($m^2\,g^{-1}$)	$D_{Scherrer}$ (nm)	STEM cluster size WO_3 (nm)
sol-gel TiO_2	3.17	56	28	--
2% WO_3/TiO_2	3.11	99	23	1.1 ($\pm$ 0.2)

The Nitrogen-doped Zinc Oxide (N-ZnO) was prepared by the sol-gel method and used for the visible photocatalytic degradation of picloram (4-amino-3,5,6-trichloropyridine-2-carboxylic acid) and 2,4-D herbicides in a mixture (20 mg L^{-1} 2,4-D and 5 mg L^{-1} picloram). Both herbicides are present in a commercial formulation and are water-soluble, producing groundwater contamination. The incorporation of nitrogen into ZnO produced particles with a nanometric size of 28.84 nm and decreased the E_g value (2.87 eV) compared to bare ZnO also synthesized via sol-gel (36.2 nm crystallite size and E_g = 3.01 eV). These modifications enhance the ability of the N-ZnO to absorb in the visible region (25 W tungsten-halogen lamp, maximum emission at 540–620 nm), and decrease the recombination rate of photogenerated electrons and holes. Therefore, higher mineralization (57%) of the herbicide mixture was attained after 270 minutes of reaction generating short-chain carboxylic acids (Macías-Sánchez et al. 2015). In contrast, bare ZnO was unable to mineralize the herbicides under the same operating conditions.

Other ZnO-based photocatalysts that have demonstrated excellent photocatalytic properties are ZnO coupled with Fe_2O_3 (ZnO-Fe_2O_3) and ZnO doped with nanometric zerovalent iron (ZnO-Fe^0) (Maya-Treviño et al. 2014, Maya-Treviño et al. 2015). Dicamba (3,6-dichloro-2-methoxybenzoic acid) and 2,4-D are active ingredients in water-soluble formulations as Fortune® (480 g L^{-1}, dicamba) and Hierbamina® (479.5 g L^{-1} 2,4-D), which are also extensively used in agriculture. The photocatalytic activity of ZnO-Fe_2O_3 (0.5 wt.%) was evaluated in the degradation and mineralization of commercial formulations. Although formulations contain additives, complete degradation of both herbicides was accomplished in 300 minutes of reaction under simulated solar light. Oxalic, formic and acetic acids were identified as byproducts during the degradation of both herbicides. Dehalogenation was complete for dicamba, and 70% was attained for 2,4-D at the end of the treatment (Maya-Treviño et al. 2014). This result, along with the mineralization percentages achieved for dicamba (57%) and 2,4-D (34%), indicated that the mineralization of organic compounds by photocatalysis depends on the chemical structure of the pollutant.

The behavior of ZnO-Fe^0 and ZnO–Fe_2O_3 nanomaterials was investigated in the degradation of 2,4-D in a pilot plant to evaluate the applicability of solar photocatalysis for wastewater treatment. The experiments were conducted in a Compound Parabolic Collector (CPC) reactor operated in a recirculation mode at 21 L min^{-1} under natural solar light (Maya-Treviño et al. 2015). ZnO-Fe^0 exhibited better photocatalytic activity than the mixed oxide (ZnO-Fe_2O_3) on the removal of the herbicide achieving rapid ($t_{1/2}$ = 38 minutes) and complete degradation of the pollutant (10 mg L^{-1}) reaching near 50% of 2,4-D mineralization while using the ZnO-Fe_2O_3 photocatalyst (E_g = 3.1 eV) only 45% degradation and 15% mineralization was attained under the same conditions. The incorporation of nanometric Fe^0 (0.5 wt.%) into the ZnO structure produced a displacement of the absorption edge towards the visible region, decreasing the bandgap to 2.8 eV, which corresponds to 442 nm. Furthermore, Fe^0 acts as a trap for the photogenerated electrons avoiding the e^-/h^+ recombination and enhancing the photocatalytic process (Maya-Treviño et al. 2015).

Similarly, ZnO doped with noble metal nanoparticles (~ 8 wt. % Ag or Pd) has been applied to methyl parathion, pendimethalin and trifluralin degradation. Individual solutions of each pollutant (100 mL at 0.001 mg L^{-1}, using 5 mg of photocatalyst) were degraded, obtaining better results with the noble metal-doped ZnO compared with ZnO and TiO_2. Similar behavior was exhibited by both modified photocatalysts achieving almost complete degradation of the herbicides in 200 minutes. The prepared ZnO based photocatalyst presented particles of star shape morphology (nanostar, NSt) with the Ag and Pd nanoparticles (Ag ~ 20 ± 2 nm, Pd ~ 5–6 nm) decorated on the ZnONSt surface, which acts as an electron trap, promoting the interfacial charge-transfer between the metal and ZnONSt enhancing the photocatalytic process (Veerakumar et al. 2021).

In another study related to 2,4-D removal, heterostructured WO_3/$NaNbO_3$ photocatalysts (Hernández-Moreno et al. 2020a) were synthesized at different WO_3 mass ratios (15, 85 and

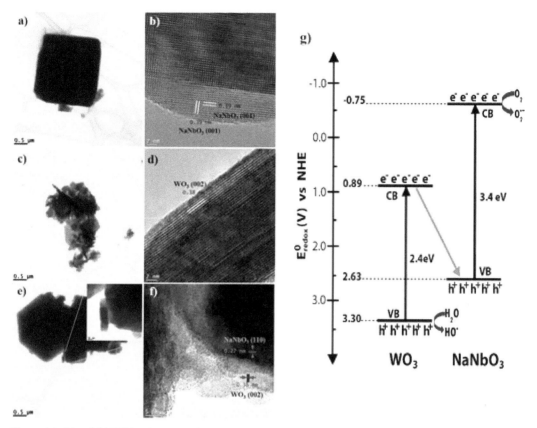

Figure 6.4. (a) and (b) TEM and HR-TEM images of NaNbO$_3$ NPs, (c) and (d) TEM and HR-TEM images of WO$_3$ NPs, and (e) and (f) TEM and HR-TEM images of WO$_3$/NaNbO$_3$, and (g) proposed Z-scheme heterojunction (Reprinted with permission from Hernández-Moreno et al. 2021).

95 wt.%) and tested under visible light to evaluate their photocatalytic performance. For comparison, bare WO$_3$ and NaNbO$_3$ were also prepared and characterized. The obtained WO$_3$(95)/NaNbO$_3$, with an average crystallite size of 28 nm, and an E_g value of 2.6 eV was the nanomaterial with lower E_g and crystallite size compared with bare NaNbO$_3$. The heterojunction between WO$_3$/NaNbO$_3$ was confirmed by TEM and HR-TEM techniques (Fig. 6.4a–f). In these figures, the crystalline phases of individual semiconductors and the heterojunction structure were identified, especially in Fig. 6.4(f), the interface generated between the (002) plane of WO$_3$ and the (110) plane of NaNbO$_3$ was demonstrated.

The WO$_3$(95)/NaNbO$_3$ showed higher photocatalytic activity under visible light than pure WO$_3$ and NaNbO$_3$ even with the 15 and 85 wt.% coupled materials. This behavior was attributed to the VB and CB alignments in the semiconductor structures, improving the charge separation and promoting the accumulation of photogenerated electrons in the CB of NaNbO$_3$ and holes in the VB of WO$_3$ due to the proposed Z-scheme-type heterojunction (Fig. 6.4g). Trapping experiments demonstrated that HO$^{\cdot}$ and O$_2^{\cdot-}$ radicals were the main reactive species that allowed the degradation of the 2,4-D herbicide (10 mg L^{-1}) during the photocatalytic process. The generation of O$_2^{\cdot-}$ radicals is only possible due to the Z-scheme construction because the potential level of electrons in the CB of WO$_3$ (0.89 V) is unable to reduce the dissolved oxygen (O$_2$/O$_2^{\cdot-}$ = −0.33 V) (Hernández-Moreno 2021). Thus, the WO$_3$(95)/NaNbO$_3$ heterojunction is a suitable alternative as an efficient photocatalyst for photocatalytic applications under visible light.

4. Nanoparticles with Photocatalytic Activity for the Degradation of Chemical Additives

Plastic materials use additives that are incorporated during the plastic compound formulation. These additives are chemical compounds that are added during the processes of polymer shaping, injection molding, extrusion, blow molding and vacuum molding, to improve the performance, functionality and properties of the polymer. The most used additives in different polymeric packaging materials are plasticizers, flame retardants, antioxidants, acid scavengers, light and heat stabilizers, lubricants, pigments, antistatic agents, slip compounds and thermal stabilizers (Hahladakisa et al. 2018, Singh et al. 2012). Except for flame retardants that polymerize with plastic molecules becoming a part of the polymeric chain, most additives are not chemically bonded into the plastic polymer. The excessive use of plastic nowadays is a concern because plastic can easily release these toxic compounds showing potential for food, soil, air and water contamination during their use or disposal (Campanale et al. 2020).

For example, the released level of some phthalates from baby bottles was in the range of $50–150$ µg kg^{-1} of food content after the contact time of 120 minutes at 70°C (Simoneau et al. 2012). The released level of bisphenol A (BPA) from food packaging items was estimated to be in the range of $100–800$ ng L^{-1}, while the values were in the range of $1–60$ µg L^{-1} for some phthalates under the same conditions (Fasano et al. 2012, Padervand et al. 2020). Furthermore, BPA was also detected in freshwater and wastewater samples at concentrations of up to 4 µg L^{-1} (Walpitagama et al. 2019).

The toxicity of these additives can cause harmful effects. These effects can strike from a single cell to an organ system. Some harmful chemicals can cause cancer and mutations to DNA, have toxic reproductive effects and are recalcitrant into the environment (OHB/CDPH 2021). Thus, it is very necessary and significant to develop effective processes to remove these additives from water.

Salazar-Beltrán et al. (2019) reported the photocatalytic degradation under UV-Vis irradiation of dimethyl phthalate (DMP) and diethyl phthalate (DEP) in a mixture (5 mg L^{-1} each), comparing three TiO$_2$ based materials, Degussa P25 (82%/18%, anatase/rutile, Aeroxide®), a Hombikat material (76%/24% anatase/rutile) and synthesized anatase TiO$_2$ by the sol-gel method. Degussa P25 showed the highest activity at a pH value and catalyst amount of 10 and 1.5 g L^{-1}, respectively. The complete degradation and 92% mineralization were achieved after 300 minutes of reaction. The higher DMP and DEP degradation percentages with Degussa P25 were attributed to its phase composition, prolonging the lifetime of the e$^-$/h$^+$ pairs due to the band alignment of anatase and rutile. The TiO$_2$ synthesized by the sol-gel method was only in the anatase phase, while Hombikat presented a higher percentage of the rutile phase (24%) than Degussa P25 (18%) which may favor the recombination of e$^-$/h$^+$ pairs (Salazar-Beltrán et al. 2019).

Wang et al. (2020) reported porous ZnO photocatalyst synthesis via a hydrothermal route with a core-shell structure for the efficient photocatalytic degradation of BPA. The materials were prepared by the hydrothermal technique, and the samples were calcined at various temperatures (350, 450, 550 and 650°C). The calcination temperature had a significant effect on the morphology and structure of ZnO. The plate structure formation with porosity was observed for ZnO samples calcined at 350 and 450°C, and ZnO exhibited the porous framework structure under calcination temperatures of 550 and 650°C. For the ZnO sample calcined at 550°C, core-shell hexagonal crystal nanoparticles were observed by HR-TEM (Fig. 6.5) in the range of 20–50 nm in length, and the shell thickness was ~ 4 nm. The ZnO nanoparticles showed the lattice spacing of 0.245 nm and 0.257 nm corresponding to (101) and (002) planes of hexagonal ZnO. The surface area was 31.4 m^2 g^{-1} and the average pore size of ~ 18 nm. The photocatalytic performance of porous ZnO photocatalysts was evaluated under simulated sunlight irradiation (Xenon lamp, 300 W). The porous core-shell ZnO calcined under 550°C exhibited the maximum photocatalytic activity, obtaining 99% degradation of BPA (30 mg L^{-1}) in 1 h (1 g L^{-1} ZnO and pH 6.5). The superior photocatalytic activity observed was due to the large surface area of the porous core-shell ZnO structure, improving the mass-transfer efficiency, promoting light absorption and exposing more active sites.

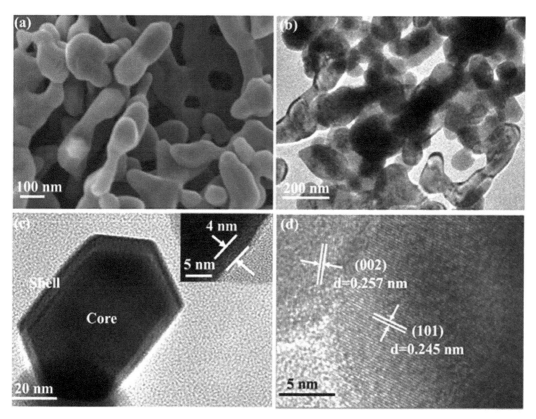

Figure 6.5. (a) SEM; (b, c) TEM (Inset was the shell thickness); (d) HR-TEM of porous ZnO calcined under 550°C (Reprinted with permission from Wang et al. 2020).

Regarding BPA degradation, Blanco-Vega et al. (2017) reported the effect of Ni^{2+} doping on the properties and photocatalytic performance of TiO_2 examined under visible radiation for the degradation of BPA (10 mg L^{-1} solution). The TiO_2 and nickel-doped TiO_2 (0.5 and 1.0 wt. %) materials were prepared using a microwave-assisted sol-gel technique. The doping with Ni ions inhibited the crystallite growth and promoted the anatase TiO_2 crystalline phase formation. The surface area of TiO_2 was 60.3 $m^2\,g^{-1}$ and increased to 83.5 $m^2\,g^{-1}$ for 1.0 wt.% $Ni\text{-}TiO_2$ material. The bandgap energy values decreased from 3.08 eV in TiO_2 to 2.72 eV in 1.0 wt.% $Ni\text{-}TiO_2$-MW. By XPS measurement, the substitutional incorporation of Ni^{2+} in the TiO_2 crystal lattice was corroborated and attributed to the formation of Ni-O-Ti bonds in the Ni-doped TiO_2, due to the similar ionic radius of Ni^{2+} and Ti^{4+} ions (0.72 and 0.68 Å, respectively). The best photocatalyst was 1.0% $Ni\text{-}TiO_2$ MW, which allowed 93.1% degradation and 77% mineralization of BPA at pH 9 within 210 minutes. The performance was superior to those allowed with TiO_2-MW (60.1% degradation and 63.8% TOC).

The *p*-cresol is used as an intermediate to prepare other additives as butylated *p*-cresol, widely used as an additive in the food industry, rubber, polymerization inhibitor and UV-absorber (Elavarasan et al. 2019). For instance, Vigil-Castillo et al. (2019) synthesized a Bi_2O_3/TiO_2 photocatalyst by the sol-gel method to degrade *p*-cresol under visible light. The sol-gel method allowed to control of the bismuth incorporation in the Bi_2O_3/TiO_2 mixed oxides. The amount of Bi_2O_3 incorporated on TiO_2 was determined by microwave plasma atomic emission spectroscopy, obtaining a result of 1.7 and 14.3 wt.% of the theoretical amount added of 2 and 20 wt.%, respectively. The E_g values estimated from the Kubelka-Munk function were 3.20, 3.12, 2.89 and 2.74 eV for TiO_2 P25, TiO_2 sol-gel, 1.7 wt.% Bi_2O_3/TiO_2 (TB2%) and 14.3 wt.% Bi_2O_3/TiO_2 (TB20%), respectively. The particle size decreased slightly from 24 nm in TiO$_2$ sol-gel to 18 and 15 nm, in TB2% and TB20%, respectively; the particle agglomeration was also reduced as Bi$_2$O$_3$ was incorporated into TiO$_2$. This trend was supported by the increase in the surface area from 98 m^2 g^{-1} in TiO$_2$ sol-gel to 108 m^2 g^{-1} in TB20%. From XPS results, it could be inferred that Bi$_2$O$_3$ was present in the modified photocatalysts; however, it was not detected by XRD and Raman spectroscopy because it can be embedded on TiO$_2$ in a nanometric size. Photocatalytic degradation efficiencies of *p*-cresol (pH 7.75) under visible light were higher in TB2% (48%) and TB20% (58%) than that of TiO$_2$ P25 (37%) within 180 minutes. This effect was attributed to the effective bandgap narrowing of TiO$_2$, by incorporating Bi$_2$O$_3$. The photogenerated holes can be transferred from the Bi$_2$O$_3$ VB to the TiO$_2$ VB to produce HO$^•$ radicals from the oxidation of H$_2$O on the photocatalyst surface, reducing the recombination process.

BPA has also been removed using a Fe-BiOBr-N photocatalyst prepared by a microwave-assisted solvothermal method. E_g values were 2.87, 2.84, 1.98 and 1.92 eV for BiOBr, BiOBr-N, Fe-BiOBr and Fe-BiOBr-N, respectively. The modification of BiOBr with Fe (1 wt.%) and N (10 wt.%) increased the surface area and pore volume, mainly when Fe is incorporated (Fe-BiOBr and Fe-BiOBr-N samples). The surface area of the materials was 7.09, 16.26, 10.97 and 16.10 m^2 g^{-1} for BiOBr, Fe-BiOBr, BiOBr-N and Fe-BiOBr-N, respectively. These properties may favor photocatalytic activity compared to pristine BiOBr due to more accessible active sites. Photocatalytic degradation of BPA (pH 6.5) under visible light (430–650 nm, 650 W m^{-2}) was 43, 62, 99 and 100% for BiOBr, Fe-BiOBr, BiOBr-N and Fe-BiOBr-N, respectively. A similar trend was observed in the mineralization process allowing 20, 31, 52 and 65% of TOC removal for the photocatalysts listed earlier. The Fe–BiOBr–N photocatalyst was the most efficient in the degradation and mineralization of BPA in 240 minutes under visible radiation, where O$_2^{•-}$ and h$^+$ were the major oxidizing species, and the photocatalyst showed excellent structural and photocatalytic stability after four reuse cycles (López-Velázquez et al. 2021).

Metal-free semiconductors, such as g-C$_3$N$_4$ are considered good candidates for the photocatalytic degradation of pollutants. In this context, Wu et al. (2020) reported the synthesis of g-C$_3$N$_4$ (obtained by calcination of melamine activated by oxalic acid) for the degradation of BPA under visible light irradiation. Different oxalic acid: melamine molar ratios (0.5:1, 1.0:1, 1.5:1 and 2.0:1) were used, and the samples were calcined to obtain CN, C$_{0.5}$CN, C$_{1.0}$CN, C$_{1.5}$CN and C$_{2.0}$CN, respectively. The bandgap energies of C$_{1.0}$CN and CN were also determined to be 2.57 and 2.69 eV, respectively. The C$_{1.0}$CN material showed a higher valence band (+1.79 eV) than that of CN (+1.70 eV), which indicated that the holes from C$_{1.0}$CN had a stronger oxidation capability for easier photo-degradation of BPA. The conduction band of C$_{1.0}$CN is lower than that of CN, indicating that CN electrons have a higher reduction potential for easier O$_2$ capture to form O$_2^{•-}$. The photocatalytic activities of samples were evaluated under visible light irradiation ($\lambda > 420$ nm cut-off filter and a 300 W Xenon lamp) for photocatalytic degradation of BPA (100 mL of solution at 10 mg L^{-1}) using 20 mg of the photocatalyst. The higher photocatalytic activity was allowed with C$_{1.0}$CN, which completely degraded BPA in solution within 90 minutes. However, ~ 22% BPA was degraded by CN under the same conditions. The mineralization rate of BPA by C$_{1.0}$CN reached 60.6% during 90 minutes, which was much higher than that of CN (4.5%).

The use of metal-organic frameworks (MOFs), a novel category of porous and well crystalline materials, has gained significant interest in current years in photocatalysis for increasing the active centers and modifying the performance of metal oxides semiconductors. Besides, substantial progress was made in producing various MOF composites for photocatalytic applications (Alhumaimess 2020). Numerous MOF materials have attracted great interest owing to their large surface area and high thermal and chemical stability. Tang et al. (2020) synthesized a novel heterostructure TiO$_2$@MIL-101(Cr) as a potential candidate for efficient BPA photocatalytic degradation. A series of X% TiO$_2$@MIL-101(Cr) composites with different TiO$_2$ content (42, 59, 70 and 85 wt.%) was synthesized using the solvothermal method. The bandgap value of MIL101(Cr) and TiO$_2$ were approximately 2.80 and 3.25 eV, respectively. The bandgap value of X% TiO$_2$@MIL-101(Cr) composites was 2.92, 2.95, 3.12 and 3.15 eV, respectively. The surface area of X%

TiO$_2$@MIL-101(Cr) composites was 3018.47, 2046.97, 858.55 and 416.12 m^2 g^{-1}, respectively. The TiO$_2$ introduction into MIL-101(Cr) facilitated the separation of charge carriers at the interface and adjusted the surface area of the composites. Photocatalytic experiments were carried out under UV irradiation (365 nm, 125 W mercury lamp), BPA solution (10 mg L^{-1}) at pH 3, and photocatalyst dosage of 0.5 g L^{-1}. The degradation efficiencies were 29.12, 46.33, 64.27, 71.45, 82.23 and 92.14% for TiO$_2$, TiO$_2$@MIL-101(Cr), 85% TiO$_2$@MIL-101(Cr), 70% TiO$_2$@MIL-101(Cr), 42% TiO$_2$@MIL-101(Cr) and 59% TiO$_2$@MIL-101(Cr) within 240 minutes of reaction. The increased surface area of the heterostructure benefits the adsorption and restrains the aggregation of TiO$_2$, resulting in an excellent photocatalytic performance.

5. Nano-Photocatalysts for Water Disinfection

The incidence of pathogenic microorganisms in drinking and surface water joined to the antimicrobial resistance (AMR) problem is a worldwide concern resulting in the risk of producing several infectious diseases. In 2015, the World Health Organization (WHO) created an action plan worldwide about AMR (WHO 2016), and in 2017 published a global priority list of antibiotic-resistant bacteria, which included 12 of the most dangerous bacteria families to human health, including *E. coli*, *Enterococcus* sp. and *Salmonella* (WHO 2017). Additionally, these kinds of bacteria have been frequently detected in wastewater treatment plants (WWTPs) in both influents and effluents. Therefore, it is necessary to develop sustainable and environmentally friendly technologies. In this context, heterogeneous photocatalysis is currently recognized as an effective technique for mitigating environmental pollution as the inactivation of pathogenic microorganisms in aqueous media (Pino-Sandoval et al. 2020).

Onkundi-Nyangaresi et al. (2019) conducted photocatalytic disinfection of *Escherichia coli* (*E. coli*, initial concentration of approximately 10^7 CFU mL^{-1}) using a UV-LED. The photocatalyst used was commercial anatase TiO$_2$ (Shanghai Macklin Biochemical Co., Ltd., China) with a density of 4.2 g cm^{-3}, a surface area of 50 m^2 g^{-1} and an average diameter size of 20–30 nm. A photocatalyst loading used was between 0.50 and 1.25 g L^{-1}. The increase of the irradiance enhanced the *E. coli* inactivation in both photolysis and photocatalysis systems, especially the former. For a given irradiance of 0.49 mW cm^{-2} from the 365 nm UV-LEDs, 1.0 g L^{-1} was found to be the optimal TiO$_2$ concentration. Inactivation in photolytic disinfection was efficient when 265 and 275 nm UV-LEDs were used. Although TiO$_2$ in suspension had a detrimental effect on the inactivation at 265 and 275 nm, its use significantly improved the inactivation efficiency using 310 and 365 nm UV-LEDs (Onkundi-Nyangaresi et al. 2019).

Zyoud et al. (2019) prepared ZnO nanoparticles by the precipitation method (27.2 nm based on XRD patterns), which formed large agglomerates (~ 200 nm). The nanophotocatalyst was evaluated in the photocatalytic inactivation of Gram-negative and Gram-positive bacteria, *Proteus mirabilis* and *Enterococcus faecium*, respectively (5 × 10^5 CFU mL^{-1}), using simulated solar light. The ZnO structure with a ~ 3.23 eV band gap value exhibited oxygen vacancies and surface defects according to PL results. ZnO nanoparticles partly inactivated *Enterococcus faecium* and *Proteus mirabilis* bacteria in the dark. Under solar simulated UV radiation, a complete bacterial inactivation occurred in 30 minutes, and complete bacterial and organic matter mineralization was achieved in 240 minutes. The performance of the ZnO photocatalyst under different experimental conditions makes it a valuable alternative for natural-water disinfection (Zyoud et al. 2019).

Regarding the strategies to improve photocatalyst activity, the semiconductor modification with transition metals (Ag, Cu) and non-metals (B) have been evaluated in the elimination/inactivation of bacteria (Núñez-Salas et al. 2021b, Pino-Sandoval et al. 2020).

Pino-Sandoval et al. (2020) evaluated the photocatalytic activity of Ag-Cu/TiO$_2$ materials on the inactivation of *E. coli* and *Salmonella typhimurium* (*S. typhimurium*), using an initial concentration of each bacterium about 10^8 CFU mL^{-1}. The incorporation of Cu^{2+} and Ag^0 had a negligible influence on the size and crystalline structure of TiO$_2$ (anatase). However, it caused a shift in the optical absorption of the codoped material towards the visible region, causing an increase in the number of oxygen vacancies on the surface of the material, generating a possible decrease in the e^-/h^+ recombination rate and thus, improving photocatalytic activity. The degree of agglomeration, particle size and homogeneity varied depending on the material: TiO_2 and Ag-TiO_2 materials showed less agglomeration of the particles, with average sizes of 52 and 22 nm, respectively. In contrast, Cu-doped materials exhibited a higher degree of particle agglomeration with calculated particle sizes of 64 and 23 nm for Cu-TiO_2 and Ag-Cu/TiO_2 photocatalysts, respectively. Copper incorporation and photocatalyst loading were significant in the inactivation of both selected bacteria. Ag-Cu/TiO_2 1.2:1.2 wt.% and Ag-Cu/TiO_2 2.1:2.1 wt.% photocatalysts showed higher photocatalytic activity, achieving a complete inactivation of both microorganisms in 10 minutes (6-log reduction) using a photocatalyst load of 0.5 g L^{-1} under simulated solar radiation. The codoped photocatalysts also showed high antimicrobial activity in the absence of light, allowing the treatment to remain effective even in dark conditions. Heterogeneous photocatalysis with Ag-Cu/TiO_2 material was more effective in eliminating Gram-negative bacteria from water than solar disinfection and photocatalysis using undoped or individually doped TiO_2 photocatalysts (TiO_2, Ag-TiO_2, and Cu-TiO_2) (Fig. 6.6).

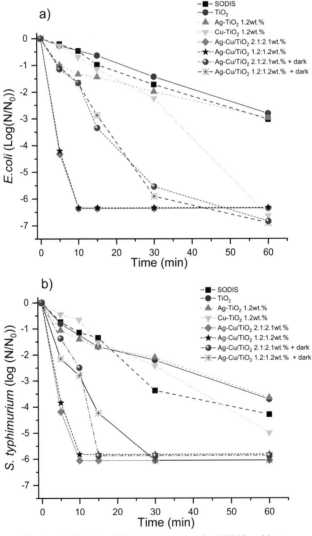

Figure 6.6. Disinfection profiles for (a) *E. coli* and (b) *S. typhimurium* by SODIS and heterogeneous photocatalysis using 0.5 g L^{-1} of photocatalyst. Reprinted with permission from Pino-Sandoval et al. 2020.

According to studies with scavengers, the generated hydrogen peroxide from oxygen reduction and the photogenerated holes mainly attack the bacterial cell membrane and subsequently, the intracellular content. Simultaneously, the metal nanoparticles and metal ions can penetrate the cell and cause internal damage (Pino-Sandoval et al. 2020).

Núñez-Salas et al. (2021b) synthesized ZnO-B (0.07 wt.% B) by a sol-gel method. The photocatalytic performance of this material was evaluated on the inactivation of *E. coli* and *Enterococcus* sp. wild strains (initial concentration of 10^6 CFU mL^{-1} from each bacterium) and was compared with those of ZnO (J.T. Backer, USA) and TiO$_2$ (Degussa P25, USA). The incorporation of B^{3+} improved the structural and optical properties of B-ZnO by decreasing the crystallite size (33.7 nm), cell parameters (a = 3.246 Å, c = 5.198 Å), cell volume (47.463 Å^3) and narrowing the bandgap (3.1 eV) compared with the properties of ZnO (45.6 nm, a = 3.251 Å, c = 5.215 Å, 47.743 Å, 3.2 eV, respectively). The crystalline lattice distortion after B^{3+} ion incorporation into the ZnO structure substituting Zn^{2+} was due to the difference in ionic radii and the charge difference between Zn^{2+} (0.074 nm) and B^{3+} (0.027 nm). The B-ZnO photocatalyst exhibited two distributions of particle size: one with a particle size between 93 and 160 nm (Fig. 6.7b) and the other one (Fig. 6.7c) with particles between 22 and 28 nm. The bigger particles may correspond to unmodified ZnO, which presented hexagonal bars that agglomerate (characteristic of ZnO synthesized by sol-gel method). The smaller particles exhibited heterogeneous morphology due to the presence of B dopant. The charge difference between Zn^{2+} and B^{3+} caused a decrease of the Zn^{2+} interstitial concentration on B-ZnO, inhibiting the grain growth. The material characterization results suggested effective doping by B incorporation, where B is replacing Zn^{2+} ions or occupying O vacancies. A reduction of 1-log unit of *E. coli* was observed in the processes with ZnO and B-ZnO in the dark. However, *Enterococcus* sp. was only inactivated in the dark using B-ZnO, reducing only 1.5-log units. The increase of the B-ZnO antimicrobial effect was attributed to the Zn^{2+} release and the formation of acid sites promoted by the B^{3+} ions in the B-ZnO photocatalyst. Under solar simulated irradiation, *E. coli* and *Enterococcus* sp. achieved an inactivation of 6-log and 3-log, respectively during 180 minutes of exposure without photocatalyst. The photocatalytic inactivation of *E. coli* was negligible using TiO$_2$, while heterogeneous photocatalysis using ZnO and B-ZnO reached 6-log inactivation during 30 and 60 minutes of lighting, respectively. The photocatalytic inactivation of *E. coli* using ZnO exhibited a higher kinetic constant value (1.12 min^{-1}) than that obtained with doped B-ZnO (0.23 min^{-1}). In the case of the photocatalytic inactivation of *Enterococcus* sp., reduction of 3-log in 180 minutes was achieved using B-ZnO (0.22 min^{-1}) while 1-log was attained using ZnO (0.02 min^{-1}) and TiO$_2$ (0.04 min^{-1}). To determine the best material to simultaneously inactivate *E. coli* and *Enterococcus* sp., a statistical approach as the desirability function was applied. From this analysis, the maximization of the kinetic constants of inactivation of *E. coli* and *Enterococcus* sp. allowed the selection of B-ZnO as the best material due to its enhanced photocatalytic activity against both strains (Fig. 6.7e–f) (Núñez-Salas et al. 2021b).

Recently, Núñez-Salas (2021) studied ZnO/MnS (1, 5, and 10 wt.% MnS) synthesized by a microwave-assisted chemical precipitation method on the photocatalytic inactivation of *E. coli* in an aqueous media (initial concentration of 10^6 CFU mL^{-1}) under solar natural light. The incorporation of MnS in ZnO caused a slight red shift in its optical absorption edge corresponding to E_g values of 2.9–3.1 eV, the active materials being in the visible region, and decreasing the e$^-$/h$^+$ recombination rate (PL analysis). The ZnO/MnS presented morphology of hexagonal bars with particle sizes in the ranges of 80–180, 80–160 and 100–200 nm for 1, 5 and 10% of MnS, respectively. In photocatalytic tests using 1.0 g L^{-1} of materials in darkness, the content of MnS in the photocatalyst notably increased the bactericidal activity, reaching its maximum efficiency with ZnO/MnS 5% that achieved greater inactivation (3.68-log) than ZnO (1.30-log), ZnO/MnS 1% (2.33-log) and ZnO/MnS 10% (2.18-log). Complete photocatalytic inactivation of *E. coli* was achieved using all materials in 30 minutes; however, the kinetic rate was improved with the modified ZnO as follows: ZnO (0.34 min^{-1}), ZnO/MnS 1% (0.35 min^{-1}), ZnO/MnS 5% (0.37 min^{-1}) and ZnO/MnS 10% (0.33 min^{-1}) (Núñez-Salas 2021).

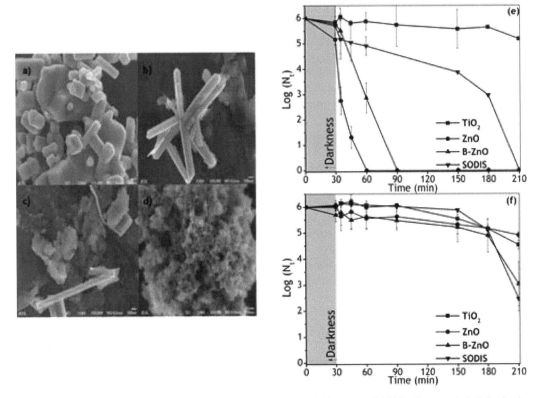

Figure 6.7. SEM images of (a) commercial ZnO, (b), (c) B-ZnO, and (d) commercial TiO$_2$. Photocatalytic behavior by solar disinfection (SODIS), TiO$_2$, ZnO and B-ZnO on inactivation of (e) *E. coli* and (f) *Enterococcus* sp. Reprinted with permission from Núñez-Salas et al. 2021b.

6. Recent Advances in Photocatalytic Nanomaterials for NO$_X$ and CO$_2$ Abatement

Atmospheric pollution is one of the most severe problems responsible for greenhouse effect due to toxic gases like Nitrogen Oxides (NO$_x$) and carbon dioxide (CO$_2$). NO$_x$ emission, composed of NO and NO$_2$, is one of the major gases generated through the combustion of N-containing sources such as automobile exhausts (55%), chemical industry and fossil fuel combustion in power plants (45%) (Nikokavoura and Trapalis 2018, Nguyen et al. 2020, Patil et al. 2019). The guidelines for ambient levels of NO$_2$ were defined by the WHO to determine the air quality (WHO Guideline Values for Health 2021). The NOx emissions contribute to acid rain, ozone accumulation, photochemical smog and global warming. Furthermore, human exposure to NO$_x$ could contribute to respiratory problems, lung and heart diseases and an increased response to allergens. Therefore, developing efficient and economical techniques for abating NO$_x$ in ambient air is required. Photocatalytic technologies for NO$_x$ oxidation constitute the most convenient way to remove these kind of air pollutants even at parts per billion (μg L^{-1}) levels with an efficient NO$_x$ conversion at ambient temperature and pressure cost-effectiveness. The NO$_x$ photocatalytic oxidation on the semiconductor surface activated by UV or visible radiation involves some steps with NO$_2$ intermediate formation and its conversion to nitrate ion (NO$_3^-$) as feasible raw material for fertilizers (Nikokavoura and Trapalis 2018, Nguyen et al. 2020).

For instance, in 2016, Hernández-Rodríguez et al. (2016) compared eight commercial TiO$_2$ based photocatalysts supported on borosilicate glass plates by the dip-coating technique on the NO$_x$ photocatalytic oxidation under UV and visible irradiation sources. Among the evaluated photocatalysts, the best results were obtained by Hombikat UV-100, Kronos vlp700 and

CristalACTIV PC500, which showed 100% anatase phase, the smallest crystal size (7–9 nm) and the largest surface area (202–358 $m^2 g^{-1}$). The increase in illumination time (24 hours) decreased the efficiency of the photocatalyst in terms of removal percentage ([NO], %) and NO_x selectivity (S). The least affected photocatalysts were CristalACTIV PC500 and Hombikat UV-100, with respective values of percentage of NO removal of 88.1 and 87.4 and selectivity of 0.75 and 0.64, respectively. The performance and selectivity decreased when the humidity increased from 25 to 65%, showing differences in the adsorbed products detected. The photocatalyst stability was demonstrated for Hombikat UV-100, Kronos vlp70 and CristalACTIV PC500 in 10 reuse cycles, allowing removals higher than 95% (Rodríguez et al. 2016).

The alternative photocatalyst of TiO_2, ZnO ($E_g > 3.1$ eV), has attracted attention in the NO_x photocatalytic removal. Luévano-Hipólito et al. (2017) prepared the nanometric ZnO wurtzite photocatalyst by a sol-gel method with preferential orientation in the crystalline plane (002), a specific surface area of 29.8 $m^2 g^{-1}$, and spherical morphology. The synthesized material showed higher photocatalytic NO conversion (> 95%) than P25 TiO_2 (65%) and commercial ZnO (53%) under UV irradiation. The enhanced performance was due to the adsorption of large amounts of hydroxyl ions that participate in the NO_x oxidation reaction.

On the other hand, Pd/TiO_2 photocatalysts were prepared by modification of commercial TiO_2 with Pd by photodeposition and impregnation to extend the TiO_2 photoresponse to the visible region and overcome the e^-/h^+ recombination. The materials were evaluated on NO_x removal under UV-Vis and visible irradiation (Hernández-Rodríguez et al. 2020). According to XPS results, the $O_{lattice}$/Ti ratio remained close to the theoretical value of ~ 2 in the photocatalysts, indicating that the modification of TiO_2 with Pd was superficial. The enhanced activity on NO_x removal was observed in the Pd/TiO_2 sample prepared by deposition due to the high Pd^0 amount on the TiO_2 surface (37% removal) compared with Pd/TiO_2 prepared by the impregnation method (35%) and the P25 sample (26%). The photocatalytic performance increase was due to the presence of Pd species that favors the NO_2 adsorption on the material surface. Moreover, Pd particles improved the charge separation and enhanced the visible-light harvesting ability of TiO_2.

Mendiola-Alvarez et al. (2021) evaluated the photocatalytic activity of mixed oxide α-Fe_2O_3/TiO_2 doped with P prepared by the microwave-assisted sol-gel route on NO_x oxidation under UV and visible light irradiation. For comparative purposes, the photocatalytic performance was compared with TiO_2 doped with P, α-Fe_2O_3/TiO_2, and pure TiO_2. As shown in Fig. 6.8(a–d), the enhanced activity of α-Fe_2O_3/TiO_2 was related to the assembling of α-Fe_2O_3 nanoparticles with TiO_2 to form an α-Fe_2O_3/TiO_2 heterostructure with values of the position of the electronic bandgap that promotes charge separation. Moreover, the activity was related to the presence of Lewis basic sites that increased the surface basicity as correlated with the highest point of zero charge (pzc) value of α-Fe_2O_3/TiO_2 ($pH_{pzc} = 6.8$). The NO_x oxidation under UV light followed the trend α-Fe_2O_3/TiO_2 > TiO_2 > TiO_2/P > α-Fe_2O_3/TiO_2/P (Fig. 6.8e), while the specific surface area values were in the opposite direction (53.2, 73.9, 173.8 and 157.7 $m^2 g^{-1}$, respectively). As can be seen in Fig. 6.8f, the α-Fe_2O_3/TiO_2 showed the lowest selectivity to generate NO_2 ($S_{NO2} = 0.67$), suggesting that this photocatalyst converts NO to less toxic species than NO_2, like NO_3^- and NO_2^-.

As described in this chapter, bismuth-based layered materials BiOX (X = I, Br) represent an alternative to TiO_2 for photocatalytic degradation of emerging pollutants in an aqueous medium. In this sense, bismuth oxybromide (BiOBr) nanoparticles were synthesized in an aqueous medium by the coprecipitation method at room temperature using EDTA as a structure-directing agent. This photocatalyst showed a 94% NO conversion degree, a 0.98 selectivity to nitrate ions and good stability after four irradiation cycles under 70% relative humidity. The enhanced activity of BiOBr was related to the crystal preferential growth that allows an efficient charge separation on (110) crystallographic exposed facets, the high surface area value (36.5 $m^2 g^{-1}$) and the presence of residual EDTA on the photocatalyst surface as a hole scavenger, which favored the $O_2^{\cdot-}$ generation, as demonstrated with a scavenger study (Montoya-Zamora et al. 2020).

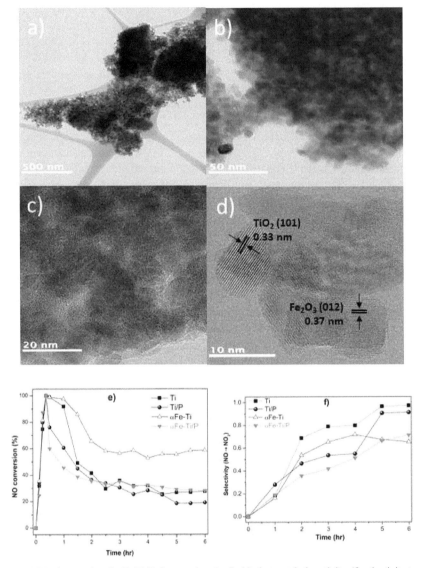

Figure 6.8. HR-TEM micrographs of α-Fe-Ti/P photocatalyst (a–d), (e) photocatalytic activity, (f) selectivity toward NO_2 of Ti, Ti/P, α-Fe-Ti, and α-Fe-Ti/P photocatalysts during NO conversion under UV radiation (Reprinted with permission from Mendiola-Alvarez et al. 2021).

BiOI with the morphology of 3D flower-like microspheres was synthesized via the microwave method. The photocatalytic activity of BiOI was evaluated in the oxidation of NO under visible-light irradiation. The material showed high NO removal activity (83%), maintaining its performance after four photocatalytic cycles with only a slight decrease of approximately 7%. Besides the morphology of BiOI samples, the high photocatalytic activity was associated with the large surface area (57 $m^2 g^{-1}$) and the low rate of charge recombination, as demonstrated by PL spectroscopy. The holes, electrons and superoxide ions were the predominant species participating in the mechanism of NO oxidation (Núñez et al. 2019).

The 2D/2D $Bi_2O_2CO_3/Bi_4O_5Br_2$ heterostructure as a direct Z-scheme photocatalyst was synthesized via a one-step hydrothermal method and applied for the NO_x oxidative removal under simulated solar light (Zhu et al. 2019). The 30% $Bi_2O_2CO_3/Bi_4O_5Br_2$ exhibited higher photocatalytic activity (53.2%) for NO_x removal than that of the single-phase $Bi_2O_2CO_3$ (20.4%) and $Bi_4O_5Br_2$

(37.9%). Hydroxyl and superoxide radicals were the main active species during the photocatalytic reaction process. The Z-scheme heterojunction formation between $Bi_2O_2CO_3$ and $Bi_4O_5Br_2$ was corroborated by DFT theoretical calculations where photoinduced e^- in $Bi_2O_2CO_3$ are combined with h^+ in $Bi_4O_5Br_2$. This process can effectively hinder the recombination of photoinduced charge carriers in the $Bi_2O_2CO_3/Bi_4O_5Br_2$ nanocomposites, enhancing photocatalytic NO_x removal (Zhu et al. 2019).

The use of photocatalysts in construction materials have been evaluated for practical applications to reduce NO_x levels from vehicle gaseous exhaust emissions. Beyond the intrinsic photocatalytic properties of the materials, such as crystalline and electronic structure, the self-cleaning property should be considered (Luévano-Hipólito et al. 2018, Pérez-Nicolás et al. 2017). Nava-Núñez et al. (2020) prepared photocatalytic bismuth-based layered material BiOI and BiOBr mortars using a cement (cem) matrix. The photocatalytic performance of mortars was evaluated by the photocatalytic NO_x conversion rate and selectivity to nitrates. The NO_x removal efficiency was BiOCl-cem > BiOI-cem > TiO_2-cem. The highest efficiency under visible light of BiOCl-cem might be ascribed to oxygen vacancies and strong oxidation ability to generate hydroxyl radicals due to its valence band potential (3.53 V). Thus, BiOCl-cem resulted in a safer and environmentally sustainable option in construction materials.

On the other hand, the increase of CO_2 concentration in the atmosphere mainly due to fossil fuel combustion, such as petroleum, natural gas and coal, has led to severe environmental problems. CO_2 is the most stable global warming and greenhouse gas, causing climate change and desertification (Sohn et al. 2017). Furthermore, alternative energy sources to reduce the energy crisis caused by intense fuel consumption have been addressed for sustainable development. Photocatalytic CO_2 reduction using solar energy that directly converts CO_2 and H_2O into value-added hydrocarbon fuels is a promising approach to address these issues (Nahar et al. 2017). The photocatalytic CO_2 reduction involves three stages, i.e., (1) CO_2 adsorption, (2) redox reactions initiated by the photocatalyst light activation where the e^-/h^+ pairs generated can be efficiently separated, suppressing the recombination of charge pairs and improving their generation rate and transport to the photocatalyst surface, (3) product desorption. Thus, the photocatalyst structure should be carefully designed to enhance the photocatalytic efficiency including an increased surface area with more active sites to ensure the rapid and strong adsorption of CO_2 molecules, quick electron transfer from the photocatalyst to CO_2 molecules and timely desorption of generated products to avoid poisoning of the material. CO_2 is a chemically stable compound due to its carbon-oxygen bonds; thus, conversion to carbon-based fuels requires substantial energy input for bond cleavage.

The reduction capacities of photogenerated electrons at the photocatalyst surface are closely related to the conduction band position of semiconductor photocatalysts. The better separation of photogenerated electrons and holes can bring higher surface photogenerated electron density of photocatalysts, accelerating the multielectron reduction reaction and obtaining higher reduced state products; however, this affects the product selectivity (Fu et al. 2020). Various photocatalysts based on the use of the semiconductors TiO_2 (Low et al. 2017), Cu_2O (Zhang et al. 2021), metal sulfides (Albero et al. 2020, Wang et al. 2021), g-C_3N_4 (Ye et al. 2015), WO_3 (Murillo-Sierra et al. 2021b) and bismuth-based compounds (Ye et al. 2019), among others, have been developed for effective CO_2 reduction, allowing a high-efficient CO_2 conversion, controlled product selectivity, low cost and high stability.

Wang et al. (2019) described a hydrothermal procedure and *in situ* chemical reduction approach to synthesize a nanometric 0D/2D Au/TiO_2 plasmon heterojunction photocatalyst. The primary CO_2 reduction product of the 0D/2D Au/TiO_2 heterojunction containing 3 wt.% Au was CH_4, obtained at the rates of 70.34 μmol $g^{-1}h^{-1}$, respectively, with 80% CH_4 selectivity. The Local Surface Plasmon Resonance (LSPR) effect of AuNPs enhanced the light absorption of TiO_2 and acted as a capturing center of photogenerated electrons. The formation of an ohmic junction between Au and TiO_2 could boost the photogenerated charge separation and promote the multielectron reaction of CO_2 and H_2O, resulting in high selectivity of CH_4 generation.

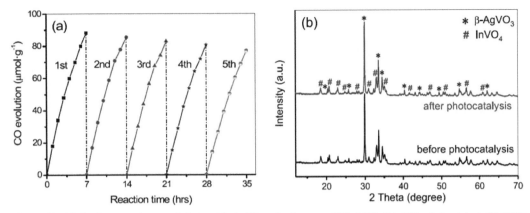

Figure 6.9. (a) Cycling experiments of photocatalytic CO_2 reduction over 20% $InVO_4/\beta AgVO_3$ photocatalyst. (b) XRD patterns of 20% $InVO_4/\beta AgVO_3$ before and after the photocatalytic reaction (Reprinted with permission from Yang et al. 2019, Copyright {2022} American Chemical Society).

Some visible-light-responsive Z-scheme photocatalysts for the reduction of CO_2, like $TiO_2/CuInS_2$ (Xu et al. 2018), g-C_3N_4/ZnO (Nie et al. 2018), $CdS/CdWO_4$ (Li et al. 2019), $InVO_4/\beta$-$AgVO_3$ (Yang et al. 2019) and Bi_2S_3 Quantum Dots (QDs)/g-C_3N_4 (Guo et al. 2020) have been reported as an effective strategy to promote superior spatial charge separation, resulting in high photocatalytic CO_2 reduction and product selectivity. For instance, Yang et al. (2019) described the enhanced photoconversion of CO_2 to CO under visible-light illumination using the direct Z-scheme $InVO_4/\beta$-$AgVO_3$ heterojunction. The composite consisted of β-$AgVO_3$ nanoribbons and $InVO_4$ nanoparticles prepared via a facile hydrothermal method and subsequent *in situ* growth process. The prepared $InVO_4/\beta$-$AgVO_3$ composite with a 20% In/Ag molar ratio exhibited enhanced photocatalytic activity for CO_2 reduction allowing a CO evolution rate of 12.61 μmol g^{-1} h^{-1} without a cocatalyst or sacrificial agent, which was approximately 11 times larger than that yielded by pure $InVO_4$ (1.12 μmol g^{-1} h^{-1}). Although undesired H_2 evolution from water vapor reduction was also detected, the CO selectivity was more than 93%. Moreover, the $InVO_4/\beta$-$AgVO_3$ retained more than 87% of the original efficiency for CO production (Fig. 6.9a), suggesting the stable heterostructure of $InVO_4/\beta AgVO_3$ after five consecutive cycles as demonstrated by DRX analysis (Fig. 6.9b).

Conclusion

The development of photocatalytic nanomaterials is of great significance for implementing clean technologies to improve the quality of the environment. Doping semiconductors with metal and non-metal, sensitization of the semiconductors with dyes or metallic complexes, MOFs incorporation and heterojunctions formation are some strategies to prepare more efficient nano-catalysts for environmental applications.

Depending on the contaminants to be degraded and the matrix where they are found, the photocatalyst must be developed with specific characteristics. In the case of water disinfection, for example, the catalyst must have bactericide properties in addition to being activated by sunlight or visible light. While for the elimination of NO_x or conversion of CO_2, the materials must present a large surface area and an improved photocatalytic activity.

The nanoscale dimensions play an important role in the electronic and textural properties that determine the performance of the photocatalytic materials for applications related to the environment. Although some photocatalysts are commercially available for applications such as air disinfection, self-cleaning surfaces, and NO_x abatement, more research is necessary on producing emerging nanomaterials with favorable surface reaction sites and long-lasting stability for decomposition of a variety of organic and inorganic pollutants and gas conversion.

Acknowledgments

The authors acknowledge Facultad de Ciencias Químicas of the Universidad Autónoma de Nuevo León and to Consejo Nacional de Ciencia y Tecnología (CONACyT) projects No. A1-S-40260 and No. 1727980 for the funding. They also thank Dr. Iliana Medina-Ramírez for the review of the final draft.

References

Aba-Guevara, C.G., I.E. Medina-Ramírez, A. Hernández-Ramírez, J. Jáuregui-Rincón, J.A. Lozano-Álvarez and J.L. Rodríguez-López. 2017. Comparison of two synthesis methods on the preparation of Fe, N-Co-doped TiO$_2$ materials for degradation of pharmaceutical compounds under visible light. Ceram. Int. 43: 5068–5079.

Ahern, J.C., R. Fairchild, J.S. Thomas, J. Carr and H.H. Patterson. 2015. Characterization of BiOX compounds as photocatalysts for the degradation of pharmaceuticals in water. Appl. Catal., B 179: 229–238.

Albero, J., Y. Peng and H. García. 2020. Photocatalytic CO$_2$ reduction to C^{2+} products. ACS Catal. 10: 5734−5749.

Alhumaimess, M.S. 2020. Metal-organic frameworks and their catalytic applications. J. Saudi Chem. Soc. 24: 461−473.

Blanco-Vega, M.P., J.L. Guzmán-Mar, M. Villanueva-Rodríguez, L. Maya-Treviño, L.L. Garza-Tovar, A. Hernández-Ramírez and L. Hinojosa-Reyes. 2017. Photocatalytic elimination of bisphenol A under visible light using Ni-doped TiO$_2$ synthesized by microwave assisted sol-gel method. Mater. Sci. Semicond. Process. 71: 275−282.

Calvete, M.J.F., G. Piccirillo, C.S. Vinagreiro and M.M. Pereira. 2019. Hybrid materials for heterogeneous photocatalytic degradation of antibiotics. Coord. Chem. Rev. 395: 63–85.

Campanale, C., C. Massarelli, I. Savino, V. Locaputo and V.F. Uricchio. 2020. A detailed review study on potential effects of microplastics and additives of concern on human health. Int. J. Environ. Res. Public. Health. 17: 1212.

Conde-Morales, I.I., L. Hinojosa-Reyes, J.L. Guzmán-Mar, A. Hernández-Ramírez, I.C. Sáenz-Tavera and M. Villanueva-Rodríguez. 2020. Different iron oxalate sources as catalysts on pyrazinamide degradation by the Photo-Fenton process at different pH values. Water, Air, Soil Pollut. 231: 425.

Cordero-García, A., J.L. Guzmán-Mar, L. Hinojosa-Reyes, E. Ruiz-Ruiz and A. Hernández-Ramírez. 2016. Effect of carbon doping on WO$_3$/TiO$_2$coupled oxide and its photocatalytic activity on diclofenac degradation. Ceram. Int. 42: 9796–9803.

Cordero-García, A., G. Turnes Palomino, L. Hinojosa-Reyes, J.L. Guzmán-Mar, L. Maya-Teviño and A. Hernández-Ramírez. 2017. Photocatalytic behaviour of WO$_3$/TiO$_2$-N for diclofenac degradation using simulated solar radiation as an activation source. Environ. Sci. Pollut. Res. 24: 4613–4624

Coronado-Castañeda, R.R.S., M.L. Maya-Treviño, E. Garza-González, J. Peral, M. Villanueva-Rodríguez and A. Hernández-Ramírez. 2020. Photocatalytic degradation and toxicity reduction of isoniazid using β-Bi$_2$O$_3$ in real wastewater. Catal. Today 341: 82–89.

Diaz-Angulo, J., J. Porras, M. Mueses, R.A. Torres-Palma, A. Hernandez-Ramirez and F. Machuca-Martinez. 2019. Coupling of heterogeneous photocatalysis and photosensitized oxidation for diclofenac degradation: role of the oxidant species. J. Photochem. Photobiol., A. 383: 112015.

Diaz-Angulo, J., J. Lara-Ramos, M. Mueses, A. Hernández-Ramírez, G. Li Puma and F. Machuca-Martínez. 2020. Enhancement of the oxidative removal of diclofenac and of the TiO$_2$ rate of photon absorption in dye-sensitized solar pilot scale CPC photocatalytic reactors. Chem. Eng. J. 381: 122520.

Elavarasan, P., K. Kondamudi and S. Upadhyayula. 2009. Statistical optimization of process variables in batch alkylation of p-cresol with tert-butyl alcohol using ionic liquid catalyst by response surface methodology. Chem. Eng. Technol. 155: 355−360.

EU Air Quality Policy and WHO Guideline Values for Health. https://www.europarl.europa.eu/RegData/etudes/STUD/2014/536285/IPOL_STU(2014)536285_EN.pdf (consulted on 14/03/2021).

Fasano, E., F. Bono-Blay, T. Cirillo, P. Montuori and S. Lacorte. 2012. Migration of phthalates, alkylphenols, bisphenol A and di (2-ethylhexyl) adipate from food packaging. Food Control 27: 132–138.

Fu, J., K. Jiang, X. Qiu, J. Yu and M. Liu. 2020. Product selectivity of photocatalytic CO$_2$ reduction reactions. Mater. Today 32: 222–243.

Gao, P., Y. Yang, Z. Yin, F. Kang, W. Fan, J. Sheng, L. Feng, Y. Liu, Z. Du and L. Zhang. 2021. A critical review on bismuth oxyhalide based photocatalysis for pharmaceutical active compounds degradation: Modifications, reactive sites, and challenges. J. Hazard. Mat. 412: 125186.

Góngora, J.F., P. Elizondo and A. Hernández-Ramírez. 2017. Photocatalytic degradation of ibuprofen using TiO$_2$ sensitized by Ru(II) polyaza complexes. Photochem. Photobiol. Sci. 16: 31–37.

Guevara-Almaraz, E., L. Hinojosa-Reyes, A. Caballero-Quintero, E. Ruiz-Ruiz, A. Hernández-Ramírez and J.L. Guzmán-Mar. 2015. Potential of multisyringe chromatography for the on-line monitoring of the photocatalytic degradation of antituberculosis drugs in aqueous solution. Chemosphere 121: 68–75.

Guo, R.T., X.Y. Liu, H. Qin, Z.Y. Wang, X. Shi, W.G. Pan and J.W. Gu. 2020. Photocatalytic reduction of CO_2 into CO over nanostructure Bi_2S_3 quantum dots/g-C_3N_4 composites with Z-scheme mechanism. Appl. Surf. Sci. 500: 144059.

Hadei, M., A. Mesdaghinia, R. Nabizadeh, M.A. Hossein, S. Rabbani and K. Naddafi. 2021. A comprehensive systematic review of photocatalytic degradation of pesticides using nano TiO_2. Environ. Sci. Pollut. Res. 28: 13055–13071.

Hahladakisa, J.N., C.A. Velis, R. Weber, E. Iacovidou and P. Purnell. 2018. An overview of chemical additives present in plastics: Migration, release, fate and environmental impact during their use, disposal and recycling. J. Hazard. Mater. 344: 179–199.

Hayati, P., Z. Mehrabadi, M. Karimi, J. Janczak, K. Mohammadi, G. Mahmoudi, F. Dadi, M.J.S. Fard, A. Hasanzadeh and S. Rostamnia. 2021. Photocatalytic activity of new nanostructures of an Ag(I) metal–organic framework (Ag-MOF) for the efficient degradation of MCPA and 2,4-D herbicides under sunlight irradiation. New J. Chem. 45: 3408–3417.

Hernández-Rodríguez, M.J., E. Pulido-Melián, J. Araña, J.A. Navío, O.M. González-Díaz, D.E. Santiago and J.M. Doña-Rodríguez. 2020. Influence of water on the oxidation of no on Pd/TiO_2 photocatalysts. Nanomater. 10: 2354.

Hernández-Rodríguez, M., E.P. Melián, O.G. Díaz, J. Araña, M. Macías, A.G. Orive and J.D. Rodríguez. 2016. Comparison of supported TiO_2 catalysts in the photocatalytic degradation of NOx. J. Mol. Catal. A Chem. 413: 56–66.

Hernández-Coronado, R. 2020. Efecto sinérgico de los procesos de fotocatálisis y foto-fenton heterogéneos en la degradación de cefuroxima bajo radiación solar simulada con el sistema α-Fe_2O_3-TiO_2/H_2O_2. M.Sc. Thesis, Universidad Autónoma de Nuevo León, Nuevo León, México.

Hernández-Moreno, E.J., A. Martínez de la Cruz, L. Hinojosa-Reyes, J.L. Guzmán-Mar, M.A. Gracia-Pinilla and A. Hernández-Ramírez. 2021. Synthesis, characterization, and visible light–induced photocatalytic evaluation of WO_3/$NaNbO_3$ composites for the degradation of 2,4-D herbicide. Mater. Today Chem. 19: 100406.

Hernández-Moreno, E.J. 2021. Actividad Fotocatalítica de WO_3/$NaNbO_3$ en la eliminación de NO_x y en la degradación de 2,4-D en medio acuoso. Ph.D. Thesis, Universidad Autónoma de Nuevo León, N.L. Mexico.

Hernández-Tenorio, R., J.L. Guzmán-Mar, L. Hinojosa-Reyes, N. Ramos-Delgado and A. Hernández-Ramírez. 2021. Determination of pharmaceuticals discharged in wastewater from a public hospital using LC-MS/MS technique. J. Mex. Chem. Soc. 65(1): 94–108.

Islam, M.R., J.B. Islam, M. Furukawa, I. Tateishi, H. Katsumata and S. Kaneco. 2020. Photocatalytic degradation of a systemic herbicide: Picloram from aqueous solution using titanium oxide (TiO_2) under sunlight. Chem. Eng. 4: 58.

Li, Y.Y., Z.H. Wei, J.B. Fan, Z.J. Li and H.C. Yao. 2019. Photocatalytic CO_2 reduction activity of Z-scheme CdS/$CdWO_4$ catalysts constructed by surface charge directed selective deposition of CdS. Appl. Surf. Sci. 483: 442–452.

Lin, L., H. Wang and P. Xu. 2017. Immobilized TiO_2-reduced graphene oxide nanocomposites on optical fibers as high performance photocatalysts for degradation of pharmaceuticals. Chem. Eng. J. 30: 389–398.

Lofrano, G., R. Pedrazzani, G. Libralato and M. Carotenuto. 2017. Advanced oxidation processes for antibiotics removal: A review. Current Organic Chemistry 21: 1–14.

López-Velázquez, K., J.L. Guzmán-Mar, A. Hernández-Ramírez, E. González-Juárez and M. Villanueva-Rodríguez. 2021. Synthesis of Fe–BiOBr–N by microwave-assisted solvothermal method: Characterization and evaluation of its photocatalytic properties. Mater. Sci. Semicond. Process. 123: 105499.

Low, J., B. Cheng and J. Yu. 2017. Surface modification and enhanced photocatalytic CO_2 reduction performance of TiO_2: a review. Appl. Surf. Sci. 392: 658–686.

Luévano-Hipólito, E., A. Martínez-de la Cruz and E.L. Cuéllar. 2017. Performance of ZnO synthesized by sol-gel as photocatalyst in the photooxidation reaction of NO. Environ. Sci. Pollut. Res. 24: 6361–6371.

Luévano-Hipólito, E. and A. Martínez-de la Cruz. 2018. Photocatalytic stucco for NOx removal under artificial and by real weatherism. Constr. Build. Mater. 174: 302–309.

Luja-Mondragón, M., L.M. Gómez-Oliván, N. SanJuan-Reyes, H. Islas-Flores, J.M. Orozco-Hernández, G. Heredia-García and O. Dublán-García. 2019. Alterations to embryonic development and teratogenic effects induced by a hospital effluent on *Cyprinus carpio* oocytes. Sci. Total Environ. 660: 751–764.

Macías, R., M. Villanueva-Rodríguez, N. Ramos-Delgado, L. Maya-Treviño and A. Hernández-Ramírez. 2017. Comparative study of the photocatalytic degradation of the herbicide 2,4-D Using WO_3/TiO_2 and Fe_2O_3/TiO_2 as catalysts. Water, Air, Soil & Pollut. 228: 379.

Macías-Sánchez, J.J., L. Hinojosa-Reyes, A. Caballero-Quintero, W. de la Cruz, E. Ruiz-Ruiz, A. Hernández-Ramírez and J.L. Guzmán-Mar. 2015. Synthesis of nitrogen-doped ZnO by sol-gel method: characterization and its application on visible photocatalytic degradation of 2,4-D and picloram herbicides. Photochem. Photobiol. Sci. 14: 536–42.

Maya-Treviño, L., N.A. Ramos-Delgado, L. Hinojosa-Reyes, J.L. Guzmán-Mar, M. Ignacio Maldonado and A. Hernández-Ramírez. 2014. Activity of the $ZnO-Fe_2O_3$ catalyst on the degradation of dicamba and 2,4-D herbicides using simulated solar light. Ceram. Int. 40: 8701–8708.

Maya-Treviño, M.L., M. Villanueva-Rodríguez, J.L. Guzmán-Mar, L. Hinojosa-Reyes and A. Hernández-Ramírez. 2015. Comparison of the solar photocatalytic activity of $ZnO-Fe_2O_3$ and $ZnO-Fe^\circ$ on 2,4-D degradation in a CPC reactor. Photochem. Photobiol. Sci. 14: 543–549.

Mendiola-Alvarez, S.Y., J.L. Guzmán-Mar, G. Turnes-Palomino, F. Maya-Alejandro, A. Hernández-Ramírez and L. Hinojosa-Reyes. 2017. UV and visible activation of Cr(III)-doped TiO_2 catalyst prepared by a microwave-assisted sol-gel method during MCPA degradation. Environ. Sci. Pollut. R. 24: 12673–12682.

Mendiola-Alvarez, S.Y., A. Hernández-Ramírez, J.L. Guzmán-Mar, M.L. Maya-Treviño, A. Caballero-Quintero and L. Hinojosa-Reyes. 2019a. A novel P-doped $Fe_2O_3-TiO_2$ mixed oxide: Synthesis, characterization and photocatalytic activity under visible radiation. Catal. Today 328: 91–98.

Mendiola-Alvarez, S.Y., A. Hernández-Ramírez, J.L. Guzmán-Mar, L.L. Garza-Tovar and L. Hinojosa-Reyes. 2019b. Phosphorous-doped TiO_2 nanoparticles: synthesis, characterization, and visible photocatalytic evaluation on sulfamethazine degradation. Environ Sci. Pollut. Res. 26: 4180–4191.

Mendiola-Alvarez, S.Y., J. Araña, J.M.D. Rodríguez, A. Hernández-Ramírez, G.T. Palomino, C.P. Cabello and L. Hinojosa-Reyes. 2021. Comparison of photocatalytic activity of $\alpha Fe_2O_3-TiO_2/P$ on the removal of pollutants on liquid and gaseous phase. J. Environ. Chem. Eng. 9: 104828.

Montoya-Zamora, J.M., A. Martínez-de la Cruz, E. López-Cuéllar and F.P. González. 2020. BiOBr photocatalyst with high activity for NOx elimination. Adv. Powder Technol. 31: 3618–3627.

Murillo-Sierra, J.C., A. Hernández-Ramírez, Z.Y. Zhao, A. Martínez-Hernández and M.A. Gracia-Pinilla. 2021a. Construction of direct Z-scheme WO_3/ZnS heterojunction to enhance the photocatalytic degradation of tetracycline antibiotic. J. Environ. Chem. Eng. 9: 105111.

Murillo-Sierra, J.C., A. Hernández-Ramírez, L. Hinojosa-Reyes and J.L. Guzmán-Mar. 2021b. A review on the development of visible light-responsive WO_3-based photocatalysts for environmental applications. Chem. Eng. J. Adv. 5: 100070.

Nahar, S., M.F.M. Zain, A.A.H. Kadhum, H.A. Hasan and M. Hasan. 2017. Advances in photocatalytic CO_2 reduction with water: a review. Mater. 10: 629.

Nava-Núñez, M.Y., E. Jimenez-Relinque, M. Grande, A. Martínez-de la Cruz and M. Castellote. 2020. Photocatalytic BiOX mortars under visible light irradiation: Compatibility, NOx efficiency and nitrate selectivity. Catal. 10: 226.

Nguyen, V.H., B.S. Nguyen, C.W. Huang, T.T. Le, C.C. Nguyen, T.T.N. Le and Q. Van Le. 2020. Photocatalytic NOx abatement: Recent advances and emerging trends in the development of photocatalysts. J. Cleaner Prod. 121912.

Nie, N., L. Zhang, J. Fu, B. Cheng and J. Yu. 2018. Self-assembled hierarchical direct Z-scheme g-C_3N4/ZnO microspheres with enhanced photocatalytic CO_2 reduction performance. Appl. Surf. Sci. 441: 12–22.

Nikokavoura, A. and C. Trapalis. 2018. Graphene and g-C_3N_4 based photocatalysts for NOx removal: a review. Appl. Surf. Sci. 430: 18–52.

Núñez, M.N., A. Martínez-de la Cruz and E. López-Cuéllar. 2019. Preparation of BiOI microspheres in 2-propanol/ ethylene glycol by microwave method with high visible-light photocatalytic activity. Res. Chem. Intermed. 45: 1475−1492.

Núñez-Salas, R.E. 2021. Eliminación de bacterias patógenas y degradación de ciprofloxacino utilizando el fotocatalizador $ZnO/FeTiO_3$. Ph.D. Thesis, Universidad Autónoma de Nuevo León, Monterrey, México.

Núñez-Salas, R.E., A. Hernández-Ramírez, V. Santos-Lozano, L. Hinojosa-Reyes, J.L. Guzmán-Mar, M.A. Gracia-Pinilla and M.L. Maya-Treviño. 2021a. Synthesis, characterization, and photocatalytic performance of $FeTiO_3/ZnO/MnS$ on ciprofloxacin degradation. J. Photochem. Photobiol. A 411: 113186.

Núñez-Salas, R.E., J. Rodríguez-Chueca, A. Hernández-Ramírez, E. Rodríguez and M.L. Maya-Treviño. 2021b. Evaluation of B-ZnO on photocatalytic inactivation of *Escherichia coli* and *Enterococcus* sp. J. Environ. Chem. Eng. 9: 104940.

Occupational Health Branch California Department of Public Health. Understanding Toxic Substances. An Introduction to Chemical Hazards in the Workplace; State of California Department of Public Health Department of Industrial Relations. Online: https://www.cdph.ca.gov/Programs/CCDPHP/DEODC/OHB/HESIS/CDPH%20 Document%20Library/introtoxsubstances.pdf (accessed on 9 March 2021).

Onkundi-Nyangaresi, P., Y. Qin, G. Chen, B. Zhang, Y. Lu and L. Shen. 2019. Comparison of UV-LED photolytic and UV-LED/TiO_2 photocatalytic disinfection for *Escherichia coli* in water. Catal. Today 335: 200–207.

Orimolade, B.O., B.A. Koiki, M.G. Peleyeju and O.A. Arotiba. 2019. Visible light driven photoelectrocatalysis on a FTO/BiVO$_4$/BiOI anode for water treatment involving emerging pharmaceutical pollutants. Electrochim. Acta 307: 285–292.

Padervand, M., E. Lichtfouse, D. Robert and C. Wang. 2020. Removal of microplastics from the environment. A review. Environ. Chem. Lett. 18: 807–828.

Patil, S.B., P.S. Basavarajappa, N. Ganganagappa, M.S. Jyothi, A.V. Raghu and K.R. Reddy. 2019. Recent advances in non-metals-doped TiO$_2$ nanostructured photocatalysts for visible-light driven hydrogen production, CO$_2$ reduction and air purification. Int. J. Hydrog. Energy 44: 13022–13039.

Pérez-Nicolás, M., I. Navarro-Blasco, J.M. Fernández and J.I. Alvarez. 2017. Atmospheric NOx removal: Study of cement mortars with iron- and vanadium-doped TiO$_2$ as visible light-sensitive photocatalysts. Constr. Build. Mater. 149: 257–271.

Pino-Sandoval, D., M. Villanueva-Rodríguez, M.E. Cantú-Cárdenas and A. Hernández-Ramírez. 2020. Performance of Ag-Cu/TiO$_2$ photocatalyst prepared by sol-gel method on the inactivation of *Escherichia coli* and *Salmonella typhimurium*. J. Environ. Chem. Eng. 8: 104539.

Radhika, N.P., R. Selvin, R. Kakkar and A. Umar. 2016. Recent advances in nano-photocatalysts for organic synthesis, Arab. J. Chem. 12: 4550–4578.

Ramos-Delgado, N. 2013. Activación solar del catalizador WO$_3$/TiO$_2$ para la degradación fotocatalítica de los plaguicidas malatión y 2,4-D. PhD Thesis, Universidad Autónoma de Nuevo León, N.L. Mexico.

Ramos-Delgado, N.A., L. Hinojosa-Reyes, J.L. Guzmán-Mar, M.A. Gracia-Pinilla and A. Hernández-Ramírez. 2013a. Synthesis by sol–gel of WO$_3$/TiO$_2$ for solar photocatalytic degradation of malathion pesticide. Catal Today 209: 35–40.

Ramos-Delgado, N.A., M.A. Gracia-Pinilla, L. Maya-Treviño, L. Hinojosa-Reyes, J.L. Guzmán-Mar and A. Hernández-Ramírez. 2013b. Solar photocatalytic activity of TiO$_2$ modified with WO$_3$ on the degradation of an organophosphorus pesticide. J. Hazard. Mater. 263: 36–44.

Rodríguez, M.H., E.P. Melián, O.G. Díaz, J. Araña, M. Macías, A.G. Orive and J.D. Rodríguez. 2016. Comparison of supported TiO$_2$ catalysts in the photocatalytic degradation of NOx. J. Mol. Catal. A: Chem. 413: 56–66.

Rueda-Salaya, L., A. Hernández-Ramírez, L. Hinojosa-Reyes, J.L. Guzmán-Mar, M. Villanueva-Rodríguez and E. Sánchez-Cervantes. 2020. Solar photocatalytic degradation of diclofenac aqueous solution using fluorine doped zinc oxide as catalyst. J. Photochem. Photobiol., A 391: 112364.

Salazar-Beltrán, D., L. Hinojosa-Reyes, F. Maya-Alejandro, G. Turnes-Palomino, C. Palomino-Cabello, A. Hernández-Ramírez and J.L. Guzmán-Mar. 2019. Automated on-line monitoring of the TiO$_2$-based photocatalytic degradation of dimethyl phthalate and diethyl phthalate. Photochem. Photobiol. Sci. 18: 863–870.

Simoneau, C., L. Van den Eede and S. Valzacchi. 2012. Identification and quantification of the migration of chemicals from plastic baby bottles used as substitutes for polycarbonate. Food. Addit. Contam. 29: 469–480.

Singh, P., S. Saengerlaub, A.A. Wani and H.C. Langowski. 2012. Role of plastics additives for food packaging. Pigment. Resin Technol. 41: 368–379.

Sohn, Y., W. Huang and F. Taghipour. 2017. Recent progress and perspectives in the photocatalytic CO$_2$ reduction of Ti-oxide-based nanomaterials. Appl. Surf. Sci. 396: 1696–1711.

Tang, Y., X. Yin, M. Mu, Y. Jiang, X. Li, H. Zhang and T. Ouyang. 2020. Anatase TiO$_2$@MIL-101(Cr) nanocomposite for photocatalytic degradation of bisphenol A. Colloids Surf. A 596: 124745.

Veerakumara, P., A. Sangili, K. Saranya, A. Pandikumar and K.C. Lin. 2021. Palladium and silver nanoparticles embedded on zinc oxide nanostars for photocatalytic degradation of pesticides and herbicides. Chem. Eng. J. 410: 128434.

Vigil-Castillo, H.H., A. Hernández-Ramírez, J.L. Guzmán-Mar, N.A. Ramos-Delgado and M. Villanueva-Rodríguez. 2019. Performance of Bi$_2$O$_3$/TiO$_2$ prepared by sol-gel on p-Cresol degradation under solar and visible light. Environ. Sci. Pollut. Res. 26: 4215–4223.

Villanueva-Rodríguez, M., R. Bello-Mendoza, A. Hernández-Ramírez and E.J. Ruiz-Ruiz. 2019. Degradation of anti-inflammatory drugs in municipal wastewater by heterogeneous photocatalysis and electro-Fenton process. Environ. Tech. 40: 2436–2445.

Walpitagama, M., M. Carve, A.M. Douek, C. Trestrail, Y. Bai, J. Kaslin and D. Wlodkowic. 2019. Additives migrating from 3D-printed plastic induce developmental toxicity and neuro-behavioral alterations in early life zebrafish (*Danio rerio*). Aquat. Toxicol. 213: 105227.

Wang, J., S. Lin, N. Tian, T. Ma, Y. Zhang and H. Huang. 2021. Nanostructured metal sulfides: Classification, modification strategy, and solar-driven CO$_2$ reduction application. Adv. Funct. Mater. 31: 2008008.

Wang, R., J. Shen, K. Sun, H. Tang and Q. Liu. 2019. Enhancement in photocatalytic activity of CO$_2$ reduction to CH$_4$ by 0D/2D Au/TiO$_2$ plasmon heterojunction. Appl. Surf. Sci. 493: 1142–1149.

Wang, Y., K. Hu, Z. Yang, C. Ye, X. Li and K. Yan. 2020. Facile synthesis of porous ZnO nanoparticles efficient for photocatalytic degradation of biomass-derived bisphenol A under simulated sunlight irradiation. Front. Bioeng. Biotechnol. 8: 616780.

Wang, X., W. Bi, P. Zhai, X. Wang, H. Li, G. Mailhot and W. Dong. 2016. Adsorption and photocatalytic degradation of pharmaceuticals by BiOClxIy nanospheres in aqueous solution. Appl. Surf. Sci. 360: 240–251.

Wen, X.J., C.G. Niu, L. Zhang and G.M. Zeng, 2017. Fabrication of SnO_2 Nanopaticles/BiOI n−p heterostructure for wider spectrum visible-light photocatalytic degradation of antibiotic oxytetracycline hydrochloride. ACS Sustainable Chem. Eng. 5: 5134–5147.

World Health Organization, WHO, Global priority list of antibiotic-resistant bacteria to guide research, discovery and development of new antibiotics, Geneva, Switzerland, 2017. Available at http://www.who.int/medicines/publications/WHO-PPL-Short_Summary_25Feb-ET_NM_WHO.pdf?ua=1 (accessed 15 March 2021).

World Health Organization, WHO, Plan de acción mundial sobre la resistencia a los antimicrobianos, Document Production Services, Geneva, Switzerland, 2016. Available at https://apps.who.int/iris/bitstream/handle/10665/255204/9789243509761-spa.pdf?sequence=1 (accessed 15 March 2021).

Wu, M., X. He, B. Jing, T. Wang, C. Wang, Y. Qin, Z. Ao, S. Wang and T. An. 2020. Novel carbon and defects co-modified g-C_3N_4 for highly efficient photocatalytic degradation of bisphenol A under visible light. J. Hazard. Mater. 384: 121323.

Xu, F., J. Zhang, B. Zhu, J. Yu and J. Xu. 2018. $CuInS_2$ sensitized TiO_2 hybrid nanofibers for improved photocatalytic CO_2 reduction. Appl. Catal. B 230: 194–202.

Yang, J., J. Hao, S. Xu, Q. Wang, J. Dai, A. Zhang and X. Pang. 2019. $InVO_4$/β-$AgVO_3$ nanocomposite as a direct Z-scheme photocatalyst toward efficient and selective visible-light-driven CO_2 reduction. ACS Appl. Mater. Interfaces 11: 32025–32037.

Ye, L., Y. Deng, L. Wang, H. Xie and F. Su. 2019. Bismuth-based photocatalysts for solar photocatalytic carbon dioxide conversion. Chem. Sus. Chem. 12: 3671–3701.

Ye, S., R. Wang, M.Z. Wu and Y.P. Yuan. 2015. A review on g-C_3N_4 for photocatalytic water splitting and CO_2 reduction. Appl. Surf. Sci. 358: 15–27.

Zhang, Y., M.M. Liu, J. Chen, S. Fang and P. Zhou. 2021. Recent advances in Cu_2O-based composites for photocatalysis: A review. Dalton Trans. 50: 4091–4111.

Zhu, G., S. Li, J. Gao, F. Zhang, C. Liu, Q. Wang and M. Hojamberdiev. 2019. Constructing a 2D/2D $Bi_2O_2CO_3$/$Bi_4O_5Br_2$ heterostructure as a direct Z-scheme photocatalyst with enhanced photocatalytic activity for NOx removal. Appl. Surf. Sci. 493: 913–925.

Zyoud, A., R. Alkowni, O. Yousef, M. Salman, S. Hamdan, M.H. Helal, S.F. Jaber and H.S. Hilal. 2019. Solar light-driven complete mineralization of aqueous gram-positive and gram-negative bacteria with ZnO photocatalyst. Sol. Energy. 180: 351–359.

Chapter 7

Photocatalytic and Photoelectrocatalytic Treatment of Water and Wastewater

Amit Kumar Singhal, Pilar Fernández-Ibáñez and *John Anthony Byrne**

1. Introduction

Most of the industries are posing a threat to life for all living species on this planet as they generate a large amount of contaminated water containing organic pollutants such as dyes, surfactants, phenolic compounds, hydrocarbons, organohalides, plasticizers, etc. Their chemical stability, water solubility, toxicity, carcinogenicity or difficulty to be degraded, make them a serious health hazard. Therefore, the removal of these pollutants from industrial wastewater is a major concern for industries. Nanotechnology has the potential to eliminate these contaminants through photocatalysis or photoelectrocatalysis in a more efficient and cost-effective manner, which can be a great help to industries.

Heterogeneous photocatalysis has been demonstrated to be a promising and low-cost method for water remediation. In this photochemical process, when a photocatalyst, a nanomaterial that is a semiconductor in nature, is irradiated by photons having energy equal to or greater than its bandgap, electrons (e^-) are excited to the conduction band, leaving holes (h^+) in the valence band. Thus, produced e^-/h^+ pairs can either recombine, releasing the energy as heat or radiation or they can react at the surface interface in redox reactions.

The terms *photokatalyse* and *photokatalytisch* were first coined in 1910 in the "Textbook on Photochemistry" by J. Plotnikow. The word photocatalyst was later introduced in France in 1913 by Landau (Serpone et al. 2012). Photocatalysis, more specifically photoelectrocatalysis, came to the fore when Fujishima and Honda in 1972 reported in Nature the photodriven water splitting on a TiO_2 electrode under UV irradiation (Fujishima and Honda 1972). In 1977, Frank and Bard reported the photocatalytic decomposition of cyanide in the presence of aqueous TiO_2 powder (Frank and Bard 1977). Later, Matsunaga et al. (1985) reported the photocatalytic inactivation of microorganisms using platinum-doped TiO_2. Since then, photocatalysis and photoelectrocatalysis have been widely studied and explored with emphasis on energy and environmental applications including water splitting, CO_2 reduction, organic pollutant degradation, microbial inactivation, etc.

Nanotechnology and Integrated Bio-Engineering Centre, School of Engineering, Faculty of Computing and Engineering, University of Ulster, Newtownabbey, Northern Ireland, UK.
Emails: singhal-a@ulster.ac.uk; p.fernandez@ulster.ac.uk
* Corresponding author: j.byrne@ulster.ac.uk

There are thousands of published articles on the subject. A search on Web of Science with the term 'photocatalysis' yields more than 58,000 returns. While there are commercial examples of photocatalysis for self-cleaning glass, paints and concrete, and some examples of air purifiers, there are few if any real examples of water or wastewater treatment systems using photocatalysis at an industrial scale. There are several reasons for this. The water industry is extremely conservative in terms of the use of new technologies. Water and wastewater treatment processes mainly consist of the transfer of pollutants from one phase to another. For example, adsorption is a phase transfer method for removing substances from gases or liquid phases that is commonly utilized in wastewater treatment methods. Photocatalysis and other advanced oxidation processes are degradative processes, whereby organic pollutants are broken down by a series of intermediate steps towards carbon dioxide and water. The problem is that the intermediate compounds may be more toxic than the initial compounds. Therefore, the photocatalytic (or photoelectrocatalytic) treatment of water or wastewater is very difficult to control because different compounds can exhibit various degradation kinetics, etc. It can be argued that photocatalysis will never really find large-scale applications for water or wastewater treatment, but it is an area for exciting research and perhaps, combined with online monitoring and feedback control, photocatalysis will find niche applications at an industrial scale.

2. Mechanisms of Photocatalysis and Photoelectrocatalysis

In this chapter, the use of heterogeneous photocatalysis using a solid photocatalyst suspended in water will be described. The photocatalyst is a semiconductor characterized by its energy band structure and consists of a valence band (VB) and a conduction band (CB). At low temperature, a semiconductor has a fully occupied valence band and an empty conduction band, acting as an insulator at 0 K. The energy difference between the lowest energy level of the conduction band (U_{CB}) and the uppermost energy level of the valence band (U_{VB}) is termed as the bandgap (E_g). In photocatalysis, the semiconductor is excited by the absorption of electromagnetic radiation with energy greater than or equal to the bandgap energy. The absorption of this photon energy results in the excitation of an electron from the valence band to the conduction band, leaving a positive hole in the valence band. Thus, one photon can generate one electron-hole pair. The electron and hole have different electrochemical potentials, which can drive redox reactions at the surface of the semiconductor (Fig. 7.1).

For water and wastewater treatment by photocatalysis, the generation of Reactive Oxygen Species (ROS) is needed. In the past few decades, many materials such as ZnO, TiO_2, SnO_2, CdS, Fe_2O_3, C_3N_4, Cu_2O, $BiVO_4$, Bi_2WO_6, WO_3, etc., have been investigated for their efficacy for various photocatalytic applications. The most commonly studied photocatalyst for water treatment applications is titanium dioxide (TiO_2), because of its efficient photoactivity, chemical stability, cost effectiveness and as it is environmentally friendly. The conduction band electron has a negative electrochemical potential (versus the normal hydrogen scale) and can reduce molecular oxygen via a one-electron reduction to yield the superoxide radical anion ($O_2^{\cdot-}$). Further electron transfer (reduction reactions) can generate hydrogen peroxide (H_2O_2), hydroxyl radical (HO$^{\cdot}$) and water. The valence band hole for TiO_2 has a very positive electrochemical reduction potential and can oxidize water to yield hydroxyl radicals (HO$^{\cdot}$). With a photocatalyst under bandgap irradiation in oxygenated water, a complex redox process takes place, leading to the formation of various ROS. In particular, the hydroxyl radical is a powerful and unselective oxidizing species that can attack and degrade a wide range of organic compounds found in water. There are thousands of published research papers reporting the degradation of a wide range of problematic pollutants including pesticides and emerging contaminants; the inactivation of disinfection-resistant microorganisms is also reported. The most interesting aspect of photocatalysis is that it only requires photons (in most cases UV), the photocatalyst, water and oxygen. If driven by solar irradiation (comprising around 4% UV), it can be considered a truly clean technology for water and wastewater treatment. TiO_2 is a wide bandgap semiconductor (E_g = 3.2 eV for anatase) and requires UV excitation (λ < 400 nm).

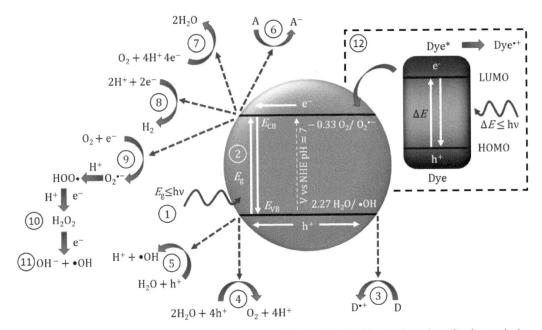

Figure 7.1. Scheme of the photocatalytic mechanism on a TiO$_2$ particle. (1) Photon absorption, (2) photoexcitation and recombination, (3) donor electron transfer, (4) oxygen evolution reaction, (5) oxidation of water to form hydroxyl radical, (6) electron transfer to an electron acceptor, (7) oxygen and proton reduction to water, (8) proton reduction to hydrogen, (9) oxygen reduction to superoxide, (10) formation of hydrogen peroxide, (11) formation of hydroxyl radical, (12) dye sensitization (*excited state) and electron transfer to the conduction band. Reprinted with permission from McMichael et al. 2021.

Hundreds of research groups around the world are searching for the best photocatalyst that can effectively utilize visible photons and there are thousands of papers published on the topic (Pelaez et al. 2012). While many papers (e.g., Li et al. 2021, Reddy et al. 2020, Zheng et al. 2019) claim novel materials with dramatically enhanced or extremely efficient photocatalytic efficiency, few (if any) will outperform TiO$_2$ under solar or solar simulated irradiation for the degradation of organic pollutants or the inactivation of microorganisms.

With photoelectrocatalysis, the photocatalyst is used to fabricate a photoelectrode. The electrode is generally produced by immobilizing a semiconductor material onto an electrically conductive supporting substrate such as metals, carbon or transparent films. Optically transparent electrodes are produced by the deposition of indium tin oxide or fluorine tin oxide onto glass. In photoelectrochemical cells, the photoanode is normally an n-type semiconductor linked to a metallic counter electrode. An alternative arrangement is a p-type photocathode linked to a metallic anode. Due to the poor stability of p-type materials in water, the former configuration of photoanode is more common.

The mechanism of photoelectrocatalysis with an n-type photoanode linked to a metallic counter electrode is shown in Fig. 7.2.

The mechanism at the photoanode is similar to that of photocatalysis. The main difference is that the application of an external electrical bias (potential) induces band bending at the photoanode causing conduction band electrons to migrate towards the conducting support electrode and holes to move towards the solution interface, thereby increasing the lifetime of the electron-hole pairs, which results in improved reaction rates. At the counter electrode, additional ROS can be produced, enhancing the process. The use of an external electrical bias to assist the process means that n-type materials that do not have the necessary conduction band edge potential to drive the oxygen reduction process ($E^0_{(O_2/O_2^{-\bullet})} = -0.33$ V) can be utilized in photoelectrocatalytic processes.

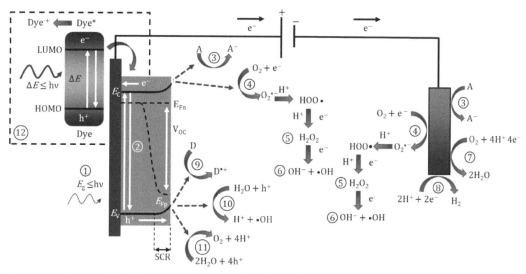

Figure 7.2. Diagram of the PEC process and pathways for radical production using a photoanode and a non-semiconducting counter electrode. (1) Photon absorption, (2) photoexcitation and recombination, (3) electron transfer to an electron acceptor, (4) oxygen reduction to superoxide, (5) formation of hydrogen peroxide, (6) formation of hydroxyl radical, (7) oxygen reduction to water, (8) proton reduction to hydrogen, (9) donor electron transfer, (10) oxidation of water to form hydroxyl radical, (11) oxygen evolution reaction, (12) dye sensitization (Dye*: excited state) and electron transfer to the conduction band. Reprinted with permission from McMichael et al. 2021.

For further reading on the mechanism of photocatalysis and photoelectrocatalysis, the reader is referred to the many reviews on the subject, e.g., Pelaez et al. (2012), Byrne et al. (2015), and McMichael et al. (2021).

3. Nanophotocatalytic Materials

The use of nanostructured materials in photocatalysis has advantages, such as the increase in surface area to volume ratio in relation to micromaterials. Other effects may be observed, e.g., size quantization, resulting in a blue shift of the bandgap and an increase of the absorption coefficient. A wide range of nanomaterials has been reported for application in photocatalysis. One of the most commonly employed commercially available photocatalysts is Evonik Aeroxide P25 TiO_2. This is a mixture of anatase and rutile (ca. 80:20), with a primary particle size of ca. 25 nm. However, the use of nanoparticles in free suspension for the treatment of water or wastewater is not convenient due to the need of removing the particles after the treatment. Various approaches have been investigated for the immobilization of P25 on different supporting substrates to give mesoporous thick films. These immobilized films have been used at different scales for investigating the degradation of model organic pollutants in water and the inactivation of microorganisms. For example, Byrne et al. (1998) examined a range of different approaches to the immobilization of TiO_2 including spray and electrophoretic coating (on electrically conducting supporting substrates). The photocatalytic efficiency of the TiO_2 coatings was compared using the degradation of phenol in an aqueous solution as a standard test system. There was no significant difference observed between spray or electrophoretic coating of the P25 powder, although the electrophoretic coating was more reproducible. Thick films of P25 on a Ti alloy were also used as photoanodes in a photoelectrochemical system for the combined degradation of organic pollutants and the recovery of metals by electrochemical reduction at the counter electrode (Byrne et al. 1999). Nanoparticle P25 TiO_2 has also been immobilized on tin oxide coated glass (optically transparent electrodes) for use in photoelectrocatalytic applications. The use of such electrodes allows the irradiation of the photocatalyst avoiding losses by absorption of photons in the water matrix, resulting in increased photoelectrocatalytic efficiency due to the efficient harvesting of photons. The ease of reusability of the material is another advantage of photoelectrocatalytic degradation over traditional photocatalysis. These photoelectrocatalytic systems show an improvement in efficiency with the application of an external electrical bias, and they can be used to combine the oxidative degradation of organic contaminants at the photoanode with the electrochemical reduction and recovery of metals at the cathode. P25 has been used by many research groups as a benchmark comparison for photocatalytic activity under UV or solar/solar simulated irradiation. Nanoengineering of TiO_2 has been explored for improving photocatalytic efficiency. For example, Wadhwa et al. (2011) studied the formation of dispersed TiO_2 nanotubes formed by the hydrothermal treatment of TiO_2. These nanotubes had a very high surface area (400 m^2 g^{-1}) but were amorphous and not photocatalytically active. Annealing to 400°C in air resulted in tube collapse to nanorods, with 30% remaining as tubes. This sample showed a greater UV photocatalytic activity than P25 for the oxidation of formic acid in water. However, nanocrystals obtained by annealing at 700°C, following the collapse of the tubes and rods, resulted in the best UV photocatalysts for the degradation of phenol. This latter finding highlights that some photocatalysts appear better than others depending on the model pollutant employed in experiments. Ryu and Choi (2008) reported a detailed comparison of different photocatalytic materials for the degradation of a range of different pollutants, however not being able to find an easy correlation between the physicochemical properties and efficiencies of the photocatalysts. They proposed that the photocatalytic activity assessment should be carried out using the following four substrates: phenol, dichloroacetic acid, tetramethylammonium and trichloroethylene.

Another interesting approach to nanoengineering of TiO_2 is the formation of self-organized aligned TiO_2 nanotubes by anodizing titanium metal in the presence of fluoride. If anodized in acid or basic electrolytes, a compact oxide layer will form on the titanium surface. In the presence of fluoride ion, a titanium fluoride complex is formed that is more soluble than the hydroxide, and the oxide layer grows in a tubular shape (see Fig. 7.3).

Dale et al. (2010) compared the efficiency of the photoelectrocatalytic oxidation of phenol and formic acid as model pollutants with different photoanodes including aligned titania nanotubes, compact oxide and nanoparticulate P25 films. With an applied electrochemical bias, the titania nanotube electrodes were the most efficient. The immobilization of nanoparticulate TiO_2 produces a porous, disordered film with a high number of particle-to-particle boundaries, which restrict the electron transfer and result in higher levels of charge carrier recombination. Aligned self-organized nanotube films have a specific nanoarchitecture that aids electron transfer, providing the tubes

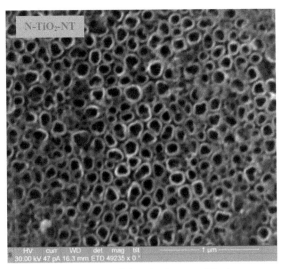

Figure 7.3. SEM of titania nanotubes formed by anodizing titanium metal in the presence of fluoride. Reprinted with permission from Pablos et al. 2017.

are straight. The tubular design of the TiO$_2$ film has a hollow structure that gives the electrolyte interface a large surface area, and the ordered form of nanotubes reduces the charge carrier recombination. If the tubes have a minimum wall thickness, band bending may be achieved aiding in charge separation and vectorial charge transfer. Pablos et al. (2017) studied the use of aligned titania nanotube electrodes for the electrochemically assisted photocatalytic disinfection of water using *E. coli* as a model microorganism. In an attempt to achieve visible light activity, the nanotubes were doped with nitrogen by annealing in NH$_3$. They also compared the nanotube electrodes with P25 nanoparticulate electrodes and N, F codoped TiO$_2$ materials, which have been reported to show visible light activity. Figure 7.4 shows the photocurrent response for the different electrodes under visible irradiation. It would appear that all electrodes, including the P25 nanoparticulate electrode, show some small visible light photocurrent. The P25 result may be explained by the fact that it is composed of 20% rutile, which has an optical gap absorbing up to 413 nm. The N-TiO$_2$ nanotube electrode gave the highest visible light photocurrent, but it should be noted that in all cases the visible light response is a mere fraction of the UV photocurrent response. For the inactivation of *E. coli*, they found that none of the materials showed any significant visible light activity, even under an applied bias, but the N-TiO$_2$ electrodes gave the best disinfection rate under UV-Vis irradiation and applied bias.

McMichael et al. (2021) utilized aligned titania nanotubes on a titanium mesh (having open area of 64%) in a cylindrical photoelectrochemical reactor, with a carbon felt counter electrode and a Compound Parabolic Collector (CPC) for the disinfection of harvested rainwater under real sun conditions in South Africa. Figure 7.5 shows the reactor design.

In this design, they utilized titania nanotubes on the Ti mesh to allow for irradiation of the photoanode and applied an external electrical potential between the photoanode and the carbon felt counter electrode. This carbon felt counter electrode is much cheaper than a platinized Ti mesh and also facilitates the formation of hydrogen peroxide by the two-electron reduction of molecular oxygen. The CPC acts to reflect diffuse and direct solar irradiation into the cylindrical reactor, effectively providing one sun irradiance over the total surface of the reactor. Culture-based and viability-based methods were used to evaluate the performance of the reactor under solar irradiation with a low average UV intensity in South African wintertime (10.7–13.3 Wm^{-2}). Disinfection was tested using *E. coli* and *P. aeruginosa* in low conductivity rainwater (70 μS cm^{-1}). The PEC

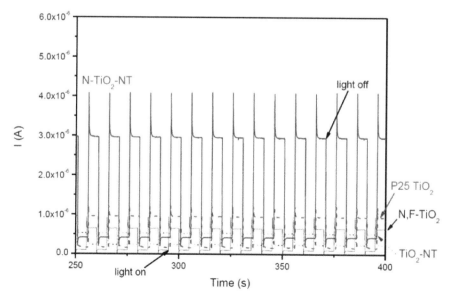

Figure 7.4. Current-time response of different photoanodes under chopped visible irradiation. N-TiO$_2$ NT = nitrogen-doped titania nanotubes, TiO$_2$-NT = titania nanotubes. Reprinted with permission from Pablos et al. 2017.

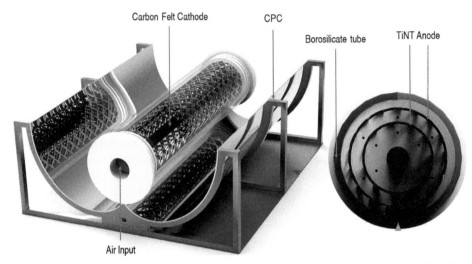

Figure 7.5. Photoelectrochemical reactor with aligned titania nanotubes on Ti mesh as the photoanode, carbon felt cathode and compound parabolic collector. Reprinted with permission from McMichael 2021.

reactor demonstrated a significant improvement in the reduction of both organisms compared to photoinactivation (solar disinfection) alone. Using culture-based methods, the PEC reactor achieved an average 5.5 log reduction for *E. coli* and 5.8 log reduction in *P. aeruginosa*. More importantly, the viability-based methods showed that the microorganisms could still be potentially viable after solar disinfection even when the culture-based methods indicate complete inactivation. However, the PEC reactor showed a much greater reduction in viability following the treatment. While a PEC reactor is more complicated and costly than a solar disinfection reactor, it will provide a faster and more effective disinfection treatment, particularly for organisms that have a greater resistance to solar disinfection.

Another approach to improve the photocatalytic activity is to combine materials to make composites or heterostructures that may improve the light-harvesting properties, the charge carrier separation or the electron transfer at the interface. One example is to combine TiO_2 with graphene to create a composite. The preparation of graphene, a one-atom-thick sheet of sp^2- bonded carbon atoms in a hexagonal two-dimensional lattice, was first reported in 2004 (Novoselov et al. 2004). Graphene is a 2D material with a very high surface area ($2600 \text{ m}^2 \text{ g}^{-1}$), ballistic electronic conduction, exceptional thermal conduction and good chemical and thermal stability. The utilization of graphene as a two-dimensional support to anchor catalyst nanoparticles (NPs) and facilitate electron transport opens up new possibilities for designing the next-generation catalysts. Graphene is commonly made using Graphene Oxide (GO) as a precursor. GO is prepared by oxidation of graphitic carbon using Hummers' method (Hummers and Offeman 1958) along with an exfoliating step, e.g., high-intensity ultrasound. GO is then reduced back towards single or multilayer graphene using thermal or chemical reduction. This reduction of GO to graphene is never complete and the partially reduced GO is referred to as reduced Graphene Oxide (rGO). Williams et al. (2008) reported that GO can be photocatalytically reduced to rGO by TiO_2 under UV irradiation in the presence of a hole scavenger. Furthermore, the interaction between TiO_2 particles and graphene sheets prevents the re-stacking of exfoliated sheets of graphene. Therefore, the photocatalytic reduction of GO by TiO_2 presents a clean and safe route to the formation of TiO_2-RGO composites.

TiO_2-RGO composites provide potential advantages for photocatalysis, e.g., conduction band electrons can be rapidly transferred to RGO, therefore inhibiting charge carrier recombination. RGO may act as an electrocatalyst for the oxygen reduction reaction and RGO may present a high surface area for adsorption of pollutants, possibly yielding a synergistic effect for pollutant removal. Furthermore, visible light activity could be enhanced because the GO/RGO material absorbs

UV/Vis, producing ROS. Fernandez et al. (2015) investigated the use of TiO_2-rGO composites for the disinfection of water under real sun irradiation and found that the rate of inactivation of *E. coli* and *F. solani* spores was dramatically enhanced as compared with solar disinfection alone. There was a small improvement in the rate of solar photocatalytic inactivation of *E. coli* with TiO_2-rGO as compared with P25, but there was no real difference observed between the inactivation rate for *F. solani* spores when comparing TiO_2 with TiO_2-rGO. There was some evidence of singlet oxygen formation with the TiO_2-rGO composite under visible irradiation. Further research under lower light intensities in the laboratory determined that the main ROS involved in the disinfection process are hydrogen peroxide, hydroxyl radicals and singlet oxygen under UV–Vis irradiation, and only singlet oxygen under visible irradiation. Chlorine can be produced by oxidizing Cl by mid-gap states in the presence of chloride. Following this, the chlorine might react with H_2O_2 to produce singlet oxygen (Cruz-Ortiz et al. 2017). The different mechanisms involved are shown schematically in Fig. 7.6.

The search for new and more efficient photocatalytic materials continues. Some interesting candidate materials for water treatment and disinfection include tungsten trioxide (WO_3), graphitic carbon nitride (g-C_3N_4) and bismuth vanadate ($BiVO_4$). WO_3 is an n-type transition metal oxide semiconductor attracting more attention in recent decades due to its narrow bandgap (E_g) of ~ 2.7 eV. This makes it capable of utilizing approximately 12% of the solar spectrum with an absorption band extended up to 460 nm. It is an attractive candidate for water treatment and disinfection due to its non-toxicity, photochemically and thermally stable nature and good stability in acidic solutions. However, just like other semiconducting photocatalysts, WO_3 has disadvantages including slow hole migration kinetics, slow interfacial charge transfer at the semiconductor/electrolyte interface and therefore, faster recombination rate. Despite a favorable overall bandgap of ~ 2.7 eV, it has an unfavorable conduction band edge potential that gives poor photoreduction properties. The valence band edge of WO_3 is more positive than the oxidation potential (H_2O/O_2) of water, which shows a very high activity to oxidize water to yield oxygen and photocatalytically oxidize organic compounds (via hydroxyl radical formation). Different strategies have been adopted to improve the WO_3 photocatalytic activity, such as enhancing the charge transfer rate and charge separation. This can be accomplished by controlling the morphology of the material through size quantization or by elemental doping/cocatalyst loading and also by the construction of heterojunctions with other semiconducting materials. Hamilton et al. (2008) investigated WO_3 thin films prepared from different tungsten precursors for their photoelectrochemical performance

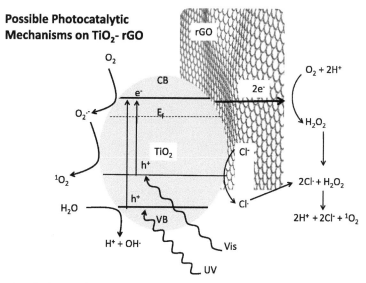

Figure 7.6. Proposed mechanism of photocatalytic generation of ROS on TiO_2-rGO under UV and visible irradiation, and in the presence of chloride. Reprinted with permission from Cruz-Ortiz et al. 2017.

under simulated solar irradiation in comparison with a sol-gel synthesized TiO_2 thin film. However, both WO_3 and TiO_2 thin films were reported to exhibit comparable efficiency (IPCE%) of 7% under UV-Vis irradiation but at different peak wavelengths (TiO_2 at 355 nm and WO_3 at 380 nm) with WO_3 showing some visible IPCE (~ 1%) (Hamilton et al. 2008).

To demonstrate the size quantization effect, single-crystal WO_3 nanosheets were engineered by exfoliating tetragonal $Bi_2W_2O_9$ to W_2O_6 and further to WO_3 nanosheets, which leads to changes in crystal structure and optical properties. The bandgap (E_g) shifts from 2.68 eV (bulk-WO_3) to 2.81 eV in $Bi_2W_2O_9$, consisting of W_2O_6 layers with 0.75 nm thickness. Further exfoliation of $Bi_2W_2O_9$ to WO_3 nanosheets shifts the bandgap to 2.88 eV with changes in the crystal structure, i.e., formation of monoclinic WO_3 from tetragonal $Bi_2W_2O_9$ (Waller et al. 2012). It is known that the surface atomic and related electronic structure of a photocatalyst has a significant impact on its reactivity. Nanostructures with specified exposed facets are believed to be more active than other facets that can be obtained through crystal facet engineering; this is becoming a popular technique for optimizing the reactivity of the photocatalysts for specific reactions. Xie et al. (2012) investigated the different crystal facets of WO_3 as photocatalyst for CO_2 photoreduction to CH_4 in the presence of H_2O vapor and O_2 evolution using $AgNO_3$ as electron acceptor.

Catalyst doping can significantly improve the photocatalytic properties of a nanomaterial. WO_3 has also been engineered with various cocatalysts, e.g., Pt, Mo, Fe, Pd, Ag, etc. For example, Zhang et al. (2014) reported that 2% platinum-doped WO_3 gave approximately four times the photocatalytic activity than bare WO_3. Reyes-Gil and Robinson (2013) coupled WO_3 with TiO_2 nanotubes to get composite heterostructures to enhance solar photoconversion reactions. TiO_2 nanotubes were used as a substrate to develop tubular WO_3 nanostructures with enhanced photoelectrochemical properties through electrodeposition. They were able to obtain a significant increase in IPCE from 30% (for bare WO_3) to 50% (for tubular WO_3/TiO_2 composites). They employed tubular WO_3/TiO_2 composites as photoanodes in a photoelectrochemical cell for organic pollutant degradation and simultaneous photocurrent generation for hydrogen production. Under visible light, the WO_3/TiO_2 electrode exhibited a six time increase in photocurrent density by the addition of methanol to the electrolyte while TiO_2 nanotubes did not display any noticeable change in the photocurrent density by the addition of methanol (see Fig. 7.7).

Graphitic carbon nitride (g-C_3N_4) is another photocatalytic material that has been given considerable attention as a promising photocatalytic material. It is a visible-light-driven photocatalyst

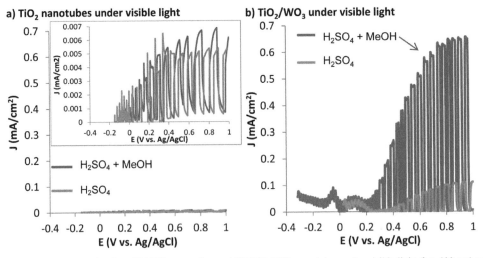

Figure 7.7. Photocurrent density of (a) TiO_2 nanotubes and (b) WO_3/TiO_2 nanotubes under visible light ($\lambda > 400$ nm) using 0.1 M H_2SO_4 as electrolyte (red curve) and adding 1 M methanol (blue curve). Reprinted with permission from Reyes-Gil and Robinson 2013.

with a narrow bandgap (~ 2.7 eV), presenting good thermal and chemical stability. Since Wang et al. (2009) reported g-C_3N_4 as a metal-free semiconductor photocatalyst for water splitting under visible light irradiation, there has been extensive research for photocatalytic applications including CO_2 reduction, organic pollutant degradation and microbial inactivation. g-C_3N_4 can be easily synthesized by thermal polymerization of various precursors such as cyanamide, melamine, dicyandiamide, thiourea, ammonium thiocyanate and urea. The photocatalytic efficiency of the bulk g-C_3N_4 is reported to be poor due to the fast recombination of photogenerated charge carriers and its small surface area (ca. ~ 10 m^2 g^{-1}). To overcome these obstacles, several approaches such as nano-engineering, metallic/non-metallic heteroatom doping, crystal structure engineering and heterostructure formation have been adopted.

Preparation of 2D nanosheets of g-C_3N_4 by exfoliation of their bulk counterparts has stimulated much interest due to their improved electrical, optical and physicochemical properties. Huang et al. (2014) reported for the first time the bacterial inactivation properties of g-C_3N_4 on *E. coli* K-12. Silica template and cyanamide as precursor were used to prepare mesoporous g-C_3N_4 whose surface area was 20 times (230 m^2 g^{-1}) larger compared to the bulk g-C_3N_4 (12 m^2 g^{-1}). They reported that the mesoporous g-C_3N_4 inactivated 100% *E. coli* (2.5×10^6 CFU mL^{-1}) after visible light irradiation for 4 hours while the bulk counterpart only inactivated 50% bacterial under the same conditions. A word of caution here is that, normally, disinfection experiment results are reported in terms of percent inactivation, which is not really appropriate and hard to analyze the actual efficiency of a photocatalyst because 99, 99.9 and 99.99% inactivation reflect only 2, 3, 4-log reduction in colony-forming units, respectively. Another novel approach 'alternated cooling and heating' was used by Kang et al. (2018) to fabricate highly crystalline g-C_3N_4 nanosheets showing high crystallinity, reduced bandgap (2.62 eV), enlarged specific surface area (88.59 m^2 g^{-1}) and enhanced electron transport ability. As a result, complete photocatalytic inactivation of 2.5×10^7 CFU mL^{-1} of *E. coli* within 120 minutes and 65% humic acid degradation after 70 minutes were observed (Kang et al. 2018).

To improve the photocatalytic properties of g-C_3N_4, several studies have been published concerning nonmetal doping (O, C, P, S, B, I, F, etc.) and metal doping (Fe, Cu, Zn, Ni, Ag, etc.) or their combination. For example, Cui et al. (2018) described the formation of sulfur-doped g-C_3N_4 microspheres using *in situ* solvothermal condensation process and trithiocyanuric acid as a sulfur source and tested its photocatalytic activity for Cr (VI) reduction. Sulfur doping was confirmed by the formation of C-S and N-S bonds into the hybrid structure suggesting improved conjugated heterocyclic structure, increase in visible light harvesting and elevated conduction band potential. As a result, sulfur-doped graphitic carbon nitride exhibited a 24 times higher reduction rate of Cr (VI) than un-doped g-C_3N_4.

In recent decades, it has been discovered that linking other semiconductors to g-C_3N_4 is an effective approach for boosting the photocatalytic activity of g-C_3N_4. Various semiconductors such as TiO_2, ZnO, Ag_3PO_4, $BiVO_4$, MoS_2, $SmVO_4$, WO_3, GO, etc. have been coupled with g-C_3N_4 to form heterostructures. For example, Du et al. (2021) recently reported a novel Z-scheme Ag_3PO_4/g-C_3N_4 heterostructure photocatalyst for *E. coli* inactivation and degradation of organic pollutants under visible light irradiation. When compared with g-C_3N_4 and Ag_3PO_4, the Ag_3PO_4/g-C_3N_4 composite with distinct concentrations of Ag_3PO_4 showed a significant bactericidal impact against *E. coli* inactivation and degradation of organic pollutants. The composite with 8 wt.% Ag_3PO_4 was superior. To investigate the potential mechanism, radical quenching experiments and Electron Spin Resonance (ESR) measurements were used, which unveiled the formation of a Z-scheme heterostructure between g-C_3N_4 and Ag_3PO_4, resulting in enhanced visible light absorption, rapid charge separation and high redox capacity. According to these findings, hole, electron and hydroxyl radicals were the most active species during *E. coli* inactivation. Figure 7.8 shows the proposed schematic mechanism.

While there are many papers on the photocatalytic properties of g-C_3N_4, more work is required to confirm its photostability. Pomilla et al. (2018) conducted an in-depth analysis of g-C_3N_4 for

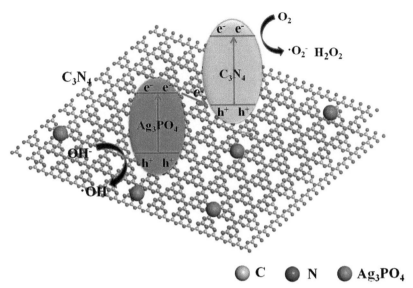

Figure 7.8. *E. coli* inactivation mechanism in the presence of Ag_3PO_4/g-C_3N_4 photocatalysts under visible light. Reprinted with permission from Du et al. 2021.

photocatalytic CO_2 reduction and discovered that C_3N_4 was photo-corroded, indicating that it was not truly catalytic. Bismuth vanadate ($BiVO_4$) has also been reported to be a photocatalytic material candidate due to its good characteristics, such as low bandgap (2.4 eV), non-toxicity, corrosion resistance and ability of visible light activity. $BiVO_4$ shows three crystal structures namely the tetragonal zircon phase (t–z), monoclinic scheelite phase (m–s) and tetragonal scheelite phase (t–s). Among them, the monoclinic phase shows a better photocatalytic activity because of lone pair distortion of the Bi 6s orbital in the $BiVO_4$ semiconductor. The first photocatalytic report on $BiVO_4$ was published by Kudo et al. (1999) and, since then, it has been extensively explored. $BiVO_4$ has been widely recognized as one of the best photoanode materials for water splitting, but it shows a very limited activity for organic degradation and water disinfection. Researchers are working to improve/align its properties so that it can be used for photocatalytic water remediation. The photocatalytic efficiency of $BiVO_4$ is well recognized to be influenced by its morphology. It is known that the crystalline phase and morphology of a material are significantly dependent on many reaction conditions such as the precursor concentration, pH, the nature of the surfactant, hydrothermal temperature and time. It has been observed that the pH of the precursor solution can alter the phase and morphology of $BiVO_4$. Tan et al. (2013) reported the effect of pH on the crystal structure and morphology of $BiVO_4$. Various structures of $BiVO_4$ were prepared by the microwave hydrothermal method at different pH values, exploring a wide range of pH from highly acidic (0.59) to highly basic (12.93) (Fig. 7.9).

The authors reported that monoclinic $BiVO_4$ crystals with octahedron and decahedron morphologies can be prepared at pH 0.59 whereas spherical and polyhedral structures can be obtained under strong acid conditions (pH 0.70–1.21), having mixed crystal phases (tetragonal and monoclinic). A pure $BiVO_4$ monoclinic phase having rodlike and dendritic structures can be prepared at a wide range of pH (4.26–9.76). At pH 3.65, a transformation from the tetragonal phase to the monoclinic one occurs. At pH above 9.76, nonstoichiometric crystals are produced, and they can be the mixed-phase $BiVO_4$ or non-$BiVO_4$ crystals (i.e., Bi_2O_3). The m–s$BiVO_4$ phase obtained at pH 7.81 exhibited a high surface area and a better photocatalytic activity than other materials for Rhodamine B (RhB) degradation (Tan et al. 2013). Again, the use of dyestuffs for the determination of photocatalytic activity may be not adequate because most dyes are photosensitive.

Like other photocatalytic materials, $BiVO_4$ has also been engineered with other metals/non-metals/compounds to improve charge transfer. For example, Zhao et al. (2016)

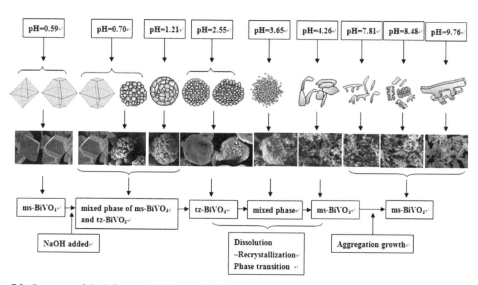

Figure 7.9. Summary of the influence of different pH values on the crystalline phase and morphology of the as-prepared BiVO₄. Reprinted with permission from Tan et al. 2013.

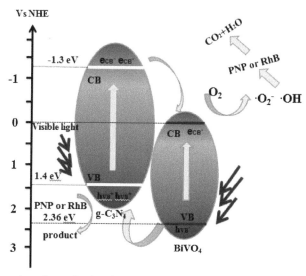

Figure 7.10. Proposed reaction scheme for the photocatalytic process of BiVO₄/g-C₃N₄ composite under visible-light irradiation. Reprinted with permission from Zhao et al. 2016.

synthesized BiVO₄/g-C₃N₄ composite by an ultrasonic method, finding enhanced photocatalytic activities of BiVO₄/g-C₃N₄ composites towards RhB and p-nitrophenol degradation compared with pure g-C₃N₄. A possible photocatalytic mechanism of the BiVO₄/g-C₃N₄ composites is proposed in Fig. 7.10 (Zhao et al. 2016).

4. Summary and Conclusions

Photocatalysis and photoelectrocatalysis are active research areas for applications including water and wastewater treatment, air remediation, carbon dioxide reduction and water splitting to yield hydrogen. While this is a very interesting and challenging area for research, there are few examples of photocatalysis being utilized for water or wastewater treatment on a large scale. There are various reasons for this, but probably the key factor is that photocatalysis is a degradative process and not a simple phase transfer. Many researchers are exploring the design and development of novel photocatalytic materials; however, it is nearly impossible to compare results between various laboratories due to the differences in reaction parameters and the substrates used for testing. There is a real need for different laboratories to try and adopt some standard approach for the comparison of the photocatalytic activity either by reporting photonic efficiencies or formal quantum efficiencies. While it is interesting to show that photocatalysis can degrade problematic pollutants, e.g., contaminants of emerging concern, and this gives novelty for papers to be published, it is very difficult to repeat such experiments if complex analytical equipment is required to monitor the reactions. Researchers should also include a model pollutant relatively easy to be analyzed by others, and they should report quantum efficiencies, along with detailed reaction conditions. In addition, it would be very useful if different laboratories include a benchmark photocatalyst in their studies for relative comparisons. Photocatalysis is an exciting area of research, inter- and even multidisciplinary, in which early career researchers can develop important skills to become the research leaders of the future.

References

Byrne, J.A., B.R. Eggins, N.M.D. Brown, B. McKinney and M. Rouse. 1998. Immobilisation of TiO$_2$ powder for the treatment of polluted water. Appl. Catal. B: Environ. 17: 25–36.

Byrne, J.A., B.R. Eggins, W. Byers and N.M.D. Brown. 1999. Photoelectrochemical cell for the combined photocatalytic oxidation of organic pollutants and the recovery of metals from waste waters. Appl. Catal. B: Environ. 20: L85.

Byrne, J.A., P.S.M. Dunlop, J.W.J. Hamilton, P. Fernandez-Ibanez, I. Polo-Lopez, P.K. Sharma et al. 2015. A review of heterogeneous photocatalysis for water and surface disinfection. Molecules 20: 5574–615.

Cruz-Ortiz, B.R., J.W.J. Hamilton, C. Pablos, L. Díaz-Jiménez, D.A. Cortés-Hernández, P.K. Sharma et al. 2017. Mechanism of photocatalytic disinfection using titania-graphene composites under UV and visible irradiation. Chem. Eng. J. 316: 179–186.

Cui, Y., M. Li, H. Wang, C. Yang, S. Meng and F. Chen. 2018. *In-situ* synthesis of sulfur doped carbon nitride microsphere for outstanding visible light photocatalytic Cr (VI) reduction. Sep. Purif. Technol. 199: 251–259.

Dale, G.R., J.W.J. Hamilton, P.S.M. Dunlop and J.A. Byrne. 2010. Electrochemically assisted photocatalysis on anodic titania nanotubes. Current Topics in Electrochemistry 14: 89–97.

Du, J., Z. Xu, H. Li, H. Yang, S. Xu, J. Wang et al. 2021. Ag$_3$PO$_4$/g-C$_3$N$_4$ Z-scheme composites with enhanced visible-light-driven disinfection and organic pollutants degradation: Uncovering the mechanism. Appl. Surf. Sci. 541: 148487.

Fernández-Ibáñez, P., M.I. Polo-López, S. Wadhwa, J.W.J. Hamilton, P.S.M. Dunlop, R. D'Sa et al. 2015. Solar photocatalytic disinfection of water using titanium dioxide graphene composites. Chem. Eng. J. 261: 36–44.

Frank, S.N. and A.J. Bard. 1977. Heterogeneous photocatalytic oxidation of cyanide ion in aqueous solutions at titanium dioxide powder. J. Am. Chem. Soc. 99(1): 303–304.

Fujishima, A. and K. Honda. 1972. Electrochemical Photolysis of water at a semiconductor Electrode. Nature 238: 37–38.

Hamilton, J.W.J., J.A. Byrne, P.S.M. Dunlop and N.M.D. Brown. 2008. Photo-oxidation of water using nanocrystalline tungsten oxide under visible light. Int. J. Photoenergy. Article ID 185479.

Huang, J., W. Ho and X. Wang. 2014. Metal-free disinfection effects induced by graphitic carbon nitride polymers under visible light illumination. Chem. Commun. 50: 4338.

Hummers, W.S. and R.E. Offeman. 1958. Preparation of graphitic oxide. Journal of the American Chemical Society 80(6): 1339.

Kang, S., L. Zhang, M. He, Y. Zheng, L. Cui, D. Sun et al. 2018. Alternated cooling and heating strategy enables rapid fabrication of highly-crystalline g-C$_3$N$_4$ nanosheets for efficient photocatalytic water purification under visible light irradiation. Carbon 137: 19–30.

Kudo, A., K. Omori and H. Kato. 1999. A novel aqueous process for preparation of crystal form-controlled and highly crystalline BiVO$_4$ powder from layered vanadates at room temperature and its photocatalytic and photophysical properties. J. Am. Chem. Soc. 121: 11459–11467.

Li, Z., Z. He, H. Lai, Y. He, Z. Zhu, Y. Chen et al. 2021. A novel high-efficiency photocatalyst Ta$_2$O$_5$/PtCl$_2$ nanosheets for benzotriazole degradation. J. Environ. Chem. Eng. 9(6): 106345.

Matsunaga, T., R. Tomoda, T. Nakajima and H. Wake. 1985. Photoelectrochemical sterilization of microbial cells by semiconductor powders. FEMS Microbiology Letters 29: 211–214.

McMichael, S., P. Fernández-Ibáñez and J.A. Byrne. 2021. A review of photoelectrocatalytic reactors for water and wastewater treatment. Water. 13(9): 1198.

McMichael, S., M. Waso, B. Reyneke, W. Khan, J.A. Byrne and P. Fernandez-Ibanez. 2021. Electrochemically assisted photocatalysis for the disinfection of rainwater under solar irradiation. Appl. Catal. B: Environ. 281: 119485.

Novoselov, K.S., A.K. Geim, S.V. Morozov, D. Jiang, Y. Zhang, S.V. Dubonos et al. 2004. Electric field effect in atomically thin carbon films. Science 306: 666–669.

Pablos, C., J. Marugán, R. van Grieken, P.S.M. Dunlop, J.W.J. Hamilton, D.D. Dionysiou et al. 2017. Electrochemical enhancement of photocatalytic disinfection on aligned TiO_2 and nitrogen doped TiO_2 nanotubes. Molecules 22: 704.

Pelaez, M., N.T. Nolan, S.C. Pillai, M.K. Seery, P. Falaras, A.G. Kontos et al. 2012. A review on the visible light active titanium dioxide photocatalysts for environmental applications. Appl. Catal. B: Environ. 125: 331–349.

Pomilla, F.R., M.A.L.R.M. Cortes, J.W.J. Hamilton, R. Molinari, G. Barbieri, G. Marcì et al. 2018. An investigation into the stability of graphitic C_3N_4 as a photocatalyst for CO_2 reduction. J. Phys. Chem. C. 122: 28727–28738.

Reddy, C.V., I.N. Reddy, K. Rabindranadh, K.R. Reddy, D. Kim and J. Shim. 2020. Ni-dopant concentration effect of ZrO_2 photocatalyst on photoelectrochemical water splitting and efficient removal of toxic organic pollutants. Sep. Purif. Technol. 252: 117352.

Reyes-Gil, K.R. and D.B. Robinson. 2013. WO_3-enhanced TiO_2 nanotube photoanodes for solar water splitting with simultaneous wastewater treatment. ACS Appl. Mater. Interfaces 5: 12400–12410.

Ryu, J. and W. Choi. 2008. Substrate-specific photocatalytic activities of tio_2 and multiactivity test for water treatment application. Environ. Sci. Technol. 42: 294–300.

Serpone, N., A.V. Emeline, S. Horikoshi, V.N. Kuznetsov and V.K. Ryabchuk. 2012. On the genesis of heterogeneous photocatalysis: a brief historical perspective in the period 1910 to the mid-1980s. Photochem. Photobiol. Sci. 11: 1121.

Tan, G., L. Zhang, H. Ren, S. Wei, J. Huang and A. Xia. 2013. Effects of pH on the hierarchical structures and photocatalytic performance of $BiVO_4$ powders prepared via the microwave hydrothermal method. ACS Appl. Mater. Interfaces 5: 5186–5193.

Wadhwa, S., J.W.J. Hamilton, P.S.M. Dunlop, C. Dickinson and J.A. Byrne. 2011. Effect of post-annealing on the photocatalytic activity of hydrothermally synthesised titania nanotubes. J. Adv. Oxid. Technol. 14(1): 147–157.

Waller, M.R., T.K. Townsend, J. Zhao, E.M. Sabio, R.L. Chamousis, N.D. Browning et al. 2012. Single-crystal tungsten oxide nanosheets: photochemical water oxidation in the quantum confinement regime. Chem. Mater. 24: 698–704.

Wang, X., K. Maeda, A. Thomas, K. Takanabe, G. Xin, J.M. Carlsson et al. 2009. A metal-free polymeric photocatalyst for hydrogen production from water under visible light. Nat. Mater. 8: 76–80.

Williams, G., B. Seger and P.V. Kamat. 2008. TiO_2-graphene nanocomposites. UV-assisted photocatalytic reduction of graphene oxide. ACS Nano. 2(7): 1487–1491.

Xie, Y.P., G. Liu, L. Yin and H.M. Cheng. 2012. Crystal facet-dependent photocatalytic oxidation and reduction reactivity of monoclinic WO_3 for solar energy conversion. J. Mater. Chem. 22: 6746–6751.

Zhang, G., W. Guan, H. Shen, X. Zhang, W. Fan, C. Lu et al. 2014. Organic additives-free hydrothermal synthesis and visible-light-driven photodegradation of tetracycline of WO_3 nanosheets. Ind. Eng. Chem. Res. 53: 5443–5450.

Zhao, J., J. Yan, H. Jia, S. Zhong, X. Zhang and L. Xu. 2016. $BiVO_4$/g-C_3N_4 composite visible-light photocatalyst for effective elimination of aqueous organic pollutants. J. Mol. Catal. A: Chem. 424: 162–170.

Zheng, Z., N. Zhang, T. Wang, G. Chen, X. Qiu, S. Ouyang et al. 2019. $Ag_{1.69}Sb_{2.27}O_{6.25}$ coupled carbon nitride photocatalyst with high redox potential for efficient multifunctional environmental applications. Appl. Surf. Sci. 487: 82–90.

Chapter 8

Nanophotocatalysts for Water Remediation
Treatment of Industrial Wastewaters

Maria Antonopoulou[1,2,]* and *Anastasia Hiskia*[2]

1. Introduction

Industrial wastewaters are the undesired aqueous discards generated through various industrial activities (e.g., chemical, electronic, petrochemical and pharmaceutical manufacturing as well as food and beverage processing and production) in nearly all phases of production. Industrial wastewaters have variable quality and quantity depending on the type of industrial activity, the applied technology, amounts of process water and types and doses of raw materials. They generally contain high concentrations of toxic or non-biodegradable pollutants. Industrial processes discharge large amounts of complex wastewater with high concentrations of Chemical Oxygen Demand (COD), Biochemical Oxygen Demand (BOD), Total Solids (TS), Total Dissolved Solids (TDS), Total Suspended Solids (TSS), inorganic ions and other pollutants (Mutamim et al. 2012, Zhang et al. 2014). The major characteristics of representative wastewaters from different industrial activities are summarized in Table 8.1.

Due to the variable quality of industrial wastewaters as well as the high concentration of toxic and non-biodegradable pollutants that they may contain, the treatment of these types of wastewaters is a big challenge. In addition, growing public awareness about the protection of the environment and increasing regulatory constraints established by governmental authorities render the development of new and more effective technologies a continuous demand (Cardoso et al. 2016, Zhang et al. 2014).

The conventional technologies currently used to treat industrial wastewater are chemical oxidation, coagulation, electrochemical separation, membrane separation and anaerobic and aerobic microbial degradation. However, most of them cannot effectively remove various non-biodegradable organic compounds and heavy metals or metalloids such as chromium and arsenic. Moreover, these methods have inherent limitations such as high operational cost, long reaction time, complicated technical requirements and, in some cases, secondary waste generation by transfer of the non-biodegradable content into sludge, which restricts their application (Al-Mamun et al. 2019, Hermosilla et al. 2015, Singh et al. 2019, Zhang et al. 2014).

[1] Department of Environmental Engineering, University of Patras, Seferi 2 Str, 30100, Agrinio, Greece.
[2] Institute of Nanoscience and Nanotechnology, NCSR "Demokritos", Patr. Gregoriou E & 27 Neapoleos Str, 15341 Agia Paraskevi, Greece.
Email: a.hiskia@inn.demokritos.gr
* Corresponding author: mantonop@upatras.gr

Table 8.1. Major characteristics of wastewaters from different industrial activities (Ambaye and Hagos 2020, Bordes et al. 2015, Talwar et al. 2018, Yuan et al. 2007).

Textile wastewater	
Parameter	**Value**
pH	8.1
COD (mg L^{-1})	1250.5
BOD (mg L^{-1})	250.5
TSS (mg L^{-1})	1150.5
TDS (mg L^{-1})	3000.2
Color (PCU)	75.0
Tannery wastewater	
Parameter	**Value**
pH	7.3
Conductivity (μS cm^{-1})	5240
Turbidity (NTU)	46
Filtered turbidity (NTU)	2.6
TSS (mg L^{-1})	34
TOC (mg L^{-1})	43.4
COD (mg L^{-1})	162
BOD_5 (mg L^{-1})	37
TN (mg L^{-1})	41
Phenol index (mg L^{-1})	6
Paper mill wastewater	
Parameter	**Value**
pH	8.42
COD (mg L^{-1})	215
BOD_5 (mg L^{-1})	52
Color (mg Pt/L^{-1})	650
Suspended solids (SS) (mg L^{-1})	3.5
Adsorbable Organic Halides (AOX) (mg L^{-1})	5.1
Pharmaceutical industry wastewater	
Parameter	**Value**
pH	5.8
COD (mg L^{-1})	12425
BOD (mg L^{-1})	1727
TSS (mg L^{-1})	3180
TDS (mg L^{-1})	1600
TS (mg L^{-1})	4780
BOD_5/COD	0.178

In recent decades, Advanced Oxidation Processes (AOPs) including heterogeneous photocatalysis, photo-Fenton, ultraviolet (UV) irradiation coupled with H_2O_2 (UV/H_2O_2) and ozone-based processes have been studied for wastewater treatment. The effectiveness of these processes is based on the production of various Reactive Oxygen Species (ROS), mainly hydroxyl radicals (HO·), which can rapidly react with the pollutants without generating toxic byproducts. HO· radicals have a high redox potential (2.8 eV), are non-selective and capable of attacking organic compounds through hydrogen abstraction, addition of radicals and electron transfer reactions (Al-Mamun et al. 2019, Garrido-Cardenas et al. 2020, Konstantinou and Albanis 2004, Rueda-Marquez et al. 2020).

Heterogeneous photocatalysis, a representative AOP, has been proven to be an alternative wastewater treatment process (Al-Mamun et al. 2019, Konstantinou and Albanis 2004, Tan et al. 2020). As an important AOP, the reactive species generated during the process, have the potential of oxidizing organic compounds, leading to their complete mineralization, as well as reducing toxic metals converting them into their less toxic or non-toxic metallic states (Kabra et al. 2004, Nogueira et al. 2018).

Various semiconducting metal oxides can be used in heterogeneous photocatalysis (Table 8.2). From a fundamental and practical perspective, TiO_2 has gained much attention for wastewater treatment, mainly due to its chemical and thermal stability, non-toxicity and low cost (Tan et al. 2020). Recent advances in the field of nanotechnology resulted in the generation of tuned catalysts in the nanometer region and have explored their applications in water remediation and wastewater treatment (Litter et al. 2012). The synthesis of appealing nanocatalysts with improved properties is of paramount importance for both the scientific community as well as large applications. During the last decades, a wide variety of TiO_2-based photocatalysts have been developed, including doped, composite and immobilized TiO_2 (Al-Mamun et al. 2019, Antonopoulou et al. 2021, Pillai et al. 2017, Saikumari et al. 2021, Tan et al. 2020).

This chapter provides a general overview of some of the recent advances in the application of heterogeneous photocatalysis using different types of TiO_2-based nanocatalysts for the treatment of industrial wastewaters. Representative applications of TiO_2-mediated photocatalysis for treating different kinds of real industrial wastewaters (textile, tannery, pulp and paper mill and pharmaceutical industry wastewaters) are also provided. Finally, the future perspectives of the photocatalytic process in this field are outlined.

Table 8.2. Semiconducting metal oxides with their bandgap energies commonly used as photocatalysts (Pawar et al. 2018).

Photocatalyst	Bandgap energy (eV)
ZrO_2	5.0
TiO_2 (anatase)	3.2
TiO_2 (rutile)	3.0
WO_3	2.8
SnO_2	2.5
Fe_2O_3	2.3
V_2O_5	2.6
CuO	1.7
MnO_2	0.25
CeO_2	2.94

2. Heterogeneous Photocatalysis

2.1 Mechanism of TiO₂-mediated photocatalysis

Heterogeneous photocatalysis involves the use of a semiconducting material (photocatalyst), having the ability to be activated under light irradiation without being consumed in the process (Paumo et al. 2021, Rueda-Marquez et al. 2020). For water and wastewater remediation, the overall photocatalytic process can be described by the following five independent steps: (i) mass transfer of pollutants from the liquid phase to the surface of the photocatalyst, (ii) adsorption of the pollutants onto the surface of the photocatalyst, (iii) photocatalytic reaction in the adsorbed state, (iv) desorption of

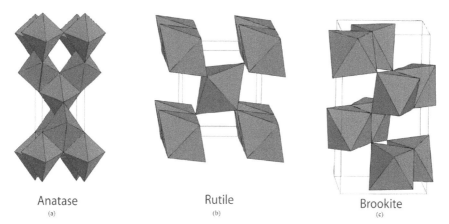

Figure 8.1. TiO$_2$ crystal structure: anatase (a), rutile (b), brookite (c). Reprinted with permission from Khaki et al. 2017.

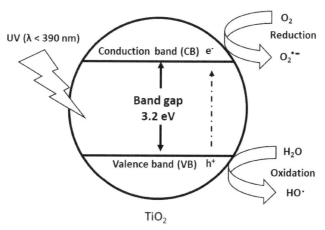

Figure 8.2. Photoexcitation mechanism of TiO$_2$ (anatase phase).

generated products from the surface of the photocatalyst, and (v) mass transfer of the products from the interface region to the bulk fluid (Al-Mamun et al. 2019).

An ideal photocatalyst should have light absorption properties, be chemically stable, easily available, economic, non-toxic, able to utilize a broad band of the solar spectrum and possess high photocatalytic activity (Pawar et al. 2018). TiO$_2$ is one of the most investigated photocatalysts for wastewater treatment having high stability, high ultraviolet absorption and cost-effectiveness. TiO$_2$ is an n-type semiconductor and exists in three forms, i.e., anatase, rutile and brookite (Fig. 8.1). Among the three phases, the anatase phase has the best photocatalytic activity (Khaki et al. 2017, Al-Mamun et al. 2019).

In a typical photocatalytic process, irradiation of TiO$_2$ with light energy equal to or greater than its bandgap leads to an electron excitation between the Valence Band (VB) and Conduction Band (CB) (Fig. 8.2). From the photoexcitation, electrons (e_{CB}^-) are promoted to the empty CB, leaving behind electron vacancies, i.e., holes in the VB (h_{VB}^+). The h_{VB}^+ and e_{CB}^- can then migrate to the TiO$_2$ surface and participate in various redox reactions. The overall process can be described by the following reactions (Antonopoulou et al. 2021, Dong et al. 2015, Legrini et al. 1993, Okamoto et al. 1985, Wang et al. 1999):

$$TiO_2 + hv \rightarrow TiO_2 \ (e_{CB}^- + h_{VB}^+) \tag{1}$$

$$TiO_2 \ (h_{VB}^+) + H_2O \rightarrow TiO_2 + H^+ + HO^\bullet \tag{2}$$

$$TiO_2 \ (h_{VB}^+) \ + OH^- \rightarrow TiO_2 + HO^\bullet \tag{3}$$

$$TiO_2 \ (e_{CB}^-) + O_2 \rightarrow TiO_2 + O_2^{\bullet -} \tag{4}$$

$$O_2^{\bullet -} + H^+ \rightarrow HO_2^\bullet \tag{5}$$

$$2HO_2^\bullet \rightarrow O_2 + H_2O_2 \tag{6}$$

$$H_2O_2 + O_2^{\bullet -} \rightarrow HO^\bullet + OH^- + O_2 \tag{7}$$

A variety of organic and inorganic pollutants present in industrial wastewater can be decomposed or transformed by the hydroxyl radicals, electrons, holes and other oxidizing species produced during the TiO_2-mediated photocatalysis. Organic pollutants can be completely degraded to CO_2 and H_2O whereas toxic metals can be transformed into less harmful species. However, many e_{CB}^- and h_{VB}^+ can recombine rapidly, which dramatically inhibits the photocatalytic activity of semiconducting materials (Nogueira et al. 2018, Younis and Kim 2020).

Although TiO_2 is a promising photocatalyst, its wide bandgap in the UV wavelength region, which contributes to 3–5% of the total solar spectrum, as well as the fast recombination of photogenerated electron-hole pairs limit its photocatalytic activity (Dong et al. 2015, Li et al. 2020). Accordingly, methods of enhancing the photocatalytic activity of TiO_2 have been studied by many researchers.

2.2 *Approaches to improve the TiO$_2$ photoactivity*

Significant properties of TiO_2 that can have an impact on the activity of the photocatalyst are pore volume, pore size, specific surface area, crystalline phase and pore structure; these properties can be improved by morphological modification during the synthesis (Nakata and Fujishima 2012). The most widely used TiO_2 form is as monodispersed nanoparticles (NPs), and to improve its activity the synthesis of small crystallite particles is favored (Al-Mamun et al. 2019). Towards this direction, various synthesis methods have been employed and optimized to obtain catalysts with a defined crystal structure, small particle sizes and high affinity to various organic pollutants. The hydrothermal method, sol-gel method, vapor deposition method and electrospinning method are some representative methods that have been used for the synthesis of TiO_2 nanostructures with various sizes and morphology types (Li et al. 2020).

One more crucial factor affecting the performance of nanophotocatalytic materials is their bandgap. To reduce the large bandgap of TiO_2, its chemical modification by doping with different metals and non-metals has been applied. Doping can lead to nanocatalysts with enhanced properties, i.e., (i) reduced bandgap energy extending the photoresponse of the catalyst to the visible light region, (ii) modified physical properties such as surface morphology, which can increase its photoactivity, and (iii) increased charge separation by the addition of dopants, which can inhibit the undesirable recombination of the photogenerated electrons and holes (Al-Mamun et al. 2019, Dong et al. 2015).

Various metal ions including V, Zn, C Mn, Al, Co, Fe, Ni, Ag, Au, Pt or Pd, and non-metals N, S, C, B, P, I or F have been utilized for doping and codoping TiO_2. The presence of dopants has been shown to increase the photocatalytic degradation efficiency of TiO_2. However, an optimum quantity of dopant(s) is required as the excess amount(s) can have detrimental effects on the photocatalytic activity (Al-Mamun et al. 2019, Dong et al. 2015, Fotiou et al. 2015, Khaki et al. 2017).

Heterojunction, i.e., the coupling of two or more semiconductors with dissimilar bandgaps, is another approach to enhance the photocatalytic performance by inhibiting the recombination of photogenerated electron-hole pairs and broaden the visible light absorption region. Semiconductor heterojunction is a good strategy that is expected to improve the photocatalytic performance of TiO_2. In TiO_2 heterojunctions, both catalysts can be excited by photons to generate charge carriers, but the charge transfer direction is mainly dependent on the relative VB and CB positions of both semiconductors (Khaki et al. 2017). In addition, multifunctional photocatalysts with excellent magnetic properties and adsorption capacity can also be fabricated (Jing et al. 2013), presenting simple and cost-effective separation advantages.

To enhance the applicability of the photocatalytic process and overcome the costly separation of the photocatalyst nanoparticles from wastewater after treatment, immobilization of TiO_2 nanoparticles on inert supports/substrates has been used. Immobilization of photocatalysts allows to avoid the possible release of nanoparticles to water and sludge generation and significantly decreases the cost of the treatment by eliminating the photocatalyst recovery step (Rueda-Marquez et al. 2020).

3. Application of TiO_2-Mediated Photocatalysis in Industrial Wastewaters Treatment

3.1 Treatment of textile wastewaters

Textile wastewater is a complex matrix consisting mainly of dyes, fibers, detergents, grease, oil, organic and inorganic salts, acids and heavy metals and is considered a major source of pollutants from the industry to the aquatic environment. Among them, dyes are the major pollutants in textile wastewater. The conventional physicochemical and biological treatment processes are not able to degrade textile dyes, failing to meet the requirement of discharge limits. The inadequate treatment of textile wastewater is a major concern and, with growing environmental awareness, there is a need for the development of more efficient technologies to remove dyes from industrial wastewater (Raman and Kanmani 2016, Zhang et al. 2014). Heterogeneous photocatalysis is considered an alternative technology to treat dye-contaminated wastewater using various kinds of photocatalysts. Diverse nanomaterials including TiO_2 have been investigated in depth (Rafiq et al. 2021).

Commercial TiO_2 photocatalysts in the form of nanoparticles have been used for the treatment of textile wastewaters under various irradiation sources (Ambaye and Hagos 2020, Bansal and Sud 2011, Cifci and Meric 2015, Gümüş and Akbal 2011, Pekakis et al. 2006, Rashidi et al. 2014, Souza et al. 2016, Vilar et al. 2011). The efficiency of the photocatalysts ranged from moderate to high for the treatment of textile wastewaters, mainly in terms of COD, BOD, color and toxicity reduction, depending on the experimental conditions. In general, the addition of H_2O_2 (Vilar et al. 2011) and the combination of TiO_2 photocatalysis with a biological process enhanced the removal (Ambaye and Hagos 2020).

As metal ion doping has invariably increased the activity, Sahoo et al. (2012a, 2012b) synthesized Ag^+ doped TiO_2 and investigated its activity for decolorization and mineralization of textile wastewaters under UVC light. High percentages of decolorization and mineralization of the five times diluted real textile wastewaters were achieved. The fabricated Ag^+-doped TiO_2 photocatalyst showed better activity than TiO_2. Reuse of the catalyst was also possible but with lesser efficiency after some cycles.

A novel end-to-end pilot-scale system involving biological and photocatalytic treatment of a real textile and dyeing industry effluent was reported by Bahadur and Bhargava (2019). Nano-TiO_2 synthesized by the sol-gel method was used as the photocatalyst. BOD and COD were reduced by 95 and 91%, respectively, and the formation of non-toxic sludge was negligible.

Although TiO_2 in the form of a nanopowder has many advantages and good photocatalytic activity for environmental application, the difficulties in separation from the suspension of a textile wastewater are still a big challenge. Alinsafi et al. (2007) and Guesh et al. (2016) reported that TiO_2 can be immobilized on various substrates such as glass slides, glass fibers and zeolites. Although the performance of the supported TiO_2 immobilized on glass slides (final TiO_2 concentration on the slides of 2.8 g m^{-2} with a specific area ~ 50 m^2 g^{-1}) and glass fibers (final TiO_2 concentration on the fibers of 20 g m^{-2} and specific area ~ 250 m^2 g^{-1}) was strongly dependent on the chemical structure of dyes and other chemicals present in the textile wastewater, decolorization with promising yields was achieved (Alinsafi et al. 2007). Ten percent loaded TiO_2 on zeolite ($SiO_2/Al_2O_3 = 60$) was tested for real textile wastewater of Ethiopia resulting in almost 84% removal of Total Organic Carbon (TOC), showing higher activity than pure TiO_2 (~ 49% TOC removal). The enhanced photocatalytic activity of the hybrid catalyst is attributed to the role of zeolites, which enhance the adsorption capacity due to their higher specific surface area as well as hinder the electron/hole recombination rate (Guesh et al. 2016).

A photocatalyst composed of TiO_2 supported on hydrotalcite (HT) and doped with iron (molar ratio of $Fe^{3+}/Fe^{2+} = 0.5$) ($HT/Fe/TiO_2$) with a Fe:Al molar ratio = 1 and a bandgap energy of 2.34 eV was also used for the treatment of a biologically treated textile effluent by heterogeneous photocatalysis under UV-visible irradiation (Arcanjo et al. 2018). The color was quantified using the ADMI (American Dye Manufacturers' Institute) scale. At a catalyst dose of 2 g L^{-1} and pH 10, $HT/Fe/TiO_2$ reduced ADMI color by 96% after 6 hours treatment, showing a higher efficiency than TiO_2 (88%).

The higher efficiency of $HT/Fe/TiO_2$ is correlated with its greater photon absorption (capable of absorbing light at wavelengths from 300 to 530 nm due to its low bandgap energy (2.34 eV)) and subsequently with enhanced $HO^{\bullet}$ production. Other advantages of $HT/Fe/TiO_2$ are its stability and magnetic nature, which would reduce the costs of photocatalyst recovery after use (Arcanjo et al. 2018).

Shivaraju et al. (2020) used Mg-doped TiO_2 coated buoyant clay hollow-spheres with well crystalline structures, adequate porosity, enough surface area, high stability and high photocatalytic activity for the treatment of textile wastewaters under different light sources. Spheres with about 10–15 mm diameter were prepared using clay and activated carbon and Mg-doped TiO_2 particles were decorated on their surface. COD reductions for the diluted (50% dilution) wastewaters of 77, 70 and 72% were observed under Light Emitting Diodes (LEDs), UV and tungsten light sources, respectively, after 6 hours. Relatively less COD removal was observed for the undiluted wastewaters, which is correlated with the numerous compounds present in the raw matrix. The recovery and reusability of the proposed photocatalyst were appreciable, as the photocatalytic activity was reported to be quite similar up to several cycles (Fig. 8.3a). The improved treatment efficiency can be explained by the spherical shape and buoyant nature (induced easy spinning) of the photocatalysts as well as by the adsorption capacity of clay and activated carbon. The proposed COD degradation mechanism is presented in Fig. 8.3b (Shivaraju et al. 2020).

Recently, Zhou et al. (2020) synthesized Europium-Nitrogen (Eu-N) codoped TiO_2/sepiolite nanocomposites via a microwave-hydrothermal method and investigated their photocatalytic activity for the treatment of real textile wastewater under visible light. The optimum Eu^{III} doped content was 0.6% for the codoped catalysts. The nanocomposites were fully characterized, and it was found that the carrier sepiolite was uniformly covered by doped-TiO_2 nanoparticles.

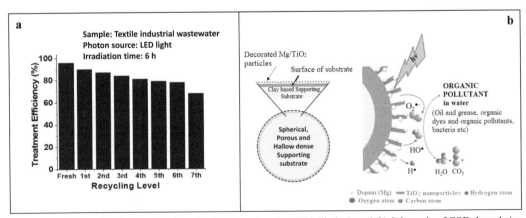

Figure 8.3. (a) Recyclability investigation with textile wastewater (50% dilution), and (b) Schematic of COD degradation mechanism. Reprinted with permission from Shivaraju et al. 2020.

N-doping extends the responsive activity of the catalysts to the visible region whereas Eu^{III}-doping promotes the growth of anatase crystallites and enhances the adsorption ability. The Eu-N codoped TiO$_2$/sepiolite effectively treated real textile wastewater, observing over 70% COD reduction in 3 hours (Zhou et al. 2020).

3.2 Treatment of tannery wastewaters

Tannery wastewaters represent a serious environmental problem causing aquatic pollution as they contain large amounts of compounds such as surfactants, acids, dyes, natural or synthetic tanning agents, sulfonated oils, salts, heavy metals, etc. Considering the large amounts and the low biodegradability of various chemicals present in tannery wastewaters as well as the fact that conventional chemical and/or biological wastewater treatment processes cannot efficiently remove the recalcitrant pollutants from these kinds of wastewaters, alternative methods are being increasingly studied (Goutam et al. 2018, Schrank et al. 2004).

UV irradiation in combination with TiO$_2$ (80% anatase and 20% rutile, specific area 59 m^2 g^{-1}) in powder form achieved a reduction of 10–12% in TOC and a slight increase in biodegradability at different pH values after 2 hours when applied for the treatment of a tannery wastewater (Schrank et al. 2004). However, the combination of TiO$_2$ photocatalysis (solar/TiO$_2$) and the Fenton process led to an enhancement of the treatment of tannery wastewaters at neutral pH conditions. The COD of tannery wastewater decreased about 96% by the combined advanced process at optimum conditions (pH 7, [TiO$_2$] = 0.2 g L^{-1}, [Fe^{2+}] = 0.5 g L^{-1}, [H$_2$O$_2$] = 1.8 g L^{-1}) after 4 hours of treatment. Under optimum experimental conditions and 60 minutes of treatment, an increase in the biodegradability of wastewater was also observed from 0.4 to 0.7 (Selvabharathi et al. 2016).

Ordered mesoporous TiO$_2$ materials with a relatively high surface area of 40.03 m^2 g^{-1} and pore diameter distribution of 13.04 nm were successfully synthesized via a sol-gel route by Zhao et al. (2017). The mesoporous TiO$_2$ catalysts were investigated to treat tannery wastewater under UVA (365 nm) and natural sunlight, and they exhibited good photocatalytic activity towards the degradation of organic pollutants present in this wastewater. Under optimum conditions ([TiO$_2$] = 1 g L^{-1}, pH 5, [H$_2$O$_2$] = 3.5%) and UVA light, color and COD reductions of 100 and 60%, respectively, were observed, showing a better photoactivity than that of commercial P25. The enhancement of the photocatalytic performance can be attributed to a larger surface area and pore volume. The ordered mesopore structure promotes the mass transport of the organic pollutants and subsequently, photoactivity. The photocatalytic activity of mesoporous TiO$_2$ was reported to be very similar even after five cycles, demonstrating its excellent photocatalytic stability. Mesoporous TiO$_2$ photocatalysts also possess superior catalytic performance under natural sunlight (Zhao et al. 2017).

TiO$_2$ NPs were synthesized using aqueous extract of leaves of the biodiesel plant, *Jatropha curcas* L. (Fig. 8.4), and tested for the simultaneous removal of COD and Cr (VI) from secondary treated tannery wastewater (Goutam et al. 2018). The average crystalline size, surface area, pore size diameter and total pore volume of the synthesized TiO$_2$ photocatalyst were 13 nm, 27.038 m^2 g^{-1}, 19.100 nm and 0.1291 cm^3 g^{-1}, respectively. The photocatalytic treatment of wastewater with synthesized TiO$_2$ NPs in a self-designed and fabricated Parabolic Trough Reactor (PTR) led to 82.26 and 76.48% removal of COD and Cr (VI) respectively. The nanosized, pure crystalline anatase phase, with a large amount of surface hydroxyl groups and high surface area of the synthesized TiO$_2$, can be considered for the observed high performance. Moreover, it was found that the phenols and tannins contained in the leaf extract used for the synthesis can contribute to the capping/stabilization of NPs (Goutam et al. 2018).

g-C$_3$N$_4$ prepared by melamine pyrolysis was combined with TiO$_2$ at different weight percentages to form TiO$_2$/g-C$_3$N$_4$ nanocomposites. Eighty percent TiO$_2$/g-C$_3$N$_4$ nanocomposite with a bandgap of 2.88 eV was investigated and compared with bare TiO$_2$ using the photocatalysts immobilized in a thin film plate reactor to treat tannery wastewaters under artificial UV light in terms of phenol, COD and color reduction (Tripathi and Narayanan 2018). Under optimum conditions (UV-irradiance time = 180 minutes, pH 5, [H$_2$O$_2$] = 0.8 g L^{-1}, flow rate = 3 L h^{-1} and UV-light intensity = 556.13 µE),

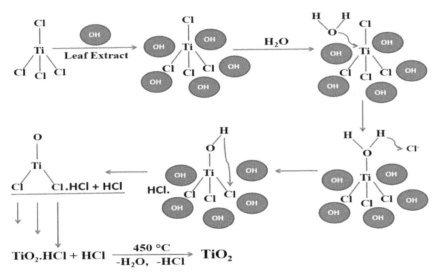

Figure 8.4. Possible reaction mechanism for the formation of TiO$_2$ NPs in the presence of hydroxyl groups (–OH) of leaf extract of *Jatropha curcas* L. as capping agent. Reprinted with permission from Goutam et al. 2018.

photocatalysis with TiO$_2$/g-C$_3$N$_4$ led to 93.06 phenol, 85.62 COD and 80.23% color reduction showing higher activity than the process with TiO$_2$ (72.39 phenol, 68.95 COD and 65.80% color reduction). The results highlight that TiO$_2$ coupled with other materials has shown an improved photocatalytic activity due to an increased electron-hole separation efficiency. Moreover, a decrease in the bandgap energy was observed, important for applications using solar light (Tripathi and Narayanan 2018).

N-doped TiO$_2$ photocatalysts were developed to increase the absorption of TiO$_2$ to visible light by Vaiano et al. (2014). Their performance was evaluated to treat tannery wastewater under different light sources and different photoreactor configurations. N-doping of TiO$_2$ leads to a reduction in the bandgap energy from 3.2 to 2.5 eV, allowing the absorption of visible radiation. The specific surface area for N-TiO$_2$ was 80 m^2 g^{-1}. As expected, N-doped TiO$_2$ showed remarkably photoactivity in the presence of visible light emitted by LEDs. In the presence of N-doped TiO$_2$, a maximum COD reduction of about 60% was reached after 5 hours using white LEDs positioned around and in contact with the photoreactor walls (Vaiano et al. 2014).

A Fe$_3$O$_4$/TiO$_2$ magnetic mesoporous composite with 8 wt.% Fe$_3$O$_4$ and a surface area of 54.36 m^2 g^{-1}, possessing rapid and efficient magnetic response, was investigated for the treatment of a tannery wastewater. The color and COD removals were 87 and 42%, respectively. After three cycles, the composite exhibited good photocatalytic performance and stability (Zhang et al. 2017).

A CuCo$_2$O$_4$/TiO$_2$ heterosystem, which is more effective than single TiO$_2$ under visible light irradiation, was studied for the removal of 2.25 mg L^{-1} of Cr (VI) from tannery industrial wastewater. Under visible light, 99.99% of Cr (VI) was removed, proving that the CuCo$_2$O$_4$/TiO$_2$ heterosystem is efficient for wastewater treatment. This heterojunction inhibits the recombination of electron-hole pairs, allowing more of these species to be available for oxidation and the reduction of pollutants (Kebir et al. 2015).

More recently, spherical TiO$_2$ catalytic materials with a hollow structure were prepared based on a facile template-free solvothermal process. TiO$_2$ hollow spheres with uniform diameter distribution and high specific surface area were fabricated and the photo electrocatalytic reduction of Cr (VI) in a real tannery wastewater collected from production lines was investigated. Complete removal of Cr (VI) by photo electrocatalysis with a TiO$_2$ spheres dosage of 0.1 g L^{-1} and 2.0 V of external potential was achieved. Considerable reduction of acute toxicity was also observed for the treated wastewater (Zhao et al. 2017a). Moreover, the electro-field-assisted-photocatalysis with hierarchical TiO$_2$ microspheres was investigated for the removal of tributyltin in real tannery wastewater. Hierarchical TiO$_2$ microspheres consisting of nanowires and possessing tunable building units, high specific surface areas and improved optical performance were fabricated via a facile self-assembly process induced by one-step microwave-assisted solvothermal reaction. The process exhibited excellent activity to remove tributyltin and reduce toxicity (Zhao et al. 2017b).

3.3 Treatment of pulp and paper mill wastewaters

The pulp and paper industry consumes enormous amounts of water, producing simultaneously almost equal amounts of wastewater. The wastewater composition is strongly dependent on the applied technology and raw materials, containing highly toxic and refractory compounds. Different pollutants can be produced during the various process steps such as hemicelluloses, pectin, lipophilic extractives (such as resin acids), terpenes, phenols, alcohols, inorganic and organic chlorine compounds, as well as chemicals used as product additives, such as fillers, whiteners, dyes, dispersion agents, surfactants and biocides (Hermosilla et al. 2015, Zhang et al. 2014). Considering that the biological treatment is not able to remove biorecalcitrant compounds present in this type of wastewater, as the biologically treated effluent still contains appreciable concentrations of COD, color and toxicity, new treatment technologies have been investigated.

Commercial TiO_2 in nanopowder form was used to remove the organic carbon content of a paper mill effluent originated from the kraft pulp bleaching process by Muñoz et al. (2006) with moderate efficiency. However, the combination of TiO_2-mediated photocatalysis with photo-Fenton significantly enhanced the removal efficiency (Muñoz et al. 2006). Considerable efficiency with TiO_2 was reported by Moiseev et al. (2004) treating biologically pretreated paper-mill effluents from a factory producing cardboard from recycled paper. In a slurry mode, Nogueira et al. (2018) investigated TiO_2-mediated photocatalysis to treat a kraft pulp mill effluent, and they found that the color, aromatic compounds, COD and toxicity could be effectively reduced. The addition of H_2O_2 led to an increased removal efficiency.

An important work was conducted by Ghaly et al. (2011), who synthesized TiO_2 nanoparticles by a conventional sol-gel process. TiO_2 nanoparticles were characterized using the X-Ray Diffraction (XRD) technique, gravimetric-differential thermal analysis (TG-DTA) and FTIR, and their performance was evaluated for the treatment of a paper mill wastewater under solar light. At optimum conditions ([TiO_2] = 0.75 g L^{-1} and pH 6.5), 75% COD removal, 80% TSS reduction as well as a substantial improvement in BOD_5/COD of the wastewater were achieved.

TiO_2 immobilization on inert support without the loss of activity has also been proposed in several studies to overcome the limitation derived from the separation of TiO_2 powder after the treatment. Yuan et al. (2007) used Activated Carbon Fibers (ACF) as support for TiO_2 and synthesized TiO_2/ACF catalysts with different surface areas. The surface area of all supported catalysts (TiO_2/ACF) decreased retaining simultaneously the good pore structure of ACF. The results showed that an increase in the surface area favored high COD removal for paper mill effluents. TiO_2/ACF with moderate surface area and pore structure (surface area of 799.4 cm^2 g^{-1}, micropore volume of 0.2250 cm^3 g^{-1}, total pore volume of 0.4485 cm^3 g^{-1}) was found to be the most effective for the removal of COD. Its photocatalytic activity was higher than suspended P25 showing also good reusability (Yuan et al. 2007). It was found that the surface area and pore structure of TiO_2/ACF materials had an important effect on their efficiency in the treatment of paper mill effluents. The optimum surface area can favor the synergistic effect between adsorption and photocatalysis promoting the adsorption of the organics present in wastewater and their transfer to the TiO_2 surface where they can degrade completely (Yuan et al. 2007).

In addition, TiO_2 immobilized on glass was also efficient to remove organic compounds present in pulp and paper bleaching effluents. The effluent was completely decolorized, and the total phenol content was reduced by 85% after 120 minutes treatment whereas 50% TOC reduction was reported (Yeber 2000).

With the aim to prepare catalysts that overcome problems related to their recovery and reuse in practical application, Subramonian et al. (2017) fabricated magnetically Fe_2O_3-TiO_2 composite with an irregular and slightly agglomerated surface morphology using a solvent-free mechanochemical process under ambient conditions. The composite material had a surface area of 58.40 m^2 g^{-1}, a pore volume of 0.29 cm^3 g^{-1}, a pore size of 18.52 nm and a bandgap energy equal to 2.95 eV. Fe_2O_3-TiO_2 showed chemical stability and resistance towards photocorrosion up to five successive cycles. The prepared composite catalysts also presented the advantage of easy recovery from an aqueous suspension. The magnetic Fe_2O_3-TiO_2 exhibited good performance for the treatment of a real industrial wastewater. A COD removal of 80.6% was achieved using initial pH, photocatalyst dosage and an airflow rate of 3.88, 1.3 g L^{-1} and 2.28 L min^{-1}, respectively. COD removal was reported to be approximately two times greater by Fe_2O_3-TiO_2 than using P25. Additionally, the presence of selected oxidants enhanced the treatment efficiency considerably (COD removal up to 93.0% using 0.625 mmol L^{-1} of HOCl). The possible photocatalytic reaction mechanism of Fe_2O_3-TiO_2 under UV irradiation is presented in Fig. 8.5. Irradiation of composite material leads to the formation of electron-hole pairs in the conduction and valence bands of TiO_2. Electrons generated in TiO_2 can be easily transferred to the CB of Fe_2O_3 due to the difference in the CB of Fe_2O_3 and TiO_2. The transferred electrons can subsequently migrate to the surface of Fe_2O_3 and promote oxygen reduction forming $O_2^{\cdot-}$. The holes produced in the VB of TiO_2 can be transferred to the surface of TiO_2 for oxidation of water or surface hydroxyl ions into $HO^{\cdot}$ radicals (Subramonian et al. 2017).

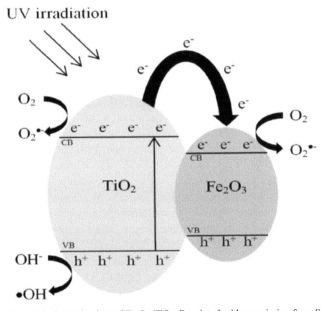

Figure 8.5. Proposed photocatalytic mechanism of Fe_2O_3-TiO_2. Reprinted with permission from Subramonian et al. 2017.

3.4 Treatment of pharmaceutical industry wastewaters

Pharmaceutical industrial facilities produce an increasing number of products for medical applications by humans and animals. Simultaneously, they produce large volumes of wastewaters with highly concentrated non-biodegradable organic compounds such as drugs, antibiotics, animal and plant steroids, hormones, anti-inflammatories, analgesics, lipid regulators, antidepressants, cytostatic agents, personal care products, other broadly used chemicals serving as reactants, reagents, catalysts and solvents in the overall process as well as their reaction residues (Alalm et al. 2016, Kumari and Tripathi 2019). Pharmaceutical industrial wastewaters contain toxic and hazardous non-biodegradable pollutants, which are ineffectively removed by conventional wastewater treatment methods (Alalm et al. 2016). Heterogeneous photocatalysis using TiO_2 and TiO_2-based photocatalysts has been investigated for the treatment of real pharmaceutical industry wastewaters, and representative scientific works are presented below.

Heterogeneous photocatalysis with TiO_2 (80% anatase and 20% rutile crystalline phases) in the form of nanoparticles was used for the degradation of ibuprofen (IBU) in a pharmaceutical industry wastewater under UV LEDs (Jallouli et al. 2018). Complete removal of IBU, as well as a mineralization percentage of 27%, was achieved with 2.5 g L^{-1} of catalyst. Microtox assay was also used to evaluate the toxicity before and after the process. The TiO_2 photocatalytic treatment was found to decrease the acute toxicity (from 73.9 to 30.3% inhibition of the bioluminescence of *Vibrio fischeri*) of the tested wastewater (Jallouli et al. 2018).

Sn-modified TiO_2 under UV irradiation was studied for the treatment of a pharmaceutical effluent taken from a pharmaceutical industry specialized in the production of nonsteroidal anti-inflammatory drugs (NSAIDs) (Gómez-Oliván et al. 2019). The main aim of the study was to evaluate the toxicity of the effluent before and after the treatment. Along with the toxicity, Sn-doped TiO_2 was evaluated in the degradation of selective drugs (paracetamol, naproxen, diclofenac, IBU) present in the effluent as well as in the overall abatement in terms of physicochemical parameters such as chlorides, fluorides, hardness, total phosphorus, total nitrogen, TSS, COD and BOD. Most of the physicochemical parameters decreased after the treatment with Sn-modified TiO_2. After the photocatalytic treatment, the toxicity of the effluent (based on testing the organism *Hyalella azteca*) and biomarkers of cell oxidation and antioxidation (oxidative stress phenomenon) were drastically reduced. A similar reduction was observed for most of the physicochemical parameters. The concentration of diclofenac, IBU, naproxen and paracetamol decreased by 78.8, 82.3, 82.7 and 86.9%, respectively, rendering the treatment with Sn-modified TiO_2 under UV irradiation an effective process for wastewater treatment (Gómez-Oliván et al. 2019).

Moreover, heterogeneous photocatalysis using commercial TiO_2 nanoparticles has been combined with other processes such as sonolysis (Verma et al. 2014) and biological treatment (Talwar et al. 2018) to treat real pharmaceutical wastewaters. In both cases, the combined processes showed a higher efficiency than the individual methods in terms of toxicity and COD reduction as well as biodegradability increase (Talwar et al. 2018, Verma et al. 2014).

Considering that the photocatalytic activity of TiO_2 can be improved by forming composites with multiwall carbon nanotubes (MWCNTs), Ahmadi et al. (2017) investigated the treatment of a real pharmaceutical wastewater using MWCNT/TiO_2 nanocomposite under UVC irradiation. COD and TOC removal of 84.9 and 82.3% were achieved, respectively (pH 5, [MWCNT/TiO_2] = 0.2 g L^{-1}, irradiation time 240 minutes). Enhancement of the biodegradability of the raw wastewater was also observed. Moreover, the formation of more simple and degradable compounds after the treatment was observed demonstrating the photocatalytic capabilities of MWCNT/TiO_2 nanocomposite. The superior photocatalytic activities of these nanocomposites are correlated with the inhibition of the recombination between electron-hole pairs, as well as the reduction of the agglomeration of TiO_2 nanoparticles that can be achieved by the addition of MWCNTs (Ahmadi et al. 2017).

A novel Fe-TiO_2 composite photocatalyst combining the *in situ* dual effect of photocatalysis and photo-Fenton has been applied for the treatment of a real pharmaceutical effluent using batch and continuous reactors (Bansal et al. 2018). In the batch mode and at the laboratory scale, the combination of coagulation and *in situ* dual process (using Fe-TiO_2 composite beads) resulted in a significant reduction in COD (89.1%) and an increase in the biodegradability index (BOD$_5$/COD) under optimum conditions (6 hours of treatment, [H_2O_2] = 1155 mg L^{-1} and a photocatalyst dose equivalent to 102% area of the reactor bed covered with Fe-TiO_2 composite beads). Successful COD reduction (~ 80%) of a real wastewater (5 L) in less than 120 minutes was also obtained when the treatment of the wastewater was carried out in a pilot-scale fixed-bed reactor using Fe-TiO_2 under natural solar irradiation. Fe-TiO_2 composite beads presented an excellent recyclability efficiency (> 70 recycles), without significant reduction in their activity. The parent chemicals present in the raw effluent were either completely removed or transformed into non-toxic intermediates after the

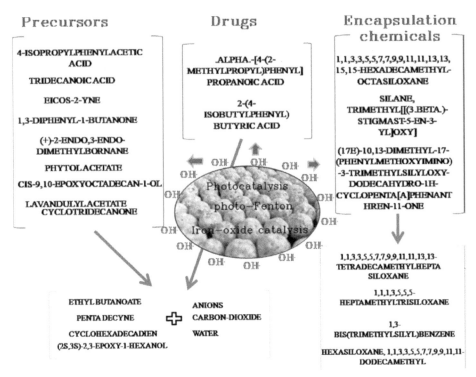

Figure 8.6. Tentative mechanism for the destruction of pollutants in real pharmaceutical effluent through *in-situ* dual process. Reprinted with permission from Bansal et al. 2018.

treatment, as depicted in Fig. 8.6 (Bansal et al. 2018). Cytotoxicity analysis and acute toxicity of the treated effluent confirmed the non-toxic nature of the treated wastewater (Bansal et al. 2018).

Mg-doped TiO_2 coated buoyant clay hollow-spheres with good photocatalytic properties reported earlier were also used for the treatment of pharmaceutical industry wastewaters under different light sources (Shivaraju et al. 2020). COD was reduced by 72, 61 and 68% from the diluted pharmaceutical wastewater under LED, UV and tungsten photo sources, respectively in 6 hours.

4. Conclusions and Future Perspectives

Environmental pollution due to industrial activities has been increasing in the last decades as industrial wastewaters have a variable composition with relatively low BOD and high COD content. Based on the complex and recalcitrant character of industrial effluents, their effective treatment is a difficult task being simultaneously time- and cost-consuming, and many times, a combination of various technologies is required. Although limited efforts have been devoted to the photocatalytic performance using real wastewaters, heterogeneous photocatalysis using TiO_2-based photocatalysts combined with UV/visible light has shown promising results for the treatment of real industrial wastewaters. Up to now, commercial, doped, hybrid and immobilized TiO_2-based photocatalysts have been used. TiO_2-based photocatalysts with improved properties have been synthesized to promote the photocatalytic activity of bare TiO_2. Based on the existing literature, high percentages of specific pollutants, COD, TOC and toxicity reduction, as well as an increase in biodegradability of the treated wastewaters, were achieved.

In general, TiO_2 photocatalysis can provide a promising pathway for the effective treatment of industrial wastewaters. However, for practical application, some challenges still need to be clarified and correlated among others with solar utilization and the separation of photocatalysts. Modification of TiO_2 by doping and the formation of nanocomposite materials are promising ways to enhance the efficiency under solar irradiation considering the low fabrication cost. More studies should be devoted to the separation and/or recovery and reuse of photocatalytic materials used for real wastewater treatment and large-scale applications.

Additionally, there is a lack of information about semiconductor modifications to achieve self-cleaning capability and reduce the risk of the deactivation of the catalyst. Thus, extensive research should be conducted on this issue. Moreover, alternative cost-effective and green methods for the synthesis of the photocatalysts as well as investigation of their efficiency at a large scale for successful field application are needed. Finally, life cycle assessment of nanocatalysts is required to address their overall benefits and environmental risks.

Acknowledgment

We acknowledge support of this work by the project "National Infrastructure in Nanotechnology, Advanced Materials and Micro-/Nanoelectronics" (MIS 5002772) which is implemented under the Action "Reinforcement of the Research and Innovation Infrastructure", funded by the Operational Programme "Competitiveness, Entrepreneurship and Innovation" (NSRF 2014–2020) and co-financed by Greece and the European Union (European Regional Development Fund).

References

Ahmadi, M., H.R. Motlagh, N. Jaafarzadeh, A. Mostoufi, R. Saeedi, G. Barzegar et al. 2017. Enhanced photocatalytic degradation of tetracycline and real pharmaceutical wastewater using MWCNT/TiO$_2$ nano-composite. J. Environ. Manage. 186: 55–63.

Alalm, M.G., A. Tawfik and S. Ookawara. 2016. Enhancement of photocatalytic activity of TiO$_2$ by immobilization on activated carbon for degradation of pharmaceuticals. J. Environ. Chem. Eng. 4: 1929–1937.

Alinsafi, A., F. Evenou, E.M. Abdulkarim, M.N. Pons, O. Zahraa, A. Benhammou et al. 2007. Treatment of textile industry wastewater by supported photocatalysis. Dyes Pigm. 74: 439–445.

Al-Mamun, M.R., S. Kader, M.S. Islam and M.Z.H. Khan. 2019. Photocatalytic activity improvement and application of UV-TiO$_2$ photocatalysis in textile wastewater treatment: A review. J. Environ. Chem. Eng. 7: 103248.

Ambaye, T.G. and K. Hagos. 2020. Photocatalytic and biological oxidation treatment of real textile wastewater. Nanotechnol. Environ. Eng. 5: 28.

Antonopoulou, M., C. Kosma, T. Albanis and I. Konstantinou. 2021. An overview of homogeneous and heterogeneous photocatalysis applications for the removal of pharmaceutical compounds from real or synthetic hospital wastewaters under lab or pilot scale. Sci. Total Environ. 765: 144163.

Arcanjo, G.S., A.H. Mounteer, C.R. Bellato, L.M.M.D. Silva, S.H. Brant Dias and P.R.D. Silva. 2018. Heterogeneous photocatalysis using TiO$_2$ modified with hydrotalcite and iron oxide under UV-visible irradiation for color and toxicity reduction in secondary textile mill effluent. J. Environ. Manage. 211: 154–163.

Bahadur, N. and N. Bhargava. 2019. Novel pilot scale photocatalytic treatment of textile & dyeing industry wastewater to achieve process water quality and enabling zero liquid discharge. J. Water Process. Eng. 32: 100934.

Bansal, P. and D. Sud. 2011. Photodegradation of commercial dye, Procion Blue HERD from real textile wastewater using nanocatalysts. Desalination 267: 244–249.

Bansal, P., A. Verma and S. Talwar. 2018. Detoxification of real pharmaceutical wastewater by integrating photocatalysis and photo-Fenton in fixed-mode. Chem. Eng. J. 349: 838–848.

Bordes, M.C., M. Vicent, R. Moreno, J. García-Montaño, A. Serra and E. Sánchez. 2015. Application of plasma-sprayed TiO$_2$ coatings for industrial (tannery) wastewater treatment. Ceram. Int. 41: 14468–14474.

Cardoso, J.C., G.G. Bessegato and M.V. Boldrin Zanoni. 2016. Efficiency comparison of ozonation, photolysis, photocatalysis and photoelectrocatalysis methods in real textile wastewater decolorization. Water Res. 98: 39–46.

Çifçi, D.I. and S. Meriç. 2015. Optimization of suspended photocatalytic treatment of two biologically treated textile effluents using TiO$_2$ and ZnO catalysts. Glob. Nest J. 17: 653–663.

Dong, H., G. Zeng, L. Tang, C. Fan, C. Zhang, X. He et al. 2015. An overview on limitations of TiO$_2$-based particles for photocatalytic degradation of organic pollutants and the corresponding countermeasures. Water Res. 79: 128–146.

Fotiou, T., T.M. Triantis, T. Kaloudis and A. Hiskia. 2015. Evaluation of the photocatalytic activity of TiO$_2$ based catalysts for the degradation and mineralization of cyanobacterial toxins and water off-odor compounds under UV-A, solar and visible light. Chem. Eng. J. 261: 17–26.

Garrido-Cardenas, J.A., B. Esteban-García, A. Agüera, J.A. Sánchez-Pérez and F. Manzano-Agugliaro. 2020. Wastewater treatment by advanced oxidation process and their worldwide research trends. Int. J. Environ. Res. Public Health 17: 170.

Ghaly, M.Y., T.S. Jamil, I.E. El-Seesy, E.R. Souaya and R.A. Nasr. 2011. Treatment of highly polluted paper mill wastewater by solar photocatalytic oxidation with synthesized nano TiO_2. Chem. Eng. J. 168: 446–454.

Gómez-Oliván, L.M., D.A. Solís-Casados, H. Islas-Flores and N.S. Juan-Reyes. 2019. Evaluation of the toxicity of an industrial effluent before and after a treatment with sn-modified TiO_2 under UV irradiation through oxidative stress biomarkers. pp. 157–175. *In*: Gómez-Oliván, L. (eds.). Pollution of Water Bodies in Latin America. Springer, Cham.

Goutam, S.P., G. Saxena, V. Singh, A.K. Yadav, R.N. Bharagava and K.B. Thapa. 2018. Green synthesis of TiO_2 nanoparticles using leaf extract of *Jatropha curcas* L. for photocatalytic degradation of tannery wastewater. Chem. Eng. J. 336: 386–396.

Guesh, K., Á. Mayoral, C. Márquez-Álvarez, Y. Chebude and I. Díaz. 2016. Enhanced photocatalytic activity of TiO_2 supported on zeolites tested in real wastewaters from the textile industry of Ethiopia. Microp. Mesop. Mat. 225: 88–97.

Gümüş, D. and F. Akbal. 2011. Photocatalytic degradation of textile dye and wastewater. Water Air Soil Pollut. 216: 117–124.

Hermosilla, D., N. Merayo, A. Gascó and Á. Blanco. 2015. The application of advanced oxidation technologies to the treatment of effluents from the pulp and paper industry: a review. Environ Sci. Pollut. Res. Int. 22: 168–191.

Jallouli, N., L.M. Pastrana-Martínez, A.R. Ribeiro, N.F.F. Moreira, J.L. Faria et al. 2018. Heterogeneous photocatalytic degradation of ibuprofen in ultrapure water, municipal and pharmaceutical industry wastewaters using a TiO_2/UV-LED system. Chem. Eng. J. 334: 976–984.

Jing, L., W. Zhou, G. Tian and H. Fu. 2013. Surface tuning for oxide-based nanomaterials as efficient photocatalysts. Chem. Soc. Rev. 42: 9509.

Kabra, K., R. Chaudhary and R.L. Sawhney. 2004. Treatment of hazardous organic and inorganic compounds through aqueous-phase photocatalysis: a review. Ind. Eng. Chem. Res. 43: 7683–7696.

Kebir, M., M. Trari, R. Maachi, N. Nasrallah, B. Bellal and A. Amrane. 2015. Relevance of a hybrid process coupling adsorption and visible light photocatalysis involving a new hetero-system $CuCo_2O_4/TiO_2$ for the removal of hexavalent chromium. J. Environ. Chem. Eng. 3: 548–559.

Khaki, M.R.D., M.S. Shafeeyan, A.A.A. Raman and W.M.A.W. Daud. 2017. Application of doped photocatalysts for organic pollutant degradation—A review. J. Environ. Manage. 198: 78–94.

Konstantinou, I.K. and T.A. Albanis. 2004. TiO_2-assisted photocatalytic degradation of azo dyes in aqueous solution: kinetic and mechanistic investigations. A review. Appl. Catal. B Environ. 49: 1–14.

Kumari, V. and A.K. Tripathi. 2019. Characterization of pharmaceuticals industrial effluent using GC–MS and FT-IR analyses and defining its toxicity. Appl. Water Sci. 9: 185.

Legrini, O., E. Oliveros and A.M. Braun. 1993. Photochemical processes for water treatment. Chem. Rev. 93: 671–698.

Li, D., H. Song, X. Meng, T. Shen, J. Sun, W. Han et al. 2020. Effects of particle size on the structure and photocatalytic performance by alkali-treated TiO_2. Nanomaterials 10: 546.

Litter, M.I., W. Choi, D.D. Dionysiou, P. Falaras, A. Hiskia, G. Li Puma et al. 2012. Nanotechnologies for the treatment of water, air and soil. J. Hazard. Mater. 211–212: 1–2.

Moiseev, A., H. Schroeder, M. Kotsaridou-Nagel, S.U. Geissen and A. Vogelpohl. 2004. Photocatalytical polishing of paper-mill effluents. Water Sci. Technol. 49: 325–330.

Muñoz, I., J. Rieradevall, F. Torrades, J. Peral and X. Domènech. 2006. Environmental assessment of different advanced oxidation processes applied to a bleaching Kraft mill effluent. Chemosphere 62: 9–16.

Mutamim, N.S.A., Z.Z. Noor, M.A.A. Hassan and G. Olsson. 2012. Application of membrane bioreactor technology in treating high strength industrial wastewater: a performance review. Desalination 305: 1–11.

Nakata, K. and A. Fujishima. 2012. TiO_2 photocatalysis: design and applications. J. Photochem. Photobiol. C Photochem. Rev. 13: 169–189.

Nogueira, V., I. Lopes, T.A.P. Rocha-Santos, F. Gonçalves and R. Pereira. 2018. Treatment of real industrial wastewaters through nano-TiO_2 and nano-Fe_2O_3 photocatalysis: case study of mining and kraft pulp mill effluents. Environ. Technol. 39: 1586–1596.

Okamoto, K., Y. Yamamoto, H. Tanaka, M. Hanaka and A. Itaya. 1985. Heterogeneous photocatalytic decomposition of phenol over TiO_2 powder. Bull. Chem. Soc. Jpn. 58: 2015–2022.

Paumo, H.K., S. Dalhatou, L.M. Katata-Seru, B.P. Kamdem, J.O. Tijani, V. Vishwanathan et al. 2021. TiO_2 assisted photocatalysts for degradation of emerging organic pollutants in water and wastewater. J. Mol. Liq. 331: 115458.

Pawar, M., S.T. Sendoğdular and P. Gouma. 2018. A brief overview of TiO_2 photocatalyst for organic dye remediation: case study of reaction mechanisms involved in Ce-TiO_2 photocatalysts system. J. Nanomater. 2018: 5953609.

Pekakis, P.A., N.P. Xekoukoulotakis and D. Mantzavinos. 2006. Treatment of textile dyehouse wastewater by TiO_2 photocatalysis. Water Res. 40: 1276–1286.

Pillai, S.C., N.B. McGuinness, C. Byrne, C. Han, J. Lalley, M. Nadagouda et al. 2017. Photocatalysis as an effective advanced oxidation process. *In*: Stefan, M.I. (ed.). Advanced Oxidation Processes for Water Treatment: Fundamentals and Applications. IWA Publishing.

Rafiq, A., M. Ikram, S. Ali, F. Niaz, M. Khan, Q. Khan et al. 2021. Photocatalytic degradation of dyes using semiconductor photocatalysts to clean industrial water pollution. J. Ind. Eng. Chem. (in press).

Raman, C.D. and S. Kanmani. 2016. Textile dye degradation using nano zero valent iron: A review. J. Environ. Manage. 177: 341–355.

Rashidi, S., M. Nikazar, A.V. Yazdi and R. Fazaeli. 2014. Optimized photocatalytic degradation of Reactive Blue 2 by TiO_2/UV process. J. Environ. Sci. Health A 49: 452–462.

Rueda-Marquez, J.J., I. Levchuk, P. Fernández Ibañez and M. Sillanpää. 2020. A critical review on application of photocatalysis for toxicity reduction of real wastewaters. J. Clean. Prod. 258: 120694.

Sahoo, C., A.K. Gupta and I.M.S. Pillai. 2012a. Heterogeneous photocatalysis of real textile wastewater: Evaluation of reaction kinetics and characterization. J. Environ. Sci. Health A 47: 2109–2119.

Sahoo, C., A.K. Gupta and I.M.S. Pillai. 2012b. Photocatalytic degradation of methylene blue dye from aqueous solution using silver ion-doped TiO_2 and its application to the degradation of real textile wastewater. J. Environ. Sci. Health A 47: 1428–1438.

Saikumari, N., S.M. Dev and S.A. Dev. 2021. Effect of calcination temperature on the properties and applications of bio extract mediated titania nano particles. Sci. Rep. 11: 1734.

Schrank, S.G., H.J. José, R.F.P.M. Moreira and H.F. Schröder. 2004. Elucidation of the behavior of tannery wastewater under advanced oxidation conditions. Chemosphere 56: 411–423.

Selvabharathi, G., S. Adishkumar, S. Jenefa, G. Ginni, J.R. Banu and I.T. Yeom. 2016. Combined homogeneous and heterogeneous advanced oxidation process for the treatment of tannery wastewaters. J. Water Reuse Desalination 6: 59–71.

Shivaraju, H.P., S.R. Yashas and R. Harini. 2020. Application of Mg-doped TiO_2 coated buoyant clay hollow-spheres for photodegradation of organic pollutants in wastewater. Mater. Today: Proc. 27: 1369–1374.

Singh, M., J. Singh, M. Rawat, J. Sharma and P.P. Singh. 2019. Enhanced photocatalytic degradation of hazardous industrial pollutants with inorganic–organic TiO_2–SnO_2–GO hybrid nanocomposite. J. Mater. Sci. Mater. Electron. 30: 13389–13400.

Souza, R.P., T.K.F.S. Freitas, F.S. Domingues, O. Pezoti, E. Ambrosio, A.M. Ferrari-Lima et al. 2016. Photocatalytic activity of TiO_2, ZnO and Nb_2O_5 applied to degradation of textile wastewater. J. Photochem. Photobiol. A Chem. 329: 9–17.

Subramonian, W., T.Y. Wu and S.P. Chai. 2017. Photocatalytic degradation of industrial pulp and paper mill effluent using synthesized magnetic Fe_2O_3-TiO_2: Treatment efficiency and characterizations of reused photocatalyst. J. Environ. Manage. 187: 298–310.

Talwar, S., V.K. Sangal and A. Verma. 2018. Feasibility of using combined TiO_2 photocatalysis and RBC process for the treatment of real pharmaceutical wastewater. J. Photochem. Photobiol. A Chem. 353: 263–270.

Tan, Y., C. Li, Z. Sun, C. Liang and S. Zheng. 2020. Ternary structural assembly of $BiOCl/TiO_2$/clinoptilolite composite: Study of coupled mechanism and photocatalytic performance. J. Colloid. Interface Sci. 564: 143–154.

Tripathi, A. and S. Narayanan. 2018. Impact of TiO_2 and TiO_2/g-C_3N_4 nanocomposite to treat industrial wastewater. Environ. Nanotechnol. Monit. Manage. 10: 280–291.

Vaiano, V., O. Sacco, M. Stoller, A. Chianese, P. Ciambelli and D. Sannino. 2014. Influence of the photoreactor configuration and of different light sources in the photocatalytic treatment of highly polluted wastewater. Int. J. Chem. React. Eng. 12: 63–75.

Verma, A., I. Chhikara and D. Dixit. 2014. Photocatalytic treatment of pharmaceutical industry wastewater over TiO_2 using immersion well reactor: synergistic effect coupling with ultrasound. Desalination Water Treat. 52: 6591–6597.

Vilar, V.J.P., L.X. Pinho, A.M.A. Pintor and R.A.R. Boaventura. 2011. Treatment of textile wastewaters by solar-driven advanced oxidation processes. Sol. Energy 85: 1927–1934.

Wang, K.H., Y.H. Hsieh, M.-Y. Chou and C.Y. Chang. 1999. Photocatalytic degradation of 2-chloro and 2-nitrophenol by titanium dioxide suspensions in aqueous solution. Applied Catalysis B: Environmental 21: 1–8.

Yeber, M.C., J. Rodríguez, J. Freer, N. Durán and H.D. Mansilla. 2000. Photocatalytic degradation of cellulose bleaching effluent by supported TiO_2 and ZnO. Chemosphere 41: 1193–1197.

Younis, S.A. and K.H. Kim. 2020. Heterogeneous photocatalysis scalability for environmental remediation: opportunities and challenges. Catalysts 10: 1109.

Yuan, R., R. Guan, P. Liu and J. Zheng. 2007. Photocatalytic treatment of wastewater from paper mill by TiO_2 loaded on activated carbon fibers. Colloids Surf. A Physicochem. Eng. Asp. 293: 80–86.

Zhang, T., X. Wang and X. Zhang. 2014. Recent progress in TiO_2-mediated solar photocatalysis for industrial wastewater treatment. Int. J. Photoenergy. 2014: 607954.

Zhang, H., X. He, W. Zhao, Y. Peng, D. Sun, H. Li et al. 2017. Preparation of Fe_3O_4/TiO_2 magnetic mesoporous composites for photocatalytic degradation of organic pollutants. Water Sci. Technol. 75: 1523–1528.

Zhao, W., X. He, Y. Peng, H. Zhang, D. Sun and X. Wang. 2017. Preparation of mesoporous TiO_2 with enhanced photocatalytic activity towards tannery wastewater degradation. Water Sci. Technol. 75: 1494–1499.

Zhao, Y., W. Chang, Z. Huang, X. Feng, L. Ma, X. Qi et al. 2017a. Enhanced removal of toxic Cr (VI) in tannery wastewater by photoelectrocatalysis with synthetic TiO_2 hollow spheres. Appl. Surf. Sci. 405: 102–110.

Zhao, Y., Z. Huang, W. Chang, C. Wei, X. Feng, L. Ma et al. 2017b. Microwave-assisted solvothermal synthesis of hierarchical TiO_2 microspheres for efficient electro-field-assisted photocatalytic removal of tributyltin in tannery wastewater. Chemosphere 179: 75–83.

Zhou, F., H. Wang, S. Zhou, Y. Liu and C. Yan. 2020. Fabrication of europium-nitrogen co-doped TiO_2/Sepiolite nanocomposites and its improved photocatalytic activity in real wastewater treatment. Appl. Clay Sci. 197: 105791.

Chapter 9

Nano- and Microplastics in Water

Roberta Hofman-Caris[1,2,*] and *Patrick Bäuerlein*[1]

1. Introduction

The present geological age is known as the 'Anthropocene', a period in which the influence of man on the environment cannot be ignored. One of the inventions of this era is plastics. In the 20th century, plastics were developed, these materials appeared to have very interesting and useful properties like low weight, high strength, versatility and durability. This, in combination with their low production costs, led to an enormous increase in production since the 1950s. They are used in every aspect of our lives (Wang et al. 2021), for packaging, constructions, toys, personal care products, electronic devices, etc. (Kirstein et al. 2021). In 2017, global plastics production reached 348 million tons (Plastics Europe 2019). According to Wang et al. (2021), in 2018 already 359 million tons of plastics were produced, showing a rapid increase. It is predicted that by 2050 this will increase to 33 billion tons (Xu et al. 2020).

Plastics can be divided into two categories: thermosets and thermoplastics. During the production of thermosets, crosslinking is induced by heating, resulting in the formation of new and irreversible covalent bonds between the polymer chains, thus forming a three-dimensional network. On heating, they keep their shape, until the bonds are irreversibly broken at very high temperatures. As a result, these plastics are extremely stable and show enhanced chemical resistance, heat resistance and structural integrity. The disadvantages of their properties, however, are that they are hard to recycle and very persistent (Rodríguez-Narvaez et al. 2021). Examples of thermosets are polyurethane, epoxy and alkyd compounds, and are often used as adhesives, construction panels, plywood, very stable coatings and insulators. Thermoplastics, on the other hand, do not form a stable network, but the polymer chains are held together by Van der Waals forces. As a result, these plastics are very flexible, and by heating, they can adopt a new form. Therefore, they can be recycled and remolded, and thus are more widely used in consumer goods (Andrady 2017, Jiang et al. 2020). Examples of common thermoplastics are polyethylene (PE), polypropylene (PP), polystyrene (PS), polyvinylchloride (PVC) and polyesters (PES) including polyethylene terephthalate (PET). PE, PET, PP, PVC and PS together account for approximately 90% of the total global plastic production (Wang et al. 2021). PE is used in a wide variety of cheap products, depending on its density. High-Density PE (HDPE) is used in bottles, toys, chemical containers and pipe systems. Low-Density PE (LDPE) is used in trays and containers, corrosion-resistant work surfaces, soft and pliable parts, floor tiles, juice and milk cartons and plastic wraps. PP is used for bottle caps, piping systems, furniture, food containers, tableware and car bumpers. PS is, e.g., used for packaging,

[1] KWR Water Research Institute.

[2] Wageningen University and Research.

* Corresponding author: Roberta.Hofman@kwrwater.nl

surfboards, automobile parts, food containers and tableware. PVC is used for piping, blood bags and tubing, wire and cable insulation, window frames and flooring. PES/PET is the most recycled thermoplastic, and it is used for packaging (bottles), tapes, textiles, and in the electrical and electronics industry. Plastic fibers, made of PES or PP, are used in clothing, agricultural, industrial and textile products. Unlike PET, PE, PVC and PS have a relatively short service life. Another polymer, which has not yet been mentioned but also can be encountered in the environment is polyamide (PA; nylon).

During the past few years, concerns have arisen on the presence of very small plastic particles (microplastics, MPs and nanoplastics, NPs) in the environment. This is illustrated in Fig. 9.1, which shows an exponential increase in scientific publications on MPs in water since 2015.

Plastic particles and fibers can be found in the environment, with a large variety in size, as shown in Table 9.1.

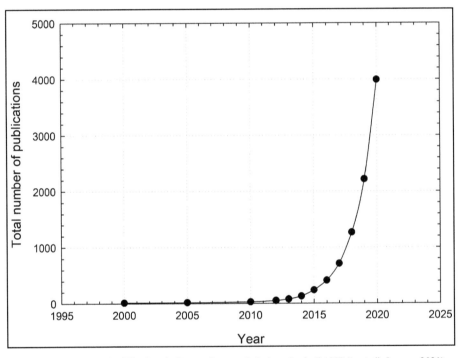

Figure 9.1. Number of publications in Scopus (keywords "microplastics" AND "water"; January 2021).

Table 9.1. Size definitions of plastics in the literature.

Plastics	Size range	Reference
Macroplastics	> 5 mm	(Hahladakis et al. 2018)
Macroplastics	> 25 mm	(Xu et al. 2020)
Mesoplastics	5 – 25 mm	(Xu et al. 2020)
Microplastics	1 μm–5 mm	(Estahbanati and Fahrenfeld 2016, Horton et al. 2017, Hahladakis et al. 2018, Enfrin et al. 2019)
Large microplastics	> 1 mm	(Enfrin et al. 2019)
Small microplastics	< 1 mm	(Enfrin et al. 2019)
Nanoplastics	< 1 μm	(Estahbanati and Fahrenfeld 2016, Horton et al. 2017, Hahladakis et al. 2018, Enfrin et al. 2019)

As plastics can have different shapes, a more detailed definition may be required. da Costa et al. (2016) defined NPs as particles with at least two dimensions < 100 nm and MPs with at least one dimension < 1 μm. Macroplastics, according to these authors, have a size ranging from 2.5–100 cm, and mesoplastics from 0.1–2.5 cm. According to Sharma et al. (2021), the fraction of MPs in the total weight of plastic accumulation around the world is predicted to be 13.2% by 2060.

The analysis of MPs and NPs is rather difficult, and only recently reliable methods have been developed, but mainly focusing on MPs. However, as it is known that NPs are formed by degradation of larger particles (like MPs), it can safely be assumed that knowledge about MPs is relevant for the presence of NPs in the environment also. MPs can be divided into primary and secondary MPs (Estahbanati and Fahrenfeld 2016). Primary MPs are manufactured with sizes in the μm- or mm-range. They are often added to personal care products or used to produce plastic products. PE microbeads are used as exfoliants in cosmetics, detergents, toothpaste, scrub and drug carriers. Glitters used in cosmetics, crafts and textiles also are primary MPs. Furthermore, they are used in preproduction pellets, air blasting and granulates (Carr et al. 2016, Jiang 2018, Jiang et al. 2020, Xu et al. 2020, Sharma et al. 2021, Wang et al. 2021). Primary MPs in general are released into industrial and domestic wastewater. The main sources of primary MPs in the environment are industrial discharge and spillage (plastic resin powder, pellet spillage from air lasting machines and raw materials from the plastic industry) and the microbeads in personal care products. Lv et al. (2019) estimated that in China alone yearly 209.7×10^{12} primary MPs enter the surface water. It is estimated that annually approximately 1.5 million tons of primary MPs are released into oceans (Xu et al. 2020, Wang et al. 2021). After a proposal was launched in 2015 to limit the use of such primary MPs, which received widespread public acclaim, the number of MPs in this type of application has significantly decreased.

Secondary MPs, which are considered to be the largest source of pollution by MPs in the aquatic environment when measured in numbers (Andrady 2017), are formed by the degradation of plastic debris (containers, nets, line fibers, films and tires) due to chemical, physical and biological processes (Jiang et al. 2020, Xu et al. 2020, Sharma et al. 2021, Wang et al. 2021). Lv et al. (2019) and Jiang (2018) also point to laundry as a source of MPs (polyesters, acrylics and polyamides) in the aquatic environment. The latter measured over 100 fibers per liter of municipal WWTP effluent. The sources for these secondary MPs, therefore, are diverse and numerous.

The degradation rate of macroplastics is affected by parameters such as temperature and sunlight, and by the properties of the plastics themselves (composition, size, density, surface roughness). Weathering and photolysis by sunlight can degrade plastics, but also mechanical forces (abrasion, fluctuation and turbulence) can produce MPs, and even biodegradation may occur (Carr et al. 2016, da Costa et al. 2016, Wang et al. 2021). However, it may take more than 50 years to reach complete degradation in the environment. In marine environments, degradation may be enhanced by abrasion, wave action and turbulence. An overview of factors affecting the degradation of polymers in the aquatic environment is given by Andrady (2017) (Table 9.2).

Table 9.2. Qualitative overview of factors affecting the degradation of MPs in different zones of a maritime and fresh water environment, based on (Andrady 2017).

Zone	UV radiation in sunlight	Temperature	Availability of O_2	Fouling
Beach	Large effects	Very large effect	Very large effect	No
Maritime surface water	Large effects	Small effect	Large effect	Yes
Fresh surface water	Large effects	Small effect	Large effect	Yes
Middle to deep marine water	No effects	Large effect	Small effect	No
Middle to deep fresh water	No effects	Large effect	Small effect	Possibly
Sediment	No effect	Very small effect	Very small effect	Yes

2. Analyses of MPs

In order to be able to study the occurrence of NPs and MPs in the aquatic system, adequate analytical techniques are required. Comparison of data often appears to be difficult as different methods are used for sampling, processing and identification. Furthermore, different reporting limits for sizes and different ways to determine shapes are applied (Koelmans et al. 2019, Barchiesi et al. 2020). Lv et al. (2019) applied stereomicroscopy to count the numbers of particles of various sizes in water. Araujo et al. (2018) applied Raman spectroscopy to identify MPs with sizes < 20 μm. Dying MPs with Nile Red to make them countable by means of fluorescence microscopy was applied by Ferraz et al. (2020). The chemical composition of MPs can be determined by means of micro-Fourier transform infrared (μFTIR) and Raman spectroscopy (Wang et al. 2021). μFTIR for the characterization of MPs has been described by various authors (Corami et al. 2021, Qin et al. 2021, Sorasan et al. 2021, Chand et al. 2022). Sorasan et al. (2021) showed that this technique also can be applied for the determination of NPs (and that NPs are generated from MPs). Raman imaging has a higher sensitivity and therefore is better suited for the identification of small size MPs (down to 1 μm), but the analysis takes more time than μFTIR. The presence of fluorescent compounds in MPs, however, can disturb Raman detection and identification. Both techniques require that the plastic particles be as pure as possible, meaning that organic and inorganic contamination, including water, need to be removed. This often requires a laborious and time-consuming work-up of environmental samples. Thermal desorption coupled with gas chromatography-mass spectrometry (TDS-GC-MS) (Dümichen et al. 2015, Campbell et al. 2020) and pyrolysis coupled with gas chromatography-mass spectrometry (Py-GC-MS) (Ahmed et al. 2021, Ainali et al. 2021, Castelvetro et al. 2021, Hermabessiere and Rochman 2021, Lauschke et al. 2021) can be used to identify the composition of MPs. As Py-GC-MS has a greater sensitivity than TDS-GC-MS, it can identify lower masses of NPs (down to about 50 μg). This technique, however, is destructive and does not allow the collection of particle numbers and characteristics. Yu et al. (2019), Kirstein et al. (2021), Liu et al. (2021) applied both μFTIR imaging and Py-GC-MS for the analysis of MPs in various matrices. They found that both techniques identified different polymer types in samples with overall low MPs content. With increasing concentrations of a given polymer type, the data generated by both techniques showed better comparability. Another technique, which has been employed recently in MPs research, is an infrared quantum cascade laser (Primpke et al. 2020, Mughini-Gras et al. 2021). The advantage of this technique is that it is in general faster than Raman or μFTIR, but, unfortunately, it delivers less information as it covers a narrower infrared band. To cover the whole size range and get as much information as possible, several techniques need to be applied (Mintenig et al. 2018).

As concentrations of particles in water samples often are very low, special attention should be paid to sampling and extraction or concentration of MPs in water. More consideration needs to be paid both during sampling and preprocessing to possible contamination of the samples by MPs originating from the devices used in their research (Costa et al. 2018). Estahbanati and Fahrenfeld (2016) could not consider fibers, as these appeared to occur as a tangle, which made characterization difficult. Dris et al. (2018) reported serious problems caused by the blocking of the sampling instrument by fibers or lines in the River Seine in Paris.

3. Plastics in the Environment

The lifetime of plastic products varies from one to over fifty years, after which 9% is recycled, 12% is used for energy generation, 8% becomes landfill and 71% eventually ends up in the environment (Hahladakis et al. 2018). However, these data strongly depend on the type of plastic. For PVC, it has been reported that 82% is discarded in landfills, 15% is incinerated and only 3% can be recycled (Jiang et al. 2020). The life cycle of plastic products is shown in Fig. 9.2.

A significant part of plastics produced finally ends up in the environment, either directly or via improper reuse or recycling (Jiang et al. 2020). These authors estimated that from the 1950s to 2015 about 6.3 billion tons of plastic wastes have been generated worldwide, and that, if the trend

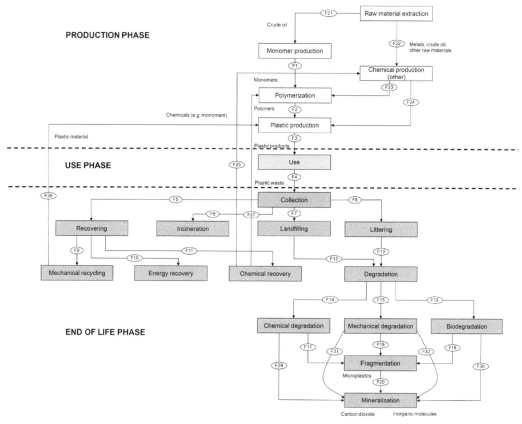

Figure 9.2. Lifecycle of plastic products (Hahladakis et al. 2018).

continues, this will increase to 26 billion tons by 2050. Of the over 300 million tons of plastics produced annually, about 13 million tons end up in rivers and oceans (Hahladakis et al. 2018, Enfrin et al. 2019, Lv et al. 2019). According to these authors, 29% of this waste material is produced in China, 18% in Europe and 18% in North America.

These days plastics are found everywhere, from the Arctic to Antarctica, in marine environments, wastewaters, surface waters, soils, sediments, food and air (Barceló and Picó 2019). In surface water, concentrations varying from 10^{-2} to 10^8 items/m^3 have been reported (Koelmans et al. 2019). In subsurface Arctic water samples, collected south and southwest of Svalbard, Norway, MP concentrations of 1–1.31 and 0–11.5 MPs/m^3 were measured. In the subwater near northeast Greenland, concentrations were shown to range from 1 to 3 MPs/m^3, and, in the vicinity of the Antarctic Peninsula, concentrations of 755–3524 MPs per m^3 were measured (Xu et al. 2020). The contribution of different regions to the MPs in oceans is as follows: 18.3% from India and South Asia, 17.2% from North America, 15.9% from Europe and Central Asia, 15.8% from China, 15.0% from East Asia and Oceania, 9.1% from South America and 8.7% from Africa and the Middle East. The average concentration of MPs in the Atlantic Ocean is 1.15/m^3, but, in the Northeastern Pacific Ocean, it ranges from 8 to 9200 per m^3. In the Carpathian Basin of Europe, MP concentrations range from 3.52 to 32.05/m^3 (Wang et al. 2021). The Danube transports MPs to the Black Sea, resulting in an MP content of about 7.5 mg/m^3/s, which corresponds to an annual load of approximately 1553 tons (Xu et al. 2020). In the northwestern Pacific Ocean, MP concentrations ranging from 640 to 42,000/km^2 have been observed, depending on the action of currents. In the surface waters of the Arabian Bay, concentrations of 4.38×10^4 to 1.46×10^6/km^2 were measured. In general, the highest concentrations of MPs can be observed in waters adjacent to highly urbanized areas. It was shown that seawater off the coast of South Korea contained 1051 particles/m³, whereas only 560 particles/m³ were found in rural areas.

Freshwater all over the world also contains significant concentrations of MPs. In China, high concentrations of MPs are found in freshwater, with concentrations varying from 1597 to 12,611 particles per m³ (Wang et al. 2021). Concentrations of 293–7924 MPs/m³ were observed in Hong Lake, Pearl River in China, and Nakdong River in South Korea (Xu et al. 2020). In Saigon River (Vietnam), even concentrations of 172,000 to 410,000 MPs/m³ have been observed. Ferraz et al. (2020) report that the Sinos River in Southern Brazil contains an average of 330.2 MPs/L, fibers being the most abundant shapes present. The authors suggest that they originate from untreated sewage from laundry machines. It should be noted that data are often difficult to compare, as MP concentrations can be expressed in numbers and masses, per volume or surface area.

In addition, for freshwater, it can be concluded that high concentrations of MPs can be found in areas with anthropogenic activities, but the actual concentrations are also affected by hydrological and meteorological conditions. With increasing residence and circulation time, plastics are broken into MPs, increasing their amounts in the water. On the other hand, strong winds may cause vertical mixing and thereby reduce the MP abundance. Freshwater sediments are major sinks of MPs, and high concentrations of MPs have been observed there.

4. Emission Routes into the Environment

Horton et al. (2017) presented an overview of sources and streams of MPs in the environment, in which disposal, degradation, accumulation in soils, soil erosion and runoff, wind dispersal, flooding and tidal deposition are taken into account (Fig. 9.3).

Several release pathways of MPs can be identified. There are human sources (e.g., textiles, personal care products), transportation sources (e.g., erosion of synthetic rubber tires), and industrial sources (e.g., pellets). According to Shen et al. (2020), the most important sources of MPs in the environment are agriculture and WWTP effluents, but Eerkes-Medrano et al. (2019) also pointed to the influence of atmospheric deposition. High concentrations of MPs in water bodies are found in areas with a high degree of urbanization, and a relation can be found with waste management (e.g., leaching of landfills) and sewage spillage. Both industrial and municipal

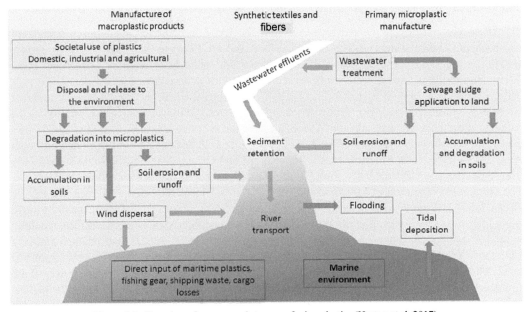

Figure 9.3. Overview of sources and streams of microplastics (Horton et al. 2017).

wastewater treatment plants (WWTPs) have been shown to release MPs to the environment (Carr et al. 2016, Estahbanati and Fahrenfeld 2016, Jiang 2018, Kay et al. 2018, Prata 2018, Yang et al. 2019). In general, WWTP processes consist of a combination of processes, including coagulation/flocculation followed by flotation or sedimentation to remove solid particles and filtration to remove suspended impurities, often followed by disinfection (Enfrin et al. 2019). More and more membrane filtration and/or oxidation processes are applied. According to Wang et al. (2021), a WWTP without a tertiary treatment step will remove on average 88% of MP, whereas WWTPs with tertiary treatment may remove over 97%. In the Netherlands and Scotland, about 98% removal is obtained (Eerkes-Medrano et al. 2019). Cheng et al. (2021) report that the influent of WWTPs contains 61 to 5,600 µg MPs per liter or 1.01 to 31,400 items/L, which is reduced to 0.5–170 µg/L, or 0.004 to 447 items/L in the effluent. Enfrin et al. (2019) described a process consisting of grids, sand filters and sedimentation tanks for the removal of "large" particles, followed by activated sludge in an aerated tank of membrane bioreactor (MBR), and finally disinfection using UV radiation, ozone or chlorine. According to these authors, WWTPs daily discharge between 9.3×10^5 and 4.0×10^9 particles, which corresponds to about 200 PET bottles. Up to 45% of the microplastics (100–5000 µm) are removed during the first treatment step, as they are trapped in precipitating particles or floating grease and oil (the latter can be removed by Dissolved Air Flotation (DAF)). Subsequently, 50% is removed in the second treatment step, as they are trapped during flocculation or precipitated by bacteria. Microplastics smaller than 500 µm can thus be removed. Less than 2% of the microplastics are found in the effluent of the third treatment step, mostly particles with a size of 20 to 100 µm. Membrane bioreactors (MBR) appear to remove 99.9% of all microplastics, rapid sand filtration 97%, DAF 95% and disk filters 40–98,5% (Talvitie et al. 2017). According to Ziajahromi et al. (2017), modern treatment processes such as membrane filtration can be very effective, but even Reverse Osmosis (RO) membranes still appear to let part of the microplastic fibers pass. However, due to the high plastic load of wastewater and the large flows, still significant amounts of MPs are released into surface water by WWTPs. These results are also affected by the analytical methods applied. Often the MP concentration is expressed in numbers per m^3. During the treatment, larger particles will be degraded and split into smaller particles, resulting in a larger number of particles. This was visualized by Enfrin et al. (2019), who showed that WWTPs cause fragmentation of 80% of all MPs, generating 10 times as many NPs. Furthermore, treatment may also change the chemical properties of MPs, e.g., by oxidation of the particle surface, thus increasing the surface roughness and porosity of the material. As a result, the adsorption capacity for micropollutants like metals may increase by a factor of 10! Besides, the biofilm at the particles will grow, resulting in increased surface hydrophobicity and roughness. When the WWTP effluent is disinfected, this process may also affect MPs. UV radiation, e.g., may result in the formation of carboxylic groups and scission of polymer chains. Exposure of PE to ozone or chlorine resulted in the formation of similar groups and emission of chlorine-containing components (Enfrin et al. 2019).

Two specific WWTP processes were described by Lv et al. (2019): MBR and an Oxidation Ditch (OD). The measurements were done at the WWTP in Wuxi, Jiangsu, China. The influent contained 47% PET, 20% polystyrene, 18% polyethylene and 15% polypropylene. Sixty five percent of the polymer consisted of fragments (secondary plastics) and 21% were fibers (in general PET). Furthermore, 12% were polymer films and 2% were foams, but real particles were hardly observed. Forty percent of all plastics were > 500 µm and 29% had a size between 62.5 and 125 µm. It was found that MBR removed 99.5% of the polymers, and the OD 97%. This leads to removal efficiencies of 821 and 53.6% respectively. In the sludge, accumulation of the particles removed from the water phase was observed.

When the MPs enter the aquatic environment, a biofilm will be formed at their surfaces within 7–14 days. This biofilm is composed of bacteria, fungi and algae (Wang et al. 2021). Thus, the physical and chemical properties of the particles (like density and surface charge) are changed, which may affect their transport and fate in the aquatic environment. In the environment, not only degradation will occur, but also accumulation and migration (Wang et al. 2021). In general, plastics like PVC, nylon and PET are more likely to sink than PP, PE and PS. The transport of MP in freshwater is affected by the water body size, wind currents and the density of the particles (Wang et al. 2021). Plastics that can float or stay suspended are eventually taken to oceans. Here, at great depths with low oxygen content and salinity, the degradation rate of the MP is significantly slowed down.

Significant amounts of MPs in the oceans originate from activities at sea such as fishing, shipping and marine industry and other MPs originate from beach littering and tourism. Atmospheric transport also contributes to the amounts of MP in the oceans (Wang et al. 2021). However, land-based plastic waste appears to contribute to 80% of the plastic waste in the ocean. Rivers serve as the primary pathway for large plastic fragments. Wang et al. (2021) estimate that yearly 1.15–2.41 million tons of plastics enter the ocean in this way; the top 20 polluting rivers are primarily located in Asia, accounting for 67% of this global total.

According to Eerkes-Medrano et al. (2019) and Barchiesi et al. (2020), it is not clear yet how and how many MPs can reach groundwater, but they have been already identified in samples related to groundwater. So far, only low particle number concentrations have been found in samples related to groundwater (Koelmans et al. 2019). However, it cannot be excluded that such particles (also) may have entered the water samples during sampling or treatment procedures, as numbers are very low.

The composition of MPs in aqueous environments, in general, is PE ≈ PP > PS > PVC > PET (Koelmans et al. 2019).

In water, various types of plastics can be found (Andrady 2017, He et al. 2019). Thermoplastics result in:

- Plastic fragments (pieces of irregular thick plastic with all three size dimensions comparable (fragments of products, virgin pellets, microbeads, fragments of thermoplastic paints, plastic packaging materials, food service items))
- Fiber fragments: elongated particles for which one dimension is much larger than both other dimensions (fishing gear, textile fibers from laundering, cigarette filter fibers)
- Plastic foams, with a spongy texture (expanded PS foam, packaging materials)

Thermosets result in:

- Composites (boat hull fragments, epoxy adhesive fragments)
- Coatings (thermoset protective coating fragments)
- Thermoset foam (polyurethane foams)

Furthermore, pellets (spherical particles) and flakes (sheets of plastic with a thickness significantly lower than both other dimensions) can be distinguished (da Costa et al. 2016). Finally, there is rubber material (e.g., balloons).

In most cases, plastics that are found in WWTP influents appear to be fragments of larger particles, which disintegrate slowly. Furthermore, many fibers can be observed. In almost 50% of all cases, PET plastics are identified, and the other 50% of plastics consist of PS, PE and PP. Yang et al. (2019) demonstrated the presence of 18 types of polymers in WWTP effluents. Over 70% of the polymers consisted of PET, PS and PP. Microfibers were encountered most frequently, their average size being 1,111 ± 863 μm. Particles only represented about 14% of the total load of polymers, with an average size of 681 ± 529 μm. These authors estimate a daily discharge of 15×10^3 to 4.5×10^6 MPs from a WWTP to surface water, regardless of the removal efficiency. In about 40% of all cases, the particles are > 500 μm, and particles with a size of 25–62.5 μm are less significantly observed. Estahbanati and Fahrenfeld (2016) studied four categories of MPs: 63–125 μm, 125–250 μm, 250–500 μm and 500–2,000 μm. They demonstrated that the amount of particles in the 135–500 and 500–2000 μm ranges significantly increased after WWTP treatment. Lv et al. (2019) noticed that the fraction of 62.5–125 μm is encountered twice as often as a fraction of 125–250 μm and 250–500 μm. However, it should be kept in mind that these data refer to numbers of particles (measured using a stereomicroscope) and that the smaller size fractions cover a smaller range than the larger size fractions. Foils and foams are not observed within the fractions with the smallest MPs. During the treatment, the primary particle content appeared to decrease downstream, probably because of the intake by organisms, dilution, precipitation, skimming during transport, etc. However, the amount of secondary particles increased due to the degradation of larger plastic particles. The lighter particles will float, and they can be removed by skimming or flotation. The heaviest particles will precipitate and end up in the sludge. During oxidation precipitation will also occur, leading to plastic particles entering the sludge. During the tertiary treatment, part of the particles will end up in the sludge, and parts may be removed using microfiltration. A complicating factor is that the density of polymers often is about 1 kg/L, similar to the density of water. This will complicate removal strategies based on density (Enfrin et al. 2019). Both MPs and NPs traveling through a WWTP can be broken into smaller particles by shears during mixing or by pumping. High concentrations of NPs and MPs may block or damage filtration units.

5. Occurrence of MPs in Drinking Water

With such large amounts of MPs being present in surface water, which is used as a source for drinking water, the question arises whether drinking water also contains MPs; if so, how many, and whether this may have negative effects on human health. Until 2019, this has hardly been a topic of research (Novotna et al. 2019), but, in very recent years, the interest in this topic has grown exponentially, as shown in Fig. 9.4. Novotna et al. (2019) reported that about 20 different materials had been found in drinking water, but that the composition and concentration of MPs strongly depended on the type of source used. Kirstein et al. (2021) measured MPs in drinking water and found on average 174 ± 405 MPs/m³. Eight different types of polymers were identified: 87% PET, 9% PE and a single PP fiber. Thirty two percent of all MPs were found to be < 20 μm, 19% being fibers and 81% particles. The removal during drinking water treatment appeared to be 70–82%. It should also be taken into account that deterioration of plastic equipment used during water treatment and distribution can be a source of MPs in drinking water, as, e.g., pipes frequently are made of PVC, PP and PE.

A similar observation was made for bottled water, which contains relatively high concentrations of MPs. In some cases, it was even observed that freshwater bodies contained fewer MPs than tap

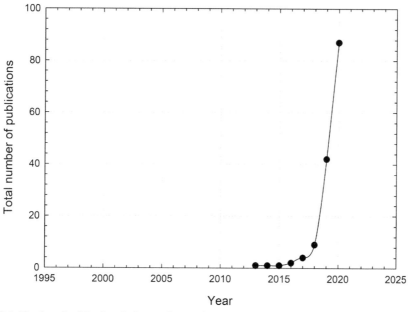

Figure 9.4. Number of publications in Scopus (keywords "microplastics" AND "drinking water"; January 2021).

water or bottled water produced from it (Zhang et al. 2020). The authors suggest that these MPs may originate from the water supply chain or product packages, such as caps and bottle walls.

Danopoulos et al. (2020) studied six types of tap water and six samples of bottled drinking water. They concluded that tap water in Europe contains a maximum of 628 MPs/L, but that the concentrations in bottled water were up to 4,889 MPs/L. Assuming an adult would consume 2 L of either tap water or bottled water daily, this would result in a yearly consumption of 458.000 MPs from tap water or 3.569.000 MPs from bottled water. Zhang et al. (2020) report concentrations in bottled water varying from 0 to 5.4×10^7 MPs/L. They found that water in returnable-used plastic bottles contained significantly more MPs compared with water in single-use bottles. The MPs in bottled water came from the bottles themselves and consisted of PET and PES (Barchiesi et al. 2020, Danopoulos et al. 2020, Shen et al. 2020, Zhang et al. 2020). However, large amounts of MPs were also found in glass bottled water, the potential source being abrasion of the plastic bottle cap against the glass bottle body. Johnson et al. (2020) studied the presence of MPs in both WWTP effluent and tap water in England and Wales. They determined the composition of particles ≥ 25 μm employing FTIR techniques, and concluded that four out of six effluents of WWTPs contained PE, PET and PP. In tap water, two out of eight sampling points showed MPs. Weber et al. (2021) carried out a similar investigation in Germany, but could not find any MPs ≥ 10 μm in tap water. This is in accordance with the research of Kirstein et al. (2021), who calculated a potential annual uptake of less than one MP per person. Tong et al. (2020), however, demonstrated that 38 tap water samples, taken in different Chinese cities, contained MPs in concentrations of 440 ± 275 MPs/L. Particles < 50 μm clearly dominated, and although 14 different materials could be identified, the majority of particles consisted of PE and PP. According to Zhang et al. (2020), 81% of 159 samples of tap water from different countries, and 93% of 259 samples of bottled water (from 11 different producers) contained MPs.

An overview of the occurrence of MPs in surface water, groundwater, tap water and bottled water in the world is given in Fig. 9.5.

Several authors have studied the fate of MPs during drinking water treatment processes. Treatment processes for the production of drinking water seem to be quite effective for the removal of MPs. Li et al. (2020) stated that MPs > 25 μm are removed for over 99.9% during drinking water treatment. However, the major part of the MPs encountered in water is < 10 μm. Particles with sizes ranging from 1 to 10 μm may be removed for over 80%. Ferraz et al. (2020) reported that the concentration of MPs decreased from 330.2/L in river water to 105.8/L in treated water. Zhang et al. (2020) suggested that particles larger than 50 μm are removed from raw water by traditional drinking water treatments with removal rates in the range of 25–90%, depending on local treatment technologies applied. This is in accordance with the findings of Ferraz et al. (2020), who reported 70% removal during drinking water production from the Sinos River in Southern Brazil. The lowest concentrations of MPs were reported in Italy and Denmark (0 MPs/L), whereas significantly higher concentrations were observed in the USA (9.2 MPs/L).

Pivokonský et al. (2020) studied two different Drinking Water Treatment Plants (DWTPs) and concluded that MPs comprised only 0.02% of all detected particles in the DWTPS. Seventy percent of the MPs appeared to be smaller than 10 μm, consisting of mostly fragments and some fibers. The most frequently occurring materials were cellulose acetate, PET, PVC, PE and PP. Two DWTPs were studied. The first one applies a simple treatment (coagulation, flocculation rapid sand filtration, softening with lime and disinfection with ClO_2); the second one uses a source containing more contaminants, and, as a consequence, it applies a more extensive treatment (pH adjustment, coagulation, flocculation, sedimentation, $KMnO_4$ oxidation, sand filtration, ozonation, filtration over Granular Activated Carbon (GAC), UV, softening with lime and disinfection with chlorine). The source of the simple treatment contained 23 ± 2 MPs/L, which was reduced to 14 ± 1 MPs/L (40% removal). The second DWTP was able to reduce the MPs content by 88%, decreasing the concentration from 1296 ± 35 MPs/L to 151 ± 4 MPs/L. Coagulation/flocculation seemed to be very effective for the removal of MPs.

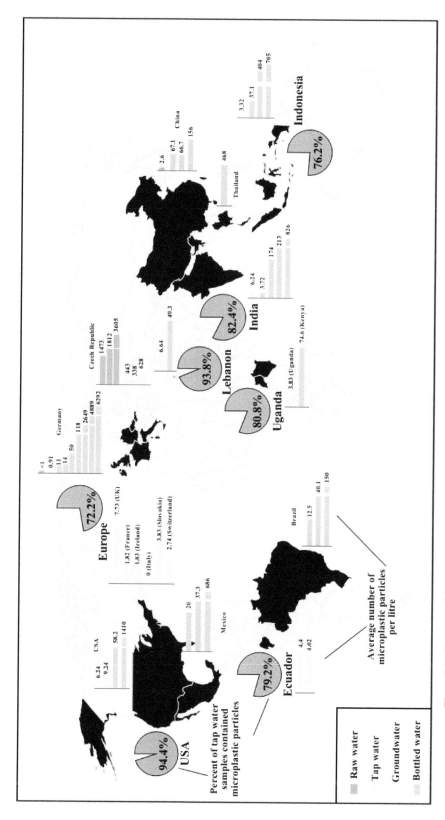

Figure 9.5. Overview of the occurrence of MPs in surface water, groundwater, tap water and bottled water in the world (Shen et al. 2020).

Coagulation/flocculation is generally used as the first treatment step in surface water treatment. A flocculant, often $FeCl_3$, $AlCl_3$ or polyaluminumchloride, is added for this purpose. The multivalent cation destabilizes the negatively charged particles present in the water, which coagulate, flocculate and subsequently can be removed utilizing rapid sand filtration of Dissolved Air Flotation (DAF). During this process, also part of the Natural Organic Matter (NOM) present in the water is captured by the flocs and removed. As MPs are often also negatively charged, they can be coagulated and flocculated as well. However, their presence during a coagulation process may result in a higher coagulant demand. How much coagulant will be required additionally is difficult to predict, as plastics are difficult to analyze and quantify. Aluminum salts appear to be more effective for the removal of PE particles than iron salts (25% of removal of microplastics during coagulation with aluminum and 15% removal with iron was reported by Enfrin et al. (2019)). Ma et al. (2019) showed that the traditional coagulation process only removed ≤ 15% of PE MPs and Wang et al. (2020) reported a removal efficiency of about 50% (mainly removal of fibers). Skaf et al. (2020) reported that using alum concentrations between 5 and 10 mg/L, turbidity in solutions containing 5 mg/L MPs was decreased from 16 NTU to < 1.0 NTU. The efficiency of coagulation/flocculation in the removal of MPs also depends on their shape. Fragments and pellets can be removed for < 40%, whereas fibers show 50–60% removal (Li et al. 2020). The efficiency of PE removal was hardly affected by water conditions like ionic strength, NOM-concentration and turbidity. Sometimes, compounds (e.g., surfactants) are added to improve coagulation, but these may have a negative effect on the removal of MPs. Sweep flocculation was identified as the dominant mechanism for the removal of MPs. The addition of anionic polyacrylamide (in combination with the coagulant) enhanced the removal efficiency to about 91% (Ma et al. 2019). The latter authors used Fe-based coagulants, because of their larger density compared to Al-based coagulants. The addition of poly (diallkyl dimethyl ammonium) chloride (DADMAC) improved the removal of MPs in the size range of 45–53 μm, partly because the particles become more sticky by adsorption of the compound, and partly because their density increases in this way (Zhang et al. 2020).

It is expected that during rapid sand filtration, MPs and NPs will quickly pass the upper (anthracite) layer because of the more open structure of these filters, but that they may cause blocking of the sand filter below. The presence of hydroxyl groups at the polymer surface, resulting from weathering, will decrease the effectiveness of backwashing for the removal of plastics, and thus may decrease the efficiency of the filter bed. Sharma et al. (2021) reported 74% removal of MPs by rapid sand filtration.

For DAF, it is important that the right ratio of air bubbles to particles and the optimum bubble size be applied. Again, in practice, this will be difficult to organize, as little is known about nano- and microplastics (Enfrin et al. 2019, Li et al. 2020, Zhang et al. 2020).

Presently, little information is available in literature on the removal of MPs by slow sand filtration, but it seems that MPs of 1–10 μm can be removed in 29–44%, whereas MPs > 10 μm are removed almost completely (Li et al. 2020). Zhang et al. (2020) state that 'large' particles (106–125 μm) and very small particles (< 1 μm) can almost completely be removed, whereas particles with sizes in between, especially 10–20 μm, can pass the filter bed. For the small particles, adsorption at the surface may play a role, whereas for large MPs the pore size of the filter bed will be crucial. The presence of a biofilm at the MP surface does not seem to influence the effectiveness of the sand filter.

Filtration over Granular Activated Carbon (GAC) is applied for the removal of NOM and organic micropollutants from water, but these granules can also filter MPs. According to Wang et al. (2020), GAC filtration can remove 60% of MPs, preferably small-sized MPs. By combining GAC filtration with ozonation, the removal efficiency could be improved by about 20%.

Biodegradation is a process that occurs in sand and GAC filters. However, for the removal of MPs, this is generally not efficient, and not fast enough to contribute to, e.g., drinking water production. In soils, the following microorganisms have been identified that can degrade MPs to some extent: *Bacillus* sp., *Rhodococcus* sp., *Pseudomonas aeruginosa*, *Aspergillus clavatus* and *Fusarium, Penicillium, Phanerochaete, Acremonium*. However, the microorganisms that may be involved depend on the MP composition. Extracellular enzymes have been shown to play a role in the degradation. Enzymes such as laccases, manganese peroxides or lignin peroxidases are responsible for hydrolyzing high-density polyethylene (HDPE), whereas hydrolases such as esterases, lipases, cutinases can induce the degradation of PE (Sharma et al. 2021).

Membrane filtration is a technique that can effectively improve water quality. Li et al. (2021) described experiments with ultrafiltration and found that membrane fouling was enhanced due to the synergistic effects of MPs and substances in raw water. MPs with an average size of 1 µm caused the most severe fouling. Fouling can be decreased by adding aluminum to the solution as a coagulant. This principle was confirmed by Ma et al. (2019), using Fe-coagulants, although these authors also noticed that less severe fouling was obtained with PE particles than without, possibly caused by a higher porosity of the cake layer. Enfrin et al. (2019) warned of deterioration of the membrane surface by frictions with the MPs. Sharma et al. (2021) reported 79% removal of MPs by means of a membrane disk filter.

In many countries, disinfection of tap water is attained by means of chlorination. Chlorine might also react with MPs because of its strong oxidizing nature. Kelkar et al. (2019) showed that PP is most resistant to chlorination, followed by HDPE and PS. Chlorination may chemically alter some MPs, which may result in the MPs becoming better carriers for contaminants. However, in drinking water treatment, chlorination will not significantly contribute to MPs degradation. Oxidation techniques, e.g., ozonation or advanced oxidation are also not effective. UV radiation may cause some photolysis, but will not be effective in quickly degrading MPs. More important than the effect that disinfection may have on MPs is the effect that MPs may have on disinfection. As MPs can react with ozone, chlorine or UV, the effective dose for disinfection may be decreased by the presence of MPs, decreasing the disinfection efficiency. Furthermore, the addition of chlorine may be less effective if microorganisms are present as flocs or suspended particles, as the MPs may 'protect' the microorganisms from, e.g., chlorine or UV, by shielding the MPs (Enfrin et al. 2019).

Experimental technologies that are being investigated for the removal of MPs from water are, e.g., electrocoagulation, magnetic extraction and some types of membrane filtration (Shen et al. 2020). A new technology may be sustainable green photocatalytic processes, as described by Uheida et al. (2021). These authors used glass fiber substrates to trap low-density MPs (like PP) and support a photocatalyst material like ZnO or TiO_2 (other semiconductors that may be used are ZnS, WO_3, ZrO_2 and g-C_3N_4 (Sharma et al. 2021)). During the process, the formation of carbonyl and hydroxyl groups and chain scissions were observed. Upon irradiation with visible light for 2 weeks, the volume of the average particles decreased by 65%. However, it will be difficult to apply such a process to large-scale drinking water treatment Ding et al. (2021).

One aspect of the presence of MPs in water not mentioned before is the fact that MPs can also adsorb certain compounds, which may be released into the water. Furthermore, they may release additives or monomers that are present in the plastics. PE and PP in general do not contain high concentrations of monomers, contrary to PS (da Costa et al. 2016, Andrady 2017, Horton et al. 2017, Jiang 2018). Additives that are often encountered in MPs are:

- plasticizers
- flame retardants
- antioxidants
- acid scavengers
- light (UV) and heat stabilizers
- anti-inflammatory agents
- lubricants
- pigments
- antistatic compounds
- 'slip compounds' (compounds that decrease the surface coefficient of friction of the polymer)

- thermic stabilizers
- surfactants
- dispersants
- nanoparticles and nanofibers
- inert fillers
- fragrances

Compounds like bisphenol A, Polycyclic Aromatic Hydrocarbons (PAHs, like phenanthrene), 4,4'-DDT, polychlorinated biphenyls (PCBs), dichlorodiphenyltrichloroethane (DDT), polybrominated diphenylethers (PBDEs) and other contaminants such as polyfluoroalkyl substances (PFAS), pharmaceuticals and personal care products, phthalates, triclosan, bisphenone, butylated hydroxytoluene, polychlorobiphenyls and organotins may enter the aqueous phase (Jiang et al. 2020, Sharma et al. 2021, Wang et al. 2021). Phthalates (like diisoheptylphthalate and dibutylphthalate) are known to be endocrine-disrupting compounds (EDCs). Besides, Cd, Zn, Ni, Pb and other heavy metals may be adsorbed on the surface of the MPs. In this way, these compounds may be brought into the water matrix, and may eventually be released.

6. Possible Health Effects

Humans are exposed to different types of fibers and particles, including MPs, by air, food and possibly drinking water: the potential health effects of MPs are largely unknown (Vethaak and Legler 2021). It is expected that both the World Health Organization (WHO) and the California government will publish a report on this topic, but, at present, little information is available. Different MPs are found in air, water, soil, indoor dust and food. Until now, MPs have been observed in more than 690 different marine species like amphipods, copepods, lugworms, barnacles, mussels, decapod crustaceans, seabirds, fish, turtles, etc. (Xu et al. 2020, Zhang et al. 2020). In literature, MPs are considered hazardous to marine species, birds, animals, soil creatures and humans. Toxic effects have been reported, like growth inhibition, oxidative damage and immune stress, and it was shown that MPs could accumulate in marine organisms. However, Mintenig et al. (2019) reported a negligible human exposure to MPs via drinking water, but, in this case, the source for the water was groundwater.

Kirstein et al. (2021) point out that, considering the annual estimated MP consumption by both drinking tap water and inhalation, the presence of MP in drinking water does not seem to be a large risk for human health. However, this largely depends on the area; Zhang et al. (2020) estimated that with a consumption of 1.4 L per day, the annual MPs intake through tap water and bottled water would be $0–2.8 \times 10^{10}$ MPs. Cox et al. (2019) estimated the annual MPs consumption from an American diet to range from 39,000 to 52,000 MPs, depending on age and sex. These estimates increased to 74,000 and 121,000 when inhalation was also considered, in accordance with the findings of Kirstein et al. (2021), who indicated that inhalation is a larger source of MPs than drinking water. Individuals who would only use bottled water may consume 90,000 MPs annually, compared to about 4,000 by someone only drinking tap water. Food also is a source of MPs for humans, especially in sea food, where MPs seem to be able to accumulate (Jiang et al. 2020).

Some health effects of MPs in humans have been described, but they strongly depend on the size and composition of the MPs, and the way they were consumed (via respiratory exposure or via digestion) (Jiang et al. 2020, Zhang et al. 2020). MPs have been observed in human feces, which means that at least part of the MPs is not absorbed in the human body but directly excreted. On the other hand, leaching of chemicals may cause health risks (Barceló and Picó 2019). Metal concentrations on plastics have been shown to be up to 800 times higher than those in the surrounding seawater (Xu et al. 2020), and, in this way, high concentrations of metals can be delivered to organisms.

7. Conclusions and Future Outlook

Although plastics have proven to be very useful materials, it is becoming clear that more plastics are found everywhere in the environment. Especially, MPs and NPs of various compositions, shapes and sizes, can be found in aqueous environments all over the world, where they are accumulating in organisms and thus in the food chain. The majority of all surface waters are contaminated with MPs, and, even in groundwater-related samples, MPs have already been observed. An important source for MPs in surface water seems to be WWTPs, but other routes can also be identified. Although WWTPs remove > 90% of all MPs, they still release significant amounts of MPs to the environment. Drinking water treatment processes show different efficiency for the removal of MPs, but in general, it has been shown that tap water only contains small amounts of MPs, which form a negligible contribution to the total MP uptake by humans. A special point of attention in this respect is the different reporting limits for MPs in water, as it has also been shown that the smaller the particle sizes, the more particles seem to be present. Furthermore, it has been seen that treatment equipment, pipes and especially bottles may increase the number of MPs in drinking water. As it is not known which health effects these MPs may have (either directly or by leaching other compounds), their presence in drinking water should be avoided.

References

Ahmed, M.B., M.S. Rahman, J. Alom, M.S. Hasan, M.A.H. Johir, M.I.H. Mondal, D.Y. Lee, J. Park, J.L. Zhou and M.H. Yoon. 2021. Microplastic particles in the aquatic environment: A systematic review. Science of the Total Environment 775.

Ainali, N.M., D.N. Bikiaris and D.A. Lambropoulou. 2021. Aging effects on low- and high-density polyethylene, polypropylene and polystyrene under UV irradiation: An insight into decomposition mechanism by Py-GC/MS for microplastic analysis. Journal of Analytical and Applied Pyrolysis 158.

Ainali, N.M., D. Kalaronis, A. Kontogiannis, E. Evgenidou, G.Z. Kyzas, X. Yang, D.N. Bikiaris and D.A. Lambropoulou. 2021. Microplastics in the environment: Sampling, pretreatment, analysis and occurrence based on current and newly-exploited chromatographic approaches. Science of the Total Environment 794.

Andrady, A.L. 2017. The plastic in microplastics: A review. Marine Pollution Bulletin 119(1): 12–22.

Araujo, C.F., M.M. Nolasco, A.M.P. Ribeiro and P.J.A. Ribeiro-Claro. 2018. Identification of microplastics using Raman spectroscopy: Latest developments and future prospects. Water Research 142: 426–440.

Barceló, D. and Y. Picó. 2019. Microplastics in the global aquatic environment: Analysis, effects, remediation and policy solutions. Journal of Environmental Chemical Engineering 7(5).

Barchiesi, M., A. Chiavola, C. Di Marcantonio and M.R. Boni. 2020. Presence and fate of microplastics in the water sources: focus on the role of wastewater and drinking water treatment plants. Journal of Water Process Engineering.

Campbell, C.G., D.J. Astorga, E. Duemichen and M. Celina. 2020. Thermoset materials characterization by thermal desorption or pyrolysis based gas chromatography-mass spectrometry methods. Polymer Degradation and Stability 174.

Carr, S.A., J. Liu and A.G. Tesoro. 2016. Transport and fate of microplastic particles in wastewater treatment plants. Water Research 91: 174–182.

Castelvetro, V., A. Corti, G. Biale, A. Ceccarini, I. Degano, J. La Nasa, T. Lomonaco, A. Manariti, E. Manco, F. Modugno and V. Vinciguerra. 2021. New methodologies for the detection, identification, and quantification of microplastics and their environmental degradation by-products. Environmental Science and Pollution Research 28(34): 46764–46780.

Chand, R., K. Kohansal, S. Toor, T.H. Pedersen and J. Vollertsen. 2022. Microplastics degradation through hydrothermal liquefaction of wastewater treatment sludge. Journal of Cleaner Production 335.

Cheng, Y.L., J.G. Kim, H.B. Kim, J.H. Choi, Y. Fai Tsang and K. Baek. 2021. Occurrence and removal of microplastics in wastewater treatment plants and drinking water purification facilities: A review. Chemical Engineering Journal 410.

Corami, F., B. Rosso, E. Morabito, V. Rensi, A. Gambaro and C. Barbante. 2021. Small microplastics (<100 μm), plasticizers and additives in seawater and sediments: Oleo-extraction, purification, quantification, and polymer characterization using Micro-FTIR. Science of the Total Environment 797.

Costa, M.F., J. Pinto da Costa and A.C. Duarte. 2018. Sampling of micro(nano)plastics in environmental compartments: How to define standard procedures? Current Opinion in Environmental Science & Health 1: 36–40.

Cox, K.D., G.A. Covernton, H.L. Davies, J.F. Dower, F. Juanes and S.E. Dudas. 2019. Human consumption of microplastics. Environmental Science and Technology 53(12): 7068–7074.

da Costa, J.P., P.S.M. Santos, A.C. Duarte and T. Rocha-Santos. 2016. (Nano)plastics in the environment—Sources, fates and effects. Science of the Total Environment 566-567: 15–26.

Danopoulos, E., M. Twiddy and J.M. Rotchell. 2020. Microplastic contamination of drinking water: A systematic review. PLoS ONE 15(7 July).

Ding, H., J. Zhang, H. He, Y. Zhu, D.D. Dionysiou, Z. Liu and C. Zhao. 2021. Do membrane filtration systems in drinking water treatment plants release nano/microplastics? Science of the Total Environment 755.

Dris, R., J. Gasperi, V. Rocher and B. Tassin. 2018. Synthetic and non-synthetic anthropogenic fibers in a river under the impact of Paris Megacity: Sampling methodological aspects and flux estimations. Science of the Total Environment 618: 157–164.

Dümichen, E., A.K. Barthel, U. Braun, C.G. Bannick, K. Brand, M. Jekel and R. Senz. 2015. Analysis of polyethylene microplastics in environmental samples, using a thermal decomposition method. Water Research 85: 451–457.

Eerkes-Medrano, D., H.A. Leslie and B. Quinn. 2019. Microplastics in drinking water: A review and assessment. Current Opinion in Environmental Science and Health 7: 69–75.

Enfrin, M., L.F. Dumée and J. Lee. 2019. Nano/microplastics in water and wastewater treatment processes—Origin, impact and potential solutions. Water Research 161: 621–638.

Estahbanati, S. and N.L. Fahrenfeld. 2016. Influence of wastewater treatment plant discharges on microplastic concentrations in surface water. Chemosphere 162: 277–284.

Ferraz, M., A.L. Bauer, V.H. Valiati and U.H. Schulz. 2020. Microplastic concentrations in raw and drinking water in the sinos river, southern brazil. Water (Switzerland) 12(11): 1–10.

Hahladakis, J.N., C.A. Velis, R. Weber, E. Iacovidou and P. Purnell. 2018. An overview of chemical additives present in plastics: Migration, release, fate and environmental impact during their use, disposal and recycling. Journal of Hazardous Materials 344: 179–199.

He, P., L. Chen, L. Shao, H. Zhang and F. Lü. 2019. Municipal solid waste (MSW)landfill: A source of microplastics? Evidence of microplastics in landfill leachate. Water Research 159: 38–45.

Hermabessiere, L. and C.M. Rochman. 2021. Microwave-assisted extraction for quantification of microplastics using pyrolysis–gas chromatography/mass spectrometry. Environmental Toxicology and Chemistry 40(10): 2733–2741.

Horton, A.A., A. Walton, D.J. Spurgeon, E. Lahive and C. Svendsen. 2017. Microplastics in freshwater and terrestrial environments: Evaluating the current understanding to identify the knowledge gaps and future research priorities. Science of the Total Environment 586: 127–141.

Jiang, B., A.E. Kauffman, L. Li, W. McFee, B. Cai, J. Weinstein, J.R. Lead, S. Chatterjee, G.I. Scott and S. Xiao. 2020. Health impacts of environmental contamination of micro- and nanoplastics: A review. Environmental Health and Preventive Medicine 25(1).

Jiang, J.Q. 2018. Occurrence of microplastics and its pollution in the environment: A review. Sustainable Production and Consumption 13: 16–23.

Johnson, A.C., H. Ball, R. Cross, A.A. Horton, M.D. Jürgens, D.S. Read, J. Vollertsen and C. Svendsen. 2020. Identification and quantification of microplastics in potable water and their sources within water treatment works in england and wales. Environmental Science and Technology 54(19): 12326–12334.

Kay, P., R. Hiscoe, I. Moberley, L. Bajic and N. McKenna. 2018. Wastewater treatment plants as a source of microplastics in river catchments. Environmental Science and Pollution Research 25(20): 20264–20267.

Kelkar, V.P., C.B. Rolsky, A. Pant, M.D. Green, S. Tongay and R.U. Halden. 2019. Chemical and physical changes of microplastics during sterilization by chlorination. Water Research 163.

Kirstein, I.V., F. Hensel, A. Gomiero, L. Iordachescu, A. Vianello, H.B. Wittgren and J. Vollertsen. 2021. Drinking plastics?—Quantification and qualification of microplastics in drinking water distribution systems by µFTIR and Py-GCMS. Water Research 188.

Koelmans, A.A., N.H. Mohamed Nor, E. Hermsen, M. Kooi, S.M. Mintenig and J. De France. 2019. Microplastics in freshwaters and drinking water: Critical review and assessment of data quality. Water Research 155: 410–422.

Lauschke, T., G. Dierkes, P. Schweyen and T.A. Ternes. 2021. Evaluation of poly(styrene-d5) and poly(4-fluorostyrene) as internal standards for microplastics quantification by thermoanalytical methods. Journal of Analytical and Applied Pyrolysis 159.

Li, J., B. Wang, Z. Chen, B. Ma and J.P. Chen. 2021. Ultrafiltration membrane fouling by microplastics with raw water: Behaviors and alleviation methods. Chemical Engineering Journal 410.

Li, Y., W. Li, P. Jarvis, W. Zhou, J. Zhang, J. Chen, Q. Tan and Y. Tian. 2020. Occurrence, removal and potential threats associated with microplastics in drinking water sources. Journal of Environmental Chemical Engineering 8(6).

Liu, Y., R. Li, J. Yu, F. Ni, Y. Sheng, A. Scircle, J.V. Cizdziel and Y. Zhou. 2021. Separation and identification of microplastics in marine organisms by TGA-FTIR-GC/MS: A case study of mussels from coastal China. Environmental Pollution 272: 115946.

Lv, X., Q. Dong, Z. Zuo, Y. Liu, X. Huang and W.M. Wu. 2019. Microplastics in a municipal wastewater treatment plant: Fate, dynamic distribution, removal efficiencies, and control strategies. Journal of Cleaner Production 225: 579–586.

Ma, B., W. Xue, Y. Ding, C. Hu, H. Liu and J. Qu. 2019. Removal characteristics of microplastics by Fe-based coagulants during drinking water treatment. Journal of Environmental Sciences (China) 78: 267–275.

Ma, B., W. Xue, C. Hu, H. Liu, J. Qu and L. Li. 2019. Characteristics of microplastic removal via coagulation and ultrafiltration during drinking water treatment. Chemical Engineering Journal 359: 159–167.

Mintenig, S.M., P.S. Bäuerlein, A.A. Koelmans, S.C. Dekker and A.P. Van Wezel. 2018. Closing the gap between small and smaller: towards a framework to analyse nano- and microplastics in aqueous environmental samples. Environmental Science: Nano 5(7): 1640–1649.

Mintenig, S.M., M.G.J. Löder, S. Primpke and G. Gerdts. 2019. Low numbers of microplastics detected in drinking water from ground water sources. Science of the Total Environment 648: 631–635.

Mughini-Gras, L., R.Q.J. van der Plaats, P.W.J.J. van der Wielen, P.S. Bauerlein and A.M. de Roda Husman. 2021. Riverine microplastic and microbial community compositions: A field study in the Netherlands. Water Research 192: 116852.

Novotna, K., L. Cermakova, L. Pivokonska, T. Cajthaml and M. Pivokonsky. 2019. Microplastics in drinking water treatment—Current knowledge and research needs. Science of the Total Environment 667: 730–740.

Pivokonský, M., L. Pivokonská, K. Novotná, L. Čermáková and M. Klimtová. 2020. Occurrence and fate of microplastics at two different drinking water treatment plants within a river catchment. Science of the Total Environment 741.

Plastics Europe. 2019. Plastics Europe—the facts 2019—An Analysis of European Plastics Production, Demand and Waste Data. Brussels.

Prata, J.C. 2018. Microplastics in wastewater: State of the knowledge on sources, fate and solutions. Marine Pollution Bulletin 129(1): 262–265.

Primpke, S., M. Godejohann and G. Gerdts. 2020. Rapid identification and quantification of microplastics in the environment by quantum cascade laser-based hyperspectral infrared chemical imaging. Environmental Science & Technology 54(24): 15893–15903.

Qin, J., B. Liang, Z. Peng and C. Lin. 2021. Generation of microplastic particles during degradation of polycarbonate films in various aqueous media and their characterization. Journal of Hazardous Materials 415.

Sharma, S., S. Basu, N.P. Shetti, M.N. Nadagouda and T.M. Aminabhavi. 2021. Microplastics in the environment: Occurrence, perils, and eradication. Chemical Engineering Journal 408.

Shen, M., B. Song, Y. Zhu, G. Zeng, Y. Zhang, Y. Yang, X. Wen, M. Chen and H. Yi. 2020. Removal of microplastics via drinking water treatment: Current knowledge and future directions. Chemosphere 251.

Skaf, D.W., V.L. Punzi, J.T. Rolle and K.A. Kleinberg. 2020. Removal of micron-sized microplastic particles from simulated drinking water via alum coagulation. Chemical Engineering Journal 386.

Sorasan, C., C. Edo, M. González-Pleiter, F. Fernández-Piñas, F. Leganés, A. Rodríguez and R. Rosal. 2021. Generation of nanoplastics during the photoageing of low-density polyethylene. Environmental Pollution 289.

Talvitie, J., A. Mikola, A. Koistinen and O. Setälä. 2017. Solutions to microplastic pollution—Removal of microplastics from wastewater effluent with advanced wastewater treatment technologies. Water Research 123: 401–407.

Tong, H., Q. Jiang, X. Hu and X. Zhong. 2020. Occurrence and identification of microplastics in tap water from China. Chemosphere 252.

Uheida, A., H.G. Mejía, M. Abdel-Rehim, W. Hamd and J. Dutta. 2021. Visible light photocatalytic degradation of polypropylene microplastics in a continuous water flow system. Journal of Hazardous Materials 406.

Vethaak, A.D. and J. Legler. 2021. Microplastics and human health. Science 371(6530): 672.

Wang, C., J. Zhao and B. Xing. 2021. Environmental source, fate, and toxicity of microplastics. Journal of Hazardous Materials 407.

Wang, Z., T. Lin and W. Chen. 2020. Occurrence and removal of microplastics in an advanced drinking water treatment plant (ADWTP). Science of the Total Environment 700.

Weber, F., J. Kerpen, S. Wolff, R. Langer and V. Eschweiler. 2021. Investigation of microplastics contamination in drinking water of a German city. Science of the Total Environment 755.

Xu, S., J. Ma, R. Ji, K. Pan and A.J. Miao. 2020. Microplastics in aquatic environments: Occurrence, accumulation, and biological effects. Science of the Total Environment 703.

Yang, L., K. Li, S. Cui, Y. Kang, L. An and K. Lei. 2019. Removal of microplastics in municipal sewage from China's largest water reclamation plant. Water Research 155: 175–181.

Yu, J., P. Wang, F. Ni, J. Cizdziel, D. Wu, Q. Zhao and Y. Zhou. 2019. Characterization of microplastics in environment by thermal gravimetric analysis coupled with Fourier transform infrared spectroscopy. Marine Pollution Bulletin 145: 153–160.

Zhang, Q., E.G. Xu, J. Li, Q. Chen, L. Ma, E.Y. Zeng and H. Shi. 2020. A review of microplastics in table salt, drinking water, and air: direct human exposure. Environmental Science and Technology 54(7): 3740–3751.

Zhang, Y., A. Diehl, A. Lewandowski, K. Gopalakrishnan and T. Baker. 2020. Removal efficiency of micro- and nanoplastics (180 nm–125 μm) during drinking water treatment. Science of the Total Environment 720.

Ziajahromi, S., P.A. Neale, L. Rintoul and F.D.L. Leusch. 2017. Wastewater treatment plants as a pathway for microplastics: Development of a new approach to sample wastewater-based microplastics. Water Research 112: 93–99.

Chapter 10

Modified Nanobentonites for Water Remediation

Estefania Bracco, Matías Butler, Roberto Candal, * *Patricio Carnelli, Lucas Guz,*
Federico Ivanic, Elsa Lopez Loveira and *José Luis Marco Brown*

1. Introduction

Bentonite is a natural clay, which consists mainly of montmorillonite (MMT), a phyllosilicate (sheet silicate minerals formed by parallel layers of silicate tetrahedra) composed of alternating tetrahedral and octahedral layers in a 2:1 ratio. MMT is a nanoclay with a layer thickness around 1 nm (Barakan and Aghazadeh 2021); the lattice has an excess negative charge, which is balanced in the interlayer space by hydrated inorganic cations, such as Na(I), Ca(II), etc. Therefore, MMT has a high Cation Exchange Capacity (*CEC*) and a large surface area, leading to high adsorption capacities for pollutants (Awad et al. 2019).

Bentonites can be—and have been—used to treat a wide variety of target pollutants, from complex organic molecules, including herbicides, fungicides and pharmaceutical drugs, to a broad range of metals with different speciation (Pandey 2017, Prabhu and Prabhu 2018, Awad et al. 2019, Yadav et al. 2019). Its extensive applicability, in addition to its low cost and easy accessibility, makes bentonite a useful alternative compared with more aggressive remediation methods. Its adsorption capacity on different types of target pollutants arises from various and accessible modifications that can be built into its structure.

The most common modification of the bentonite structure is the intercalation of cations (Fe, Al, etc.) into its layers, as this increases the interlayer spacing and therefore, the adsorption capacity, usually accentuating the hydrophilic nature of the clay by the presence of hydrated cations (Calabi Floody et al. 2009). The adsorption behavior of iron exchanged nanobentonite clay and Fe_3O_4 nanoparticles for removing NO_3^- and HCO_3^- from wastewater was compared, and a higher efficiency of the modified bentonite was observed, as a result of its higher specific surface area and increased basal spacing (Mukhopadhyay et al. 2019).

Despite bentonite having an overall neutral charge, the excess negative charge on its lattice may provide affinity to cationic species and to a lesser extent, to anionic and neutral molecules. Thus, organic modifiers are usually included in its structure for improving affinity to organic compounds. L-tryptophan (Trp) was added to the structure of a modified bentonite in order to benefit from

Instituto de Investigación e Ingeniería Ambiental (IIIA), Universidad Nacional de San Martin (UNSAM), CONICET, 3iA, Campus Miguelete, Av. 25 de Mayo y Francia, 1650 San Martín, Provincia de Buenos Aires, Argentina.
Emails: ebracco@agro.uba.ar; mbutler@unsam.edu.ar; pcarnelli@unsam.edu.ar; lguz@unsam.edu.ar; fivanic@unsam.edu.ar; elsaglopez@gmail.com; jlbrown@unsam.edu.ar
* Corresponding author: rcandal@unsam.edu.ar

the amphoteric nature of amino acids and to improve the selectivity and adsorption capacity of the bentonite for organic molecules (Gallouze et al. 2021). Comparing the achieved results among non-modified bentonite, Na-bentonite, Fe-Na-bentonite and Trp-Na-bentonite, the latter exhibited the greatest adsorption capacity for the emerging contaminant 17α-ethinylestradiol (EE2), that led to the conclusion that it may be a promising low-cost adsorbent resource, not only for EE2 and steroids but also for other organic pollutants.

In another example of removal of organic pollutants, bentonite was modified by two cationic surfactants (octadecyltrimethylammonium -C18- and dioctadecyltrimethylammonium -2C18-) to remove triclosan, an antibacterial agent, toxic to humans and other living organisms (Phuekphong et al. 2020). These modifications allowed a higher surface hydrophobicity and a greater adsorption capacity of the mentioned contaminant, especially when using 2C18 modified bentonite. Furthermore, the reusability of the latter was tested, finding only a slight decrease in the adsorption capacity after washes. Improved adsorption methods may not only be cheaper but also more ecofriendly than other removal techniques.

Chemical modifications of clays such as acid treatment, intercalation of organic compounds and pillaring by different metal polyoxycations may be combined to expand the sorption capacity and selectivity of bentonites. A montmorillonite modified with a cationic surfactant (cetyltrimethylammonium, CTMA), followed by an acid treatment and subsequent introduction of iron hydroxides was synthesized, the material showing the ability to adsorb simultaneously oxyanions, hydrophobic compounds and heavy metals in batch experiments in water (Yang et al. 2020).

Moreover, the applicability of modified and unmodified bentonites can be further expanded through coupling with other methods, thus enhancing the removal of pollutants. For example, nanoscale zerovalent iron (nZVI) has been recently used for remediation purposes, especially as a permeable reactive barrier, due to the reducing ability of Fe^0. The performance of nZVI particles attached over montmorillonite (MMT-nZVI) and monodispersed nZVI particles in removing Cr (VI), a well-studied carcinogenic element, was studied (Yin et al. 2020). It was concluded that MMT-nZVI presented a similar removal yield of Cr (VI) from water compared with monodispersed elemental Fe nanoparticles, but much higher than aggregated nZVI particles and MMT alone, as this material provides a higher porous area and prevents particle aggregation.

At the same time, the removal of contaminants and target molecules promoted by the adsorption affinity of bentonites can be followed by a complete or partial degradation using Advanced Oxidation Processes (AOPs). AOPs rely on the formation of various oxidant radicals (primarily the hydroxyl radical, HO·), which may react with the contaminant and transform it into more oxidized products. A classic example of AOPs are the Fenton and photo-Fenton processes, which consist of a pH-dependent array of reactions employing a catalytic amount of Fe (II) to react with H_2O_2 and form Fe (III) and HO·; the photo-Fenton process uses UV or solar light to regenerate the Fe (II) more efficiently (Langford and Carey 1975, Kim and Vogelpohl 1998).

Other examples show the simultaneous use of clay minerals and photocatalysis. The presence of clay seemed to affect the reaction yields and product selectivity in the oxidation of benzene (Ide et al. 2012) and 2-chlorophenol (Mogyorósi et al. 2002) in water using TiO_2 photocatalysts and UV radiation. A few other examples include: an organically modified bentonite combined with TiO_2 for the removal of toluene (Chen et al. 2011) and a smectite-TiO_2 film used for the degradation of methyl orange and methylene blue (Deepracha et al. 2019).

A common way of exploiting the combination of AOPs and adsorption techniques is the removal and storage of the pollutants using the adsorbent material followed by desorption and treatment of the sample. An advantage of bentonites, as mentioned earlier, is the wide variety of modifications that can be built into their structures, thereby facilitating the coupling and enhancing the yield of transformation. This means that the same material can be utilized to adsorb and transform the target pollutant. A modified bentonite synthesized through a pillaring process using Fe and Al cations was used to treat water samples containing an anionic dye (Congo red, CR) (Khelifi and Ayari

2019). These alterations not only enhanced the adsorbent capacity towards the dye but also made possible its use as a catalyst by the addition of UV light and H_2O_2, allowing mineralization of the adsorbed pollutant molecules and the recovery of the adsorbent through a photo-Fenton process, with a degradation percentage of 98%.

Bentonites can also be converted into catalysts for AOPs by the incorporation of nanoparticles of iron oxides. A modified organo-bentonite was prepared with CTMA, which later favored the fixation of TiO_2 and Fe_2O_3, showing promising heterogeneous photo-Fenton activity for the removal of two pharmaceutical drugs in water (Molina et al. 2020). The degradation of an organic pollutant, diethylphthalate ester (DEP) by photo-Fenton-like reactions was compared using free Fe_2O_3 nanoparticles and Fe_2O_3 nanoparticles embedded in montmorillonite clay particles (Fe-MMT). Experiments were performed at pH 5.5 in the presence of citric acid as an iron complexing agent. The Fe-MMT system exhibited a higher photocatalytic efficiency for DEP degradation than the pure Fe_2O_3 nanoparticles due to its porous structure, large surface area and more oxygen vacancies. Additionally, the agglomeration of Fe_2O_3 nanoparticles is also prevented when they are supported on MMT. Moreover, the recovery and regeneration of the material were improved, achieving a degradation efficiency of more than 50% after three recycling cycles (Sun et al. 2021).

There are also examples of novel uses that involve composites of bentonites with other adsorbents, such as: the combination of an organoclay and activated carbon to treat water containing benzene, toluene, xylene, naphthalene and various oils (Alther 2002), arsenic and lead (Mangwandi et al. 2016, Mo et al. 2018), in batch mode, and organic micropollutants such as carbamazepine, 4-tert-octylphenol, 4-nonylphenol and anthracene in a continuous mode (Kamińska 2018). A Montmorillonite-Graphene oxide Composite (MGC) was tested to treat water containing Pb^{2+} and *p*-nitrophenol (PNP), although the removal of the separated pollutants (97% for PNP and 99% for Pb^{2+}) appeared to be more efficient than when combined (98% for Pb^{2+} and 51% for PNP) due to Pb^{2+} strong competitiveness (Zhang et al. 2019).

Although membrane filtration is an established method for water remediation, filtration efficiency and fouling avoidance still need improving. Organically-modified clays can be used either in membranes made of polymer-clay nanocomposites (Alshangiti et al. 2019) or in the combination of adsorption on clay minerals with membrane filtration (Shaalan 2009).

Finally, the combination of bacteria, archaea and fungi with bentonites could be a sustainable and environmentally friendly bioremediation alternative. The adsorption properties of bentonites provide protection to microbes from toxic substances, thus enhancing their biodegradation capabilities and the mycorrhizal symbiotic associations of fungi with plants, which is of great importance for phytoremediation (Gadd 2010, Dong 2012). There are other environmental applications for the combination of bentonites with microorganisms and enzymes. For example, enzymes can be immobilized onto a bentonite support that provides protective properties allowing their use as biosensors or biocatalysts (An et al. 2015).

This chapter presents different strategies for the modification of bentonites to increase their versatility and capacity as adsorbents for pollutants and describes different mechanisms proposed for the adsorption of organic compounds. Some applications as catalysts for Fenton and photo-Fenton-like processes are also discussed.

2. Synthesis and Characterization of Modified Bentonites

2.1 Synthesis of modified clays

As indicated above, the characteristics of clays such as surface area, interlaminar space, charge or active surface sites can be modified by several methods. Depending on the target pollutant, various methodologies may be used. Moreover, the modification of clays may provide a material that can be easily removed from the aqueous medium after its use in pollutant removal processes. Here some of the most used clay modification methods will be described.

2.1.1 Synthesis of organoclays

The modification of clays by intercalation of Quaternary Ammonium Compounds (QAC) has attracted interest as a facile process for the preparation of an adsorbent material for inorganic anions. The intercalation of positively charged molecules such as QAC may change the electrical charge of MMT from negative to positive, improving the performance as adsorbent for negatively charged ions in water like PO_4^{3-}, Cr(VI) or NO_3^- (Uddin 2017, Jaworski et al. 2019). Besides, the combination of surface charge with acid-base behavior of sorbates may be used to control the adsorption or desorption with the solution pH, as will be exemplified later.

The synthesis methodology consists typically in the preparation of a suspension of the clay in a QAC aqueous solution, with constant stirring at 50–60°C, followed by washing to remove the excess of QAC. Thus, the solid is obtained after washing by successive resuspensions in deionized water before centrifugation. Organo-montmorillonites (OMMT) were recently prepared with several octadecyltrimethyl ammonium (ODTMA), hexadecyltrimethyl ammonium (HDTMA), tetradecyltrimethyl ammonium (TDTMA) and dodecyltrimethyl ammonium (DDTMA) bromides (Jaworski et al. 2019). The concentration of QAC solutions can be varied to obtain OMMT materials with several surfactant loadings. The effect of loading on nitrate adsorption was studied by exchanging MMT with 100, 200 and 400% of QAC with respect to its CEC (Jaworski et al. 2019).

2.1.2 Synthesis of iron modified clays

Clays can be modified by the addition of iron in their architecture, both in the interlayer and/or at its external surface, altering its electrical charge and improving its adsorption capacity for negatively charged molecules. An increment in the BET area (see later) of iron-modified clays is another benefit given by this methodology. These types of adsorbents were used for the removal of anionic molecules in water (organic or inorganic), as picloram (a chlorinated derivative of the picolinic acid herbicide) or As(V) adsorption (Marco-Brown et al. 2012, Iriel et al. 2020).

The synthesis is based on clay suspension in an Fe (III) solution with stirring, followed by successive steps of centrifugation and washing by resuspending in deionized water. The solvents typically used for the Fe (III) solution are acetone and water. For the acetone-based synthesis, the clay is dispersed in acetone containing $FeCl_3$ with a MMT:Fe mass ratio of 4.88:1 to ensure the exchange of Na^+ for Fe (III) in the intermediate layer (Komlósi et al. 2007). The dispersion is stirred for 2 hours and centrifuged, rinsing the solid with acetone, ethanol and water, followed by centrifugation after each step, and final lyophilization (Guz et al. 2014). In the aqueous synthesis, KOH is slowly added under vigorous stirring to a Fe (III) solution until a OH^-/Fe molar ratio of 2 is reached and kept at room temperature for 4 hours. The clay is added to the Fe (III) solution and the mixture is stirred at room temperature for 12 hours. The solid is separated by centrifugation and washed with deionized water (Iriel et al. 2020).

Supporting nZVI on clays leads to a well-dispersed and chemical stable nZVI material. This system was proposed for removal of U(VI), Cr(VI) and trichloroethylene in water (Shi et al. 2011, Sheng et al. 2014, Xu et al. 2020, Marco-Brown et al. 2021). For its preparation, the clay is dispersed in $FeCl_3$ solution, stirred for 24 hours and then reduced by adding a $NaBH_4$ solution dropwise. The solid obtained is filtered, washed with ethanol and stored under N_2 atmosphere unless the material is used in the next few days, as storing it under inert atmosphere would not be necessary due to its stability.

2.1.3 Synthesis of magnetic clays

The inclusion of iron magnetic particles into clay structures is a simple, efficient and economical method to prepare interesting adsorbent materials. Magnetic nanomaterials have attracted scientific interest in effluent treatments, for removal processes of dyes, As(III/V), heavy metals and biologically active compounds, with high separation efficiency and reusability (Barraqué et al. 2018), as well as its isolation capacity by the application of external magnetic fields.

The synthesis is based on controlled alkaline oxidation of Fe (II) sulfate in the presence of the clay, employing KNO_3 as the oxidant. The solid is recovered by magnetic separation and lyophilized (Barraqué et al. 2021).

2.2 Characterization of raw and modified clays

Several characterization techniques may be employed to determine the changes induced by the previously mentioned modification processes of the raw clays. The most frequently used characterization methods reported will be described next.

2.2.1 Electrophoretic mobility

Electrophoretic mobility is related to the zeta-potential, and can be calculated through the Smoluchowski equation (Smoluchowski 1921). Information about the electrical charge of materials can be obtained through the plotting of zeta-potential values *vs.* pH. MMT samples do not show any significant change of zeta potential over a wide pH range as a consequence of isomorphic substitutions of Si (IV) by Al (III) in the tetrahedral layer and substitutions of Al (III) by Fe (II)/Mg (II) in the octahedral layer, building a localized negative charge at the interlayer and external surface. Nonetheless, positively and negatively charged parts on the surface of clay minerals coexist simultaneously under acidic conditions.

Modifications of clays may alter the zeta-potential *vs.* pH function. The incorporation of QAC molecules and iron species caused, in most cases, the change of the zeta-potential to more positive values. Larger loadings (1 *CEC* and higher) of ODTMA and HDTMA reverse the surface charge of MMT due to the progressive covering of the edge and external surface of the clay by alkylammonium groups in bilayer arrangements (Bianchi et al. 2013). The incorporation of Fe (III) metallic centers in the clay mineral structure changes the properties of the surface and the zeta-potential *vs.* pH function has a different slope, similar to that of Fe (III) (hydr) oxides (Marco-Brown et al. 2012).

The analysis of the changes in the zeta-potential *vs.* pH plot may help to understand the adsorption mechanism of contaminants on the external surface of raw or modified clays. The formation of inner-sphere surface complexes between the charged molecules and the active surface sites of these solids produces changes in the net charge of the pollutant-adsorbent final material. Moreover, when the adsorbate has acid-base properties, the surface complexes formed may depend on the pH of the suspension. A zeta-potential analysis has been used to determine the formation of inner-sphere complexes of As (V) and picloram species (pH-dependent) on surface sites of iron-modified MMT (Marco-Brown et al. 2012, Iriel et al. 2020, Barraqué et al. 2021).

Electrophoretic mobility of clay suspensions is measured in a light-scattering zeta-potential analyzer instrument. Samples are prepared by dispersion of clay samples in an inert electrolyte (e.g., KCl) with adjusted pH (Barraqué et al. 2021).

2.2.2 X-Ray diffraction (XRD), small angle X-ray scattering (SAXS), and wide-angle X-ray scattering (WAXS) analysis

The entrance of substances into the interlayer of clays by interlayer exchange during its modification and/or pollutant adsorption leads to a basal space change of the clay mineral-based material. The thickness of the interlayer may be determined by the position of the d001 peak obtained by analysis of the sample in the range from 3 to 10°(2θ) for XRD analysis or in the range of scattering vector (*q*) from 0.25 to 6.0 nm for SAXS-WAXS (Marco-Brown et al. 2018).

The shifts of the position of the diffraction peaks with respect to the raw material provide information about the arrangements of a substance (e.g., QAC, Fe (III) oxo-hydroxides, pollutants) within the interlayer space. The dimension of a molecule can be estimated using modeling software (e.g., HyperChem) and, in this way, a model for the arrangement of molecules located in the interlaminar space can be proposed. Arrangements and rearrangements into the MMT basal space of picloram, Crystal Violet (CV, a cationic dye) and QAC molecules have been determined by this methodology (Bianchi et al. 2013, Marco-Brown et al. 2015, 2018).

Due to the adsorption of water into the interlayer space of the clay-based materials, it is important for the samples to be equilibrated at a Relative Humidity (*RH*) before the analysis, to allow comparison between the samples. On the other hand, it is recommended that the samples are oriented before their analysis. Orientation of the sample is done by placing a few drops of a suspension of the solid on a sample holder and allowing it to air (or *RH* controlled) dry. In this way, the sheets of mineral tend by gravity to orient themselves in the drop as it dries.

2.2.3 *Textural properties obtained by analysis of N_2 adsorption/desorption isotherms*

Clay minerals are porous materials; their external surface area, meso- and macroporous characteristics can be determined by the analysis of the N_2 adsorption/desorption isotherms (Wang et al. 2004, 2020). Briefly, the clay-based materials are degassed between 100 and 150°C for at least 6 hours, and N_2 adsorption/desorption isotherms are recorded at −196°C. CO_2 adsorption studies are carried out in order to characterize the micropore structures of clays. As expected, the modification of the clay or the adsorption of pollutants may lead to changes in its textural properties. The analysis of these changes may be relevant in the determination of the mechanisms involved in the modification or adsorption processes.

The analysis of the changes of total pore volume (V_{TP}), diameter pore size (W_P), and surface area determined by the BET method (S_{BET}) after iron modification of MMT or after adsorption of CV have been relevant to propose adsorption mechanisms (Guz et al. 2014, Marco-Brown et al. 2018, Iriel et al. 2020).

2.2.4 *Fourier-transform infrared (FTIR) analysis*

FTIR is a helpful tool to explore the adsorption mechanisms of substances. The presence of surface complexes can be inferred through the analysis of shifts, appearance or disappearance of IR absorption peaks of molecules or clay surface sites.

Transmission FTIR analysis can be used for the determination of surface interactions between an adsorbate and clay surface sites. Even more, FTIR has been employed for the recognition of the adsorbate chemical groups that are involved in the formation of surface complexes between the adsorbate and the clay. The study of the adsorption of picloram on MMT has suggested an inner-sphere complex formation by coordination through its pyridine nitrogen atom and carboxyl groups. Moreover, the FTIR analyses also indicated the formation of bidentate and bridge surface complexes by coordination of one picloram molecule to two external surface sites (Marco-Brown et al. 2015).

Studies of adsorption mechanisms of contaminants on clays can be performed using ATR-FTIR. Within the FTIR accessory, a thin colloidal film of clay is deposited onto an ATR crystal, a solution of the contaminant is placed on the clay film, and the ATR-FTIR spectrum is recorded as a function of time. In this technique, the IR beam passes through the ATR crystal and reflects at least once in the internal surface in contact with the sample, in this case, the Crystal Violet (CV) dye (Fig. 10.1).

The penetration depth into the sample is typically between 0.5 and 2 µm, thus allowing the detection of substances adsorbed on the clay. In this way, the adsorption kinetics of the pollutant can be studied and related to adsorption mechanisms. The study of the kinetic adsorption of CV on MMT using this technique will be described below in the text.

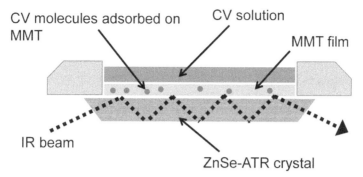

CV molecules adsorbed on MMT

CV solution

MMT film

IR beam

ZnSe-ATR crystal

Figure 10.1. Scheme of the experimental setup of the study of *in situ* CV adsorption kinetics on MMT films.

2.2.5 *Transmission (TEM) and scanning electron microscopy (SEM)*

TEM and SEM are powerful techniques to determine the formation of structures and nanostructures on the surface of clays, particularly when the structures are inorganic substances like iron oxides that display enough contrast against clay frameworks.

Analysis of samples by SEM has been extensively used on iron-modified clays as this technique can reveal the presence of iron species at the surface. SEM images of MMT typically show a face-to-edge contact between particles with random orientation and no formation of domains or clusters. A change in the roughness of the sample and the formation of prismatic shaped nanoparticles or rhombic nanoparticle structures of magnetite have been proposed to indicate the formation of iron oxides species at clay surfaces (Marco-Brown et al. 2012, Guz et al. 2014, Barraqué et al. 2018).

TEM images can reveal the growth of nanoparticles at the clay structures. nZVI supported on MMT materials were prepared through a chemical reduction procedure (Crane et al. 2011, Marco-Brown et al. 2021). Samples of nZVI supported on MMT were analyzed by TEM and SEM, and the images obtained are shown in Fig. 10.2. As it can be observed, the material obtained with the nZVI particles supported on MMT was a mixture of necklace-like aggregates and well discrete and dispersed nanoparticles (Marco-Brown et al. 2021). Similar structures were reported before (Sheng et al. 2014).

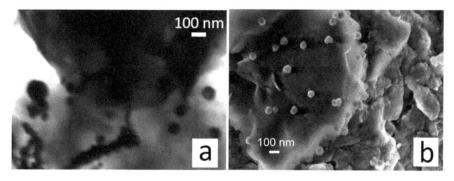

Figure 10.2. (a) TEM and (b) SEM images of nZVI supported on MMT.

3. Adsorption of Pollutants on Clays

3.1 *Kinetics of adsorption of organic molecules using ATR-FTIR spectroscopy*

Studies on the adsorption kinetics of pollutants on clays are typically carried out using batch systems. Nevertheless, a novel methodology using ATR-FTIR spectroscopy was employed to study the *in situ* kinetics of adsorption of organic molecules on MMT (Marco-Brown et al. 2014, 2018).

ATR-FTIR results obtained from the adsorption of CV on MMT helped to identify the chemical groups involved in the adsorption process and to determine the kinetics of the process through a simple and reproducible experimental setup. The schematic setup for the study of the CV adsorption kinetics on an MMT film is described in Fig. 10.1, and it may be extended to the study of adsorption of other organic molecules on clay films. Typically, a film of the clay is deposited over a commercial ZnSe-ATR crystal unit and rinsed with water to eliminate loosely deposited particles. After this, an electrolyte solution is placed on the clay film and the spectrum is recorded. Then, the electrolyte solution is withdrawn, a known volume with a given concentration of the organic substance is placed over the clay film and the spectra are recorded as a function of time until no further changes are detected. Shifts in FTIR spectra peaks of adsorbed molecules with respect to the non-adsorbed molecules allow determining the functional groups involved in the adsorption process.

In this particular case, CV solutions of varying concentration at a constant ionic strength and pH were placed on the MMT film, recording the temporal changes in absorbance (A_t) in the spectra.

The groups involved in the adsorption mechanisms of CV on MMT were determined by comparing the FTIR spectra of the CV solution with those of CV adsorbed on MMT. Figure 10.3 shows the FTIR spectra of different CV solutions and the FTIR spectra of CV adsorbed on MMT films at equilibrium time (CV@MMT film). The peaks between 1590 and 1480 cm^{-1} were assigned to the C=C stretching vibration of aromatic rings and the band centered at 1367 cm^{-1} was attributed to the N-phenyl stretching vibration. The doublet with peaks at 1176 and 1192 cm^{-1} was assigned to the N-CH$_3$ stretching vibration.

The appearance of a shoulder at 1380 cm^{-1} and the increment of the intensity of the peak at 1192 cm^{-1} observed in the CV@MMT film with respect to the CV solution for all the initial CV concentrations ([CV]$_0$) studied were attributed to the coordination of the CV molecule with surface sites through the interaction of its quaternary amine and the formation of an outer-sphere complex.

The adsorption kinetics of CV on MMT was followed *in situ* by ATR-FTIR (Fig. 10.4a) using the temporal evolution of the peaks centered at 1588, 1367, and 1184 cm^{-1}.

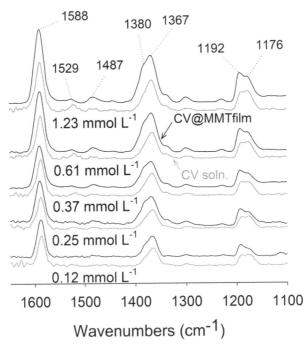

Figure 10.3. FTIR spectra of CV solutions and CV adsorbed on MMT films at equilibrium time (CV@MMTfilm). [CV]$_0$ are indicated. Reprinted with permission from Chem. Eng. J., 333(2018): 495–504.

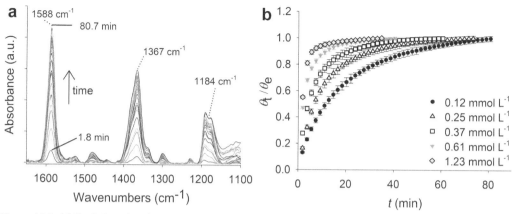

Figure 10.4. (a) Evolution of peaks at 1588, 1367 and 1184 cm⁻¹ over time in a typical experiment of CV adsorption ($[CV]_0$ = 0.12 mM) onto MMT. Time intervals are indicated in the figure. (b) Temporal evolution of θ_t/θ_e for the peak centered at 1367 cm⁻¹ (N-phenyl stretching vibration) for CV adsorption onto MMT. $[CV]_0$ is indicated in the legends. θ is the degree of surface coverage at a given time (θ_t) or at equilibrium (θ_e). Reprinted with permission from Chem. Eng. J. 333(2018): 495–504.

According to the Lambert-Beer law, the areas under the absorbance-wavelength curve at a given time (A_t) or at equilibrium (A_e), are directly proportional to the degree of surface coverage at a given time (θ_t) or at equilibrium (θ_e). Additionally, the ratio A_t/A_e is equivalent to the θ_t/θ_e ratio. Figure 10.4b shows θ_t/θ_e *vs.* time for the band centered at 1367 cm⁻¹. The evolution of θ_t/θ_e over time using the bands centered at 1588 and 1184 cm⁻¹ at all $[CV]_0$ had a similar behavior. It is necessary to relate θ_t to θ_e (θ_t/θ_e) in these kinds of systems in order to make them independent of the thickness of the clay films, allowing the comparison of results between experiments.

Several kinetic models for simulation of the experimental data such as Lagergren Pseudo-First Order (PFO), Pseudo-Second Order (PSO) and Intra-particle Diffusion Model (IDM) were tested (Areco et al. 2013, Haerifar and Azizian 2013, Marco-Brown et al. 2014). The kinetic law follows the next equations:

$$\ln(\theta_e - \theta_t) = \ln\theta_e - k_1 t \qquad \text{PFO (1)}$$

$$\frac{t}{\theta_t} = \frac{1}{k_2\theta_e^2} + \frac{t}{\theta_e} \qquad \text{PSO (2)}$$

$$\theta_t = k_{id}\sqrt{t} + C \qquad \text{IDM (3)}$$

Table 10.1 shows the parameters obtained after fitting the experimental data. For PFO, k_1 is the pseudo-first-order sorption rate constant; for PSO, k_2 is the sorption rate constant, and for IDM, k_{id} is the intraparticle diffusion constant and C is the y-intercept of the linear plots. θ_e and θ_t are the surface coverage at equilibrium and at a determined time t, respectively, and they were measured as the area under the bands centered at 1588, 1367, and 1184 cm⁻¹.

Plots of $\ln(1-\theta_t/\theta_e)$ *vs.* time (t) for bands centered at 1367 cm⁻¹, $t*\theta_e/\theta_t$ *vs.* t for the band centered at 1367 cm⁻¹, and of θ_t/θ_e *vs.* $t^{1/2}$ for the band centered at 1367 cm⁻¹ are presented in Fig. 10.5.

For PFO, the values of k_1 for each $[CV]_0$ (Table 10.1) were calculated from the slope by plotting $\ln(1-\theta_t/\theta_e)$ *vs.* t (Fig. 10.5a). The coefficients of determination (R^2) obtained using the PFO model for the evaluated bands were lower than 0.96.

For PSO, the values of k_2 and θ_e are typically obtained from the slope and y-intercept of a t/θ_t *vs.* t plot. In order to normalize the systems and make them independent of the thickness of the MMT film, $t*\theta_e/\theta_t$ *vs.* t plots were obtained. Figure 10.5b shows these plots for the band centered at 1367 cm⁻¹ at the $[CV]_0$; higher deviations from the PSO model as the $[CV]_0$ decreases were observed. The slope and the intercept differ from the expected values (1 and 0, respectively) as shown in

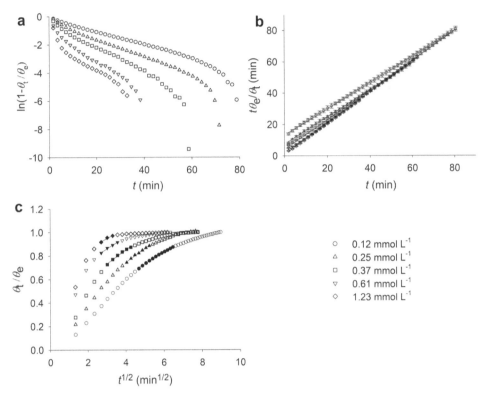

Figure 10.5. (a) Plots of $\ln(1-\theta_t/\theta_e)$ vs. time (t) for bands centered at 1367 cm^{-1}. (b) $t^*\theta_e/\theta_t$ vs. t linear plots for the band centered at 1367 cm^{-1}. The solid line was obtained from the PSO equation. (c) Plots of θ_t/θ_e vs. $t^{1/2}$ for the band centered at 1367 cm^{-1}. Full filled dots are related to the intraparticle diffusion stage and its linear regression was used to determine $k_{id}\theta_e^{-1}$ and C values. Empty dots at initial times are related to a high adsorption rate on the external surface. Empty dots at final times correspond to the final equilibrium stage. [CV]$_0$ is indicated. Reprinted with permission from Chem. Eng. J. 333(2018): 495–504.

Table 10.1. An intercept value different from zero or a deviation from the linearity at initial times was attributed to the contribution of a diffusional process at low [CV]$_0$, as will be addressed below.

The IDM was investigated to analyze the adsorption kinetics data due to the possible existence of a diffusional process contributing to the global adsorption rate. Plots of θ_t/θ_e at 1367 cm^{-1} vs. $t^{1/2}$ for CV adsorption on MMT film at different [CV]$_0$ are shown in Fig. 10.5c. After an extensive analysis of the data, it was concluded that the increase of the adsorbate concentration resulted in an increase of the concentration gradient between the porous and the solution bulk, which increased the diffusion rate of the adsorbate, k_{id}. The calculated intraparticle diffusion coefficient, k_{id}, related to θ_e ($k_{id}\theta_e^{-1}$) values at different initial dye concentrations are shown in Table 10.1. Results indicated that an increase of [CV]$_0$ caused no change in the $k_{id}\theta_e^{-1}$ value; nevertheless, θ_e value increased with [CV]$_0$; consequently, the k_{id} value should increase with [CV]$_0$. On the other hand, the C value (y-intercept of the linear plots) reflected the boundary layer effect (Weber and Morris 1963). The larger the value of the y-intercept of the plot (C), the greater is the contribution of the surface sorption in the rate-limiting step. The increase of k_{id} and $C\theta_e^{-1}$ with [CV]$_0$ indicated that the diffusion process was less important in samples with high [CV]$_0$ and the surface adsorption became the rate-limiting step. Hence, the three-stage linearity of intraparticle diffusion plots confirmed the presence of both surface adsorption and intraparticle diffusion as was proposed by PSO modeling. Thus, the adsorption process might present a complex nature, involving both surface adsorption and intraparticle diffusion.

Table 10.1. Pseudo-first order (PFO), pseudo-second order (PSO) and intraparticle diffusion model (IDM) parameters of the kinetics of CV adsorption on MMT obtained for $[CV]_0 = 0.12$; 0.25; 0.37; 0.61 and 1.23 mM.

	Kinetic Model								
	Pseudo-first order		Pseudo-second order			Intraparticle diffusion			
$[CV]_0$ (mM)	k_1 (min^{-1})	R^2	$\theta_e k_2$ (min^{-1})	M	R^2	$k_{id}\theta_e^{-1}$ (min$^{-1/2}$)	$C\theta_e^{-1}$	R^2
0.12	-0.06 ± 0.01	0.9374	0.075 ± 0.004	0.84 ± 0.02	0.9999	0.10 ± 0.01	0.24 ± 0.02	0.9944
0.25	-0.08 ± 0.01	0.9598	0.14 ± 0.01	0.89 ± 0.02	0.9994	0.12 ± 0.01	0.28 ± 0.02	0.9927
0.37	-0.08 ± 0.02	0.9611	0.24 ± 0.02	0.92 ± 0.02	0.9996	0.12 ± 0.01	0.38 ± 0.02	0.9878
0.61	-0.10 ± 0.02	0.9598	0.52 ± 0.08	0.96 ± 0.02	0.9998	0.10 ± 0.01	0.55 ± 0.02	0.9876
1.23	-0.14 ± 0.01	0.9075	1.0 ± 0.1	0.97 ± 0.02	0.9997	0.09 ± 0.01	0.68 ± 0.02	0.9608

Source: Data from Chem. Eng. J. 333(2018): 495–504, with permission.

3.2 pH-tuned sorption of imazalil on ODTMA-MMT

Intercalation of octadecyltrimethylammonium (ODTMA) in MMT can be performed by simple ionic exchange as explained above. The organoclay (OMMT) used in this work was intercalated with ODTMA equivalent to two times the Cationic Exchange Capacity (CEC). OMMT displayed different structural and surface properties compared with MMT, as determined by XRD analysis and electrophoretic mobility. Figure 10.6a shows the XRD pattern at low angle of MMT and OMMT. The ionic exchange of MMT with ODTMA produced the shift of the diffraction peak from $2\theta = 7.0°$ to $2\theta = 4.2°$. The shift to lower 2θ values indicates that the voluminous ODTMA was in the interlayer space increasing the interlayer distance from 1.24 to 2.02 nm as determined by the Bragg equation (nλ = 2d.sen (θ), where for CuKα, λ = 1.541 Å). Figure 10.6b shows the zeta-potential values calculated from electrophoretic data using the Smoluchowski equation, for MMT and OMMT at different solution pH values and 1 mM KCl. Clearly, the incorporation of ODTMA completely reverted the natural negative charge of MMT.

Imazalil (IMZ) is a well-known antifungal postharvest agent, widely used to preserve fruits before packaging for exports. IMZ belongs to the family of imidazoles, with a pK_a = 6.5; its chemical structure is shown in Fig. 10.7. IMZ is the active principle of several commercial formulations and provides approximately 50% of the Total Organic Carbon (TOC) of the product. The other components are typically surfactants and other coadjutants that help in the application of the products. IMZ is an endocrine disruptor; its presence in natural waters may affect the biota, in particular fishes and other aquatic organisms. For this reason, it is important to remove IMZ from liquid effluents (for example, wastewaters coming from fruit packaging companies) before it reaches the receptor watercourse.

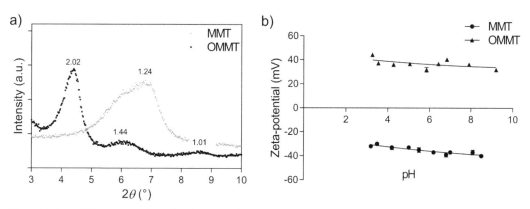

Figure 10.6. (a) XRD patterns of MMT and OMMT. (b) Zeta-potential at different pH values determined for MMT (black circles) and OMMT (black triangles).

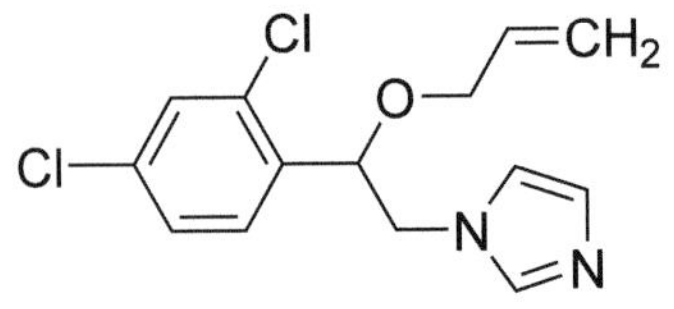

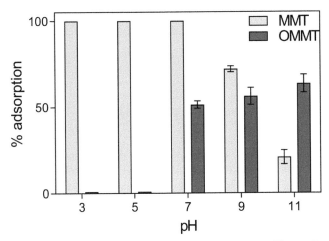

Figure 10.7. Chemical structure of imazalil.

Figure 10.8. Percentages of IMZ adsorption on MMT or OMMT at different pH values.

Based on IMZ acid-base properties, MMT surface charge and the presence of hydrophobic moieties in OMMT, a strategy for the removal of IMZ from water solution was developed (López Loveira 2017). Figure 10.8 shows the percentage of IMZ adsorbed from 100 mg L^{-1} solutions containing 1 g L^{-1} of clay (MMT or OMMT). At pH 7 or lower, IMZ was completely adsorbed by MMT, while at higher pH the percentage of adsorption decreased. These results can be explained in terms of the MMT surface charge and the IMZ pK_a. At pH values lower than pK_a (6.5), IMZ acquires a positive charge, while MMT displays a negative surface charge (see Fig. 10.6b). The electrostatic interaction between the IMZ molecules and the MMT surface led to the formation of ionic surface complexes with the consequent removal of the fungicide from the water solution. At pH higher than pK_a, the IMZ molecule is neutral and cannot be retained by the negative surface of MMT. In the case of OMMT as an adsorbent, the behavior was the opposite. At pH lower than 7, both IMZ and OMMT displayed positive charges, hindering the adsorption due to electrostatic repulsion, whereas at pH higher than 7, IMZ becomes a neutral molecule, being retained by OMMT through hydrophobic interactions.

Desorption experiments were performed to determine the possibility of reusing both sorbents several times. This is particularly important in the case of OMMT, where the incorporation of ODTMA increases the cost of the product with respect to simple MMT. Figure 10.9 shows the results obtained by resuspending the clays with adsorbed IMZ in water at different pH values, in such a proportion that allows reconstituting the original system from where IMZ was removed. Desorption of a small fraction of IMZ from MMT was accomplished only at the highest studied pH. However, in the case of OMMT, more than 40% of IMZ was desorbed at pH 6, increasing the fraction desorbed as the pH decreased. At pH 3.0, 100% of the adsorbed IMZ was desorbed. These results were likely a consequence of electrostatic repulsion between protonated IMZ and the positive surface charge of OMMT.

The results indicated that MMT is a better adsorbent for IMZ than OMMT at pH 9 or lower, but the adsorption is almost irreversible at pH 11 or lower. MMT could be used as an IMZ sorbent, but the complete MMT/IMZ system must be treated to eliminate IMZ or has to be safely disposed

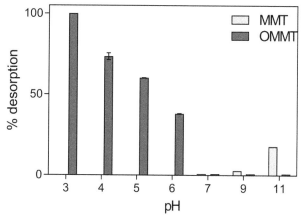

Figure 10.9. Percentages of IMZ desorbed from MMT or OMMT at different pH values.

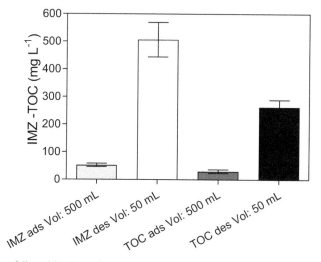

Figure 10.10. Adsorption followed by desorption and concentration to 1:10 of the original volume, from a 100 mg L⁻¹ commercial IMZ solution with 1 g L⁻¹ OMMT.

in a landfill for dangerous compounds. Adsorption-desorption experiments of IMZ were designed using OMMT as an adsorbent by adding it into a pH 7.0 water solution of IMZ from a commercial fungicide (formulated with 50% wt IMZ and coadjutants). After stirring and centrifugation, desorption was done by resuspending the clay in water at pH 3.0 (1/10 of the volume of the original solution), obtaining a solution of IMZ with higher concentration than the solution from where IMZ was adsorbed.

Figure 10.10 shows the results obtained from adsorption-desorption experiments. The concentration of IMZ removed from solution by adsorption at OMMT was 53 ± 6 mg L⁻¹ (light gray bar), while the concentration after desorption was 507 ± 63 mg L⁻¹ (white bar). The concentration factor after desorption was 10, meaning that almost all adsorbed IMZ was released to the smaller volume of water at pH 3. The concentration of TOC removed from solution at pH 7.0 was 31 ± 6 mg L⁻¹ (dark gray bar), while the TOC released at pH 3.0 was 263 ± 26 mg L⁻¹ (black bar). These concentrations of TOC are equivalent to the carbon contributed to the solution by IMZ, suggesting that coadjutants remained adsorbed at OMMT. In this way, it would be possible to obtain a concentrated solution of IMZ, which can be properly treated by different methods to eliminate the pollutant (advanced oxidation, biotreatment, among others). The remaining OMMT was used three consecutive times in adsorption experiments without a notably reduction in the performance for IMZ adsorption. However, the amount of adsorbed TOC decreased slightly, suggesting the accumulation of organics at the adsorbent.

OMMT represents an interesting alternative for the removal of IMZ (or other pollutants with acid-base characteristics), whose adsorption-desorption capacity can be tuned by changes in the solution pH. The sorbent can be recycled several times reducing the cost of the process. Due to the presence of hydrophobic moieties, coadjutants present in the formulation of most commercial pesticides are also partially adsorbed.

4. Applications as Catalysts by Coupling with Advanced Oxidation Processes

The incorporation of iron into clays is a promising strategy to treat diluted recalcitrant pollutants in water, coupling adsorption and Fenton-like processes (Herney-Ramirez et al. 2010, Guz et al. 2014, Zhu et al. 2019). As described earlier, adsorption is a fast and simple process, whilst the Fenton-like process consumes H_2O_2 to generate highly oxidative radicals, using iron incorporated to clays as a catalyst. The Fenton reaction has been known for over a century and has proven to be a very efficient technique to remove organic recalcitrant pollutants at very low concentrations (Liu et al. 2021, Oller and Malato 2021). In recent years, the number of scientific articles regarding the application of heterogeneous iron catalysts for Fenton-like processes has increased exponentially (Litter and Slodowicz 2017, Mirzaei et al. 2017, Zhu et al. 2019).

The production of hydroxyl radicals at the Fe-MMT surface by a Fenton Process can be explained by the following reactions:

$$\equiv Fe\,(III) + H_2O_2 \rightarrow \equiv FeOOH^{2+} + H^+ \tag{4}$$

$$\equiv Fe\,(III)\text{-}OOH^{2+} \rightarrow \equiv Fe\,(II) + HO_2^{\cdot} \tag{5}$$

$$\equiv Fe\,(II)\text{-}H_2O_2 \rightarrow \equiv Fe\,(III) + HO^{\cdot} + OH^- \tag{6}$$

The slowest step is the production of Fe (II) by eq. (5). By illumination of the catalyst with visible or UVA light, regeneration of Fe (II) is accelerated through reaction (7) (Litter and Slodowicz 2017):

$$\equiv Fe\,(III)\text{-}(OH)^{2+} + h\nu \rightarrow \equiv Fe\,(III)(OH)^{2+*} \rightarrow \equiv Fe\,(II) + OH^{\cdot} \tag{7}$$

Several authors have investigated the possibility of coupling adsorption with Fenton-like processes to reduce sludge generation and reuse the adsorbents (Hu et al. 2011, Guimarães et al. 2019, Xu et al. 2019, Meng et al. 2020). In some cases, the adsorption and Fenton or photo-Fenton-like processes are carried out sequentially in the same vessel. The Fenton reaction starts with the addition of hydrogen peroxide (and turning on the lamp, in the case of photo-Fenton processes) after achieving adsorption equilibrium (Hu et al. 2011, Guimarães et al. 2019, Xu et al. 2019). However, another interesting strategy is the separation of the loaded adsorbent and the regeneration or treatment of the adsorbed pollutant in a different appropriate reactor (or photoreactor). In this way, the pollutant could be easily removed from the water, since adsorption is fast and simple, and stored for a later regeneration. Besides, the adsorbent containing the pollutant could be resuspended in water at a higher solid/liquid ratio, increasing the pollutant concentration in a much smaller water volume.

For example, the iron-modified MMT previously described was employed as Fenton and photo-Fenton-like catalyst (Guz et al. 2014). The experimental setup tried to simulate the removal of low concentrations of CV from high water volumes, recovery of the loaded adsorbent/catalyst and finally redispersing the used clay in a smaller amount of water to be treated by Fenton or photo-Fenton-like oxidation. CV solutions were stirred for 80 minutes with Fe-MMT until the adsorption equilibrium was reached (Marco-Brown et al. 2018). The loaded clay was recovered and resuspended in water adjusting the pH to 3. Finally, the suspension was placed in a thermostatted reactor and H_2O_2 was added. In the case of photo-Fenton-like experiments, the same reactor was illuminated from above with a 300 W solar lamp (maximum emission at 630 nm and lower emissions between 315 and 400 nm). This preconcentration by rapid adsorption on MMT or Fe-MMT, reduces the volume of the wastewater to be treated and of the photocatalytic reactors employed. Furthermore, since there was no apparent desorption of the dye into the solution (because the dye could not be detected by spectrophotometry), it is likely that the contaminant was heterogeneously oxidized on the surface of the clay. The oxidized dye byproducts were released to the solution, as measured by TOC. In the case of the photo Fenton-like experiments, TOC values after 2 hours of treatment indicated the almost complete mineralization of the released organics.

The Fenton and photo-Fenton processes are affected by the experimental conditions in addition to the characteristics of the catalyst. Figure 10.11 shows the degradation of the adsorbed CV by the Fenton reaction using Fe-MMT as adsorbent/catalyst and varying several key parameters of the process. In order to measure the degradation of the adsorbed dye, the remaining dye was desorbed from the loaded clay at different degradation times by a 5% Sodium Dodecyl Sulfate (SDS) solution (Guz et al. 2014). The remaining CV was measured in the SDS solution by spectrophotometry.

As seen in Fig. 10.11, when the amount of clay is decreased (and therefore the concentration of iron) from 1.5 g L^{-1} to 1 g L^{-1}, the process rate decreased, reaching the maximum discoloration after 0.75 hours and 1.5 hours respectively. Despite the dye being completely adsorbed in both cases, the decrease of the clay mass while keeping the CV concentration constant, increases the adsorbed CV/Fe-MMT ratio explaining the slower initial degradation rate observed. A similar effect was reported before using goethite (α-FeOOH) as catalyst for the degradation of methyl orange by a heterogeneous Fenton process (Wang et al. 2015).

By raising the pH close to neutrality, the discoloration rate notably decreases. However, as the oxidation of the dye progresses, the pH begins to decrease, reaching values of 4.5 after 2 hours of treatment. The decrease of the pH with reaction time is likely a consequence of iron leaching to the solution from the clay and the formation of acid byproducts (Guz et al. 2014). This acidification process may explain the notable drop in the CV concentration observed after 2 hours at initial pH 6.0 (see Fig. 10.11, experiments performed at 1.5 hours: 36% remaining dye; 2.0 hours: 14% remaining dye). When working at pH 6.0, the data suggest that the dye could be completely degraded in treatment times larger than 2 hours. It is widely reported in literature that the Fenton-type process is pH-dependent, mainly due to the low solubility of iron at pH close to neutrality (Fan et al. 2009). In the case of the experiments performed with Fe-MMT at pH 6.0, the dissolved iron concentration found at the end of the process was 0.025 mM (determined by atomic absorption spectroscopy).

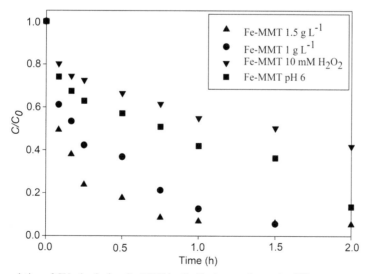

Figure 10.11. Degradation of CV adsorbed on Fe-MMT by the Fenton reaction under different process conditions. Unless otherwise specified in the legend, the conditions were: [CV] = 0.06 mM, Fe-MMT = 1.5 g L^{-1}, [H_2O_2] = 50 mM, pH 3.0.

This value is similar to the one obtained working with simple MMT as adsorbent and catalyst, where negligible degradation of the dye was observed after Fenton treatment at pH 6.0. These results indicate that Fe (III) present at the Fe-MMT surface play an important role as a catalyst for a photo-Fenton process at pH 6.0. Consequently, Fe-MMT results in an interesting alternative as adsorbent/catalyst for Fenton process at pH close to neutrality.

The concentration of H_2O_2 affects the process rate and efficiency the most: by reducing its concentration from 50 to 10 mM, the concentration of the remaining dye increased 8.5 times after 2 hours of treatment. Besides, in the case of 10 mM initial H_2O_2 concentration, 95% of this concentration is consumed during the first 15 minutes, indicating that the amount of the oxidant is not sufficient to achieve CV discoloration (data not shown). Wang et al. found similar results using goethite as heterogeneous Fenton catalyst (Wang et al. 2015). The amount of generated $HO^\bullet$ is directly dependent on the H_2O_2 concentration, and in this case, 10 mM was not sufficient to completely discolor the adsorbed dye.

Figure 10.12 shows the evolution of TOC for Fenton-like treatments. At pH 6, the oxidized organic compounds are released at a slower initial rate to the solution, as observed in the discoloration experiments. This delay in the desorption of the degradation products could be associated with the slow acidification of the suspension, as described earlier. As in the degradation experiments, the only parameter that significantly reduces the final release of TOC to the solution is the initial concentration of H_2O_2. In the rest of the experiments, the concentration of H_2O_2 employed was 5.7 times higher than the stoichiometric amount necessary to completely mineralize the CV adsorbed onto the clay (0.006 mmol), according to reaction (8), whilst using 10 mM initial concentration is equivalent to 1.1 times the stoichiometric quantity. According to these results, the concentration necessary to mineralize the dye is greater than that theoretically calculated, as already reported by other authors (Chen and Zhu 2009, Chen et al. 2009, Herney-Ramirez et al. 2010).

$$C_{25}H_{30}N_3^+ + 72\ H_2O_2 \rightarrow 25\ CO_2 + 85\ H_2O + 3\ NO_3^- + 4\ H^+ \tag{8}$$

Figure 10.13 shows the degradation of CV adsorbed onto Fe-MMT by photo-Fenton-like processes.

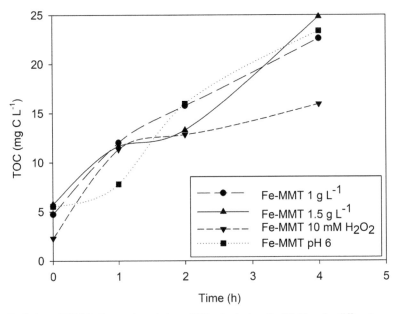

Figure 10.12. Evolution of TOC in Fenton degradation of CV adsorbed on Fe-MMT under different process conditions. Unless otherwise specified in the legend, the conditions were: [CV] = 0.06 mM, Fe-MMT = 1.5 g L⁻¹, [H₂O₂] = 50 mM, pH 3.0. Lines are only for clear visualization.

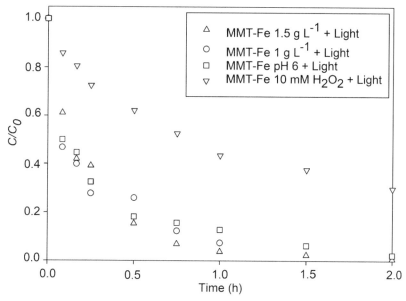

Figure 10.13. Degradation of CV adsorbed on Fe-MMT by photo-Fenton reaction under different process conditions. Unless otherwise specified in the legend, the conditions were: [CV] = 0.06 mM, Fe-MMT = 1.5 g L^{-1}, [H$_2$O$_2$] = 50 mM, pH 3.0.

The main factor that significantly influences the degradation rate of the dye is the H$_2$O$_2$ concentration. At initial pH 6.0, the reaction was slightly slower but reached a similar discoloration efficiency compared with the experiments at initial pH 3.0. As explained earlier, this could be a consequence of the slow acidification of the solution. Other authors have already demonstrated the versatility of supporting the catalyst and employing it in a heterogeneous phase, allowing its use at a pH closer to neutrality (Zhu et al. 2019). The authors conclude that the immobilization of iron within the structure of the catalysts leads to a retention of their ability to decompose H$_2$O$_2$ into HO$^{\bullet}$, preventing the leaching of iron ions and the precipitation of iron (III) hydroxide at pH higher than 4. Unlike to that observed in the case of the Fenton process (Fig. 10.11), similar degradation rates were obtained for lower catalyst concentrations. This could be a consequence of the higher degradation rate obtained through the photo-Fenton process, due to the higher regeneration rate of Fe (II) and HO$^{\bullet}$ production under illumination (Guz et al. 2014). If the catalyst were regenerated at a faster rate, the concentration of the catalyst needed in the photo-Fenton experiments would be lower than in the Fenton experiments under the same experimental conditions. This might suggest that a less catalyst/adsorbent ratio is required for the photo-Fenton experiments, representing an advantage compared with Fenton process.

Figure 10.14 shows the evolution of TOC for photo-Fenton-like treatments. As in the discoloration experiments, there are no significant differences in terms of the final TOC released to the solution by the different treatments, except when the H$_2$O$_2$ concentration is decreased. However, it can be observed that, when using a lesser amount of catalyst, the release rate of TOC to the solution is slower, reaching its maximum after 2 hours of treatment, instead of 1 hour.

As it can be seen in these results, the support of iron catalysts into nanoclays not only reduces the sludge generation of the Fenton or photo-Fenton processes and/or allows the recovering and reusability of the catalyst, as previously reported by several authors, but also gives more flexibility to the overall wastewater treatment. The loaded adsorbent/catalyst could be treated under different Fenton or photo-Fenton process conditions (pH near neutrality or higher pollutant/catalyst ratio) without affecting the final mineralization efficiency of the treatment.

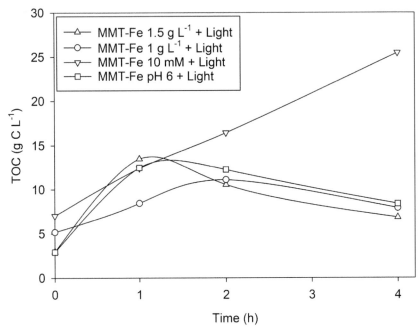

Figure 10.14. Evolution of TOC in photo-Fenton degradation of CV adsorbed on Fe-MMT under different process conditions. Unless otherwise specified in the legend, the conditions were: [CV] = 0.06 mM, Fe-MMT = 1.5 g L^{-1}, [H$_2$O$_2$] = 50 mM, pH 3.0. Lines are only for clear visualization.

5. Conclusions and Future Trends

Over the past years, the properties of bentonites to be used in adsorption and detoxification systems have been studied for their application in environmental remediation (Phuekphong et al. 2020). There has also been an interest in adsorbent regeneration, where bentonites have an advantage over other adsorbents, including activated carbons (Momina et al. 2018). For the years to come, the challenge is to develop "multifunctional" (for the simultaneous adsorption and decomposition of different contaminants) and "smart" (for sensors/detection, stimuli-responsive adsorption/desorption) clay-based adsorbents (Phuekphong et al. 2020). There are already examples of next-generation applications of bentonites in the remediation and bioremediation of complex pollutant matrices (Zhang et al. 2019, Fomina and Skorochod 2020, Phuekphong et al. 2020). Even more, there are other cutting-edge applications in several fields, such as food packaging (Chaudhary et al. 2020), transportation and aerospace industries (Mathew et al. 2019), and human and animal health (Abduljauwad et al. 2020, Baigorria et al. 2020, Fomina and Skorochod 2020, Phuekphong et al. 2020).

These examples illustrate the versatility of bentonites for environmental applications, not only in remediation processes, but also in the development of systems with lower environmental impact than the ones traditionally used. The combination of the abundance of clays, their low price and versatility make modified nano bentonites a promising material for multiple applications focused in environmental preservation.

Acknowledgments

The authors are grateful to Universidad Nacional de San Martín, and Agencia Nacional de Promoción de la Investigación, el Desarrollo Tecnológico y la Innovación, for the support given through PICTs: 2014-2386; 2014-0750, 2015-1260, 2016-2940, 2018-2788, 2019-2320 and 2019-3263. CONICET fellowships given to E.B. and F.I. are acknowledged. M.B., R.C., P.C., L.G. and J.L.M.B. are members from CONICET.

References

Abduljauwad, S.N., T. Habib and H. ur R. Ahmed. 2020. Nano-clays as potential pseudo-antibodies for COVID-19. Nanoscale Res. Lett. 15.

Alshangiti, D.M., M.M. Ghobashy, S.A. Alkhursani, F.S. Shokr, S.A. Al-Gahtany and M.M. Madani. 2019. Semi-permeable membrane fabricated from organoclay/PS/EVA irradiated by X-rays for water purification from dyes. J. Mater. Res. Technol. 8: 6134–6145.

Alther, G. 2002. Using organoclays to enhance carbon filtration. Waste Manag. 22: 507–513.

An, N., C.H. Zhou, X.Y. Zhuang, D.S. Tong and W.H. Yu. 2015. Immobilization of enzymes on clay minerals for biocatalysts and biosensors. Appl. Clay Sci. 114: 283–296.

Areco, M.M., L. Saleh-Medina, M.A. Trinelli, J.L. Marco-Brown and M. dos Santos Afonso. 2013. Adsorption of Cu(II), Zn(II), Cd(II) and Pb(II) by dead Avena fatua biomass and the effect of these metals on their growth. Colloids Surfaces B Biointerfaces 110: 305–312.

Awad, A.M., S.M.R. Shaikh, R. Jalab, M.H. Gulied, M.S. Nasser, A. Benamor et al. 2019. Adsorption of organic pollutants by natural and modified clays: A comprehensive review. Sep. Purif. Technol. 228: 115719.

Baigorria, E., L.A. Cano, L. Sánchez, V.A. Alvarez and R.P. Ollier. 2020. Bentonite-composite polyvinyl alcohol/alginate hydrogel beads: Preparation, characterization and their use as arsenic removal devices. Environmental Nanotechnology, Monitoring & Management 14: 100364.

Barakan, S. and V. Aghazadeh. 2021. The advantages of clay mineral modification methods for enhancing adsorption efficiency in wastewater treatment: a review. Environ. Sci. Pollut. Res. 28: 2572–2599.

Barraqué, F., M.L. Montes, M.A. Fernández, R.C. Mercader, R.J. Candal and R.M. Torres Sánchez. 2018. Synthesis and characterization of magnetic-montmorillonite and magnetic-organo-montmorillonite: Surface sites involved on cobalt sorption. J. Magn. Magn. Mater. 466: 376–384.

Barraqué, F., M.L. Montes, M.A. Fernández, R. Candal, R.M. Torres Sánchez and J.L. Marco-Brown. 2021. Arsenate removal from aqueous solution by montmorillonite and organo-montmorillonite magnetic materials. Environ. Res. 192.

Bianchi, A.E., M. Fernández, M. Pantanetti, R. Viña, I. Torriani, R.M.T. Sánchez et al. 2013. ODTMA+ and HDTMA+ organo-montmorillonites characterization: New insight by WAXS, SAXS and surface charge. Appl. Clay Sci. 83–84: 280–285.

Calabi Floody, M., B.K.G. Theng, P. Reyes and M.L. Mora. 2009. Natural nanoclays: applications and future trends—a Chilean perspective. Clay Miner. 44: 161–176.

Chaudhary, P., F. Fatima and A. Kumar. 2020. Relevance of nanomaterials in food packaging and its advanced future prospects. J. Inorg. Organomet. Polym. Mater. 30: 5180–5192.

Chen, J. and L. Zhu. 2009. Comparative study of catalytic activity of different Fe-pillared bentonites in the presence of UV light and H_2O_2. Sep. Purif. Technol. 67: 282–288.

Chen, Q., P. Wu, Y. Li, N. Zhu and Z. Dang. 2009. Heterogeneous photo-Fenton photodegradation of reactive brilliant orange X-GN over iron-pillared montmorillonite under visible irradiation. J. Hazard. Mater. 168: 901–908.

Chen, X., K. Chen, G. Bai, J. Liu, M. Lin and X. Shen. 2011. Photodegradation of toluene by composite photocatalysts of organobentonites and TiO_2: Influences of carbon chain length. Adv. Mater. Res. 233–235: 1958–1965.

Crane, R.A., M. Dickinson, I.C. Popescu and T.B. Scott. 2011. Magnetite and zero-valent iron nanoparticles for the remediation of uranium contaminated environmental water. Water Res. 45: 2931–2942.

Deepracha, S., S. Bureekaew and M. Ogawa. 2019. Synergy effects of the complexation of a titania and a smectite on the film formation and its photocatalyst' performance. Appl. Clay Sci. 169: 129–134.

Dong, H. 2012. Clay-microbe interactions and implications for environmental mitigation. Elements 8: 113–118.

Fan, H.J., S.T. Huang, W.H. Chung, J.L. Jan, W.Y. Lin and C.C. Chen. 2009. Degradation pathways of crystal violet by Fenton and Fenton-like systems: Condition optimization and intermediate separation and identification. J. Hazard. Mater. 171: 1032–1044.

Fomina, M. and I. Skorochod. 2020. Microbial interaction with clay minerals and its environmental and biotechnological implications. Minerals 10: 1–54.

Gadd, G.M. 2010. Metals, minerals and microbes: Geomicrobiology and bioremediation. Microbiology 156: 609–643.

Gallouze, H., D.E. Akretche, C. Daniel, I. Coelhoso and J.G. Crespo. 2021. Removal of synthetic estrogen from water by adsorption on modified bentonites. Environ. Eng. Sci. 38: 4–14.

Guimarães, V., M.S. Lucas and J.A. Peres. 2019. Combination of adsorption and heterogeneous photo-Fenton processes for the treatment of winery wastewater. Environ. Sci. Pollut. Res. 26: 31000–31013.

Guz, L., G. Curutchet, R.M. Torres Sánchez and R. Candal. 2014. Adsorption of crystal violet on montmorillonite (or iron modified montmorillonite) followed by degradation through Fenton or photo-Fenton type reactions. J. Environ. Chem. Eng. 2: 2344–2351.

Haerifar, M. and S. Azizian. 2013. Mixed surface reaction and diffusion-controlled kinetic model for adsorption at the solid/solution interface. J. Phys. Chem. C 117: 8310–8317.

Herney-Ramirez, J., M.A. Vicente and L.M. Madeira. 2010. Heterogeneous photo-Fenton oxidation with pillared clay-based catalysts for wastewater treatment: A review. Appl. Catal. B Environ. 98: 10–26.

Hu, X., B. Liu, Y. Deng, H. Chen, S. Luo, C. Sun et al. 2011. Adsorption and heterogeneous Fenton degradation of 17α-methyltestosterone on nano Fe_3O_4/MWCNTs in aqueous solution. Appl. Catal. B Environ. 107: 274–283.

Ide, Y., M. Matsuoka and M. Ogawa. 2012. Controlled photocatalytic oxidation of benzene in aqueous clay suspension. ChemCatChem 4: 628–630.

Iriel, A., J.L. Marco-Brown, M. Diljkan, M.A. Trinelli, M. Dos Santos Afonso and A. Fernández Cirelli. 2020. Arsenic adsorption on iron-modified montmorillonite: kinetic equilibrium and surface complexes. Environ. Eng. Sci. 37: 22–32.

Jaworski, M.A., F.M. Flores, M.A. Fernández, M. Casella and R.M. Torres Sánchez. 2019. Use of organo-montmorillonite for the nitrate retention in water: influence of alkyl length of loaded surfactants. SN Appl. Sci. 1.

Kamińska, G. 2018. Removal of organic micropollutants by grainy bentonite-activated carbon adsorbent in a fixed bed column. Water (Switzerland) 10.

Khelifi, S. and F. Ayari. 2019. Modified bentonite for anionic dye removal from aqueous solutions. Adsorbent regeneration by the photo-Fenton process. Comptes Rendus Chim. 22: 154–160.

Kim, S.M. and A. Vogelpohl. 1998. Degradation of organic pollutants by the photo-fenton-process. Chem. Eng. Technol. 21: 187–191.

Komlósi, A., E. Kuzmann, N.M. Nagy, Z. Homonnay, S. Kubuki and J. Kŏnya. 2007. Incorporation of Fe in the interlayer of Na-bentonite via treatment with $FeCl_3$ in acetone. Clays Clay Miner. 55: 89–95.

Langford, C.H. and J.H. Carey. 1975. The charge transfer photochemistry of the Hexaaquoiron(III) Ion, the Chloropentaaquoiron(III) ion, and the μ-dihydroxo dimer explored with tert-Butyl alcohol scavenging. Can. J. Chem. 53: 2430–2435.

Litter, M.I. and M. Slodowicz. 2017. An overview on heterogeneous Fenton and photoFenton reactions using zerovalent iron materials. J. Adv. Oxid. Technol. 20.

Liu, Y., Y. Zhao and J. Wang. 2021. Fenton/Fenton-like processes with *in-situ* production of hydrogen peroxide/ hydroxyl radical for degradation of emerging contaminants: Advances and prospects. J. Hazard. Mater. 404: 124191.

López Loveira, E.G. 2017. Tratamiento de efluentes agroindustriales a través de técnicas combinadas de adsorción, biológicas y procesos avanzados de oxidación. Ph. D. Thesis. Universidad Nacional de San Martín, Buenos Aires, Argentina.

Mangwandi, C., S.N.A. Suhaimi, J.T. Liu, R.M. Dhenge and A.B. Albadarin. 2016. Design, production and characterisation of granular adsorbent material for arsenic removal from contaminated wastewater. Chem. Eng. Res. Des. 110: 70–81.

Marco-Brown, J.L., C.M. Barbosa-Lema, R.M. Torres Sánchez, R.C. Mercader and M. dos Santos Afonso. 2012. Adsorption of picloram herbicide on iron oxide pillared montmorillonite. Appl. Clay Sci. 58: 25–33.

Marco-Brown, J.L., M.M. Areco, R.M. Torres Sánchez and M. Dos Santos Afonso. 2014. Adsorption of picloram herbicide on montmorillonite: Kinetic and equilibrium studies. Colloids Surfaces A Physicochem. Eng. Asp. 449: 121–128.

Marco-Brown, J.L., M.A. Trinelli, E.M. Gaigneaux, R.M. Torres Sánchez and M. Dos Santos Afonso. 2015. New insights on the structure of the picloram-montmorillonite surface complexes. J. Colloid Interface Sci. 444: 115–122.

Marco-Brown, J.L., L. Guz, M.S. Olivelli, B. Schampera, R.M. Torres Sánchez, G. Curutchet et al. 2018. New insights on crystal violet dye adsorption on montmorillonite: Kinetics and surface complexes studies. Chem. Eng. J. 333: 495–504.

Marco-Brown, J.L., R. Valiente, C.P. Ramos, M.A. Fernández and R. Candal. 2021. Stable nZVI-based nanocomposites for adsorption and reduction processes: The case of U(VI) removal. Environ. Nanotechnol. Monit. Manag. 16: 100563.

Mathew, J., J. Joy and S.C. George. 2019. Potential applications of nanotechnology in transportation: A review. J. King Saud Univ. - Sci. 31: 586–594.

Meng, X., C. Zhang, J. Zhuang, G. Zheng and L. Zhou. 2020. Assessment of schwertmannite, jarosite and goethite as adsorbents for efficient adsorption of phenanthrene in water and the regeneration of spent adsorbents by heterogeneous fenton-like reaction. Chemosphere 244: 125523.

Mirzaei, A., Z. Chen, F. Haghighat and L. Yerushalmi. 2017. Removal of pharmaceuticals from water by homo/heterogonous Fenton-type processes—A review. Chemosphere 174: 665–688.

Mo, W., Q. He, X. Su, S. Ma, J. Feng and Z. He. 2018. Preparation and characterization of a granular bentonite composite adsorbent and its application for Pb^{2+} adsorption. Appl. Clay Sci. 159: 68–73.

Mogyorósi, K., A. Farkas, I. Dékány, I. Ilisz and A. Dombi. 2002. TiO_2-based photocatalytic degradation of 2-chlorophenol adsorbed on hydrophobic clay. Environ. Sci. Technol. 36: 3618–3624.

Molina, C.B., E. Sanz-Santos, A. Boukhemkhem, J. Bedia, C. Belver and J.J. Rodriguez. 2020. Removal of emerging pollutants in aqueous phase by heterogeneous Fenton and photo-Fenton with Fe_2O_3-TiO_2-clay heterostructures. Environ. Sci. Pollut. Res. 27: 38434–38445.

Momina, M. Shahadat and S. Isamil. 2018. Regeneration performance of clay-based adsorbents for the removal of industrial dyes: A review. RSC Adv. 8: 24571–24587.

Mukhopadhyay, R., T. Adhikari, B. Sarkar, A. Barman, R. Paul, A.K. Patra et al. 2019. Fe-exchanged nano-bentonite outperforms Fe_3O_4 nanoparticles in removing nitrate and bicarbonate from wastewater. J. Hazard. Mater. 376: 141–152.

Oller, I. and S. Malato. 2021. Photo-Fenton applied to the removal of pharmaceutical and other pollutants of emerging concern. Curr. Opin. Green Sustain. Chem. 29: 100458.

Pandey, S. 2017. A comprehensive review on recent developments in bentonite-based materials used as adsorbents for wastewater treatment. J. Mol. Liq. 241: 1091–1113.

Phuekphong, A.F., K.J. Imwiset and M. Ogawa. 2020. Organically modified bentonite as an efficient and reusable adsorbent for triclosan removal from water. Langmuir 36: 9025–9034.

Prabhu, P.P. and B. Prabhu. 2018. A review on removal of heavy metal ions from waste water using natural/ modified bentonite. MATEC Web Conf. 144: 1–13.

Shaalan, H.F. 2009. Treatment of pesticides containing effluents using organoclays/nanofiltration systems: Rational design and cost indicators. Desalin. Water Treat. 5: 153–158.

Sheng, G., X. Shao, Y. Li, J. Li, H. Dong, W. Cheng et al. 2014. Enhanced removal of uranium(VI) by nanoscale zerovalent iron supported on na-bentonite and an investigation of mechanism. J. Phys. Chem. A 118: 2952–2958.

Shi, L. na, Y.M. Lin, X. Zhang and Z. liang Chen. 2011. Synthesis, characterization and kinetics of bentonite supported nZVI for the removal of Cr(VI) from aqueous solution. Chem. Eng. J. 171: 612–617.

Smoluchowski, M. 1921. Handbuch der Elektrizität und des Magnetismus. Vol. II. L. Grätz (ed.). Ambrosius Barth Verlag, Leipzig.

Sun, Z., L. Feng, G. Fang, L. Chu, D. Zhou and J. Gao. 2021. Nano Fe_2O_3 embedded in montmorillonite with citric acid enhanced photocatalytic activity of nanoparticles towards diethyl phthalate. J. Environ. Sci. (China) 101: 248–259.

Uddin, M.K. 2017. A review on the adsorption of heavy metals by clay minerals, with special focus on the past decade. Chem. Eng. J. 308: 438–462.

Wang, C.C., L.C. Juang, C.K. Lee, T.C. Hsu, J.F. Lee and H.P. Chao. 2004. Effects of exchanged surfactant cations on the pore structure and adsorption characteristics of montmorillonite. J. Colloid Interface Sci. 280: 27–35.

Wang, X., H. Cheng, P. Chai, J. Bian, X. Wang, Y. Liu et al. 2020. Pore characterization of different clay minerals and its impact on methane adsorption capacity. Energy and Fuels 34: 12204–12214.

Wang, Y., Y. Gao, L. Chen and H. Zhang. 2015. Goethite as an efficient heterogeneous Fenton catalyst for the degradation of methyl orange. Catal. Today 252: 107–112.

Weber, W.J. and J.C. Morris. 1963. Kinetics of adsorption carbon from solutions. Journal Sanitary Engineering Division Proceedings. American Society of Civil Engineers 89: 31–42.

Xu, B. De, D.C. Li, T.T. Qian and H. Jiang. 2020. Boosting the activity and environmental stability of nanoscale zero-valent iron by montmorillonite supporting and sulfidation treatment. Chem. Eng. J. 387.

Xu, X., W. Chen, S. Zong, X. Ren and D. Liu. 2019. Magnetic clay as catalyst applied to organics degradation in a combined adsorption and Fenton-like process. Chem. Eng. J. 373: 140–149.

Yadav, V.B., R. Gadi and S. Kalra. 2019. Clay based nanocomposites for removal of heavy metals from water: A review. J. Environ. Manage. 232: 803–817.

Yang, Y., R. Zhu, Q. Chen, H. Fu, Q. He, J. Zhu et al. 2020. A novel multifunctional adsorbent synthesized by modifying acidified organo-montmorillonite with iron hydroxides. Appl. Clay Sci. 185: 105420.

Yin, Y., C. Shen, X. Bi and T. Li. 2020. Removal of hexavalent chromium from aqueous solution by fabricating novel heteroaggregates of montmorillonite microparticles with nanoscale zero-valent iron. Sci. Rep. 10: 1–12.

Zhang, C., J. Luan, X. Yu and W. Chen. 2019. Characterization and adsorption performance of graphene oxide–montmorillonite nanocomposite for the simultaneous removal of Pb^{2+} and p-nitrophenol. J. Hazard. Mater. 378: 120739.

Zhu, Y., R. Zhu, Y. Xi, J. Zhu, G. Zhu and H. He. 2019. Strategies for enhancing the heterogeneous fenton catalytic reactivity: A review. Appl. Catal. B Environ. 255: 117739.

Chapter 11

Transparent Open Cellular Structures Formed with Silica Nanopowder for Industrial Applications in Photocatalysis

Ethel C. Bucharsky,* Frank B. Löffler and *Karl G. Schell*

1. Introduction

Glass devices of pure and clean silica have excellent properties such as a wide range of high transparency in the visible and IR range and a good resistance against a large number of chemicals. Conventional glass processing is mostly based on melting processes, and that is the reason for the use of high temperatures for processing SiO_2. In order to lower these temperatures, on one hand, network formers can be used, whereby the material composition is influenced or sintering techniques based on particulate systems can be applied. By the use of silica nanopowders, the process temperature can be reduced and the forming process is no longer limited to the melting process.

This chapter describes the potential of nanoscale silica powder, used to manufacture stable slurries, which can be used in the replica process (Schwartzwalder and Somers 1961) or in additive manufacturing. This means that various, complex-shaped structures can be created using either templating techniques or 3D printing. Using the example of open-porous structures, which can be used as a substrate for photocatalytic coatings, the present chapter describes the way to improve novel water treatment processes and other applications.

2. The Role of the Support on Photocatalysis

The photocatalyst support is not just a simple "inert" mechanical template for the photocatalytic particles, as it can affect their reactivity by chemical stabilization and modification of their geometric

Institute for Applied Materials – Ceramic Materials and Technologies, Karlsruhe Institute of Technology (KIT), 76131 Karlsruhe, Germany.
Emails: frank.loeffler@kit.edu; g.schell@kit.edu
* Corresponding author: ethel.bucharsky@kit.edu

and electronic properties. Applications of SiO_2 as a support can be found in different fields, from catalysis to composite synthesis. Some major studies are, for example:

- Employing a Pickering emulsion polymerization stabilized by silicon dioxide nanoparticles to form molecularly imprinted polymers for steroid recognition (Zhou et al. 2014).
- Using mesoporous silica as a catalyst support in the preparation of multiwalled carbon nanotubes (MWCNTs) (Cervantes-Sodi et al. 2012).
- Ring opening of tetrahydrofuran-dimethanol to form 1,2,6-hexanetriol (Buntara et al. 2013).
- Applications such as preparing membranes based on silica/brominated poly (phenylene oxide) nanocomposites for CO_2 separations (Hu et al. 2007).
- Functionalization of mesoporous silica nanoparticles for targeting, biocompatibility, combined cancer therapies and theragnosis. The high surface area of the nanoparticles contributes to the loading of active substances and their ability to be released. These are just some of the positive aspects that make this technology a great asset to the field of nanotechnology (Kuthati et al. 2013).

In all these cases, silica is employed as a support in the form of nanoparticles. A novel and most interesting field is where silica nanoparticles are used to produce porous materials with sufficient mechanical stability, optical transparency and chemical resistance. These properties are relevant in different applications, e.g., in innovative water treatment processes (Löffler et al. 2020a) or in photoreactors allowing a high light dispersion and, consequently, an increased growth due to photosynthesis (for example, when using algae) (Jacobi et al. 2012).

3. Open Cellular Structures

Porous materials or open cellular structures are recognized as solids having large amounts of pores and the fraction of pore volume to the total volume, namely porosity, is 0.2–0.95 (Ishizaki et al. 1998). According to IUPAC, porous materials are classified based on pore sizes as macropores (pore diameter with more than 50 nm), mesopores (pore diameter between 2 and 50 nm) and micropores (pores with diameter no more than 2 nm). Pores are classified according to their connectivity into two categories: open pores that are connected to the outside of the material and closed pores that are isolated from the outside of the material.

4. How Can Porous Materials be Made?

4.1 Solid space holder (or spacer) method

Porous structures as catalysis supports, with defined pore size, shape and volume fractions can be produced by using the solid space holder (or spacer) method. Placeholders, usually in powder form, are mixed evenly with the ceramic powder to be sintered. For the dry processing, this powder mixture is mechanically compressed before the green body obtained is subjected to burnout and sintering. An interconnected porous structure that mimics the structure of the space holder can then be created. The geometries and the proportion of space holders that are mixed into the glass powder generally directly reflect the pore geometries and consequently, the desired porosity of the material. The basic requirements for this process are that the space holders have sufficient strength during the entire mechanical compression process, that they are inert to the bulk material, and that they can easily be removed, e.g., thermally, without leaving any residual contamination.

4.2 Production via the sol-gel process

Silica as an inorganic catalyst support with micro- and mesoporous structures can be produced by a sol-gel process. What is interesting about these techniques is that they can also be designed to be environmentally friendly if water without co-solvents is used as the reaction medium. The sol-gel process involves initially the polymerization of monomeric alkoxide precursors; for example, tetraethylorthosilicate (TEOS) is used in the manufacture of silicon dioxide. The hydrolysis of TEOS leads to ethanol and water, creating a so-called "sol", followed by the formation of a homogeneous dispersion of polymers as clusters. The solid nanoparticles are transformed into a gel through links between the individual nanoparticulate clusters. Depending on the pH value, the basic structure and morphology of the solid phase can range from discrete colloidal particles to continuous particle collections. At an acidic pH, the formation of linear molecules is favored, and their crosslinking leads to network growth. In contrast, polymerization is preferred at basic pH, whereby individual nanoclusters are linked one to another and thus promote gel formation. After removal of the solvent by evaporation, the resulting material is called xerogel, which is compressed by the action of capillary forces during the removal of the solvent. The size, shape and regularity of the particles can be influenced by different variables such as temperature, molar composition of the mixture, solvent and co-solvent, duration of the hydrothermal treatment and the silica source (Brinker et al. 1982, Dislich and Hinz 1982, Sakka and Kamiya 1982).

4.3 Hydrothermal sintering treatment

Due to their simplicity and lower temperatures compared with conventional sintering, hydrothermal processes are the most frequently used method to obtain mesoporous structures. SiO_2 can be solidified, for example, by hydrothermal sintering at 350°C and 27 MPa. These conditions lead to sintering without decomposition, high open porosity and small pores. Control parameters are, for example, pressure, sintering temperature and reaction time in order to obtain a certain pore size, pore distribution, porosity and specific surface area (Yanagisawa et al. 1993).

4.4 Freeze-casting process

The freeze-casting process is an interesting and well-distinct process for manufacturing of porous materials. This method consists of freezing a suspension, mainly preserving the frozen structure. Therefore, the particles dispersed in an aqueous medium are ordered according to the direction of ice crystals, which normally solidify in a hexagonal shape. Water crystals grow and suspension particles are released from the frozen medium until the sample is completely frozen during the freezing step. The application of different freezing conditions such as temperature, particle concentration in the used slurry, solvent type and freezing direction can influence the resulting pore structure of the materials (Deville 2010). In this way, the directional freezing and growth of solvent crystals (ice crystals) can be controlled to determine the orientation of the pores. The latter step is the freeze-drying process, which creates a macroporous material that retains its original size and shape.

The direction of freezing is one of the most interesting observations. In general, three different zones can be clearly distinguished during the freezing process, which are characterized by a certain pore shape and dimension (Fig. 11.1). Almost no porosity is observed in zone 1, which is closest to the initial cold finger, resulting in a nearly dense material. In zone 2, the material shows a lamellar morphology. Finally, the ceramic, metal or glass show a lamellar structure with typically long parallel pores that are aligned in the direction of movement of the ice front. Zone 3 is marked by the beginning of freezing, whereby the particle lamellae to be formed are not yet solidified.

Freeze-drying usually needs the longest time in the overall process, as the time required is directly related to the rate of sublimation of the solvent (ice) and is determined by factors such as vacuum, temperature, particle size and concentration that affect the exposed surface.

Many parameters can be varied to control the pore size. The most common procedure is an adjustment of the freezing rate; for example, faster freezing leads to smaller pores, while lower freezing rates are required to achieve larger pores, of a few hundred micrometers (Deville 2006, Song et al. 2006, Yoon et al. 2007).

In recent times, the use of other solvents has become more interesting. Mediums such as camphene or camphene mixtures are the most common alternatives (Koh et al. 2006, Rahaman and

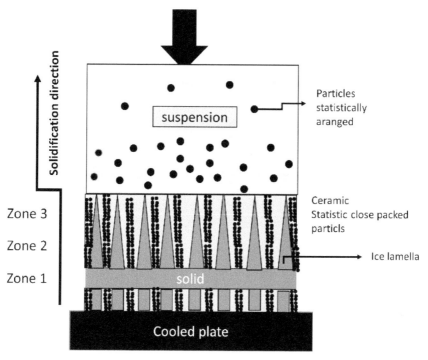

Figure 11.1. Schematic illustration of microstructure development (modified from Waschkies et al. 2009).

Fu 2008). The reason is that, by solidification of camphene, a partial to total crystallization occurs, resulting in a template for the formation of large pores.

Due to the advantages associated with the avoidance of toxic organic solvents and the low temperature during the freeze-drying processes, it is presumed that applications of this method will increasingly come into focus in the future. Also, the combination of porous materials and their functionality offers an immense field of application in a variety of novel applications such as purification of pollutants, e.g., biosensors and biocatalytic systems, cell technologies and, in general, biomedicine.

4.5 Glass foams obtained from replication process

The polymer sponge replication method fulfills the requirement of a good control of cell size distributions. The process consists of only three steps: (1) coating of a polyurethane (PU) template with an aqueous slurry; (2) burnout of the PU templates and binder; (3) final sintering (thermal processing) of the foams, which is shown in Fig. 11.2. Those foams have controllable pore size, a high porosity (> 90%) and a high surface-area-to-volume ratio.

This technique is widely applied in the fabrication of ceramic or metal foams, and allows to obtain high porosities (Manonukul et al. 2016, Wang et al. 2017, Sutygina et al. 2020).

Commercial glass foams are generally obtained by foaming methods using the action of a gas-generating agent. As the cell size distribution for the formation of open cells is hardly adjustable, this method leads mainly to closed cells, which is not suitable for the designated application.

However, it is a big challenge to obtain foams with transparent struts by this process (Fig. 11.2). Transparency requires the elimination of pores that cause scattering effects within the struts. Especially, pores ranging from 50 to 5000 nm have a detrimental effect on transparency, while pores smaller than 100 nm do not contribute significantly to the scattering. The macroporous polymer plays the role of the preform, which determines the morphology of the resulting structure. After removal of the polymer, the material retains the structural properties of the original template (Lange and Miller 1987, Colombo et al. 1998, Colombo and Modesti 2004, Fu et al. 2008).

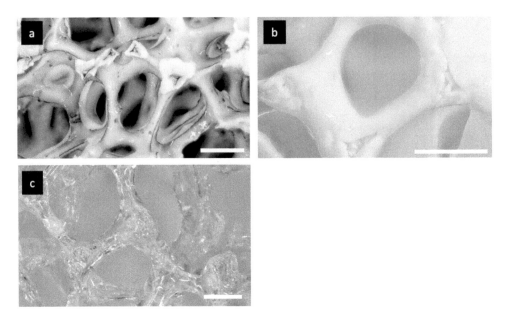

Figure 11.2. Different steps to manufacturing transparent glass structures: (a) coated template, (b) green body, i.e., the structure obtained after burning-out the polymer, and (c) sintered glass sponge (scale bar 500 μm) (modified from Löffler et al. 2018).

Transparent glass sponges are a new class of materials that may be used as membranes, catalyst carriers, in furnace equipment or biomedical or biological devices. Glass sponges offer several advantages over other open-pore structures, such as good thermal properties, resistance to chemical attack and interesting optical properties. For example, when installed in photobioreactors, they can contribute to an improved light dispersion and thus increase efficiency. This application can come into effect when the production of transparent or light-conducting sponges with open pores becomes economically feasible.

Producing transparent glass sponges presented a great challenge in terms of combining transparency with open cellular structures. In the field of technical ceramics such as Al_2O_3 or spinels, transparent, polycrystalline materials can be produced if it is possible to manufacture defect-free and high-density components. This often requires complex procedures, for example, the use of high pressures to obtain samples with a density > 99.9% of the theoretical density. For this purpose, hot isostatic pressing is used, one prerequisite being that there is no open porosity. Obviously, this condition is not applicable to obtain open cellular structures. Bucharsky et al. proposed a conventional replica method utilizing dispersed SiO_2 nanoparticles in an aqueous suspension for coating a polymer (Bucharsky et al. 2010). The properties of glass sponges produced by this method could be adjusted by varying the viscosity of the slurry and the polymeric foam characteristics. The quality of the SiO_2 coating on the polymeric sponge is largely dependent on the solids loading and thus on the viscosity.

4.5.1 Process description

4.5.1.1 Powder selection

Natural silica is abundant in the terrestrial system, with O and Si being the two most abundant elements on the earth (46.6% oxygen and 27.7% silicon). Natural silica exists most commonly as quartz found in sand, and also in various living organisms. However, SiO_2 has also been shown to pose a significant risk to human health, because, when inhaled as fine dust, it may lead to severe respiratory infections and systemic autoimmune diseases (Adams et al. 2006, Yamashita et al. 2011, Ghio et al. 2014).

A stable form under ambient conditions is alpha quartz, in which crystalline silicon dioxide is usually encountered. In nature and in labor, impurities in crystalline α-quartz can give rise to colors. For optimal transparency, the selection of the powder to be used in the manufacturing process is very important. The powder must be pure, without contaminants that can influence color or opacity.

The different modifications of silicon dioxide can be classified into natural and synthetic. For natural forms, they can be divided into crystalline, such as quartz powder and amorphous as kieselguhr (diatomaceous earth); on the other hand, synthetic forms can be precipitated (CAS Nr: 112926-00-8) or pyrogenic (CAS Nr: 112945-52-5 and 60842-32-2) silicon dioxide.

To produce pyrogenic silicon dioxide (also called fumed silica), sand is first reduced with carbon to obtain silicon, which then reacts with chlorine to form silicon tetrachloride. The last step is the high-temperature pyrolysis of silicon tetrachloride with oxyhydrogen ($H_2 + O_2$), where HCl is formed as a byproduct.

In these processes, silicon dioxide particles are formed first droplet-like directly in the flame, and they later join other particles to form a three-dimensional network, schematically represented in Fig. 11.3.

As a result, an extremely pure, white, SiO_2 powder with particle sizes of typically 5 to 100 nm is obtained. A high Specific Surface Area (SSA) of 50 to 600 m^2/g and a density of 160 to 190 kg/m^3 are typically measured. Interesting characteristics are non-porosity and exceptionally low (bulk) density. Its three-dimensional structure makes it largely thickening, thixotropic and protects against sedimentation. Possibly, because of a strong adsorption on the silicon dioxide particles, the mobility of the polymer clusters is reduced. Due to the formation of hydrogen bonds between primary particles, a thixotropy is observed, which can be temporarily wrecked by shear.

Important differences between the products obtained by different preparation methods, such as precipitated and pyrogenic silicon dioxide can be due to the size of the particles and their purity: precipitated silica leads to particle sizes in the μm range, whereas fumed silica lies in the nm range. It is also noteworthy that precipitated silica consists of 93% SiO_2 and therefore, it contains more impurities than fumed silica (99% SiO_2). Due to all these properties, the use of this type of silica is recommended.

Important manufacturers of synthetic silicon dioxide are Evonik Industries (formerly Degussa, trade name: Aerosil), Wacker Chemie (HDK), Rhodia, and W. R. Grace.

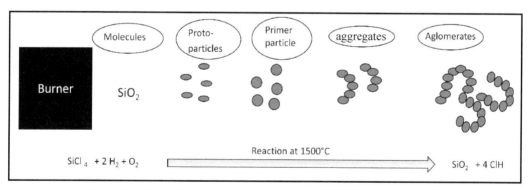

Figure 11.3. Schematic representation for the production of pyrogenic silicon dioxide.

4.5.1.2 Preparation and characterization of silica suspension

It is common to see that the suspensions begin to sediment after a long time after preparation, which is undesirable. There are several ways to improve the suspension stability, for example, by preventing coagulation through inter-particle repulsion, by retarding sedimentation by reducing the particle size, by increasing the density of the external phase or by increasing the viscosity.

Particles in suspensions can form either aggregates or floccules. Floccules form when weak van der Waals forces are holding the particles together, but they are easy to break. If floccules sediment, they trap the solvent and are easier to be resuspended. Nevertheless, the particles tend to clump together, which characterizes a flocculated system and causes it to sediment quickly. The control of the velocity of flocculation and sedimentation is fundamental for the development of a stable suspension. In order to obtain a well-stabilized suspension, particles dispersed in the suspending medium must exhibit sufficient repulsive forces to overcompensate the van der Waals attraction. These stabilizing forces originate from either electrostatic or steric interaction mechanisms. Repulsive forces are created by either a surface charge being developed at the solid-liquid interface (electrostatic stabilization) or by adsorbed polymer chains on the ceramic particle surfaces (steric stabilization), as well as by a combination of both mechanisms (electrosteric stabilization). The charge developed at the solid-liquid interface leads to the formation of a diffuse electrical double-layer surrounding each particle. The interactions of these electrical double layers give rise to mutually repulsive forces which keep the particles separated (Hyatt 1986, Roosen and Bowen 1988). One of the helpful parameters is the zeta-potential (ζ-potential), which is per definition the potential difference between the dispersion medium and the stationary layer of fluid close to the dispersed particle.

In the case of a silica suspension used for coating the polymer, it is necessary first to identify a suitable pH range for the electrostatic stabilization of the silica particles; therefore, a potential measurement is suggested. A typical dependence of the ζ-potential on the pH value of a SiO_2 slurry is shown in Fig. 11.4 (Bucharsky et al. 2011, Löffler et al. 2018) for the slurry with 10 vol.% solid content and 3 wt.-% PVA (related to the powder). Together with an increasing electrolyte concentration, a pH value of about 11 has to be considered as an upper limit for electrostatic stabilization. Thus, a water-based SiO_2 suspension can be electrostatically stabilized at pH 10 according to the measured high ζ-potential and the comparatively good flow behavior.

Coming from a strong acid medium (pH 2), silanol groups dissociate and ionize with increasing pH values forming negatively charged $=SiO-$ and $-Si(OH)_2-$ groups (Lange and Miller 1987, Bucharsky et al. 2010). In consequence, the absolute value of the measured ζ-potential grows with the increasing pH. As it can be seen in Fig. 11.4, the maximum ζ-potential is reached in the neutral

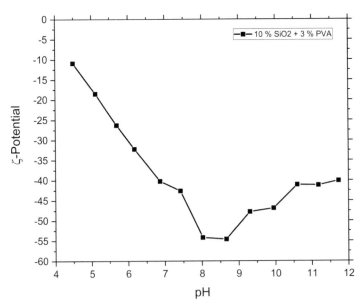

Figure 11.4. Dependence of the ζ-potential as a function of pH of a suspension with 10 vol.% of silica and 3 wt.% of PVA (modified from Bucharsky et al. 2011 and Löffler et al. 2018).

range. However, from about pH 4 on, hydrogen bonds between neutral and negative charged silanol groups are developed. This leads to gelation, i.e., to a strong increase of the suspension viscosity. With regard to this behavior, the weak acid and neutral pH range seems to be inappropriate for dip coating processes.

At higher pH values, the negatively charged groups are predominant and thus, there are fewer neutral groups available for binding and gelation. Besides, due to the repulsive electrostatic interaction between silica particles at higher pH values, agglomeration and bonding effects are prevented more and more (Mayo 1996). Therefore, the gelation process and consequently, the viscosity of the suspension decreases. However, it must be noted that the solubility of SiO_2 grows with increasing pH values, and, therefore, the stabilization of the slurries is affected. For that reason, the pH of the slurry should not be too high. Not only the pH value and the electrolyte concentration, but also the amount of SiO_2 influences the viscosity. Accordingly, a higher solid loading leads to a higher viscosity. Nevertheless, for well-stabilized slurries, the increase in viscosity is lower than that for slurries, which tend to gelation.

For coating the polymeric templates, low-viscosity slurries are necessary. In the SiO_2–water system, a compromise between a maximum of ζ-potential and the suitable viscosity can be obtained at a pH level higher than 10. For a better understanding, the dependence of ζ-potential and viscosity on the pH variation, for a slurry with a high loading of solids, are shown in Fig. 11.5.

Looking at the viscosity of the suspensions, the experiments showed that, at pH values below 9, the slurries presented the tendency to gelation and the viscosity increased immensely as shown in Fig. 11.5. This is probably due to the formation of hydrogen bonds between negatively charged and neutral silanol groups. This indicates that the pH of the slurry has to be in the range 9–11, where the best compromise of ζ-potential and viscosity values can be found. In that region of pH, suspensions with a solid loading less than about 34 vol.-% SiO_2 showed really low viscosities at appropriate shear rates ($\dot{y} > 10$ s^{-1}), as seen in Fig. 11.5. For a dip-coating process, it can be assumed that the flow and drop properties of a suspension can be described by the viscosity at a shear rate equal to 50 s^{-1}.

As shown in Fig. 11.5, the difference in viscosity is higher for a solid loading of 35.7 vol.-% than for 25 vol.-%. This suggests that, even at higher solid loadings, the viscosity exceedingly increased the desired values for dip-coating. From these experiments, the authors suggest that the optimal viscosity values for coating are about 3–7 Pa·s.

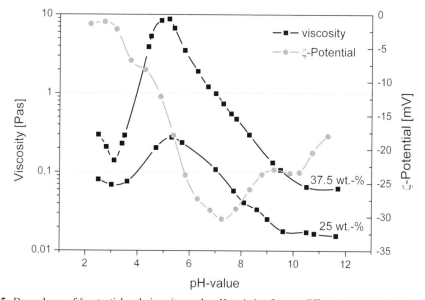

Figure 11.5. Dependence of ζ-potential and viscosity on the pH variation for two different concentrations of SiO_2 slurries (modified from Bucharsky et al. 2011).

4.5.1.3 Burning out the polymer template

The thermal decomposition of the polymeric template is a critical step for obtaining a stable green body suitable for sintering. An adjusted temperature regime and a sufficient concentration of SiO_2 on the struts are necessary to prevent the framework from collapsing. Therefore, the solid loading of the slurry should be high. In order to find out which solid loading is necessary, slurries with different SiO_2 concentrations, with solid loadings varying from 50 to 60 wt.-% at two pH values, were prepared. To obtain such high solid loadings, further mixing was done in a silica glass bottle with the help of an ultrasonic bath, using a power output of 75 watts for 5 minutes. The dependence of the amount of SiO_2 on a given volume of template on the solid loading is depicted in Fig. 11.6. As expected, higher solid loadings result in a greater amount of SiO_2 being brought onto the support. However, the solid loading and the pH values influence the viscosity and therefore, the impregnation; particularly, the removal of excess slurry is hindered. As a result, the error bars in Fig. 11.6 increase with higher solid loadings.

Regardless of the used slurry, the recommended amount of SiO_2 should be higher than approximately 0.1 grams per cubic centimeter of template. Below this value, for the polymer employed in this study, it was not possible to get the desired structure through the burning-out process. However, if the thermal treatment is successful, there are no significant changes in the structure visible to the naked eye, and the color of the samples remains white.

It is worth noting that the geometrical density of the shown green body was > 0.1 g/cm^3. Figure 11.7 depicts a coated sample before (a) and after burning out the polymeric framework (b). In the case of a density of 0.08 g/cm^3 (Fig. 11.7 (c)), it can be seen that the framework was pulverized during the burning-out step.

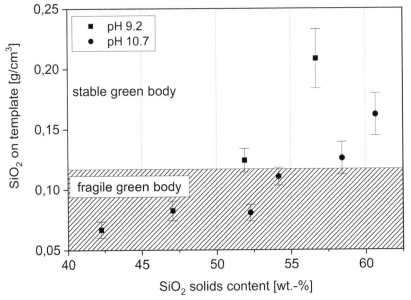

Figure 11.6. Dependence of the amount of SiO_2 on a given volume of template on the content of solids (modified from Bucharsky et al. 2011).

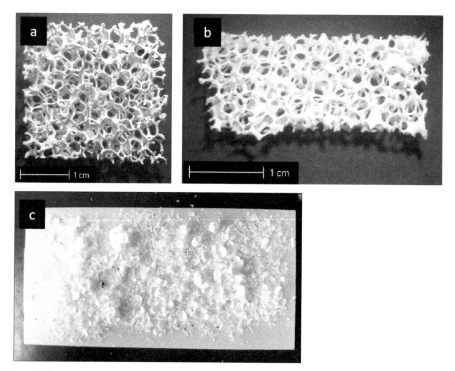

Figure 11.7. (a) SiO_2 coated polymeric template; (b) resulting structure after the burning-out process of green bodies with densities > 0.1 g cm³; (c) sample with a density of 0.08 g/cm³ leading to a pulverized framework (modified from Jacobi et al. 2012).

4.5.1.4 Sintering and its problems because of crystallization behavior

After the burning-out step, the obtained green bodies can be densified by viscous-flow sintering (Poppe 2003). This leads to an overall shrinkage of about 20% in all directions; this observation was determined experimentally from the geometrical size and confirmed by the ¹HNMR technique and by microcomputer tomography (micro CT) (Jacobi et al. 2012, Löffler et al. 2020a). By carefully selecting the sintering regime, i.e., an appropriate heating rate, sintering temperature and dwell time, the initial green bodies can be transformed into transparent open-celled foams. Transparency indicates that the amorphous character of the initial powder is still present. Accordingly, the sintered foams appear transparent as long as amorphous SiO_2 is present. Otherwise, the foam is white in color. There are two reasons for the white color: either sintering was incomplete or the amorphous structure has changed into a crystalline one. In the latter case, the white color is probably caused by the scattering of light at grain boundaries.

Monitoring of phase composition has been performed by analyzing X-ray spectra recorded after sintering at temperatures varying up to 1400°C. In Fig. 11.8, typical XRD-diffraction patterns of amorphous SiO_2 after sintering at 1330°C are shown. The broad peak near 22° is attributed to the short-range order in amorphous silica. Above this temperature, SiO_2 foams commence to crystallize. Hence, after sintering at temperatures higher than 1400°C, the sharp peaks of the XRD pattern correspond to crystalline cristobalite as the only detectable phase.

In addition to the XRD patterns, changes in structure and consequently on transparency can be explained considering the apparent density, defined as the relationship between the mass and volume of the material, including pores (apparent volume) (Fig. 11.9). For sintering temperatures below 1150°C, the relative apparent density is low as expected for a non-compacted green body. Only in the interval between 1200 and 1350°C, the relative apparent density is close to the density of amorphous silica. At temperatures above 1350°C, the measured density exceeds this value due to the formation of cristobalite. Therefore, this temperature can be interpreted as an upper-temperature limit.

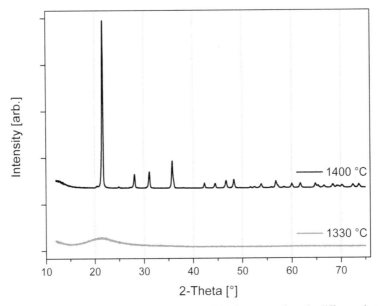

Figure 11.8. XRD patterns of sponges sintered at 1330°C and 1400°C. The spectra show the difference between amorphous (1330°C) and crystalline (1400°C) structures (modified from Bucharsky et al. 2011).

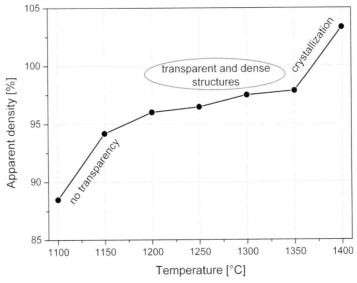

Figure 11.9. Dependence of apparent density of sintered samples on the sintering temperature in relation to the density of amorphous silica (2.2 g/cm³) (modified from Bucharsky et al. 2011).

SEM analysis can be used to map the surfaces of transparent and non-transparent foams, which provides very interesting findings. A micrograph of a non-transparent strut is depicted in Fig. 11.10(a). The surface is flaky, probably due to the crystallization process. This coarse surface contributes to the loss of transparency. In contrast, in Fig. 11.10(b), for a foam sintered at 1330°C, a smooth surface can be observed with only a few defects and consequently, the strut appears transparent.

 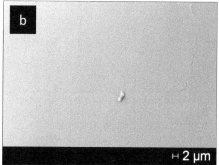

Figure 11.10. SEM micrographs of (a) a non-transparent foam sintered at 1400°C in comparison with (b) a transparent foam sintered at 1330°C (modified from Bucharsky et al. 2011).

4.5.1.5 Volume imaging analysis

The starting point for the image analysis is 3-D images of sponges obtained from ^{1}HNMR or micro CT, consisting of gray values for the 2,563 voxels of the scan domain. Using this 3-D volume data of transparent sponge models of the solid, surfaces can be extracted by applying digital filter routines for noise reduction and an individual threshold value to separate the solid from the void. The selected threshold directly influences the calculated specific surface area and thus, simply selecting a value is somewhat vague as this threshold additionally depends on the operational parameters when acquiring the raw data. Therefore, an automatic selection procedure was used, which is described in detail in Löffler et al. 2020a. The whole procedure was programmed using a freely available visualization toolkit, and there are now various suitable programs for this, e.g., VTK or Python algorithm based on scikit-image, see (Yen et al. 1995, Oliphant 2007, Millmann and Aivazis 2011, van der Walt et al. 2014). The results of the procedure are discretized surfaces represented by typically 0.5–1 million connected triangles, with the exact number depending on the surface area and resolution. After discretization, the calculation of the surface areas is reduced to a simple addition of these triangle areas (Lewiner et al. 2003).

In order to determine the porosity, the void space on one side of the interface area can be filled with tetrahedral volume elements and again summed up to result in volumes of the void phase (Saito and Toriwaki 1994). The tetrahedral meshes for this purpose were generated in this work using the commercially available meshing tool ANSYS ICEMCFD, as these meshes can be easily imported by many simulation toolkits. In order to give an overview, in Fig. 11.11, an image of the resulting geometry is shown.

Additionally, the polymer template was imaged to compare it with the sintered glass sponge, as shown in Fig. 11.12(c). Using the mathematical method described above, the pore average of the polymer is 3.12 mm whereas that of the glass sponge is 2.21 mm. These data lead to a shrinkage of about 30%, coinciding with the geometrical data. The stronger reduction in pore size cannot be explained by sintering effects alone, but by the combination of sintering shrinkage and an increase in strut thickness, as can be seen by comparing Fig. 11.11(a) and (b), from sintered glass foams and reconstructed coated polymer structure.

A very important characteristic parameter of the material is the Specific Surface Area (SSA) due to its relevance to the catalytic activity. From micro CT data, the SSA can be calculated for sintered glass foams. In this case, the sponge volume for the SSA was calculated as the volume of the measurement convex hull for different ppi values (Löffler et al. 2020a). The calculated SSAs for different ppi values are depicted in Fig. 11.12.

An exponential growth of the SSA as a function of the ppi values is clearly observed up to 60 ppi, but, from 60 to 80 ppi, the presence of many closed pores produces a deviation of this behavior, which is probably also the reason for the large error bars observed for the 60 ppi sample.

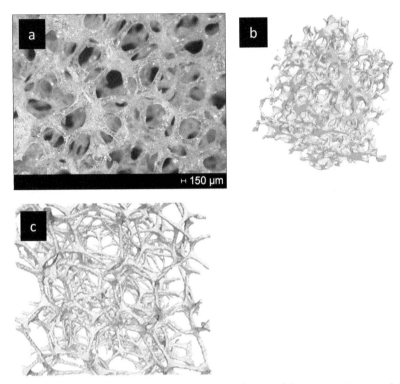

Figure 11.11. (a) Real and (b) reconstructed structure from ¹HNMR images of the transparent sponge sintered at 1300°C; (c) a reconstructed structure from an uncoated 20 pores per inch (ppi) polymer template (modified from Bucharsky et al. 2011).

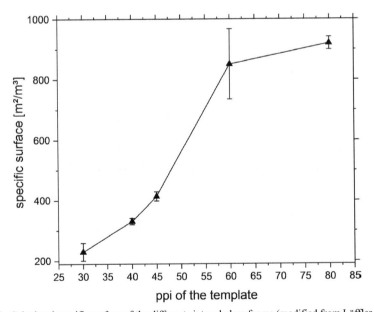

Figure 11.12. Calculated specific surface of the different sintered glass foams (modified from Löffler et al. 2020a).

Next we will describe the application of the prepared sponges in biophotoreactors and as supports of photocatalysts. In these examples, the available surface of the cellular structures used plays a prominent role, as the photocatalytically active materials are deposited on them.

5. Application of Transparent Glass Sponges in Biophotoreactors and as Supports for Photocatalysts

Transparent silica sponges are a new class of materials that, when coated with TiO_2, are capable of removing pollutants from water and air by heterogeneous photocatalysis (Löffler et al. 2020a). If transparent structures are used for this purpose, additional advantages come into effect, such as the effective distribution of light, which promotes photocatalytic activity. Two examples are intended to demonstrate the suitability of this concept in an application-oriented form, for which experimental results are used: (1) with regard to the improvement of the photocatalytic activity, (2) the effective light distribution in biophotoreactors that increases the growth rate of algae can be used as a producer of various substances. Due to the multiple and complex light reflection within the porous glass structures, algae can obtain an optimal amount of light for growth at any point of the total volume.

The basic requirement for a high photoactivity of the materials is a proper distribution of light within the whole structure. Furthermore, a large surface of the material is needed to get a good contact of the pollutant with the photocatalyst, assuring both a fast mass transfer and a proper light distribution. Therefore, only open-cell materials are suitable for these photo applications, as closed pores only contribute superficially (and slightly) to the activity. However, open-porous glass sponges are not commercially available at present.

5.1 Photocatalytic activity

Here the comparison of the photocatalytic activity of TiO_2 nanoparticulate layers applied on slides and glass sponges are described. The photocatalytic activity of TiO_2 in the anatase form is employed for coating the transparent glass structures. A dispersion of P25 TiO_2 (mainly anatase, the most used photocatalyst) nanoparticles in water, was prepared. The photocatalytic activity of the layers was determined by the measurement of the decay of Cr (VI) (starting from $K_2Cr_2O_7$), which was tested under a UV excitation source (Löffler et al. 2020a). Finally, the relationship of sponge geometry parameters such as ppi and porosity and the amount of TiO_2 was examined.

The photocatalytic activity of the SiO_2 sponges coated with TiO_2 was tested with the Cr (VI)/EDTA system ([Cr (VI)]$_0$ = 0.8 mmol/L, using [EDTA]/[Cr (VI)] = 1.25 (molar ratio)) at pH 2. The mean UV irradiance incident on the surface of the solution was E_0 = 3,520 µW cm^{-2}, measured at 365 nm (Fredrich et al. 1995, Bucharsky et al. 2010, Löffler et al. 2018). EDTA was used as a sacrificial synergetic agent to improve Cr (VI) transformation (Löffler et al. 2020a).

Figure 11.13 shows the results of normalized time profiles of Cr (VI) concentration using the TiO_2-coated sponges with different ppi values, compared with the coated glass slide containing 2.1 mg TiO_2, together with the results in the absence of the photocatalyst obtained under similar conditions (blank).

The experimental points for both, the homogeneous and the heterogeneous reactions, could be adjusted with Eq. (1):

$$C/C_0 = e^{-kt} \tag{1}$$

where C is the Cr (VI) concentration in solution, C_0 is the Cr (VI) concentration at the beginning of the photocatalytic test, and k is the pseudo-first order kinetic constant. The fitting curves using Eq. (1) show a good agreement ($R^2 > 0.97$) with the experimental points. From the data depicted in Fig. 11.13, it can be observed that the photocatalytic Cr (VI) reduction with all the coated samples is faster than the reduction in the absence of TiO_2 and that all coated sponges exhibit a higher photocatalytic activity than the P25 coated glass slide sample.

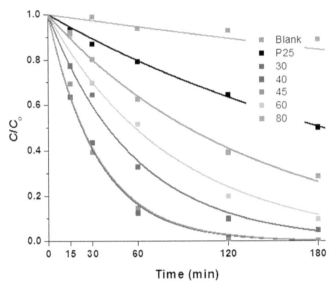

Figure 11.13. Evolution of the normalized Cr (VI) concentration (C/C_0) with the irradiation time for the photocatalytic Cr (VI) reduction in the presence of EDTA. Experimental conditions: $[K_2Cr_2O_7]$ = 0.4 mmol/L, [EDTA] = 1 mmol/L, pH 2, E_0 = 3,500 µW/cm² (measured at 365 nm), T = 25°C. The lines represent the adjustment of the curves fitted with Eq. (1) (modified from Löffler et al. 2020b).

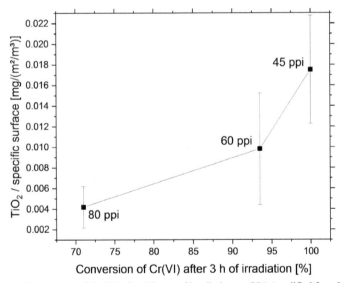

Figure 11.14. Percent of conversion of Cr (VI) after 3 hours of irradiation vs. SSA (modified from Löffler et al. 2020a).

Another point of interest is the relationship between photocatalytic efficiency and the foam geometry, namely the ppi values that determine the specific surface area (SSA). Figure 11.14 shows the percentage of conversion of Cr (VI) after 3 hours of irradiation as a function of SSA.

It can be observed that the photocatalytic activity of the coated sponges for the Cr (VI) conversion finds the optimal behavior with the 45 ppi sample. These photocatalytic results may be related to the accessibility of light and reagents in the samples: the accessibility would increase up to a certain ppi value, but then the accessibility would be impeded, as shown in Fig. 11.14, and the Cr (VI) conversion increases accordingly. This combination of conversion efficiency and mechanical stability at 45 ppi or more is a very promising result for further photocatalytic applications. In addition, it was found that after being reused three times (not shown) the sponges still exhibited a good efficiency of about 70% of the transformation.

5.2 Biophotoreactors

The cultivation of the microalgae *Chlamydomonas reinhardtii* wild type wt 13 derived from the original Ebersold/Levine line 137c was used as a model organism (Jacobi et al. 2012). The main problem of phototrophic processes is that light is needed as an energy source but light cannot simply be mixed in the system like other components. If there is insufficient light supply, only low biomass concentrations can be achieved. The light attenuation shows a very rapid exponential decrease over the layer thickness due to the light absorption and scattering by the algae. Large volume elements only provide light limiting conditions that lead to less or no photosynthetic growth, while light saturation produces the same effect on the illuminated surface, i.e., photo inhibition.

With regard to the process efficiency, for most applications, the illuminated surface should not be overexposed and the gradient in the bulk phase should be reduced. These requirements can be fulfilled by using the new transparent glass foams described above. In the results presented here, porous miniplate reactors have been designed to test these effects. The transparent glass foams can provide an enlargement of the inner surface area of the reactor and therefore, they contribute to illumination (light intensities up to 1,500 $\mu E\ m^{-2}\ s^{-1}$ were applied), light distribution and light supply (Lehr and Posten 2009, Posten 2009, Morweiser et al. 2010).

The algae grow in suspension within the pores of the sponge as schematically shown in Fig. 11.15, which compares the empty reactor with that of the sponges. The sponge structure distributes the light into the deeper, dark parts of the reactor by conduction the light within the struts, as well as by reflection and refraction at the surfaces due to the refractive index of glass (which is higher than that of water). Therefore, an amorphous SiO_2 glass structure has to be employed. Ideally, sunlight intensity should be adjusted to the optimal value in the whole cultivation volume. An additional effect of the use of sponges is the reduction of light gradients inside the whole reactor volume. Based on these concepts, photobioreactors have quite promising potential for further developments. Thereby, the efficiency of production scale reactors could be improved by maximizing overall growth rates.

Figure 11.15 depicts schematic conditions in plate reactors illuminated from one side with changing cell densities. The maximum growth rate (μ_{max}) is only achieved at low cell concentrations, when all algae receive sufficient light for growth. With an increasing optical density of the suspension volume elements, light-limiting conditions (μlim) or even the absence of cell growth ($\mu = 0$) become apparent. The μ_{max} and the photoconversion efficiency (PCE), i.e., the efficiency of the biomass buildup can be determined from the captured light. The relevant parameters for

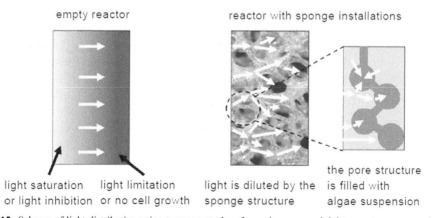

Figure 11.15. Schema of light distribution using transparent glass foams in porous miniplate reactors compared with the empty reactor (modified from Jacobi et al. 2012).

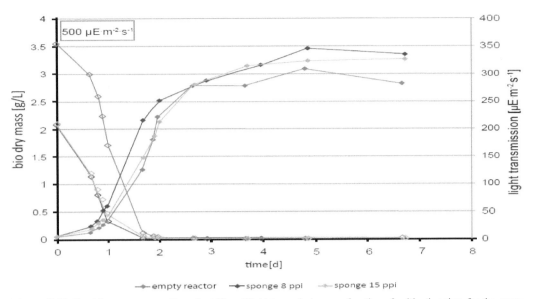

Figure 11.16. Dry biomass concentration of wt 13 and light transmission as a function of cultivation time for the empty bioreactor and the bioreactor equipped with glass sponges of 8 and 15 ppi as light dispersion agents (modified from Jacobi et al. 2012).

estimating the growth rate (μ) in the exponential growth phase are low cell densities and more importantly, the photoconversion efficiency (PCE) at high cell densities. The reference system is the "empty" reactor for growth experiments without any light distributing installations inside. To increase the final cell densities reached in the reactor, concentrated, Tris-Phosphate (TP) medium and CO_2 supply by aeration (5%) are used. Typical growth curves of batch experiments at a selected light intensity (500 μE m^{-2} s^{-1}) and the corresponding light transmittance behind the reactor for 8 and 15 ppi glass sponges and the empty reactor are shown in Fig. 11.16.

A characteristic growth curve of microalgae batch cultures can be seen in Fig. 11.16. A very short exponential phase with low cell densities < 1 g L^{-1} is observed up to the first day. Already at day one, light becomes limiting but, in contrast to other substrates such as flat plate or tubular reactors or airlift systems, light is continuously supplied, which leads to a transition into a linear growth phase. When ammonium is limited in the medium (ion chromatography measurements, not shown), a maximum biomass concentration of about 3–3.5 g L^{-1} is reached during the third day. The light transmission at the beginning of the cultivation is lower with reactors filled with sponges compared with the empty ones. This is possibly due to a better light distribution inside the reactor caused by refraction and scattering due to the sponge structure.

The measured reflection at the illuminated surface is less than < 10% of the incident light intensity and no difference between the different configurations (empty or sponge-filled reactor) can be observed. The influence of the reflection at the reactor surface (typically 4% for air-glass) on the incident light energy is eliminated by calibrating inside the reactor behind the first glass plate. The absorption of light by glass is only in the range of < 0.1% per meter of light path.

Cultivation experiments showed an increase in the growth rate (about more than 25%) at low cell densities and the PCE was increased from 4.9% for the empty reactor to 5.6 and 5.9% for the reactor filled with sponges of 8 and 15 ppi, respectively. The results show a first positive effect, but further research is necessary. The effect will be more apparent at higher light intensities and higher cell densities. Therefore, experiments to increase the biomass concentrations by concentrating the TP medium (2.5 fold) and with appropriate feeding are in progress. Furthermore, a scale-up to larger volumes would be necessary to mimic conditions in production reactors, which is limited to the producible sponge size now due to the complexity of the process.

6. 3D-Structures by Additive Manufacturing: Development of Silica-Based Organic Slurries for Stereolithographic Printing Process

6.1 Stereolithography

In the process of Additive Manufacturing (AM), the structures are created by applying a material layer by layer (layer construction principle), which means that very complex structures can be obtained with fewer design restrictions (Löffler et al. 2020b).

Special attention is paid to the use of designed cellular structures. These structures can have a variable cell size and their component geometry can be adapted to the intended situation. One advantage of using materials with cellular structures is the reduction in the amount of building material while maintaining a certain overall mechanical strength. Cellular structures are a particular strength of AM processes because the entire structure can be designed using software applications, optimized to achieve the most efficient strength-to-weight properties. Objects are digitally created by Computer-Aided Design (CAD) software. Besides, various software solutions are also available for topology optimization, for example, Abaqus from Dassault Systèmes and OptiStruct from Altair.

Stereolithography offers great benefits in the field of rapid prototyping due to important factors that can be linked together: free choice of shape, simultaneous production of several parts, few or no post-processing steps and very high resolution of the parts (Hull 1986). It has been shown that the stereolithographic process can also be used to manufacture glass parts (Kotz et al. 2016, Kotz et al. 2017).

If the desired CAD data for the component to be printed is available, the creation process generally includes the following steps:

i)   Preparation of glass suspensions

ii) Printing of green body

iii) Thermal treatment (debindering/burn out and sintering)

6.2 Preparation of glass suspensions for stereolithography

Photocurable slurries based on a variety of different materials such as alumina, zirconia or tricalcium phosphate have been developed and marketed, but no commercially available suspension based on high-purity SiO_2 is currently available (Schwentenwein et al. 2014, Schwentenwein and Homa 2015). Moreover, there are very few works on printing glass and glass structures, meaning that a photocurable suspension based on silicon dioxide, suitable for the desired printing process (Halloran 2016, Kotz et al. 2017) must first be developed. Acrylate-based slurries are developed considering parameters such as the content of solids and its influence on the viscosity and its critical shear rate. This aspect is very important for printing samples with precise structures. Other properties affecting the suitability of the suspension in terms of curability, such as the polymerization time and the curability kinetics are normally investigated using time dependent attenuated total reflection infrared spectroscopy (ATR-IR) (Löffler et al. 2020b).

An acrylate-based slurry with a certain photo initiator content gives the starting composition of most photosensitive slurries used in stereolithography. In 2009, Wozniak et al. investigated the rheological behavior of slurries based on hydrophilic and non-hydrophilic monomers such as butylacrylate and 4-hydroxybutylacrylate using amorphous silica as a glass former in the composition (Wozniak et al. 2009). The authors showed that the viscosity of silica-based slurries strongly depends on the particle size of the used silica powder. Furthermore, a certain critical shear rate was observed, which depends mainly on the loading of solids. As an important result, it is reported that, when using non-polar butylacrylate, even high solid loadings lead to comparatively low viscosities. This is attributed to a solvation layer around the particles formed by hydroxyl groups, which avoids attractive van der Waals forces in the organic substance. Thus, high particle concentrations can be dispersed, while the viscosity of the slurry is kept low. The shear forces

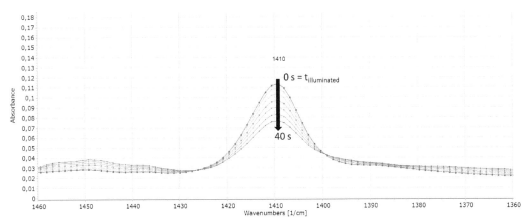

Figure 11.17. IR spectra of a 40 vol.-% silica slurry illuminated with an intensity of 42 mW cm⁻² for 40 s (modified from Löffler et al. 2020b).

cause the slurry to reduce its viscosity until a critical shear rate is reached. Then, the solvation layer collapses and hydration bonds form between the silica nanoparticles. This results in gelation of the slurry and an increase in viscosity (Raghavan and Khan 1997, Raghavan et al. 2000, Wozniak et al. 2009, Wozniak et al. 2011).

As photo initiators, camphorquinone and ethyl-4(dimethylamino) benzoate (EDMAB) were used. Photopolymerization is a time depending process, which is initiated by the radicalization of the photo initiators. These photo initiators absorb photons when irradiated under a certain wavelength and form reactive species from the excited state, which initiate consecutive reactions. The kinetic parameters can be monitored using time-dependent IR.

In Fig. 11.17, time-resolved IR spectra at a wavenumber of 1410 cm⁻¹, where the acrylate double bond absorbs, are used as a measure for monitoring the ongoing polymerization progress. A new IR spectrum during illumination was recorded every four seconds. Changes in the absorbance intensity during illumination are observed, indicating the progress of the polymerization reaction. The peak with the maximum absorbance represents the slurry that was not illuminated at the beginning, while the peak with the lowest absorbance intensity represents the slurry that was illuminated for 40 seconds. With these time depending measurements, the polymerization kinetics of the slurry can be evaluated.

A very important parameter to consider in the printing process is the slurry viscosity. It is important to maintain the flow properties required for the printing process so that excess suspension can flow off and thus complex structures and sharp edges can be printed. When the viscosity is low enough, which is up to 10 Pa·s, the slurry that is not cured can run off after the printing step. On the other hand, high viscosities of more than 30 Pa·s can lead to the curing of the rest on the surfaces near the printed layers.

6.3 Printing of green bodies, debindering, and sintering

In order to evaluate printing parameters, a commercial printer was used. The slurries containing SiO_2 nanoparticles were printed using a Cerafab 7500 printer (Lithoz GmbH, Austria) to test their applicability. The used method was bottom-up and operated at a wavelength of 460 nm. The light is split and focused by a digital micromirror device, allowing a layer-by-layer curing. The printer is able to print layer thicknesses between 25 and 100 µm with a lateral resolution of 40 µm minimum per pixel. The slurries were placed in the vat and pre-sheared a few minutes before starting the printing process. Evaluated printing parameters were used to print high-resolution printing samples with the developed silica slurry.

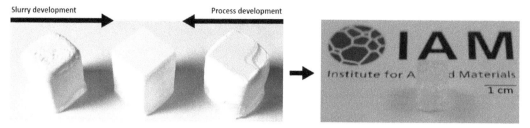

Figure 11.18. Stereolithographically printed cubes, which represent the results of the investigations on printing, debindering conditions and the sintering parameters (modified from Löffler et al. 2020b).

Figure 11.18 shows stereolithographically printed cubes. On the left-hand side of (Fig. 11.18a), the results can be seen as printed structures when the suspension was modified in terms of the concentration of solids, the amount of photo initiator and the addition of co-photo initiator (Löffler et al. 2020b). The printed structures were debindered in a multi-stage temperature process up to 800°C and then brought into a glassy structure by sintering at 1300°C in air (Fig. 11.18 (b)).

7. Conclusions

This chapter examines different processes for the production of catalytic or photocatalytic materials by various methods, with the focus on the creation of light-guiding structures. The use of nanoscale starting powders is common to all processes. The illustrated methods overcome most of the manufacturing difficulties and disadvantages usually encountered. After a brief introduction to the various possibilities, chemical mechanisms of the processes, the specific challenges in the manufacture of ceramic materials are presented. Finally, results from the field of photocatalysis, bioreactors and organic slurries for stereolithographic printing processes are presented as application examples of the cellular transparent substrates. The versatility, the wide-ranging applicability of the carriers with low manufacturing costs and the possible material applications are demonstrated.

Acknowledgments

The authors would like to thank the German Research Foundation (DFG) for funding the Re-search Group FOR 583 "Solid Sponges - Application of monolithic network structures in process engineering". Thanks to SPUK–Strategic Partnership between the National University of San Martin, Argentina and the Karlsruhe Institute of Technology, Germany.

References

Adams, L.K., Y.D.Y. Lyon and P.J.J. Alvarez. 2006. Comparative eco-toxicity of nanoscale TiO$_2$, SiO$_2$, and ZnO water suspensions. Water Res. 19: 3527–3532.

Brinker, C.J., K.D. Keefer, D.W. Schaefer and C.S. Ashley. 1982. Sol-gel transition in simple silicates. J. Non-Cryst. Solids. 1: 47–64.

Bucharsky, E.C., K.G. Schell, R. Oberacker and M.J. Hoffmann. 2010. Preparation of transparent glass sponges via replica method using high-purity silica. J. Am. Ceram. Soc. 1: 111–114.

Bucharsky, E.C., K.G. Schell, P. Habisreuther, R. Oberacker and M.J. Hoffmann. 2011. Preparation of optically transparent open-celled foams and its morphological characterization employing volume image analysis. Adv. Eng. Mater. 11: 1060–1065.

Buntara, T., I. Melian-Cabrera, Q. Tan, J.L.G. Fierro, M. Neurock, J.G. deVries and H.J. Heeres. 2013. Catalyst studies on the ring opening of tetrahydrofuran-dimethanol to 1,2,6-hexanetriol. Catal. Today. 210: 106–116.

Cervantes-Sodi, F., J.J. Vilatela, J.A. Jimenez-Rodriguez, L.G. Reyes-Gutierrez, S. Rosas-Melendez, A. Iniguez-Rabago, M. Ballesteros-Villarreal, E. Palacios, G. Reiband and M. Terrones. 2012. Carbon nanotube bundles self-assembled in double helix microstructures. Carbon. 10: 3688–3693.

Colombo, P., M. Griffoni and M. Modesti. 1998. Ceramic foams from a preceramic polymer and polyurethanes: preparation and morphological investigations. J. Sol-Gel Sci. Techn. 1: 195–199.

Colombo, P. and M. Modesti. 2004. Silicon oxycarbide ceramic foams from a preceramic polymer. J. Am. Ceram. Soc. 3: 573–578.

Deville, S. 2006. Freezing as a path to build complex composites. Science. 5760: 515–518.

Deville, S. 2010. Freeze-casting of porous biomaterials: structure, properties and opportunities. Materials. 3: 1913–1927.

Dislich, H. and P. Hinz. 1982. History and principles of the sol-gel process, and some new multicomponent oxide coatings. J. Non-Cryst. Solids. 1: 11–16.

Fredrich, J.T., B. Menéndez and T.F. Wong. 1995. Imaging the pore structure of geomaterials. Science 5208: 276–279.

Fu, Q., M.N. Rahaman, B.B. Sonny, R.F. Brown and D.E. Day. 2008. Mechanical and *in vitro* performance of 13–93 bioactive glass scaffolds prepared by a polymer foam replication technique. Acta Biomater. 6: 1854–1864.

Ghio, A.J., S.T. Kummarapurugu, H. Tong, J.M. Soukup, L.A. Dailey, E. Boykin, M.I. Gilmor, P. Ingram, V.L. Roggli, H.L. Goldstein and R.L. Reynolds. 2014. Biological effects of desert dust in respiratory epithelial cells and a murine model. Inhal. Toxicol. 5: 299–309.

Halloran, J.W. 2016. Ceramic stereolithography: additive manufacturing for ceramics by photopolymerization. Ann. Rev. Mater. Res. 1: 19–40.

Hu, X., H. Cong, Y. Shen and M. Radosz. 2007. Nanocomposite membranes for CO_2 separations: silica/brominated poly(phenylene oxide). In. Eng. Chem. Res. 5: 1547–1551.

Hull, C.W. 1986. Apparatus for Production of Three-Dimensional Objects by Stereolithography. U.S. Patent # 4,575,330.

Hyatt, E.P. 1986. Making thin, flat ceramics—a review. Am. Ceram. Soc. Bull. 4: 637–638.

Ishizaki, K., S. Komarneni and M. Nanko. 1998. Porous Materials: Process Technology and Applications. Springer, Boston.

Jacobi, A., E.C. Bucharsky, K.G. Schell, P. Habisreuther, R. Oberacker, M.J. Hoffmann, N. Zarzalis and C. Posten. 2012. The application of transparent glass sponge for improvement of light distribution in photobioreactors. J. Bioproces. Biotechniq. 1: 1000113.

Koh, Y.-H., E.-J. Lee, B.-H. Yoon, J.-H. Song, H.-E. Kim and H.-W. Kim. 2006. Effect of polystyrene addition on freeze casting of ceramic/camphene slurry for ultra-high porosity ceramics with aligned pore channels. J. Am. Ceram. Soc. 89: 3646–3653.

Kotz, F., K. Plewa, W. Bauer, N. Schneider, N. Keller, T. Hergang, D. Helmer, K. Sachsenheimer, M. Schäfer, M. Worgull, C. Greiner, C. Richter and B.E. Rapp. 2016. Liquid glass: a facile soft replication method for structuring glass. Adv. Mater. 28: 4646–4650.

Kotz, F., K. Arnold, W. Bauer, D. Schild, N. Keller, K. Sachsenheimer, T.M. Nargang, C. Richter, D. Helmer and B.E. Rapp. 2017. Three-dimensional printing of transparent fused silica glass. Nature 544: 337–339.

Kuthati, Y., P.-J. Sung, C.-F. Weng, C.-Y. Mou and C.-H. Lee. 2013. Functionalization of mesoporous silica nanoparticles for targeting, biocompatibility, combined cancer therapies and theragnosis. J. Nanosci. Nanotechnol. 13: 2399–2430.

Lange, F.F. and K.T. Miller. 1987. Open-cell, low-density ceramics fabricated from reticulated polymer substrates. Adv. Ceram. Mater. 2: 827–831.

Lehr, F. and C. Posten. 2009. Closed photo-bioreactors as tools for biofuel production. Curr. Opin. Biotech. 20: 280–285.

Lewiner, T., H. Lopes, A.W. Vieira and G. Tavares. 2003. Efficient implementation of marching cubes' cases with topological guarantees. J. G. T. 8: 1–11.

Löffler, F.B., E.C. Bucharsky, K.G. Schell and M.J. Hoffmann. 2018. Preparation and characterization of transparent SiO_2 sponges for water treatment. CFI-Ceram. Forum. Int. 85: E33–E37.

Löffler, F.B., F.J. Altermann, E.C. Bucharsky, K.G. Schell, M.L. Vera, H. Traid, A. Dwojak and M.I. Litter. 2020a. Morphological characterization and photocatalytic efficiency measurements of pure silica transparent open-cell sponges coated with TiO_2. Int. J. Appl. Ceram. Tech. 17: 1930–1939.

Löffler, F.B., E.C. Bucharsky, K.G. Schell, S. Heißler and M.J. Hoffmann. 2020b. Development of silica based organic slurries for stereolithographic printing process. J. Eur. Ceram. Soc. 40: 4556–4561.

Manonukul, A., P. Srikudvien, M. Tange and C. Puncreobutr. 2016. Geometry anisotropy and mechanical Property isotropy in Titanium foam fabricated by replica impregnation method. Mat. Sci. Eng. A-Struct. 655: 388–395.

Mayo, M.J. 1996. Processing of nanocrystalline ceramics from ultrafine particles. In. Mater. Rev. 41: 85–115.

Millman, K.J. and M. Aivazis. 2011. Python for scientists and engineers. Comput. Sci. Eng. 13: 9–12.

Morweiser, M., O. Kruse, B. Hankamer and C. Posten. 2010. Developments and perspectives of photobioreactors for biofuel production. Appl. Microbiol. Biot. 87: 1291–1301.

Oliphant, T.E. 2007. Python for scientific computing. Comput. Sci. Eng. 9: 10–20.

Poppe, T. 2003. Sintering of highly porous silica-particle samples: analogues of early Solar-System aggregates. Icarus. 164: 139–148.

Posten, C. 2009. Design principles of photo-bioreactors for cultivation of microalgae. Eng. Life Sci. 9: 165–177.

Raghavan, S.R. and S.A. Khan. 1997. Shear-thickening response of fumed silica suspensions under steady and oscillatory shear. J. Colloid Interf. Sci. 185: 57–67.

Raghavan, S.R., H.J. Walls and S.A. Khan. 2000. Rheology of silica dispersions in organic liquids: new evidence for solvation forces dictated by hydrogen bonding. Langmuir 16: 7920–7930.

Rahaman, M.N. and Q. Fu. 2008. Manipulation of porous bioceramic microstructures by freezing of suspensions containing binary mixtures of solvents. J. Am. Ceram. Soc. 91: 4137–4140.

Roosen, A. and H.K. Bowen. 1988. Influence of various consolidation techniques on the green microstructure and sintering behavior of alumina powders. J. Am. Ceram. Soc. 71: 970–977.

Saito, T. and J.-I. Toriwaki. 1994. New algorithms for euclidean distance transformation of an n-dimensional digitized picture with applications. Pattern Recogn. 27: 1551–1565.

Sakka, S. and K. Kamiya. 1982. The sol-gel transition in the hydrolysis of metal alkoxides in relation to the formation of glass fibers and films. J. Non-Cryst. Solids. 48: 31–46.

Schwartzwalder, K. and A.V. Somers. 1961. Method of Making Porous Ceramic Articles. U.S. Patent # 3,090,094.

Schwentenwein, M., P. Schneider and J. Homa. 2014. Lithography-based ceramic manufacturing: a novel technique for additive manufacturing of high-performance ceramics. Adv. Sci. Tech. 88: 60–64.

Schwentenwein, M. and J. Homa. 2015. Additive manufacturing of dense alumina ceramics. Int. J. Appl. Ceram. Tec. 12: 1–7.

Song, J.-H., Y.-H. Koh, H.-E Kim, L.-H. Li and H.-J. Bahn. 2006. Fabrication of a porous bioactive glass–ceramic using room-temperature freeze casting. J. Am. Ceram. Soc. 89: 2649–2653.

Sutygina, A., U. Betke and M. Scheffler. 2020. Open-cell aluminum foams by the sponge replication technique: a starting powder particle study. Adv. Eng. Mater. 22: 1901194.

van der Walt, S., J.L. Schönberger, J. Nunez-Iglesias, F. Boulogne, J.D. Warner, N. Yager, E. Gouillart and T. Yu. 2014. scikit-image: image processing in Python. PeerJ. 2: e453.

Wang, C., H. Chen, X. Zhu, Z. Xiao, K. Zhang and X. Zhang. 2017. An improved polymeric sponge replication method for biomedical porous titanium scaffolds. Sci. Eng. C. 70: 1192–1199.

Waschkies, T., R. Oberacker and M.J. Hoffmann. 2009. Control of lamellae spacing during freeze casting of ceramics using double-side cooling as a novel processing route. J. Am. Ceram. Soc. 92: S79–S84.

Wozniak, M., T. Graule, Y. de Hazan, D. Kata and Jerzy Lis. 2009. Highly loaded UV curable nanosilica dispersions for rapid prototyping applications. J. Eur. Ceram. Soc. 29: 2259–2265.

Wozniak, M., Y. de Hazan, T. Graule and D. Kata. 2011. Rheology of UV curable colloidal silica dispersions for rapid prototyping applications. J. Eur. Ceram. Soc. 31: 2221–2229.

Yamashita, K., Y. Yoshioka, K. Higashisaka, K. Mimura, Y. Morishita, M. Nozaki et al. 2011. Silica and titanium dioxide nanoparticles cause pregnancy complications in mice. Nat. Nanotechnol. 6: 321–328.

Yanagisawa, K., M. Nishioka, K. Ioku and N. Yamasaki. 1993. Densification of silica gels by hydrothermal hot-pressing. J. Mater. Sci. Lett. 12: 1073–1075.

Yen, J.-C., F.-J. Chang and S. Chang. 1995. A new criterion for automatic multilevel thresholding. IEEE Trans. Image Process. 4: 370–378.

Yoon, B.-H., Y.-H. Koh, C.-S. Park and H.-E. Kim. 2007. Generation of large pore channels for bone tissue engineering using camphene-based freeze casting. J. Am. Ceram. Soc. 90: 1744–1752.

Zhou, T., X. Shen, S. Chaudhary and L. Ye. 2014. Molecularly imprinted polymer beads prepared by pickering emulsion polymerization for steroid recognition. J. Appl. Polym. Sci. 131: 39606.

Chapter 12

A Review of the Structure and Metal(loid) Adsorption Reactivity of Nanoscale Fe(III) and Mn(IV) (Oxyhydr)oxides for Industrial Application

Case M. van Genuchten

1. Introduction

The use of nanoscale and poorly-crystalline iron (Fe) and manganese (Mn) (oxyhydr)oxide minerals to immobilize and remove toxic oxyanions, such as arsenic (As), antimony (Sb) and heavy metals (e.g., cadmium (Cd) and lead (Pb)) (Wang et al. 2012) has become one of the most widely applied methods in water and soil remediation (Hering et al. 2017). The application of nanoscale Fe and Mn oxides, hydroxides and oxyhydroxides (herein oxides, for brevity) has become widespread due to the inexpensive synthesis of these minerals, their non-toxic nature and their remarkable reactivity with a diverse range of contaminants (Tan et al. 2010, Wang et al. 2012). In particular, two phases of nanoscale Fe and Mn oxides have received the most attention in the context of industrial application for remediation purposes: two-line ferrihydrite (2LFh) and layered-type MnO_2, which is represented by the synthetic Mn oxide, δ-MnO_2 (Dixit and Hering 2003, Gude et al. 2017). Both 2LFh and δ-MnO_2 have unique structures which give rise to exceptional adsorption reactivity with inorganic oxyanion and bivalent cation contaminants (Pena et al. 2010, van Genuchten and Pena 2016a, van Genuchten and Pena 2016b).

This chapter centers on describing the structure and adsorption reactivity of 2LFh and δ-MnO_2, two nanoscale minerals commonly used for pollutant remediation. The unique crystal structures of 2LFh and δ-MnO_2 will be summarized, including the key particle properties that contribute to their adsorption reactivity. The macroscopic (i.e., adsorption isotherms) and molecular-scale (surface speciation and contaminant uptake mode) oxyanion and bivalent cation adsorption reactivity of 2LFh and δ-MnO_2 will then be described using As and Sb as probe metalloids and Cd and Pb as probe metals. Arsenic and Sb were selected as environmentally relevant probe metalloids because they are both negatively charged oxyanions, but their pentavalent forms have different coordination

Department of Geochemistry, Geological Survey of Denmark and Greenland, Oester Voldgade 10, Copenhagen K 1350, Denmark.
Email: cvg@geus.dk

geometries (i.e., As(V) is tetrahedrally coordinated, Sb(V) is octahedrally coordinated). Lead and Cd were selected as environmentally relevant probe cations because they are both positively-charged heavy metals, but they fall on opposite ends of the adsorption selectivity sequence for metal oxides (Kinniburgh et al. 1976) because of their significantly different ionic potentials (i.e., Cd(II) displays hard-shell behavior and Pb(II) displays soft-shell behavior) (Sposito 2008). In addition, As, Sb, Cd and Pb all have high atomic numbers, which ensures high-quality differential pair distribution function data, which is the novel surface structural characterization technique discussed mainly in this chapter. It is important to acknowledge that many industrial applications of Mn oxides, in particular, exploit the high redox potential of these minerals and their unique ability to oxidize recalcitrant organic pollutants (Remucal and Ginder-Vogel 2014). However, since this chapter is constrained to the structure and metal(loid) adsorption reactivity of nanoscale Fe and Mn oxides, the redox reactivity of Mn oxides will not be discussed.

2. Methods to Determine the Structure and Sorption Reactivity of Nanoparticles

The most common traditional method of investigating the adsorption reactivity of any type of solid particle is to perform adsorption isotherm experiments. Adsorption isotherms typically involve the addition of a dissolved adsorbate, often a heavy metal(loid) or contaminant of concern, to a suspension of a solid adsorbent (Sposito 2008). The adsorbate and adsorbent are then allowed to react for a defined time, with the suspension maintained at a selected pH, ionic strength, temperature and pressure. After the defined reaction time, the solids are separated from the solution and the final concentration (i.e., equilibrium concentration, C_{eq}) of the adsorbate is measured. An adsorption isotherm is generated by performing several of these experiments across a range of either adsorbate or adsorbent concentrations and plotting the surface excess or surface loading (q, in mol adsorbate:mol adsorbent) as a function of the C_{eq}.

While adsorption isotherms provide some insight into the macroscopic reactivity of the adsorbent, this method cannot unambiguously determine the sorption mechanism. To identify how exactly the adsorbate interacts with the absorbent (e.g., outer-sphere adsorption, inner-sphere adsorption, ternary complexation, incorporation), techniques that directly probe the chemical environment of the surface species are required. In addition, poorly-ordered Fe(III) and Mn(IV) nanoparticles lack the long-ranged structural order necessary for structural determination using standard X-ray diffraction (Perret et al. 2000, Morin et al. 2003, Voegelin et al. 2013). This chapter includes data from two synchrotron-based characterization techniques that meet the challenges of these complex systems: X-ray absorption spectroscopy and differential pair distribution function analysis of high-energy X-ray scattering data.

X-ray Absorption Spectroscopy (XAS) is based on scanning the energy of incident X-rays across the binding energy of a core-shell electron of a specific element (e.g., Fe, Mn, As, Pb), which excites a photoelectron from the absorbing atom (Kelly et al. 2008). The excited photoelectron interacts with neighboring atoms, creating oscillations in the absorption spectrum, which is also known as the extended X-ray Absorption Fine Structure (EXAFS) spectrum. Fourier transformation of the EXAFS spectrum produces a series of peaks that correspond to atomic pairs (i.e., absorbing atom and neighboring atoms) at a given radial distance. The amplitude and position of these peaks can then be modeled to derive structural information in the form of the Coordination Number (CN) and interatomic distance (R). Since X-ray absorption edge energies depend on the atomic number of the element, XAS is an element-specific technique that can probe dilute species in complex matrices, such as adsorbed contaminants at trace levels. However, due to the finite lifetime of the excited photoelectron, XAS can only probe the short-ranged environment (i.e., < 6 Å) around the absorbing atom.

Differential Pair Distribution Function (d-PDF) analysis of high-energy X-ray scattering data is a novel approach to determine surface speciation that offers a key advantage over XAS: this technique is not limited to the local bonding environment of surface species. Fourier-transformation of the total structure function ($S(Q)$), which is obtained from high-energy X-ray scattering data of solid samples, yields the pair distribution function (PDF). The PDF itself yields data in the form of atomic pairs separated by a given radial interatomic distance, which is similar to XAS data, but the PDF is not element-specific and thus yields information on all atomic pairs in the sample (Egami and Billinge 2003). The d-PDF is generated by subtracting the PDF of a mineral adsorbent blank (i.e., host) from the PDF of the adsorption sample (i.e., host + guest). The peaks in the d-PDF that remain after subtraction of the host mineral PDF from the adsorption sample PDF arise only from atomic pairs from the surface species (Harrington et al. 2010, Li et al. 2011). The peaks in the d-PDF can then be analyzed by comparison with theoretical (i.e., calculated) d-PDFs using hypothetical models of metal(loid) adsorption geometries. The peaks in the d-PDF extend to much larger interatomic distances ($R > 10$ Å) than possible with XAS, which yields more robust conclusions based on peak analyses. The novel d-PDF approach to determine surface speciation will be the primary method examined in this chapter.

3. Crystal Structures of Nanoscale Fe(III) and Mn(IV) Oxides Relevant to Industry

The sorption and redox reactivity of nanoscale Fe(III) and Mn(IV) oxides is intimately related to their unique structures (Borch et al. 2010). Two Fe(III) and Mn(IV) oxide mineral nanoparticles in particular, 2LFh and birnessite, have found widespread application across a variety of industries, ranging from water treatment and soil remediation to catalysis and energy storage (Post 1999, Okkenhaug et al. 2012, Okkenhaug et al. 2013, Remucal and Ginder-Vogel 2014). Here the key structural properties of each mineral will be briefly summarized.

3.1 Two-line ferrihydrite

Two-line ferrihydrite (2LFh) is the synthetic analog of poorly-ordered Hydrous Ferric Oxide (HFO) often found in nature and gets its name from the two characteristic broad peaks in its X-ray diffraction pattern. The most common formation pathways of 2LFh are by rapid Fe(II) oxidation or by the addition of ferric salts to solutions at circumneutral pH, which is often practised at conventional water treatment plants. Two structural models have been highly cited for 2LFh: the "Michel model" proposed by Michel et al. (2007a) and the "Drits model" proposed by Drits et al. (1993). The two models differ primarily in that the Michel model describes 2LFh as a single phase consisting approximately of 20% four-fold coordinated tetrahedral Fe ([IV]Fe), whereas the Drits model for 2LFh consists of three separate phases of entirely six-fold coordinated octahedral Fe ([VI]Fe). The confirmation of [IV]Fe in 2LFh has been reported by independent research groups using Fe K-edge XAS, Fe L-edge XAS, neutron scattering and X-ray magnetic circular dichromatism (Harrington et al. 2011, Maillot et al. 2011, Peak and Regier 2012), although these studies have themselves been disputed (Manceau 2011, Manceau 2012, Manceau and Gates 2013, Paktunc et al. 2013). However, there seems to be a growing consensus of the use of the Michel model for 2LFh (Cismasu et al. 2011, Cismasu et al. 2012, Dyer et al. 2012, Toner et al. 2012). This chapter adopts the Michel model for the crystal structure of 2LFh, which is shown in Fig. 12.1.

The Michel model of 2LFh consists of Fe atoms in different coordination environments occupying multiple sites. Edge-sharing FeO$_6$ octahedra ([VI]Fe$_E$, site Fe1: $R_{Fe-Fe} = 2.94$–3.02 Å) form an octahedral sheet, with each edge-sharing sheet containing a six-membered ring opening. Either a second FeO$_6$ octahedron ([VI]Fe$_C$, site Fe2: $R_{Fe-Fe} = 3.50$ Å) or an FeO$_4$ tetrahedron ([IV]Fe$_C$, site Fe3: $R_{Fe-Fe} = 3.39$ Å) sits on top of the opening in the sheet and shares corners with multiple [VI]Fe$_E$ octahedra of the six-membered ring. The Fe polyhedra that sit atop the opening in the sheet link together separate octahedral sheets by sharing either edges ($R_{Fe-Fe} = 3.13$ Å) or corners ($R_{Fe-Fe} = 3.31$ Å) with [VI]Fe$_E$ octahedra. These different polyhedral linkages are evident in the 1–5 Å region of the 2LFh PDF (Fig. 12.1), which clearly shows peaks due to Fe–O (2.0 Å),

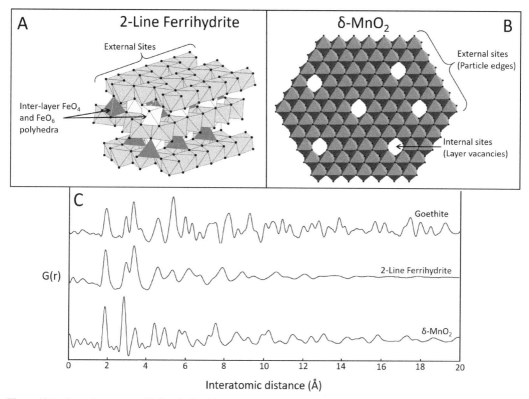

Figure 12.1. Crystal structures of 2-line ferrihydrite (2LFh, A) and δ-MnO$_2$ (B). Pair distribution functions (*C*) of crystalline goethite (α-FeOOH), 2LFh, and δ-MnO$_2$. This figure is a modified version of Fig. 1 in van Genuchten (2016b).

edge-sharing Fe–Fe (3.0 Å) and corner-sharing Fe–Fe (3.4 Å) atomic pairs. What should be clear from the description of the Michel model, and in the visual representation of the Michel model in Fig. 12.1, is the high degree of complexity and disorder in the 2LFh structure. This complexity and structural disorder results in a Svanishingly small metastable crystal size of 2.0 nm for 2LFh (Michel et al. 2007b). The PDF data of 2LFh given in Fig. 12.1 shows that the amplitude of PDF peaks decays to 0 near 2.0 nm, consistent with the diameter of the 2LFh particle. This decay at 2.0 nm in the 2LFh PDF is neither observed for crystalline goethite particles nor even for nanocrystalline δ-MnO$_2$, indicating the poor structural coherence of 2LFh. In fact, 2LFh is considered a metastable nanomineral because it does not have a crystalline counterpart. As 2LFh grows and becomes more crystalline, the mineral transforms and reorganizes into more stable Fe(III) oxide phases, such as goethite and hematite (Zhao et al. 1994, Gilbert et al. 2004, Pedersen et al. 2005, Waychunas et al. 2005, Liu et al. 2007). The inherent poorly-ordered nature of 2LFh is what gives this mineral a large reactive surface area per mass of roughly 300 m^2/g, depending on synthesis conditions (Schwertmann and Cornell 1991). The surface area of 2LFh is composed of purely external surface sites (Fig. 12.1), a distinct difference from the reactive sites of birnessite minerals, such as δ-MnO$_2$, which will be discussed next.

3.2 δ-MnO$_2$

δ-MnO$_2$ is the synthetic analog of the natural-occurring mineral, birnessite, which is found in the environment primarily as a result of biogenic oxidation of aqueous Mn(II) by a range of microorganisms (e.g., fungi, bacteria) (Learman et al. 2011, Droz et al. 2015). In industrial applications, birnessite minerals can form by the reduction of permanganate (Ahmad et al. 2019) and by controlled biogenic Mn(II) oxidation, which occurs in rapid sand filters in water treatment plants (Gude et al. 2017). While still complex, the structure of birnessite is debated in literature far less than the structure of 2LFh in part because birnessites have (slightly) more structural order than 2LFh. In general, birnessite minerals typically consist of randomly stacked sheets of edge-sharing MnO_6 octahedra having diameters of approximately 2–10 nm in the ab-plane and varying Mn(III) content ranging from 0–35% (Lanson et al. 2002, Manceau et al. 2013). For δ-MnO_2, the particle diameter is often around 4–5 nm and the ideal δ-MnO_2 crystal contains 100% Mn(IV) (i.e., no significant fraction of Mn(III) (Zhu et al. 2012)). As shown in the δ-MnO_2 PDF in Fig. 12.1, the Mn–O (1.9 Å) and Mn–Mn (2.9 Å) atomic pairs from edge-sharing MnO_6 octahedra that make up individual sheets in the δ-MnO_2 particle are the defining features of the first 1–5 Å of the PDF. Furthermore, in contrast to 2LFh, the PDF peaks for δ-MnO_2 do not decay to 0 at 2.0 nm, indicating a larger crystal size than 2LFh. However, the peaks in the δ-MnO_2 PDF are broader and have less amplitude than those of goethite at $R > 10$ Å. This relatively poor crystallinity can lead to large surface areas for δ-MnO_2, which have been measured by BET N_2 adsorption to exceed 100 m²/g.

One of the key features of birnessite, and specifically δ-MnO_2, is the presence of internal metal sorption sites in the form of MnO_6 cation vacancies in the crystal structure (Manceau et al. 2002). Therefore, in contrast to 2LFh, metal sorption by δ-MnO_2 can occur both by complexation to internal surfaces (MnO_6 vacancies) and external surfaces (lateral edges of the MnO_2 sheets). The MnO_6 vacancies of some birnessites can reach tens of mol%, which can be a substantial part of the entire reactive surface, in addition to external sites on the particle edges. However, for the δ-MnO_2 particles described later in this chapter, the expected vacancy content is roughly 10 mol%.

4. Bond Strength Analysis of Surface Sites and Particle Surface Charge for 2LFh and δ-MNO_2

A complete description of the adsorption reactivity of 2LFh and δ-MnO_2 requires a detailed examination of their corresponding reactive surface sites (both external and internal) since these surface sites largely control the particle surface charge and potential for chemical bonding. To examine the reactive external surface sites of both 2LFh and δ-MnO_2, bond strength analysis was performed on the surface oxygen atoms of each mineral. Bond strength distributions were calculated for the singly-, doubly- and triply-coordinated (where applicable) oxygen atoms, though this chapter focuses primarily on singly-coordinated surface oxygen atoms (Fig. 12.2) since these atoms are the least saturated with respect to bond strength and are thus the most likely to engage in bonding. Bond strength distributions for 2LFh and δ-MnO_2 external surfaces were compared with those of a roughly 3.0 nm crystal of gibbsite (Al(OH)$_3$), which has a well-characterized point of zero charge (PZC) (Kosmulski 2009). For δ-MnO_2, MnO_6 cation vacancy sites (i.e., internal sites), which comprise three localized doubly-coordinated oxygen atoms (Fig. 12.2) and contribute to the surface charge characteristics of birnessite, were not included in the analysis.

The following empirical formula reported by Brown and Altermatt (1985) was used for bond strength calculations, which was derived from a systematic analysis of thousands of mineral structures: $s = \exp(27.03(R_0 - R))$, where s is the bond strength, R_0 is the bond valence parameter and R is the metal-oxygen interatomic distance. In these calculations, R-values were derived from the crystal structures of 2LFh (Michel et al. 2007a), δ-MnO_2 (Lanson et al. 2002), and Al(OH)$_3$ (Megaw 1934). The Me–O distances from ideal crystal structures were used to approximate those at mineral surfaces, but it is acknowledged that Me–O bonds at mineral surfaces can vary due to surface structural modifications and interactions with the bulk solution (Rustad 2001).

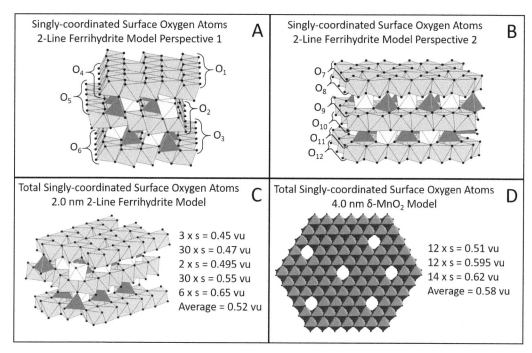

Figure 12.2. Visual representation of the Singly-coordinated Surface Oxygen (SSO) atoms for 2LFh and δ-MnO$_2$. Panels A and B show different visual perspectives of the same 2.0 nm 2LFh particle to highlight the different SSO atoms. Each group of SSO atoms highlighted in panels A and B has identical bond strengths. Panels C and D list the bond strength of all SSO atoms for 2LFh and δ-MnO$_2$, respectively.

4.1 Two-line ferrihydrite

Figure 12.2 highlights the positions of singly-coordinated surface oxygen (SSO) atoms for different crystal faces of a 2.0 nm 2LFh particle. Figure 12.2C summarizes the bond strengths for all SSO atoms of the 2LFh particle. Since a stable SSO atom would have a bond strength near 2 (equal to the magnitude of the charge of O atoms, which is 2 in oxide minerals), the bond strength distribution of 2LFh indicates that surface O atoms are highly undersaturated. The bond strength for all SSO atoms of 2LFh falls within the range of 0.44 to 0.56 valence units (vu), with an average of 0.52 vu, which is slightly larger than that of Al(OH)$_3$. The slightly larger bond strength distribution for SSO atoms of 2LFh compared with Al(OH)$_3$ can explain the slightly lower PZC for 2LFh (PZC = 7 to 8) compared with that of Al(OH)$_3$ (PZC = 8 to 9) since H$^+$ adsorption, which increases the surface charge, is more likely to occur with SSO atoms having lower bond strengths (Sposito 2004, Sposito 2008, Kosmulski 2009). Taken together, the undersaturation of SSO atoms for 2LFh combined with its small particle size (i.e., large surface area per mass ratio) explains why 2LFh has a relatively high adsorption reactivity with a variety of metal(loids).

4.2 δ-MnO$_2$

One of the major barriers to reach a complete understanding of the adsorption reactivity of δ-MnO$_2$ is the experimental difficulties in determining the PZC of the external sites. Since this mineral contains as much as 10 mol% MnO$_6$ cation vacancies, the surface charge is dominated by the contribution of these vacancies. As such, the PZC for δ-MnO$_2$ is extremely low, with most estimates indicating a PZC below 3 (Appelo and Postma 1999). However, the external sites of δ-MnO$_2$ can play a critical role in the adsorption because they are often higher in abundance than the vacancies (Villalobos et al. 2014). Therefore, the bond strength analysis for δ-MnO$_2$ focused on the external surface sites (Fig. 12.2 D). Figure 12.3 shows the bond strength distribution of SSO atoms for δ-MnO$_2$ external

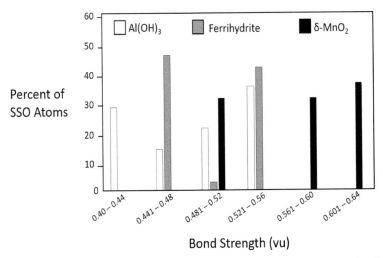

Figure 12.3. Distribution of the bond strengths of singly-coordinated surface oxygen (SSO) atoms for gibbsite (Al(OH)$_3$), 2-line ferrihydrite (2LFh), and δ-MnO$_2$. This figure is a modified version of Fig. 8 from van Genuchten and Pena (2016b).

surface sites compared with those of 2LFh and Al(OH)$_3$. The range of bond strength for δ-MnO$_2$ (0.48 to 0.64 vu) and the average value (0.56 vu) was similar but slightly larger than that of 2LFh. This result suggests that the external surfaces of δ-MnO$_2$ can be expected to have a similar, but slightly lower PZC than that of 2LFh (7 to 8). Therefore, at pH 5.5, the net δ-MnO$_2$ surface charge would be negative, which is due to the overwhelming impact of MnO$_6$ cation vacancies, but the external surface sites are expected to be neutral or positive. These different localized charging properties of δ-MnO$_2$ surfaces can give this mineral a unique adsorption reactivity, depending on the solution pH, with the potential for adsorption of negatively-charged species to external surface sites despite highly negative overall particle surface charge.

5. Oxyanion Sorption Reactivity

The structure and surface properties of 2LFh and δ-MnO$_2$ described earlier will now form the basis for interpreting the adsorption reactivity of these minerals. The interaction between As(V) and Sb(V) oxyanions with 2LFh and δ-MnO$_2$ will be examined using macroscopic (adsorption isotherms) and molecular-scale (synchrotron-based characterization) approaches.

5.1 As(V) and Sb(V) adsorption isotherms for 2LFh and δ-MnO$_2$

Figure 12.4 shows As(V) and Sb(V) sorption isotherms for 2LFh and δ-MnO$_2$ at pH 5.5. The surface excess (q, mol:mol) has been normalized by the concentration of the added adsorbent. In the 2LFh system, the steep initial slopes of the As(V) and Sb(V) isotherms follow H-curve isotherm behavior (Sposito 2008). As the initial slope of the sorption isotherms is related to sorption affinity (Hinz 2001), we concluded that As(V) and Sb(V) have similar high affinities for 2LFh surfaces. However, 2LFh showed a moderately higher sorption capacity for Sb(V) compared to As(V). Arsenate sorption to 2LFh plateaus at a solids ratio near 0.18 mol As:mol 2LFh, which is consistent with the previously reported As(V) sorption capacities (Raven et al. 1998), whereas Sb(V) sorption reaches a loading above 0.22 mol Sb:mol 2LFh.

Similar to oxyanion sorption behavior in the 2LFh system, δ-MnO$_2$ shows a larger sorption capacity for Sb(V) compared with that of As(V). The initial slope of the Sb(V) sorption isotherm for δ-MnO$_2$ is steeper than that of As(V), but both are significantly less steep than those for 2LFh. This result indicates that As(V) and Sb(V) have a lower affinity for δ-MnO$_2$ surfaces than for 2LFh, which is consistent with the net negative charge of δ-MnO$_2$ (PZC ≈ 3.0) at the pH of the experiments (5.5), whereas 2LFh is positively charged (PZC ≈ 7) (Kosmulski 2009). The maximum As(V)

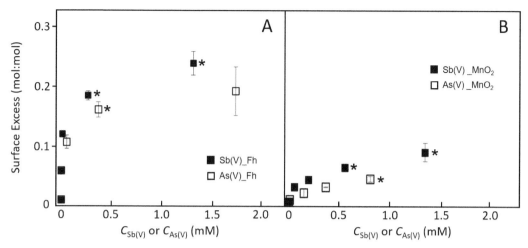

Figure 12.4. Sorption isotherms of As(V) (empty squares) and Sb(V) (filled squares) for 2LFh (A) and δ-MnO₂ (B). The * symbols indicate samples for which characterization data are reported next. This figure is a modified version of Fig. 1 in van Genuchten and Pena (2016a).

surface excess measured is less than 0.06 mol As:mol MnO$_2$, which is consistent with the values reported previously (Villalobos et al. 2014), whereas the surface excess for Sb(V) is slightly larger than 0.1 mol Sb:mol MnO$_2$.

5.2 Oxyanion sorption mechanisms

5.2.1 As(V) sorption on 2LFh

Figure 12.5 compares the Fourier-transformed EXAFS spectrum of As(V) adsorbed onto 2LFh with the experimental and calculated d-PDFs of As(V) and Sb(V) adsorbed onto 2LFh.

The Fourier-transformed EXAFS spectrum (Fig. 12.5B) of As(V) adsorbed on 2LFh is primarily characterized by two peaks: an intense peak at a low interatomic distance, which is indicative of As–O atomic pairs and a second broad peak at higher R that is due to As–Fe backscattering. Comparison of the Fourier-transformed EXAFS spectrum of this sample with the d-PDF of the identical sample clearly reveals the strengths and weaknesses of both methods. The d-PDF of As(V)-adsorbed 2LFh shows two symmetric peaks at 1 to 5 Å, which is similar to the XAS data, but the second-shell peak for the d-PDF data has a substantially larger amplitude. The larger amplitude of these peaks in the d-PDF compared with the Fourier-transformed EXAFS spectrum facilitates a more robust analysis of peak position. Gaussian fits of these short-ranged d-PDF peaks indicated R-values of 1.70 ± 0.01 and 3.29 ± 0.01 Å, respectively, which is consistent with As–O and As–Fe scattering due to As(V) bound to 2LFh in the DCS 2C geometry (Sherman and Randall 2003, Harrington et al. 2010). Apart from the 2C geometry, the d-PDF data showed no evidence for additional As(V) coordination geometries, such as the Single Edge-Sharing (SES) 1E complex ($R_{As–Fe} = 2.9$ Å) nor the Single Corner-Sharing (SCS) 1C geometry ($R_{As–Fe} = 3.6$ Å) (Waychunas et al. 1993, Fendorf et al. 1997). The absence of other prominent peaks at $R > 4$ Å is consistent with a previous d-PDF study at similar As(V) surface excess (Harrington et al. 2010), although a broad peak with low amplitude does appears at 4.6 Å.

The calculated d-PDF based on the model of As(V) bound to 2LFh in the 2C geometry, which is given in Fig. 12.5C below the experimental d-PDF data, reproduces the positions well of the As–O and As–Fe peaks near 1.7 and 3.3 Å, respectively. Although some DFT studies suggest that the interatomic distance of 1C complexes can match those of the 2C complex (Loring et al. 2009, Otte et al. 2013), the good agreement in relative intensities of the As–O and As–Fe peaks between our model and data argues against this 1C geometry, which would display an As–Fe peak with roughly half the amplitude. Therefore, we concluded that As(V) binds primarily in the 2C geometry to 2LFh.

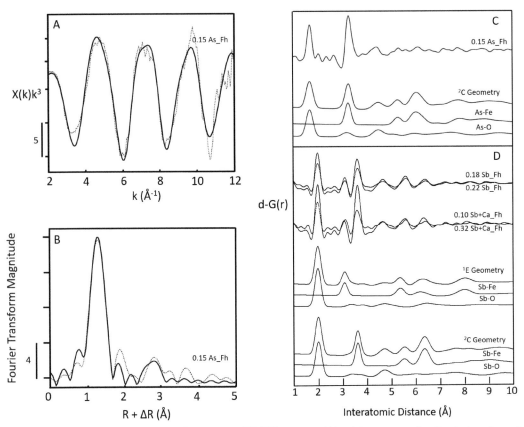

Figure 12.5. Extended X-ray Absorption Fine Structure (EXAFS) spectrum (A) and the corresponding Fourier transform of As(V) adsorbed to 2LFh compared with the d-PDFs of As(V) (C) and Sb(V) (D) adsorbed to 2LFh. Theoretically calculated d-PDFs of As(V) and Sb(V) adsorbed to 2LFh in the ²C and ²E geometries are presented below the experimental data.

At $R > 4$ Å, the calculated As(V) d-PDF reproduces the position of the peak near 4.6 Å, which is shown in the partial d-PDFs to appear from As–O atomic pairs. However, the calculated d-PDF overpredicts the amplitude of this peak and the amplitude of the two As–Fe peaks near 5.5 and 6.2 Å. This overprediction can be explained by differences in bonding disorder between the model and experimental data. The calculated d-PDF was generated using disorder terms derived from the average Fh structure (Michel et al. 2007a, van Genuchten et al. 2014). Thus, the overpredicted amplitudes of the As–Fe peaks at 5.5 and 6.2 Å indicate that intermediate-ranged Fe(III) polyhedra (i.e., next-nearest neighbor Fe atoms) located near adsorbed As(V) are more disordered than in the average 2LFh structure. The bonding of Tetrahedral Oxyanions (TO_4) to adjacent FeO_6 octahedra in the 2C geometry has been shown to induce FeO_6 octahedral distortions to accommodate the shorter and more rigid T–O and O–O bond lengths (Kwon and Kubicki 2004). As a consequence, the distortion of surface Fe(III) polyhedra on As(V) adsorption likely propagates through the particle and leads to increased structural disorder at $R > 4$ Å.

5.2.2 Sb(V) sorption to 2LFh

Figure 12.5D compares the experimental d-PDFs of Sb(V) adsorbed to Fh with surface loadings ranging from 0.10 to 0.32 mol Sb:mol 2LFh. Although the positions of the d-PDF peaks do not change significantly with the expanding Sb surface loading, the absolute magnitudes of the peaks grow due to the increasing fraction of Sb atoms in the solid phase. In contrast to the two major peaks at $R < 4$ Å in the As(V) d-PDF, all Sb(V) d-PDFs display three symmetric peaks centered at 1.98, 3.05–3.10, and 3.60–3.61 Å, which indicates the presence of at least two Sb(V) adsorption

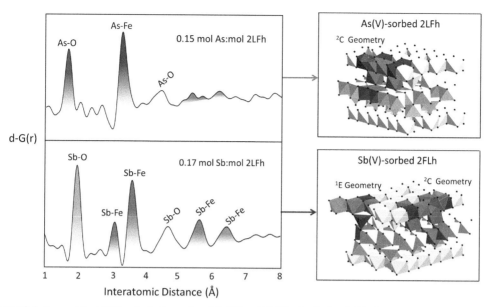

Figure 12.6. Zoomed in d-PDFs of select samples of As(V) and Sb(V) adsorbed to 2LFh showing the major atomic pair correlations (A) and the corresponding structural models of As(V) and Sb(V) adsorbed to 2LFh derived from the d-PDF data.

geometries. These peak positions are in excellent agreement with Sb–O and Sb–Fe interatomic distances previously attributed to the SES 1E ($R_{Sb–Fe} \approx 3.1$ Å) and double corner-sharing (DCS) 2C ($R_{Sb–Fe} \approx 3.6$ Å) sorption geometries to Fe(III) oxides (Mitsunobu et al. 2013). At $R > 4$ Å, the Sb(V) d-PDFs exhibit broad, but well-resolved next-nearest-neighbor peaks near 4.8, 5.6, and 6.4 Å with a larger amplitude than the corresponding peaks in the As(V) d-PDF data. Figure 12.6 highlights the difference between the experimental d-PDFs of As(V) and Sb(V) adsorption to 2LFh.

The calculated Sb(V) d-PDFs derived from the 1E and 2C models closely reproduce the positions of the nearest-neighbor Sb–O and Sb–Fe peaks at $R < 4$ Å (Fig. 12.5D). The 2C model provides a better fit than the 1E model to the next-nearest-neighbor peaks at 4.8, 5.6 and 6.4 Å, which are shown in the partial component d-PDFs to arise from Sb–O, Sb–Fe and Sb–Fe atomic pairs, respectively. Better agreement between the model and experimental Sb(V) d-PDFs at $R < 4$ Å is observed relative to the As(V) d-PDFs. This closer agreement between the experimental and model Sb(V) d-PDFs, which were derived using the same bulk 2LFh disorder parameters as the model As(V) d-PDF, suggests that Sb(V) adsorption introduces less disorder to 2LFh than As(V) adsorption. This result is consistent with the similar size and shape of SbO_6 ($R_{Sb–O} = 1.98$ Å) and FeO_6 ($R_{Fe–O} = 2.00$ Å) octahedra and the ability of Sb(V) to become incorporated into Fe(III) oxide minerals (Mitsunobu et al. 2010, Mitsunobu et al. 2013).

5.2.3 As(V) and Sb(V) sorption to δ-MnO₂

Figure 12.7 shows the experimental and calculated d-PDFs of As(V) and Sb(V) adsorbed to δ-MnO$_2$.

In contrast to the As(V) d-PDF in the 2LFh system, the d-PDF of As(V) adsorbed to δ-MnO$_2$ (Fig. 12.3A) lacks prominent peaks, which prevented reliable Gaussian peak fits and additional analysis by comparison with calculated d-PDFs. The poor signal of As(V) likely arises from the insufficient As(V) surface loading of this sample (0.05 mol As:mol δ-MnO$_2$ compared with 0.15 mol As:mol 2LFh), which is one of the weaknesses of the d-PDF analysis. However, based on previous spectroscopic investigations, including XAS (Foster et al. 2003, Villalobos et al. 2014), the small peaks near 1.7 and 3.2 Å in the d-PDF can be attributed to As–O and As–Mn atomic pairs, respectively, from As(V) adsorbed to δ-MnO$_2$ in the SCS 2C geometry. The d-PDFs of Sb(V) adsorbed to δ-MnO$_2$, which have surface loadings from 0.06 to 0.10 mol Sb:mol δ-MnO$_2$, display three well-resolved peaks at $R < 4$ Å. These short-ranged peaks located at 1.97–1.98, 3.09 and

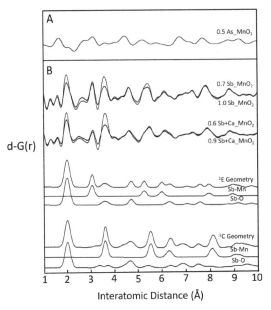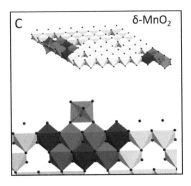

Figure 12.7. Differential pair distribution functions (d-PDFs) of As(V) (A) and Sb(V) (B) adsorbed onto δ-MnO$_2$. Theoretically calculated d-PDFs of As(V) and Sb(V) adsorbed onto δ-MnO$_2$ in the 2C and 2E geometries are presented below the experimental data. The structural model of Sb(V) adsorbed to δ-MnO$_2$ is given in panel C.

3.59–3.62 Å are analogous to those in the d-PDFs of Sb(V) adsorbed to 2LFh and are consistent with Sb(V) adsorbed in the SES 1E and DCS 2C geometries. The presence of well-resolved peaks in the d-PDF of the sample with the lowest loading in this series (0.06 mol Sb:mol δ-MnO$_2$), which is only slightly larger than the loading of the As(V)-adsorbed δ-MnO$_2$ d-PDF (0.05 mol As:mol δ-MnO$_2$), can be explained by the larger atomic number of Sb ($Z = 51$) relative to As ($Z = 33$) since the amplitude of peaks in the d-PDF is proportional to the scattering power of the atoms involved. Beyond 4 Å, the d-PDFs of Sb(V) adsorbed to δ-MnO$_2$ display well-defined intermediate-ranged peaks at 4.6, 5.2 and 6.2 Å, which reflects an ordered bonding environment that extends beyond the Sb-Mn second-shell.

The calculated d-PDFs of Sb(V) adsorbed to δ-MnO$_2$ particle edges in the 1E and 2C geometries are shown in Fig. 12.7B. The 1E and 2C models reproduce well the positions of the Sb–O and Sb–Mn nearest-neighbor peaks at 2.0, 3.1 and 3.6 Å. Good agreement in peak position and amplitude between the experimental and model d-PDFs at $R < 4$ Å suggests that the disorder parameters taken from bulk δ-MnO$_2$ (Lanson et al. 2002, Manceau et al. 2013) can be used to describe adequately the atomic environment of Sb(V) adsorption sites on δ-MnO$_2$ particle edges. However, at $R > 7$ Å, the calculated d-PDF reproduces the peak positions in the experimental data less accurately, which may be rationalized by the contribution of inter-sheet atomic pairs not accounted for in the single-sheet δ-MnO$_2$ model used in the d-PDF calculations. The calculated d-PDF of Sb(V) bound in a Triple Corner-Sharing (TCS) 3C configuration to MnO$_6$ layer vacancies contains Sb–Mn peaks at similar positions as in the 2C model, but with considerably more amplitude. Although the 3C model reproduces the peak positions of the experimental d-PDFs, we argued against the formation of the Sb(V) 3C complex to layer vacancies for several reasons. First, although ions with octahedral coordination, such as Sb(V), can sorb onto MnO$_6$ vacancy sites, the negative charges of vacancy sites and Sb(V) oxyanions should promote strong electrostatic repulsion. Second, the low initial slope of the Sb(V) adsorption isotherm on δ-MnO$_2$ is consistent with a low extent of ion sorption to vacancies since vacancies are often considered high-affinity sites (Villalobos et al. 2014). Last, similar 2:1 and 3:1 peak amplitude ratios are observed in the d-PDFs of Sb(V) adsorbed onto δ-MnO$_2$ and 2LFh, where 2LFh has no vacancies and thus adsorbs Sb(V) in 1E and 2C geometries.

Therefore, we concluded that Sb(V) binds to δ-MnO$_2$ particle edges (i.e., external sites) in the same 1E and 2C geometries (Fig. 12.7C) as was observed for Sb(V) adsorption to 2LFh external sites.

6. Heavy Metal Sorption Reactivity

6.1 Cadmium(II) and Pb(II) adsorption isotherms for 2LFh and δ-MnO$_2$

Figure 12.8 compares the adsorption isotherms measured for the heavy metals, Cd(II) and Pb(II), reacting with 2LFh and δ-MnO$_2$ at the same pH value (5.5) as that used in the As(V) and Sb(V) adsorption experiments.

Cadmium(II) and Pb(II) showed distinct sorption behavior on 2LFh (Fig. 12.8A), which is consistent with their opposite positions in the selectivity sequence of 2LFh (Kinniburgh et al. 1976) and their different hard and soft-shell character. In particular, Cd(II) sorption to 2LFh was minimal, attaining surface loadings less than 0.04 mol Cd:mol 2LFh. These low surface loadings rendered d-PDF analysis of Cd(II)-sorbed ferrihydrite impossible. Conversely, Pb(II) sorbed extensively onto 2LFh. The roughly linear initial slope of the Pb(II) sorption isotherm for 2LFh reached a maximum loading of approximately 0.17 mol Pb:mol 2LFh, which is in excellent agreement with the isotherm shape and capacity reported for Pb(II) adsorption onto 2LFh in previous work (Trivedi et al. 2003). The enhanced uptake of Pb(II) relative to Cd(II) by 2LFh is consistent with the Linear Free Energy Relationship (LFER) documented previously for metal sorption by metal oxides, clays and natural sediments (Dzombak and Morel 1990, Wang et al. 1997). Despite the enhanced sorption of Pb(II) compared with Cd(II) by 2LFh, both metals were removed from solution far less effectively by 2LFh than As(V) and Sb(V) oxyanions. Considering that the surface charge of 2LFh is positive at pH 5.5, this result can be explained by the electrostatic repulsion of positively charged Cd(II) and Pb(II) and electrostatic attraction of negatively charged Sb(V) and As(V) on the 2LFh surface.

The sorption isotherms in the δ-MnO$_2$ system (Fig. 12.8B) differ considerably relative to those in the 2LFh system. Both Cd(II) and Pb(II) sorption isotherms for δ-MnO$_2$ are characterized by steep initial slopes typical of H-curve isotherms, which indicates a strong affinity for δ-MnO$_2$ surfaces (Sposito 2008). In the H-curve portion of the isotherms, surface loadings of up to 0.12 mol

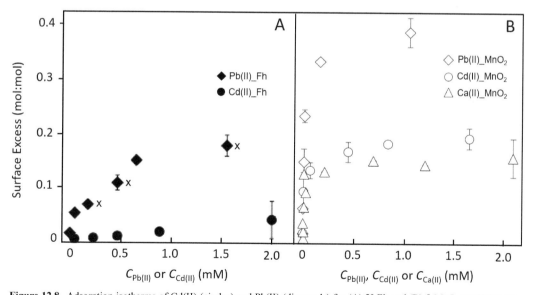

Figure 12.8. Adsorption isotherms of Cd(II) (circles) and Pb(II) (diamonds) for (A) 2LFh and (B) δ-MnO$_2$ at pH 5.5. The adsorption isotherm of calcium(II) for δ-MnO$_2$ (triangles) is also given for comparison. The × symbols indicate samples for which surface structural characterization is reported. This figure is a modified version of Fig. 2 in van Genuchten and Pena (2016b).

Cd:mol δ-MnO$_2$ ($C_{Cd(II)}$ = 4 μM) and 0.24 mol Pb:mol δ-MnO$_2$ ($C_{Pb(II)}$ = 15 μM) were reached. With increasing equilibrium concentrations of aqueous Cd(II), the slope of the sorption isotherm leveled off, with the highest measured surface loading never exceeding 0.20 mol Cd:mol δ-MnO$_2$. This loading is similar to the highest loadings reported previously for Cd(II) (0.1 mol:mol at pH 6) and Ni(II) (0.18 mol:mol at pH 6.6) on δ-MnO$_2$ (Gadde and Laitinen 1974, Simanova et al. 2015). We also noted that the shape of the corresponding isotherm of calcium (Ca(II)) adsorbed onto δ-MnO$_2$ was an excellent match to that of Cd(II) (Fig. 12.8B), which is consistent with the similar ionic potentials (IP, charge/ionic radius) and the hard-shell character of Ca(II) and Cd(II) (Sposito 2008). By contrast, the slope of the Pb(II) sorption isotherm decreased more gradually than those of Cd(II) and Ca(II), with Pb(II) sorption achieving a maximum loading of about 0.40 mol Pb:mol δ-MnO$_2$. The 2-fold difference between the highest measured Cd(II) and Pb(II) surface excess suggests that Cd(II) and Pb(II) sorb onto different surface sites of δ-MnO$_2$. When compared to the oxyanion experiments, both heavy metals displayed a higher affinity for δ-MnO$_2$ surfaces than As(V) and Sb(V) and a higher sorption capacity. This observation suggests that, for a given pH and identical adsorbent mass, δ-MnO$_2$ would be a more useful mineral nanoparticle than 2LFh for heavy metal removal from polluted soils, industrial wastewaters and drinking water.

6.2 Cation sorption mechanisms

6.2.1 Pb(II) adsorption onto 2LFh

Since Cd(II) sorption onto 2LFh was minimal, determining the mechanism of Cd(II) uptake by 2LFh using the d-PDF technique was not possible. Therefore, this chapter will only use Pb(II) as a probe species to identify the modes of heavy metal uptake by 2LFh. The experimental and calculated d-PDFs of Pb(II) adsorbed onto 2LFh at surface loadings from 0.06 to 0.17 mol% are shown in Fig. 12.9. In this figure, the experimental d-PDFs are scaled to the most intense peak and offset vertically to facilitate comparison among samples. The first shell of Pb–O atoms was fitted with a Gaussian peak at 2.36–2.40 Å, consistent with Pb(II) in a trigonal pyramidal coordination (Bargar et al. 1998, Tiberg et al. 2013). The dominant peak in the experimental d-PDFs is centered at 3.47 Å, which is longer than the Pb–Fe distances attributed to edge-sharing Pb(II) sorption complexes with 2LFh (3.29–3.41 Å) and goethite (3.27–3.43 Å) (Takahashi et al. 2007, Tiberg et al. 2013, Xu et al. 2006). With increasing Pb:Fe surface excess, a shoulder located near 4.0 Å emerges in this major peak (Fig. 12.9), indicating that Pb(II) binds in multiple coordination environments. The position of this shoulder is nearly identical to the Pb–Fe distance assigned to mononuclear corner-sharing adsorption complexes on 2LFh and goethite (Bargar et al. 1998, Ostergren et al. 2000, Elzinga et al. 2001, Trivedi et al. 2003, Tiberg et al. 2013). Although broad peaks extend to about 11 Å in the d-PDFs of Pb(II)-adsorbed 2LFh, Gaussian functions could only be fitted to two peaks at $R > 4$ Å, which were located near 6.0 and 6.6–6.7 Å, due to the low amplitude of peaks at $R > 7$ Å.

Calculated d-PDFs derived from different models of Pb(II) bound to 2LFh surfaces in various single and double edge- and corner-sharing complexes are given in Fig. 12.9C. The models used to calculate the d-PDFs overestimate the relative amplitude of the first-shell Pb–O peak. The discrepancy in relative peak intensity is likely due to poorly-modeled static disorder in the average Pb–O bonding environment when Pb is in trigonal pyramidal coordination. At R-values greater than 2.4 Å, Pb(II) bound in both Edge-Sharing (ES) and SCS configurations were required to reproduce the dominant Pb–Fe peak and shoulder from 3.4–4.0 Å. However, no single combination of ES and SCS models provided a satisfactory fit to the d-PDF features at $R > 5.0$ Å. The model of Pb(II) bound in SES complexes to the top octahedral sheet of ferrihydrite (SES A) reproduced the position of the Pb–Fe peak at 6.0 Å, but only the SES B and Double Edge-Sharing (DES) models matched the position of the Pb–Fe peak near 6.7 Å. The inability of a single model to reproduce the experimental data suggests that Pb(II) does not bind preferentially to any one 2LFh sorption site that can accommodate ES complexes. In Fig. 12.9C, a model where Pb(II) is bound in multiple coordination geometries to different sorption sites on 2LFh is given. The calculated d-PDF

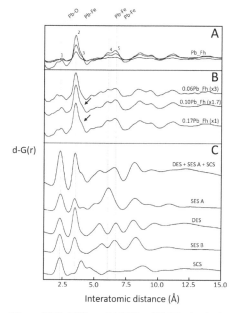

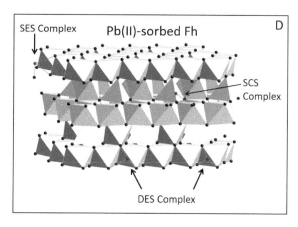

Figure 12.9. Differential PDFs of Pb(II) adsorbed onto 2LFh (A–C) and visual representation of Pb(II) adsorption to 2LFh based on the d-PDF analysis (D).

derived from a combination of 50% DES, 25% SES A and 25% SCS models reproduces well the relative amplitude and position of the major Pb–Fe peak and shoulder near 3.5–4.0 Å and matches the positions of peaks near 6.7 and 8.1 Å. Therefore, we concluded that Pb(II) binds in multiple different inner-sphere adsorption modes to a variety of surface sites on 2LFh, which is shown in a visual representation in Fig. 12.9D.

6.2.2 Cd(II) sorption to δ-MnO₂

The experimental d-PDFs of Cd(II) sorbed to δ-MnO_2 at surface loadings from 9 to 16 mol Cd:mol δ-MnO_2 are given in Fig. 12.10. The short-ranged ($R < 4$ Å) region of the d-PDFs is dominated by two peaks: a relatively short peak near 2.27–2.28 Å and an intense peak centered at 3.66 Å. The positions of these d-PDF peaks are in excellent agreement with the Cd–O ($R_{Cd–O} = 2.24$ Å) and Cd–Mn ($R_{Cd–Mn} = 3.65$ Å) interatomic distances of Cd(II) sorbed as a multidentate complex ($N_{Cd–Mn} = 5$) to internal sites of a Mn(IV) oxide with a tunnel structure, cryptomelane (Randall et al. 1998). Beyond the first 4 Å, the experimental d-PDFs display well-resolved peaks out to $R > 10$ Å (Fig. 12.10). These long-range atomic pairs reflect both a highly-ordered Cd(II) coordination environment and a rigid, coherently-scattering sorbent structure, both of which are consistent with metal sorption at MnO_6 vacancy sites in the δ-MnO_2 crystal.

The calculated d-PDFs based on Cd(II) sorbed in TCS geometries to one end (TCS) of a MnO_6 layer vacancy is given in Fig. 12.10A. Below the calculated TCS d-PDF, Fig. 12.10A plots the TCS partial component d-PDFs, which separate the contribution of individual atomic pairs (i.e., Cd–O, Cd–Mn, Cd–Cd) to the complete d-PDF. Excellent agreement in peak position and relative peak amplitude between the TCS model and the experimental data at $R < 10$ Å indicates that Cd(II) sorbs dominantly onto internal sites of δ-MnO_2. Combining the excellent fit between the TCS model and the experimental data with the sorption isotherm data presented in the preceding section reveals a key conclusion regarding Cd(II) sorption to δ-MnO_2. The sorption isotherms revealed an approximately 20 mol% sorption capacity for Cd(II). However, the vacancy content of the δ-MnO_2h used in the sorption experiments was only 10 mol% (Villalobos et al. 2006, Grangeon et al. 2012, Villalobos 2015). Since no evidence of Cd(II) sorption to δ-MnO_2 particle edges was found in the d-PDF data, we concluded that Cd(II) binds to both ends of a MnO_6 cation vacancy (internal

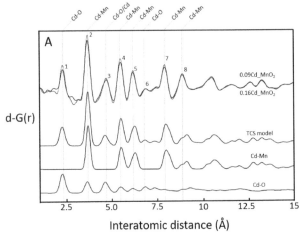
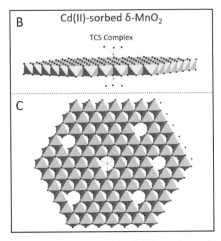

Figure 12.10. Differential PDFs of Cd(II) sorbed onto δ-MnO₂ (A) and visual representation of the Cd(II) uptake mode (B, C) based on the d-PDF analysis.

δ-MnO₂ sorption site). This conclusion is chemically reasonable based on the 4 vu charge deficit at each internal site, which is analogous to the position of Zn(II) in the stable crystal structure of chalcophanite (ZnMn₃O₇·3H₂O) (Post 1999). A visual representation of Cd(II) bound to both ends of a cation vacancy in a δ-MnO₂ sheet is shown in Fig. 12.10C.

6.2.3 Pb(II) sorption to δ-MnO₂

The experimental d-PDFs of Pb(II) sorbed to δ-MnO₂ at surface loadings from 0.14 to 0.38 mol Pb:mol δ-MnO₂ are given in Fig. 12.11A. In this figure, the experimental d-PDFs are scaled to the most intense peak and offset vertically to facilitate comparison among samples. Similar to Pb(II) adsorbed onto 2LFh, the highly distorted Pb–O first-shell is evident in the low amplitude of the first atomic pair correlation near 2.5 Å. Two peaks at 2.34–2.36 Å and 2.63–2.73 Å were modeled in the first Pb–O coordination sphere of Pb(II)-sorbed δ-MnO₂, which is consistent with previous EXAFS studies of Pb(II)-sorbed birnessite (Manceau et al. 2002). The most intense peak in the Pb(II)-sorbed δ-MnO₂ d-PDFs is positioned at 3.74–3.78 Å, which is slightly longer than the analogous peak in the d-PDFs of Cd(II)-sorbed δ-MnO₂. Gaussian fits of peaks at $R > 4$ Å in both the Cd(II)- and Pb(II)-sorbed δ-MnO₂ d-PDFs indicated that nearly every peak in the Pb(II) series is longer than the analogous peak in the Cd(II) series by about 0.1 Å. This difference reflects the difference between the first-shell Cd–O distance ($R_{Cd–O} = 2.27$ Å) and the shorter first-shell Pb–O distance ($R_{Pb–O} = 2.35$ Å). This trend indicates that the first-shell O atoms at shorter Pb–O distances are those involved in sorption, whereas the O atoms at longer Pb–O distances extend into solution (Kwon et al. 2010). Unlike the d-PDFs of Cd(II)-sorbed onto δ-MnO₂, the Pb(II) d-PDFs show a systematic decrease in the relative amplitude of the peak near 5.4–5.5 Å with increasing Pb:Mn surface excess from 0.14 to 0.38 mol Pb:mol δ-MnO₂.

The calculated d-PDFs of Pb(II) sorbed in a TCS complex to one end (TCS_1) and both ends (TCS_2) of a MnO₆ cation vacancy (internal sorption site) are given in Fig. 12.11C in dotted and solid lines, respectively, with the partial component d-PDFs of the TCS_2 model given in Fig. 12.11D. Although both TCS_1 and TCS_2 models reproduce the data, the peak centered at 4.8 Å in the experimental d-PDFs is located at lower R in the TCS_1 model and has a lower amplitude. By contrast, the TCS_2 model provides a better match to the relative amplitude of the peak at 4.8 Å and reproduces more accurately its position, which is due to the significant contribution of a Pb–Pb pair near 4.9 Å. As demonstrated in the Pb(II)-sorbed δ-MnO₂, partial component d-PDFs (Fig. 12.11D), the resolvable contribution of Pb–Pb pairs at 4.9 Å in the experimental data provides evidence that heavy metals can bind to both ends of δ-MnO₂ layer vacancies. The TCS_2 model alone, however, does not reproduce the systematic decrease in the relative amplitude of the peak near 5.4–5.5 Å with

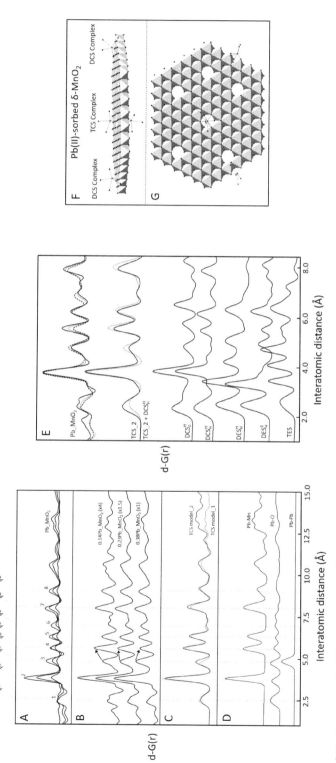

Figure 12.11. Experimental (A, B) and calculated (C–E) d-PDFs of Pb(II) sorbed to δ-MnO₂ and visual representation of the Pb(II) bonding mode to δ-MnO₂ surfaces (F, G) based on the d-PDF data. The superscripts and subscripts in panel E indicate the protonation state of different O atoms on the MnO₂ surface adjacent to the Pb(II) sorption site. A linear combination of 50% TCS and 50% DCSH H d-PDFs (dotted lines) is also overlain to the TCS_2 model (solid lines).

increasing Pb:Mn surface excess observed in the experimental d-PDFs. Thus, we investigated the formation of multiple Pb(II) surface complexes at δ-MnO$_2$ external sites (i.e., particle edges), as has been proposed earlier (Villalobos et al. 2003, Takahashi et al. 2007). In Fig. 12.11E, the experimental d-PDFs to models of edge- and corner-sharing Pb(II) sorption complexes on δ-MnO$_2$ particle edges derived from DFT calculations, performed in a separate investigation, were compared (Kwon et al. 2010). Inspection of the geometry-optimized Pb(II) sorption model d-PDFs (Fig. 12.11E) rules out significant fractions of Pb(II) bound in DES or triple edge-sharing (TES) modes due to the absence of the dominant DES or TES peak near 3.5 Å in the experimental d-PDF data. By contrast, the DCS models fit the data more closely, particularly the DCS$_H^H$ model, which corresponds to Pb(II) bound to adjacent surface MnO$_6$ octahedra with protonated singly- and doubly-coordinated O atoms (Kwon et al. 2010). Thus, a d-PDF was calculated based on a model where Pb(II) was bound in both TCS_2 (50%) and DCS$_H^H$ (50%) modes (Fig. 12.11E). This combination of Pb(II) sorption geometries at internal and external surfaces reproduced the trend of decreased relative amplitude of the peak at 5.5 Å and slight decrease in amplitude and shortening in bond length of the Pb–Mn peak near 8 Å more accurately than all other geometry-optimized sorption models.

Given the conclusion that Cd(II) sorbed dominantly in the TCS configuration, it is expected that the maximum amount of bivalent cations that can associate directly with layer vacancies in the present experiments is 20 mol% (van Genuchten and Pena, 2016b). Accordingly, the fraction of Pb(II) sorbed in modes other than the TCS geometry at the largest surface loading in our study (38 mol%) is expected to be approximately 0.5 (20 mol%:38 mol%). This conclusion is supported by the good fitting of the calculated d-PDF of the 50% TCS_2 and 50% DCS models, particularly the reproduction of the trend in peak position and relative amplitude with increasing Pb:Mn surface excess observed in the experimental d-PDFs. Finally, although it cannot be disputed that Pb(II) binds to δ-MnO$_2$ edges at surface loadings below the lowest Pb loading of 0.14 mol Pb: mol δ-MnO$_2$ for which samples were taken, the d-PDF data are consistent with an increase of Pb(II) sorption by external sites (i.e., δ-MnO$_2$ particle edges) relative to internal sites (i.e., MnO$_6$ vacancies) as the surface loading increases. This conclusion is consistent with the δ-MnO$_2$ particle edges contributing more to Pb(II) sorption as vacancies saturate. A visual representation of Pb(II) sorbed to δ-MnO$_2$ that is based on the analysis of the complete d-PDF data set and Pb(II) adsorption isotherms is given in Figs. 12.11F and 12.11G.

7. Conclusions

The crystal structures of both 2LFh and δ-MnO$_2$ contain severe to moderate amounts of disorder, which breeds a significant adsorption reactivity towards a range of oxyanion metalloids and bivalent cation heavy metals with key industrial relevance. Combining all sets of sorption data, including As(V), Sb(V), Cd(II) and Pb(II) interactions with 2LFh and δ-MnO$_2$ allows several general conclusions to be drawn regarding the reactivity of Fe(III) and Mn(IV) nanoparticles in the context of industrial application (i.e., soil and water remediation). Both 2LFh and δ-MnO$_2$ are effective adsorbents for oxyanions, but 2LFh outperforms δ-MnO$_2$ on a mol or mass basis. The enhanced uptake of oxyanions by 2LFh compared to δ-MnO$_2$ is best explained by the differences in surface charge of 2LFh, which is positively charged at pH above 5.5 and δ-MnO$_2$, which is strongly negatively charged. It is worth reiterating that although Sb(V) attains an octahedral coordination, no evidence for Sb(V) occupying MnO$_6$ vacancies in the δ-MnO$_2$ structure was found, which again can be explained by electrostatic repulsion between the negative SbO$_6$ octahedron and the negative charge of the cation vacancy.

In contrast to the behavior of oxyanion uptake by 2LFh and δ-MnO$_2$, heavy metals Cd(II) and Pb(II) were more effectively removed by δ-MnO$_2$ than 2LFh. At pH 5.5, only the internal δ-MnO$_2$ sites (i.e., MnO$_6$ vacancies) were reactive with Cd(II), a hard-shell metal with a high ionic potential, but both internal and external sites (i.e., δ-MnO$_2$ particle edges) were reactive with Pb(II), a soft-shell metal with lower ionic potential (Sposito 2008). The reactivity of δ-MnO$_2$ particle edges towards Cd(II) and Pb(II) was similar to the reactivity of 2LFh, with minimal Cd(II) uptake at pH 5.5, but effective Pb(II) adsorption. The similarity in oxyanion and heavy metal sorption reactivity between δ-MnO$_2$ particle edges and 2LFh surfaces is consistent with the similar degrees of bond strength undersaturation of their SSO atoms. Taken together, this chapter demonstrates that 2LFh and δ-MnO$_2$ particle edges will be more reactive with oxyanions, whereas heavy metals are expected to bind preferentially to MnO$_6$ cation vacancies in δ-MnO$_2$ sheets. Considering the differences in reactive surface area between 2LFh and δ-MnO$_2$ and the costs and complexity of synthesizing large quantities of these minerals for remediation purposes, it is recommended that 2LFh be implemented in treatment schemes that target oxyanion contaminants, whereas δ-MnO$_2$ is considered the optimum nanoparticle to remove heavy metals from soils, sediments, wastewater and drinking water.

References

Ahmad, A., A. van der Wal, P. Bhattacharya and C.M. van Genuchten. 2019. Characteristics of Fe and Mn bearing precipitates generated by Fe(II) and Mn(II) co-oxidation with O$_2$, MnO$_4$ and HOCl in the presence of groundwater ions. Water Res. 161: 505–516.

Appelo, C.A.J. and D. Postma. 1999. A consistent model for surface complexation on birnessite (-MnO2) and its application to a column experiment. Geochim. Cosmochim. Acta. 63: 3039–3048.

Bargar, J.R., G.E. Brown and G.A. Parks. 1998. Surface complexation of Pb(II) at oxide-water interfaces: III. XAFS determination of Pb(II) and Pb(II)-chloro adsorption complexes on goethite and alumina. Geochim. Cosmochim. Acta. 62: 193–207.

Borch, T., R. Kretzschmar, A. Kappler, P. Van Cappellen, M. Ginder-Vogel, A. Voegelin and K. Campbell. 2010. Biogeochemical redox processes and their impact on contaminant dynamics. Environ. Sci. Technol. 44: 15–23.

Brown, I.D. and D. Altermatt. 1985. Bond-valence parameters obtained from a systematic analysis of the inorganic crystal-structure database. Acta Crystallogr. Sect. B-Structural Sci. 41: 244–247.

Cismasu, A.C., F.M. Michel, A.P. Tcaciuc, T. Tyliszczak and G.E. Brown. 2011. Composition and structural aspects of naturally occurring ferrihydrite. Comptes Rendus Geosci. 343: 210–218.

Cismasu, A.C., F.M. Michel, J.F. Stebbins, C. Levard and G.E. Brown. 2012. Properties of impurity-bearing ferrihydrite I. Effects of Al content and precipitation rate on the structure of 2-line ferrihydrite. Geochim. Cosmochim. Acta. 92: 275–291.

Dixit, S. and J.G. Hering. 2003. Comparison of arsenic(V) and arsenic(III) sorption onto iron oxide minerals: Implications for arsenic mobility. Environ. Sci. Technol. 37: 4182–4189.

Drits, V.A., B.A. Sakharov, A.L. Salyn and A. Manceau. 1993. Structural model for ferrihydrite. Clay Miner. 28: 185–207.

Droz, B., N. Dumas, O. Duckworth and J. Pena. 2015. A comparison of the sorption reactivity of bacteriogenic and mycogenic Mn oxide nanoparticles. Environ. Sci. Technol. 49: 4200–4208.

Dyer, L.G., K.W. Chapman, P. English, M. Saunders and W.R. Richmond. 2012. Insights into the crystal and aggregate structure of Fe^{3+} oxide/silica co-precipitates. Am. Mineral. 97: 63–69.

Dzombak, D.A. and F.M.M. Morel. 1990. Surface Complexation Modeling: Hydrous Ferric Oxide. Wiley. New York.

Egami, T. and S. Billinge. 2003. Underneath the Bragg Peaks: Structural Analysis of Complex Materials. Pergamon. Kiddington, Oxford, UK.

Elzinga, E.J., D. Peak and D.L. Sparks. 2001. Spectroscopic studies of Pb(II)-sulfate interactions at the goethite-water interface. Geochim. Cosmochim. Acta. 65: 2219–2230.

Fendorf, S., M.J. Eick, P. Grossl and D.L. Sparks. 1997. Arsenate and chromate retention mechanisms on goethite 1. Surface structure. Environ. Sci. Technol. 31: 315–320.

Foster, A.L., G.E. Brown and G.A. Parks. 2003. X-ray absorption fine structure study of As(V) and Se(IV) sorption complexes on hydrous Mn oxides. Geochim. Cosmochim. Acta. 67: 1937–1953.

Gadde, R.R. and H.A. Laitinen. 1974. Studies of heavy-metal sorption by hydrous oxides. Abstr. Pap. Am. Chem. Soc. 142.

Gilbert, B., F. Huang, H.Z. Zhang, G.A. Waychunas and J.F. Banfield. 2004. Nanoparticles: Strained and stiff. Science. 305: 651–654.

Grangeon, S., A. Manceau, J. Guilhermet, A.C. Gaillot, M. Lanson and B. Lanson. 2012. Zn sorption modifies dynamically the layer and interlayer structure of vernadite. Geochim. Cosmochim. Acta. 85: 302–313.

Gude, J.C.J., L.C. Rietveld and D. van Halem. 2017. As(III) oxidation by MnO$_2$ during groundwater treatment. Water Res. 111: 41–51.

Harrington, R., D.B. Hausner, N. Bhandari, D.R. Strongin, K.W. Chapman, P.J. Chupas et al. 2010. Investigation of surface structures by powder diffraction: A differential pair distribution function study on arsenate sorption on ferrihydrite. Inorg. Chem. 49: 325–330.

Harrington, R., D.B. Hausner, W.Q. Xu, N. Bhandari, F.M. Michel, G.E. Brown et al. 2011. Neutron pair distribution function study of two-line ferrihydrite. environ. Sci. Technol. 45: 9883–9890.

Hering, J.G., I.A. Katsoyiannis, G.A. Theoduloz, M. Berg and S.J. Hug. 2017. Arsenic removal from drinking water: experiences with technologies and constraints in practice. J. Environ. Eng. 143.

Hinz, C. 2001. Description of sorption data with isotherm equations. Geoderma. 102: 405–406.

Kelly, S.D., D. Hesterberg and B. Ravel. 2008. Analysis of soils and minerals using X-ray absorption spectroscopy. *In*: Ulery, A.L. and L.R. Drees (eds.). Methods of Soil Analysis. Part 5. Mineralogical Methods. SSA Book Series No. 5. Soil Science Society of America.

Kinniburgh, D.G., M.L. Jackson and J.K. Syers. 1976. Adsorption of alkaline-earth, transition, and heavy-metal cations by hydrous oxide gels of iron and aluminum. Soil Sci. Soc. Am. J. 40: 796–799.

Kosmulski, M. 2009. Compilation of PZC and IEP of sparingly soluble metal oxides and hydroxides from literature. Adv. Colloid Interface Sci. 152: 14–25.

Kwon, K.D. and J.D. Kubicki. 2004. Molecular orbital theory study on surface complex structures of phosphates to iron hydroxides: Calculation of vibrational frequencies and adsorption energies. Langmuir. 20: 9249–9254.

Kwon, K.D., K. Refson and G. Sposito. 2010. Surface complexation of Pb(II) by hexagonal birnessite nanoparticles. Geochim. Cosmochim. Acta. 74: 6731–6740.

Lanson, B., V.A. Drits, Q. Feng and A. Manceau. 2002. Structure of synthetic Na-birnessite: Evidence for a triclinic one-layer unit cell. Am. Mineral. 87: 1662–1671.

Learman, D.R., B.M. Voelker, A.I. Vazquez-Rodriguez and C.M. Hansel. 2011. Formation of manganese oxides by bacterially generated superoxide. Nat. Geosci. 4: 95–98.

Li, W., R. Harrington, Y.Z. Tang, J.D. Kubicki, M. Aryanpour, R.J. Reeder et al. 2011. Differential pair distribution function study of the structure of arsenate adsorbed on nanocrystalline gamma-alumina. Environ. Sci. Technol. 45: 9687–9692.

Liu, H., P. Li, M.Y. Zhu, Y. Wei and Y.H. Sun. 2007. Fe(II)-induced transformation from ferrihydrite to lepidocrocite and goethite. J. Solid State Chem. 180: 2121–2128.

Loring, J.S., M.H. Sandstrom, K. Noren and P. Persson. 2009. Rethinking arsenate coordination at the surface of goethite. Chem. Eur. J. 15: 5063–5072.

Maillot, F., G. Morin, Y.H. Wang, D. Bonnin, P. Ildefonse, C. Chaneac et al. 2011. New insight into the structure of nanocrystalline ferrihydrite: EXAFS evidence for tetrahedrally coordinated iron(III). Geochim. Cosmochim. Acta. 75: 2708–2720.

Manceau, A., B. Lanson and V.A. Drits. 2002. Structure of heavy metal sorbed birnessite. Part III: Results from powder and polarized extended X-ray absorption fine structure spectroscopy. Geochim. Cosmochim. Acta. 66: 2639–2663.

Manceau, A. 2011. Critical evaluation of the revised akdalaite model for ferrihydrite. Am. Mineral. 96: 521–533.

Manceau, A. 2012. Comment on "Direct Observation of Tetrahedrally Coordinated Fe(III) in Ferrihydrite." Environ. Sci. Technol. 46: 6882–6884.

Manceau, A. and W.P. Gates. 2013. Incorporation of Al in iron oxyhydroxides: implications for the structure of ferrihydrite. Clay Miner. 48: 481–489.

Manceau, A., M.A. Marcus, S. Grangeon, M. Lanson, B. Lanson, A.C. Gaillot et al. 2013. Short-range and long-range order of phyllomanganate nanoparticles determined using high-energy X-ray scattering. J. Appl. Crystallogr. 46: 193–209.

Marafatto, F., M.L. Strader, J. Gonzalez-Holguera, A. Schwartzberg, B. Gilbert and J. Pena. 2015. Rate and mechanism of the photoreduction of birnessite (MnO_2) nanosheets. Proc. Natl. Acad. Sci. U. S. A. 112: 4600–4605.

Megaw, H.D. 1934. The crystal structure of hydrargillite, $Al(OH)_{(3)}$. Zeitschrift Fur Krist. 87: 185–204.

Michel, F.M., L. Ehm, S.M. Antao, P.L. Lee, P.J. Chupas, G. Liu et al. 2007a. The structure of ferrihydrite, a nanocrystalline material. Science. 316: 1726–1729.

Michel, F.M., L. Ehm, G. Liu, W.Q. Han, S.M. Antao, P.J. Chupas et al. 2007b. Similarities in 2- and 6-line ferrihydrite based on pair distribution function analysis of X-ray total scattering. Chem. Mater. 19: 1489–1496.

Mitsunobu, S., Y. Takahashi, Y. Terada and M. Sakata. 2010. Antimony(V) incorporation into synthetic ferrihydrite, goethite, and natural iron oxyhydroxides. Environ. Sci. Technol. 44: 3712–3718.

Mitsunobu, S., C. Muramatsu, K. Watanabe and M. Sakata. 2013. Behavior of antimony(V) during the transformation of ferrihydrite and its environmental implications. Environ. Sci. Technol. 47: 9660–9667.

Morin, G., F. Juillot, C. Casiot, O. Bruneel, J.C. Personne, F. Elbaz-Poulichet et al. 2003. Bacterial formation of tooeleite and mixed Arsenic(III) or Arsenic(V)-Iron(III) gels in the carnoulbs acid mine drainage, France. A XANES, XRD, and SEM study. Environ. Sci. Technol. 37: 1705–1712.

Okkenhaug, G., Y.G. Zhu, J.W. He, X. Li, L. Luo and J. Mulder. 2012. Antimony (Sb) and Arsenic (As) in Sb Mining impacted paddy soil from xikuangshan, china: differences in mechanisms controlling soil sequestration and uptake in rice. Environ. Sci. Technol. 46: 3155–3162.

Okkenhaug, G., K. Amstatter, H.L. Bue, G. Cornelissen, G.D. Breedveld, T. Henriksen et al. 2013. Antimony (Sb) contaminated shooting range soil: Sb mobility and immobilization by soil amendments. Environ. Sci. Technol. 47: 6431–6439.

Ostergren, J.D., T.P. Trainor, J.R. Bargar, G.E. Brown and G.A. Parks. 2000. Inorganic ligand effects on Pb(II) sorption to goethite (alpha-FeOOH)-I. Carbonate. J. Colloid Interface Sci. 225: 466–482.

Otte, K., W.W. Schmahl and R. Pentcheva. 2013. DFT+U study of arsenate adsorption on FeOOH surfaces: Evidence for competing binding mechanisms. J. Phys. Chem. C 117: 15571–15582.

Paktunc, D., A. Manceau and J. Dutrizac. 2013. Incorporation of Ge in ferrihydrite: Implications for the structure of ferrihydrite. Am. Mineral. 98: 848–858.

Peak, D. and T. Regier. 2012. Direct observation of tetrahedrally coordinated Fe(III) in ferrihydrite. Environ. Sci. Technol. 46: 3163–3168.

Pedersen, H.D., D. Postma, R. Jakobsen and O. Larsen. 2005. Fast transformation of iron oxyhydroxides by the catalytic action of aqueous Fe(II). Geochim. Cosmochim. Acta 69: 3967–3977.

Peña, J., K.D. Kwon, K. Refson, J.R. Bargar and G. Sposito. 2010. Mechanisms of nickel sorption by a bacteriogenic birnessite. Geochim. Cosmochim. Acta 74: 3076–3089.

Peña, J., J.R. Bargar and G. Sposito. 2015. Copper sorption by the edge surface of synthetic birnessite nanoparticles. Chem. Geol. 396: 196–207.

Perret, D., J.F. Gaillard, J. Dominik and O. Atteia. 2000. The diversity of natural hydrous iron oxides. Environ. Sci. Technol. 34: 3540–3546.

Post, J.E. 1999. Manganese oxide minerals: Crystal structures and economic and environmental significance. Proc. Natl. Acad. Sci. U. S. A. 96: 3447–3454.

Randall, S.R., D.M. Sherman and K.V. Ragnarsdottir. 1998. An extended X-ray absorption fine structure spectroscopy investigation of cadmium sorption on cryptomelane (KMn(8)O(16)). Chem. Geol. 151: 95–106.

Raven, K.P., A. Jain and R.H. Loeppert. 1998. Arsenite and arsenate adsorption on ferrihydrite: Kinetics, equilibrium, and adsorption envelopes. Environ. Sci. Technol. 32: 344–349.

Remucal, C.K. and M. Ginder-Vogel. 2014. A critical review of the reactivity of manganese oxides with organic contaminants. Environ. Sci. Impacts 16: 1247–1266.

Rustad, J.R. 2001. Molecular models of surface relaxation, hydroxylation, and surface charging at oxide-water interfaces. Mol. Model. Theory Appl. Geosci. 42: 169–197.

Schwertmann, U. and S. Cornell. 1991. Iron Oxides in the Laboratory: Preparation and Characterization. VCH, Weinheim; New York.

Sherman, D.M. and S.R. Randall. 2003. Surface complexation of arsenie(V) to iron(III) (hydr)oxides: Structural mechanism from ab initio molecular geometries and EXAFS spectroscopy. Geochim. Cosmochim. Acta. 67: 4223–4230.

Simanova, A.A., K.D. Kwon, S.E. Bone, J.R. Bargar, K. Refson, G. Sposito et al. 2015. Study on the sorption reactivity of the edge surfaces in birnessite nanoparticles using nickel as a probe ion. Geochim. Cosmochim. Acta. 164: 191–204.

Sposito, G. 2004. The Surface Chemistry of Natural Particles. Oxford University Press, Oxford; New York.

Sposito, G. 2008. The Chemistry of Soils. 2nd ed., Oxford University Press, Oxford; New York.

Takahashi, Y., A. Manceau, N. Geoffroy, M.A. Marcus and A. Usui. 2007. Chemical and structural control of the partitioning of Co, Ce, and Pb in marine ferromanganese oxides. Geochim. Cosmochim. Acta. 71: 984–1008.

Tan, H., G.X. Zhang, P.J. Heaney, S.M. Webb and W.D. Burgos. 2010. Characterization of manganese oxide precipitates from Appalachian coal mine drainage treatment systems. Appl. Geochem. 25: 389–399.

Tiberg, C., C. Sjostedt, I. Persson and J.P. Gustafsson. 2013. Phosphate effects on copper(II) and lead(II) sorption to ferrihydrite. Geochim. Cosmochim. Acta. 120: 140–157.

Toner, B.M., T.S. Berquo, F.M. Michel, J.V. Sorensen, A.S. Templeton and K.J. Edwards. 2012. Mineralogy of iron microbial mats from loihi seamount. Front. Microbiol. 3: 118.

Trivedi, P., J.A. Dyer and D.L. Sparks. 2003. Lead sorption onto ferrihydrite. 1. A macroscopic and spectroscopic assessment. Environ. Sci. Technol. 37: 908–914.

van Genuchten, C.M., J. Pena, S.E. Amrose and A.J. Gadgil. 2014. Structure of Fe(III) precipitates generated by the electrolytic dissolution of Fe(0) in the presence of groundwater ions. Geochim. Cosmochim. Acta. 127: 285–304.

van Genuchten, C.M. and J. Pena. 2016a. Antimonate and arsenate speciation on reactive soil minerals studied by differential pair distribution function analysis. Chem. Geol. 429: 1–9.

van Genuchten, C.M. and J. Pena. 2016b. Sorption selectivity of birnessite particle edges: A d-PDF analysis of Cd(II) and Pb(II) sorption by δ-MnO_2 and ferrihydrite. Environ. Sci. Process. Impacts. 18: 1030–1041.

Villalobos, M., B. Toner, J. Bargar and G. Sposito. 2003. Characterization of the manganese oxide produced by *Pseudomonas putida* strain MnB1. Geochim. Cosmochim. Acta. 67: 2649–2662.

Villalobos, M., B. Lanson, A. Manceau, B. Toner and G. Sposito. 2006. Structural model for the biogenic Mn oxide produced by *Pseudomonas putida*. Am. Mineral. 91: 489–502.

Villalobos, M., I.N. Escobar-Quiroz and C. Salazar-Camacho. 2014. The influence of particle size and structure on the sorption and oxidation behavior of birnessite: I. Adsorption of As(V) and oxidation of As(III). Geochim. Cosmochim. Acta. 125: 564–581.

Villalobos, M. 2015. The role of surface edge sites in metal(loid) sorption to poorly-crystalline birnessites. *In*: Feng, X., W. Li, M. Zhu and D. Sparks (eds.). Advances in the Environmental Biogeochemistry of Manganese Oxides. American Chemical Society.

Voegelin, A., A.C. Senn, R. Kaegi, S.J. Hug and S. Mangold. 2013. Dynamic Fe-precipitate formation induced by Fe(II) oxidation in aerated phosphate-containing water. Geochim. Cosmochim. Acta. 117: 216–231.

Wang, F.Y., J.S. Chen and W. Forsling. 1997. Modeling sorption of trace metals on natural sediments by surface complexation model. Environ. Sci. Technol. 31: 448–453.

Wang, Y., X.H. Feng, M. Villalobos, W.F. Tan and F. Liu. 2012. Sorption behavior of heavy metals on birnessite: Relationship with its Mn average oxidation state and implications for types of sorption sites. Chem. Geol. 292: 25–34.

Waychunas, G.A., B.A. Rea, C.C. Fuller and J.A. Davis. 1993. Surface-chemistry of ferrihydrite.1. Exafs studies of the geometry of coprecipitated and adsorbed arsenate. Geochim. Cosmochim. Acta. 57: 2251–2269.

Waychunas, G.A., C.S. Kim and J.F. Banfield. 2005. Nanoparticulate iron oxide minerals in soils and sediments: unique properties and contaminant scavenging mechanisms. J. Nanoparticle Res. 409–433.

Xu, Y., T. Boonfueng, L. Axe, S. Maeng and T. Tyson. 2006. Surface complexation of Pb(II) on amorphous iron oxide and manganese oxide: Spectroscopic and time studies. J. Colloid Interface Sci. 299: 28–40.

Zhao, J.M., F.E. Huggins, Z. Feng and G.P. Huffman. 1994. Ferrihydrite—surface-structure and its effects on phase-transformation. Clays Clay Miner. 42: 737–746.

Zhu, M.Q., C.L. Farrow, J.E. Post, K.J.T. Livi, S.J.L. Billinge, M. Ginder-Vogel et al. 2012. Structural study of biotic and abiotic poorly-crystalline manganese oxides using atomic pair distribution function analysis. Geochim. Cosmochim. Acta. 81: 39–55.

Chapter 13

Titanium Dioxide Photocatalysts for Environmental Applications
Metal-Doping with Cerium Ions for Visible Light Activation and Efficiency Improvement

Marcela V. Martin, Orlando M. Alfano* and *María L. Satuf*

1. Introduction

The photocatalytic mechanism with titanium dioxide (TiO_2) as a catalyst is initiated by the absorption of photons with energy equal to or greater than the bandgap of TiO_2 (~ 3.1 eV for the anatase phase), producing electron-hole pairs in the semiconductor particle, as shown schematically in Fig. 13.1. Electrons are promoted to the conduction band, while positive holes are formed in the valence band. Excited-state electrons and holes can recombine and dissipate the input energy as heat, get trapped in metastable surface states or react with electron donors and electron acceptors adsorbed at the semiconductor surface or within the surrounding electrical double layer of the charged particles. Holes can react with adsorbed water to produce hydroxyl radicals with high redox oxidizing potential (Dong et al. 2015, Gopinath et al. 2020).

TiO_2-based photocatalysis has shown great potential as a low-cost and environmentally friendly technology for water purification. Nevertheless, there are technical limitations that hinder its application at industrial scale: the inefficient exploitation of the visible spectrum of the sun, the low adsorption capacity of hydrophobic pollutants and the requirement of a post-process step to recover the TiO_2 catalyst.

A substantial amount of research has been conducted towards the enhancement of TiO_2 photocatalysis by metal, nonmetal and ion doping of the semiconductor (Grabowska et al. 2012, Pelaez et al. 2012, Ibrahim et al. 2020). The primary purposes of TiO_2 doping are to retard the fast electron-hole pair recombination and to enable visible light absorption by creating defect states in the bandgap of the semiconductor. Coating TiO_2 with dyes has also proven to produce good photocatalysts able to extend the absorption to the visible (Litter et al. 2018).

Instituto de Desarrollo Tecnológico para la Industria Química (INTEC) (CONICET and Universidad Nacional del Litoral), Ruta Nacional 168, Santa Fe (3000), Argentina.
Emails: alfano@santafe-conicet.gov.ar; mlsatuf@santafe-conicet.gov.ar
* Corresponding author: mmartin@intec.unl.edu.ar

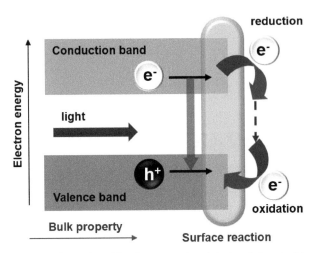

Figure 13.1. Schematic illustration of the photogeneration of electron-hole pairs (charge carriers).

This chapter summarizes and discusses the last developed strategies for improving the degradation of organic contaminants using metal-doped catalysts, with major emphasis on cerium-doped TiO_2 materials.

2. General Aspects of Metal-Doped TiO_2 Photocatalysts

Metal-doped TiO_2 photocatalysts have been tested for the degradation of a wide variety of substrates under visible light irradiation, including endocrine disruptors (Martin et al. 2019), dyes (Choi et al. 2010), phenolic compounds (Martin et al. 2015, Rossi et al. 2021), acetaldehyde (Murakami et al. 2010) and nitric oxide (Morikawa et al. 2006). However, the performances reported in these studies are dissimilar, presenting contradictory results of the effects of doping on the photoactivity of TiO_2. This variability may be due to the diversity of experimental conditions employed, which includes different catalyst preparation methods, chemical reactions, reactor configurations and irradiation sources. Therefore, the lack of a standard procedure to test the photocatalytic materials makes it difficult to compare the net effects of metal dopants on the activity of TiO_2.

3. Synthesis of Metal-Doped TiO_2

The synthesis of metal-doped TiO_2 is quite simple: TiO_2 particles can be just interstitially or substitutionally doped with distinct cations, and they can form a mixture of oxides or mixed oxides. The type of material obtained depends mainly on the following factors: the method of doping, the characteristics and amounts of dopants and the thermal treatment (Palmisano et al. 1994).

Metal-doped TiO_2 materials can be prepared in the laboratory in the form of films and powders by different synthetic routes, such as sol-gel process (Choi et al. 2010, Vera et al. 2017, Martin et al. 2019), solvothermal methods (Silva et al. 2009), Metal-Organic Chemical Vapor Deposition (MOCVD) (Zhang and Lei 2008), ion implantation (Anpo et al. 1999), impregnation and coprecipitation (Litter and Navío 1996) and solid-state reactions (Niishiro et al. 2005).

At the research scale, the sol-gel method is preferred, principally because of its simplicity and cost-effectiveness in terms of design, process, materials and implementation (Banerjee et al. 2016). In effect, the chemical and physical features of the materials synthesized by this process (e.g., particle size, crystallinity, surface area and mechanical properties) can be easily modified by changing well-defined experimental parameters: reaction time, pH, molar relation between reactants, temperature and precursors. The usual precursors are titanium tetrachloride or alkoxides (Yanagisawa and Ovenstone 1999, Falaras and Xagas 2002, Rodríguez et al. 2005).

To obtain metal-doped materials by the sol-gel method, the precursor containing the metal of interest (dissolved in water or alcohol) is generally added at the hydrolysis stage. In this way, the dopant has a better chance of remaining in the network of TiO_2 (compared with impregnation or deposition techniques) due to the formation of metal-O-Ti or metal-Ti-O bonds, and it becomes more difficult to be lost by leaching (Martin et al. 2015).

3.1 Type and amount of metal used as a dopant

Lanthanide ions are recognized for their capacity to form complex compounds with several Lewis bases (e.g., aldehydes, alcohols, amines, etc.) through the interaction of these functional groups with the *f* orbitals of lanthanides (Ranjit et al. 2001). Cerium has been identified as one of the most interesting lanthanides able to act as TiO_2 dopant because of two main features: the formation of the Ce^{3+}/Ce^{4+} redox couple (Martin et al. 2019) and the facile formation of labile oxygen vacancies as a result of the relatively high mobility of bulk oxygen species (Aman et al. 2012).

The utilization of Ce-doped TiO_2 has been reported for the degradation of several compounds: 17-α-ethinylestradiol (Martin et al. 2019), propranolol (Santiago-Morales et al. 2013), 2-mercaptobenzothiazole (Li et al. 2005), phenol (Martin et al. 2015), rhodamine B (Maarisetty and Baral 2019), formaldehyde (Xu et al. 2006), 4-chlorophenol (Silva et al. 2009), methylene blue (Aman et al. 2012), formic acid (Meksi et al. 2015) and bisphenol A (Chang and Liu 2014), among others. Although these treatments were carried out under different conditions (light source, catalyst preparation method, catalyst in powder form or immobilized, etc.), the authors agree that there is an optimal concentration of cerium for each material and each substrate. Very small amounts of dopant can be insufficient to produce a significant effect on the photocatalytic activity, but an excess can also be detrimental. This subject will be discussed in more detail later in this chapter.

3.2 Use of surfactants to generate selective catalysts

Photocatalytic reactions are governed by free radical mechanisms, especially when HO˙ is generated and is involved in nonselective oxidations. Specific pollutants should not be treated without taking into account the rest of the compounds present in the matrix (Gaya and Abdullah 2008). Industrial effluents may contain low concentrations of highly toxic substances along with high concentrations of Natural Organic Matter (NOM) that can act as scavengers of the reactive species generated on the surface of TiO_2 (Martin et al. 2012). Long et al. (2017) claimed that an important challenge for the application of TiO_2 photocatalysis in water treatment is to control the strong inhibitory effect developed by NOM.

Palacio et al. (2017) reported the synthesis of TiO_2 with a mesoporous structure to restrict the access of large NOM molecules and to achieve the preferential oxidation of phenol. They synthesized TiO_2 by the sol-gel method under two different conditions. In one of the synthesis conditions, a nonionic surfactant (Tween 80) was used as a pore-directing agent along with an acetic acid-based sol-gel route without the addition of water. These authors proposed that, by controlling the porous structure of TiO_2, the access of large size NOM molecules to TiO_2 can be suppressed, thus improving the selective oxidation of small size target contaminants. The particles of NOM with a size larger than the pore entrances mostly adsorb onto the grain boundary (external surface) of TiO_2, while small target molecules migrate into and adsorb onto its porous structure (internal surface), as schematized in Fig. 13.2.

The molecular weight of the Aldrich humic acids used in this study as an example of NOM is 3,614 g mol^{-1} (Mignone et al. 2012), while the molecular weight of phenol is 94.1 g mol^{-1}. The porous structure contributes to the majority of the surface of the materials and acts as a sorption site for the chemical reaction. Such a porous structure, if controlled properly, is advantageous to target chemicals by providing them with much more space for the reaction, but disadvantageous to NOM by restricting it to sitting on the grain boundary. Effectively, under UV irradiation, highly porous catalysts prepared with Tween 80 surfactant exhibited higher photocatalytic activity for

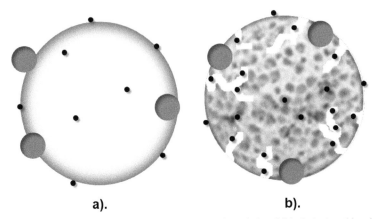

a).   **b).**

Figure 13.2. Proposed size-exclusion mechanism for the decomposition of phenol (black dots) and humic acid (gray dots) onto: (a) the control (synthesis without surfactant) TiO_2 and (b) the porous TiO_2.

the degradation of phenol in the presence of commercial humic acids than the materials prepared without this surfactant.

4. Characterization

4.1 Cristal structure and particle size

Titanium dioxide presents three forms of crystal structure: rutile, anatase and brookite. All three polymorphs can be synthesized in the laboratory. Usually, the metastable anatase and brookite can be transformed into the thermodynamically stable rutile by calcination at temperatures higher than 600°C. Rutile and anatase are the most frequent and active phases either under visible and UV light (Pelaez et al. 2012). The most common commercial form, Evonik P25, contains these two polymorphs, with anatase being the most abundant.

Martin et al. (2015) reported that increasing the level of cerium doping in TiO_2 catalysts produces four main effects: (1) it increases the transition temperature for anatase to rutile phase transformation, (2) it decreases the crystallite size of the anatase phase, (3) it decreases the crystallinity of the solid, and (4) it increases the specific surface area.

The ionic radius of Ce^{3+} is 0.097 nm, which is much larger than that of Ti^{4+} (0.061 nm) but smaller than that of oxygen (0.14 nm). Hence, it is difficult for lanthanide ions to enter the TiO_2 lattice to replace Ti^{4+} ions (Nguyen-Phan et al. 2009), but it is conceivable that Ti–O–Ce bonds are formed around the anatase crystallites during the thermal treatment process; this possibly inhibits the formation and growth of the crystal nucleus of rutile (Liqiang et al. 2004). Moreover, as the content of Ce increases in the doped samples, the XRD peaks broaden, and their intensity slightly weakens, indicating a lower crystallinity. Lanthanide ions are known to be less reactive than the titania precursor species, which also slows down the condensation and crystallization processes of the titania matrix (Aman et al. 2012). For samples calcined at 600°C, the anatase crystallite size (estimated by the Scherrer equation using the XRD line broadening) decreases from 26.32 nm (for TiO_2 doped with 0.05 atomic percent of Ce) to 9.05 nm (for TiO_2 doped with 1.0 atomic percent of Ce). The presence of relatively large Ce^{3+} ions on the particle surfaces, at grain boundaries and grain junctions, could inhibit the crystallite growth of titania through the formation of Ti–O–Ce bonds, responsible for increasing the diffusion barrier at the titania grain junctions (Sibu et al. 2002). Moreover, the specific surface area increases from 4.75 m^2/g for the undoped sample to 57.71 m^2/g for the Ce-doped samples. Considering the XRD results of the Ce-doped sample, it can be concluded that the doping with Ce ions can decrease the crystallite size, which results in the larger specific surface area of the doped samples.

4.2 Optical properties

The bandgap energy (E_g) of the photocatalytic materials can be estimated using the following equation (Zhou et al. 2011):

$$(\alpha h\nu)^n = C\,(h\nu - E_g) \tag{1}$$

where $h\nu$ is the photon energy, α is the absorption coefficient that can be obtained from the scattering and reflectance spectra according to the Kubelka–Munk theory, C is a constant intrinsic of the material, n is the value that depends on the nature of the transition: 2 for a direct allowed transition, 3/2 for a direct forbidden transition and 1/2 for an indirect allowed transition. Thus, the E_g value is calculated by extrapolation at $\alpha = 0$ when representing $(\alpha h\nu)^n$ (with $n = 1/2$) versus $h\nu$.

As mentioned earlier, one of the goals of metal-doping of TiO_2 is to enable the absorption of visible light. Cerium incorporation to TiO_2 induces a red shift of the electronic absorption band that correlates with the Ce content of the catalyst. Figure 13.3 shows the absorption spectra of undoped and Ce-doped powdered TiO_2 prepared by the sol-gel method, evaluated by UV–Vis diffuse reflectance spectroscopy (DRS) (Martin et al. 2019).

The E_g value calculated for the undoped TiO_2 is 3.06 eV and it decreases with the Ce content down to 2.43 eV for the TiO_2 catalyst with 0.8 atomic percent of Ce (Martin et al. 2019). It has been suggested that, under visible illumination, Ce 4f levels can trap the photoexcited electrons from the valence band of Ce-TiO_2 (Liu et al. 2013, Meksi et al. 2015). The trapped electrons can be subsequently transferred to the surrounding adsorbed O_2, thus enhancing the generation and separation of electron-hole pairs under visible light.

When photocatalysts are immobilized as thin films over glass plates, the fraction of radiation that is effectively absorbed by the film can be calculated from measurements of diffuse transmittance (T) and reflectance (R) of the immobilized system (photocatalytic film + support) and the inert support alone, in a spectrophotometer equipped with an integrating sphere reflectance attachment (Martin et al. 2019).

The absorbed fraction of energy of the catalytic film, $\alpha_{f,\lambda}$, at wavelength λ can be computed as:

$$\alpha_{f,\lambda} = 1 - T_{f,\lambda} - R_{f,\lambda} \tag{2}$$

where $T_{f,\lambda}$ and $R_{f,\lambda}$ represent the fraction of energy transmitted and reflected by the catalytic film at wavelength λ, respectively. $T_{f,\lambda}$ and $R_{f,\lambda}$ cannot be directly measured from a film without support, but they can be calculated by Eqs. (3) and (4). The following expressions, obtained by applying the Net-Radiation method to the photocatalytic glass plates (Siegel and Howell 2002), relate the values

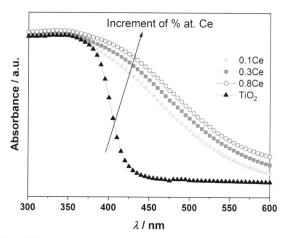

Figure 13.3. UV–Visible diffuse reflectance spectra (DRS) for the undoped and Ce-doped TiO_2.

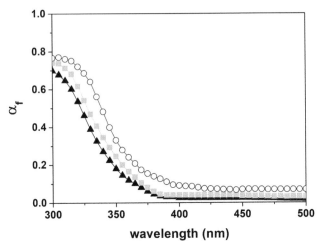

Figure 13.4. Absorbed fraction of energy by the TiO_2 catalytic films. Undoped: ▲; doped with 0.3 atomic percent Ce: ■; and 0.8 atomic percent Ce: ○.

of T and R of the catalytic film (*f*), the bare glass plate (glass, *g*) and the immobilized system (film + glass, *fg*):

$$T_{f,\lambda} = \frac{T_{fg,\lambda}}{T_{g,\lambda}}(1 - R_{f,\lambda}R_{g,\lambda}) \tag{3}$$

$$R_{f,\lambda} = \frac{R_{fg,\lambda}T_{g,\lambda}^2 - T_{fg,\lambda}^2 R_{g,\lambda}}{T_{g,\lambda}^2 - T_{fg,\lambda}^2 R_{g,\lambda}^2} \tag{4}$$

A detailed description of the methodology used in the calculation of the absorbed radiation by the films can be found in Zacarías et al. (2012).

As an example, the values of $\alpha_{f,\lambda}$ of undoped and Ce-doped TiO_2 thin films (140 nm thick), calculated with Eq. (2), are presented in Fig. 13.4.

The absorption in the UV region is expanded with the increase in the Ce content, being more evident for the 0.8 Ce sample (TiO_2 doped with 0.8 nominal atomic percent of Ce). Absorption of all samples decreases from maximum values at 300 nm to almost zero values in the visible region, except for the 0.8 Ce sample, which still exhibits a slight absorption beyond 400 nm.

5. Stability and Recovery

The effective separation of suspended photocatalysts and their reuse is still a practical challenge. One strategy involves the development of magnetic photocatalysts, which allows a quick and easy recovery of the catalyst from the treated water by using an external magnetic force (Ahmadpour et al. 2020, Narzary et al. 2020). Furthermore, the preparation of monodisperse Ce-doped TiO_2 microspheres with visible light photocatalytic activity and easy recovery from the treated water, as well as fine durability, has also been reported (Xie et al. 2010). All the Ce-doped TiO_2 microsphere samples showed higher photocatalytic activity than the undoped TiO_2.

Another approach that avoids the recovery step of catalyst particles in aqueous suspension is the immobilization of the photoactive material on different supports. For Ce-doped TiO_2 catalysts, the deposition was made on borosilicate plates (Martin et al. 2019), an indium-doped tin oxide glass (Poo-arporn et al. 2019) and glass substrates (Kayani et al. 2020). Although immobilized photocatalysts are easy to be separated and reused, this type of strategy implies a reduction in the efficiency of the materials.

Besides, the catalyst reuse has been studied by repeated photocatalytic experiments, calling them photocatalytic cycles, runs or rounds. In addition, the stability of the metal-doped catalysts has been analyzed in terms of the Ti and doping ions leached to the liquid phase, by using analytical techniques such as Inductively Coupled Plasma Mass Spectrometry (ICP-MS) and/or Atomic Absorption Spectroscopy (AAS). Martin et al. (2019) reported high stability of Ce-doped TiO$_2$ films in terms of metal leaching, and minor loss of activity after 12 hours of continuous use.

6. Photocatalytic Efficiency

As mentioned earlier, it is difficult to compare the performance of new photocatalytic materials with kinetic data reported by different research groups under different experimental conditions. An interesting alternative to standardize this information is to report efficiency parameters, which relate reaction rate data with the radiant energy provided to activate the photocatalyst.

The photonic efficiency parameter, η_{ph}, relates the photocatalytic reaction rate with the rate of incident radiation:

$$\eta_{ph} = \frac{\text{observed reaction rate}}{\text{rate of incident radiation}} \tag{5}$$

Besides, the quantum efficiency parameter, η_{rxn}, also called reaction efficiency, relates the photocatalytic reaction rate with the rate of photon (or radiation) absorption:

$$\eta_{rxn} = \frac{\text{observed reaction rate}}{\text{rate of photon absorption}} \tag{6}$$

The numerator of Eqs. (5) and (6) can be obtained from degradation experiments of a model pollutant. The rate of incident radiation, denominator of Eq. (5), can be experimentally measured by chemical actinometry or with a spectroradiometer. On the other hand, the rate of photon absorption, denominator of Eq. (6), usually involves the measurement of the optical properties of the absorbing material and the resolution of radiation models, taking into account the phenomena of absorption, reflection and scattering. Although η_{rxn} is more difficult to compute than η_{ph}, the former provides more information about the performance of new photocatalytic materials. For example, pollutant degradation experiments under the same operating conditions but using a photocatalyst with two different types of dopants (sample A and sample B) may present the same value of photonic efficiency, but the experiment with sample A may render a lower quantum efficiency. This indicates that sample A absorbs more energy than sample B but the use of this energy is not employed for the chemical reaction and it is probably lost as heat. Therefore, instead of improving the ability of sample A to absorb energy, research efforts should be conducted to reduce the undesirable recombination of charge carriers.

Another interesting parameter to characterize the ability of photocatalysts to absorb the incident radiation is the absorption efficiency:

$$\eta_{abs} = \frac{\text{rate of photon absorption}}{\text{rate of incident radiation}} \tag{7}$$

In what follows, the procedure to compute the quantum efficiency in reactors with the catalyst immobilized over a surface as thin films will be given. η_{rxn} takes the following form:

$$\eta_{rxn} = \frac{\langle r^s \rangle A_{\text{cat}}}{\langle e_f^{a,s} \rangle A_{\text{cat}}} \tag{8}$$

where $\langle r^s \rangle A_{\text{cat}}$ is the surface reaction rate averaged over the catalytic area, A_{cat}, and $\langle e_f^{a,s} \rangle A_{\text{cat}}$ is the surface rate of photon absorption, also averaged over A_{cat}. Moreover, $\langle e_f^{a,s} \rangle A_{\text{cat}}$ can be expressed as:

$$\langle e_f^{a,s} \rangle A_{\text{cat}} = \langle q_{f,in} \rangle A_{\text{w}} \, \Sigma_\lambda (\alpha_{f,\lambda} \, \varphi_\lambda) \tag{9}$$

where $\langle q_{f,in} \rangle A_w$ represents the incident radiation flux averaged over the irradiated window area A_w; $\alpha_{f,\lambda}$, as defined earlier, is the fraction of energy absorbed by the photocatalytic film; φ_λ is the normalized fraction of the radiation that reaches the film at wavelength λ; and $\Sigma_\lambda(\alpha_{f,\lambda} \varphi_\lambda)$ is the summation of the product of both quantities over the wavelength range of interest.

Another interesting approach to implement a standard comparison among photocatalysts is the use of microreactors. The advantages of employing microreactors instead of conventional laboratory-scale reactors for catalyst assessment include: more accurate and easy control of operation variables, more uniform irradiation, lower consumption of reagents, minimal waste generation and shorter test times. Next, an example of the use of a microreactor and the calculation of efficiency parameters to compare the performance of cerium doped TiO_2 is presented.

7. Applications

TiO_2 films, undoped and doped with different amounts of Ce, were tested for the degradation of 17-α-ethinylestradiol (EE2) in an aqueous solution using a microreactor under simulated solar radiation (Martin et al. 2019). EE2 is a synthetic hormone that is extensively used in formulations of oral contraceptive pills and hormone replacement therapy, and it has been detected in sewage effluents and natural waters throughout the world.

The microreactor used in the experiments is shown in Fig. 13.5. It has a planar reaction chamber of 200 µL and the photocatalyst is immobilized as a thin film on one side of a glass plate, which acts as the reactor window.

The photocatalysts were synthesized by a standard sol-gel method employing titanium tetraisopropoxide, cerium (III) nitrate hexahydrate and absolute ethanol. Immobilization of the photocatalyst over the glass plates was carried out by the dip-coating technique.

The absorption efficiency and the quantum efficiency parameters for the undoped and cerium-doped catalysts are reported in Table 13.1.

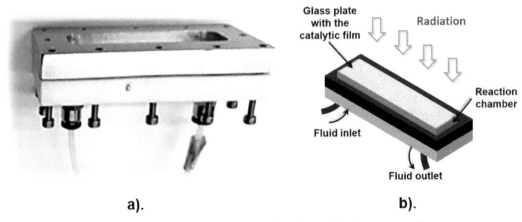

a).   b).

Figure 13.5. (a) Photograph of the assembled reactor. (b) Scheme of the microreactor main components.

Table 13.1. Efficiency parameters of undoped and Ce-doped photocatalytic films.

Ce content of the TiO_2 film (at.%)	η_{abs} (%)	$\eta_{rxn} \times 10^3$
0	5.8	1.31
0.1	6.5	1.36
0.3	7.7	1.42
0.8	12.6	0.70

Films with 0.1 and 0.3 atomic percent of Ce present a slight enhancement of the radiation absorption compared with pure TiO_2 films. The film with 0.8 atomic percent of Ce presents the most significant improvement, rendering a η_{abs} value greater than twice the η_{abs} value of pure TiO_2. Nevertheless, the film with the highest amount of dopant gave the lowest quantum efficiency, reaching almost half the η_{rxn} value of the undoped TiO_2 film. The low performance of the film with 0.8 atomic percent of Ce is not due to the lack of absorption but to the inefficient use of the absorbed energy for the degradation reaction. This effect can be attributed to the presence of Ce^{3+}/Ce^{4+} pairs in the catalyst. These species reduce the recombination of electrons and holes up to a certain concentration but when the optimal metal concentration is surpassed, the Ce^{3+}/Ce^{4+} pairs can act as recombination centers, and the efficiency of the reaction decreases. Consequently, the optimal dopant concentration of a photocatalyst, for a certain reaction, has to be found experimentally and cannot be predicted only by the absorption properties of the material. The highest quantum efficiency for EE2 degradation was obtained with TiO_2 doped with 0.3 atomic percent of Ce, indicating that the absorbed photons in this material are more efficiently employed for the chemical reaction.

8. Conclusions

Recent publications indicate significant progress in the field of metal-doped TiO_2-based photocatalysts towards a large-scale application of this type of materials. However, the strategies employed to improve the efficiency of photocatalysts also have limitations:

- The shift of the absorption spectrum of TiO_2 towards the visible region can be achieved by metal doping. Nevertheless, there is always an optimal level of the doping agent because an excess of dopant can decrease the performance of the catalyst, thus reducing the possibility of practical application. In particular, when TiO_2 is doped with the optimal amount of cerium, the radiation absorption of the photocatalyst increases, and the degradation of water pollutants is enhanced. On the contrary, when the optimal cerium concentration is surpassed, the activity of the photocatalyst is even lower than that of the undoped TiO_2.

- Immobilization of catalysts on inert supports facilitates their separation and reuse but reduces the access of active sites for the reaction of interest, rendering lower reaction rates when compared to suspended catalysts.

- The evaluation of the optical parameters of undoped and doped TiO_2 materials and mainly, the subsequent determination of the quantum efficiency is an important tool to compare and select the optimal catalyst for the photocatalytic process.

Therefore, the synthesis, preparation and evaluation of heterogeneous TiO_2-based photocatalysts with a wide absorption spectrum and high efficiency for real-scale application of water and wastewater treatment remains a relevant area of research.

Acknowledgments

The authors are grateful to Universidad Nacional del Litoral (UNL, CAI+D 2020 50620190100040LI), Consejo Nacional de Investigaciones Científicas y Técnicas (CONICET, PIP 2015 112 201501 00093), and Agencia Nacional de Promoción Científica y Tecnológica (ANPCyT, PICT 2018-0926) for financial support.

References

Ahmadpour, N., M.H. Sayadi, S. Sobhani and M. Hajiani. 2020. Photocatalytic degradation of model pharmaceutical pollutant by novel magnetic TiO_2@$ZnFe_2O_4$/Pd nanocomposite with enhanced photocatalytic activity and stability under solar light irradiation. J. Environ. Manage. 271: 110964.

Aman, N., P.K. Satapathy, T. Mishra, M. Mahato and N.N. Das. 2012. Synthesis and photocatalytic activity of mesoporous cerium doped TiO_2 as visible light sensitive photocatalyst. Mater. Res. Bull. 47: 179–183.

Anpo, M., Y. Ichihashi, M. Takeuchi and H. Yamashita. 1999. Design and development of unique titanium oxide photocatalysts capable of operating under visible light irradiation by an advanced metal ion-implantation method. Sci. Technol. Catal. 121: 305–310.

Banerjee, A.N., N. Hamnabard and S.W. Joo. 2016. A comparative study of the effect of Pd-doping on the structural, optical, and photocatalytic properties of sol-gel derived anatase TiO_2 nanoparticles. Ceram. Intern. 42: 12010–12026.

Chang, S-m. and W-s. Liu. 2014. The roles of surface-doped metal ions (V, Mn, Fe, Cu, Ce, and W) in the interfacial behavior of TiO_2 photocatalysts. Appl. Catal. B: Environ. 156–157: 466–475.

Choi, J., H. Park and M.R. Hoffmann. 2010. Effects of single metal-ion doping on the visible-light photoreactivity of TiO_2. J. Phys. Chem. C. 114: 783–792.

Dong, H., G. Zeng, L. Tang, C. Fan, C. Zhang, X. He and Y. He. 2015. An overview on limitations of TiO_2-based particles for photocatalytic degradation of organic pollutants and the corresponding countermeasures. Water Res. 79: 128–146.

Falaras, P. and A.P. Xagas. 2002. Roughness and fractality of nanostructured TiO_2 films prepared via sol-gel technique. J. Mater. Sci. 37: 3855–3860.

Gaya, U.I. and A.H. Abdullah. 2008. Heterogeneous photocatalytic degradation of organic contaminants over titanium dioxide: a review of fundamentals, progress and problems. Photochem. Photobiol. C. 9: 1–12.

Gopinath, K.P., N.V. Madhav, A. Krishnan, R. Malolan and G. Rangarajan. 2020. Present applications of titanium dioxide for the photocatalytic removal of pollutants from water: A review. J. Environ. Manag. 270: 110906.

Grabowska, E., J. Reszczynska and A. Zaleska. 2012. Mechanism of phenol photodegradation in the presence of pure and modified-TiO_2: a review. Water Res. 46: 5453–5471.

Ibrahim, N.S., W.L. Leaw, D. Mohamad, S.H. Alias and H. Nur. 2020. A critical review of metal-doped TiO_2 and its structure-physical properties-photocatalytic activity relationship in hydrogen production. Inter. J. Hydrog. Energy. 45: 28553–28565.

Kayani, Z.N., M.S. Riaz and S. Naseem. 2020. Magnetic and antibacterial studies of sol-gel dip-coated Ce doped TiO_2 thin films: Influence of Ce contents. Ceram. Int. 46: 381–390.

Li, F.B., X.Z. Li, M.F. Hou, K.W. Cheah and W.C.H. Choy. 2005. Enhanced photocatalytic activity of Ce^{3+}–TiO_2 for 2-mercaptobenzothiazole degradation in aqueous suspension for odour control. Appl. Catal. A: Gen. 285: 181–189.

Liqiang, J., S. Xiaojun, X. Baifu, W. Baiqi, C. Weimin and F. Honggang. 2004. The preparation and characterization of La doped TiO_2 nanoparticles and their photocatalytic activity. J. Solid State Chem. 177: 3375–3382.

Litter, M.I. and J.A. Navío. 1996. Photocatalytic properties of iron-doped titania semiconductors. J. Photochem. Photobiol. A. 98: 171–181.

Litter, M.I., E. San Román, M.A. Grela, J.M. Meichtry and H.B. Rodríguez. 2018. Sensitization of TiO_2 by dyes: a way to extend the range of photocatalytic activity of TiO_2 to the visible region. pp. 255–282. *In*: Ghosh, S. (ed.). Visible-light-active Photocatalysis: Nanostructured Catalyst Design, Mechanisms, and Applications. Wiley-VCH Verlag GmbH & Co. KGaA.

Liu, Y., P. Fang, Y. Cheng, Y. Gao, F. Chen, Z. Liu and Y. Dai. 2013. Study on enhanced photocatalytic performance of cerium doped TiO_2-based nanosheets. Chem. Eng. J. 219: 478–485.

Long, M., J. Brame, F. Qin, J. Bao, Q. Li and J.J. Alvarez. 2017. Phosphate changes effect of humic acids on TiO_2 photocatalysis: from inhibition to mitigation of electron-hole recombination. Environ. Sci. Technol. 51: 514–521.

Maarisetty, D. and S.S. Baral. 2019. Defect-induced enhanced dissociative adsorption, optoelectronic properties and interfacial contact in Ce doped TiO_2: Solar photocatalytic degradation of Rhodamine B. Ceram. Int. 45: 22253–22263.

Martin, M.V., G.T. Ruiz, M.C. Gonzalez, C.D. Borsarelli and D.O. Mártire. 2012. Photolytic and radiolytic oxidation of humic acid. Photochem. Photobiol. 88: 810–815.

Martin, M.V., P.I. Villabrille and J.A. Rosso. 2015. The influence of Ce doping of titania on the photodegradation of phenol. Environ. Sci. Pollut. Res. 22: 14291–14298.

Martin, M.V., M.O. Alfano and M.L. Satuf. 2019. Cerium-doped TiO_2 thin films: Assessment of radiation absorption properties and photocatalytic reaction efficiencies in a microreactor. J. Environ. Chem. Eng. 7: 103478.

Meksi, M., H. Kochkar, G. Berhault and C. Guillard. 2015. Effect of cerium content and post thermal treatment on doped anisotropic TiO_2 nanomaterials and kinetic study of the photodegradation of formic acid. J. Mol. Catal. A: Chem. 409: 162–170.

Mignone, R.A., M.V. Martin, F.M. Vieyra, V.I. Palazzi, B.L. de Mishima, D.O. Martiré and C.D. Borsarelli. 2012. Modulation of optical properties of dissolved humic substances by their molecular complexity. Photochem. Photobiol. 88: 792–800.

Morikawa, T., Y. Irokawa and T. Ohwaki. 2006. Enhanced photocatalytic activity of $TiO_{2-x}N_x$ loaded with copper ions under visible light irradiation. Appl. Catal. A: Gen. 314: 123–127.

Murakami, N., A. Ono, M. Nakamura, T. Tsubota and T. Ohno. 2010. Development of a visible-light-responsive rutile rod by site-selective modification of iron (III) ion on {1 1 1} exposed crystal faces. Appl. Catal. B: Environ. 97: 115–119.

Narzary, S., K. Alamelu, V. Raja and B.M. Jaffar Ali. 2020. Visible light active, magnetically retrievable $Fe_3O_4@SiO_2@g$-C_3N_4/TiO_2 nanocomposite as efficient photocatalyst for removal of dye pollutants. J. Environ. Chem. Eng. 8: 104373.

Nguyen-Phan, T.-D., M.B. Song, E.J. Kim and E.W. Shin. 2009. The role of rare earth metals in lanthanide-incorporated mesoporous titania. Micropor. Mesopor. Mat. 119: 290–298.

Niishiro, R., H. Kato and A. Kudo. 2005. Nickel and either tantalum or niobium-codoped TiO_2 and $SrTiO_3$ photocatalysts with visible-light response for H_2 or O_2 evolution from aqueous solutions. Phys. Chem. Chem. Phys. 7: 2241–2245.

Palacio, M., L. Rossi, M. Estefanía, F. Hermosilla, J.A. Rosso, P.I. Villabrille and M.V. Martin. 2017. Selective photodegradation of phenol in the presence of a commercial humic acid. J. Environ. Chem. Eng. 5: 5540–5546.

Palmisano, L., M. Schiavello, A. Sclafani, C. Martin, I. Martin and V. Rives. 1994. Surface properties of iron-titania photocatalysts employed for 4-nitrophenol photodegradation in aqueous TiO_2 dispersion. Catal. Lett. 24: 303–315.

Park, H., Y. Park, W. Kim and W. Choi. 2013. Surface modification of TiO_2 photocatalyst for environmental applications. J. Photochem. and Photobiol. C 15: 1–20.

Pelaez, M., N.T. Nolan, S.C. Pillai, M.K. Seery, P. Falaras, A.G. Kontos, P.S.M. Dunlop, J.W.J. Hamilton, J.A. Byrne, K. O'Shea, M.H. Entezari and D. Dionysiou. 2012. A review on the visible light active titanium dioxide photocatalysts for environmental applications. Appl. Catal. B: Environ. 125: 331–349.

Poo-arporn, Y., S. Kityakarn, A. Niltharach, M.F. Smith, S. Seraphin, M. Wörner and A. Worayingyong. 2019. Photocatalytic oxidation of thiophene over cerium doped TiO_2 thin film. Mater. Sci. Semicond. Process. 93: 21–27.

Ranjit, K.T., I. Willner, S.H. Bosmann and A.M. Braun. 2001. Lanthanide oxide-doped titanium dioxide photocatalysts: novel photocatalysts for the enhanced degradation of *p*-chlorophenoxyacetic acid. Environ. Sci. Technol. 35: 1544–1549.

Rodríguez, J., R.J. Candal, J. Solís, W. Estrada and M.A. Blesa. 2005. El fotocatalizador: síntesis, propiedades y limitaciones. pp. 135–152. *In*: Blesa, M. and Blanco Gálvez, Solar Safe Water. La Plata, Argentina: ByToner.

Rossi, L., M. Palacio, P.I. Villabrille and J.A. Rosso. 2021. V-doped TiO_2 photocatalysts and their application to pollutant degradation. Environ. Sci. Pollut. Res. (in press).

Santiago-Morales, J., A. Agüera, M. Gómez, A.R. Fernández-Alba, J. Giménez, S. Esplugas and R. Rosal. 2013. Transformation products and reaction kinetics in simulated solar light photocatalytic degradation of propranolol using Ce-doped TiO_2. Appl. Catal. B: Environ. 129: 13–29.

Sibu, C.P., S. Rajesh Kumar, P. Mukundan and K.G.K. Warrier. 2002. Structural modifications and associated properties of lanthanum oxide doped sol-gel nanosized titanium oxide. Chem. Mater. 14: 2876–2881.

Siegel, R. and J. Howell. 2002. Thermal Radiation Heat Transfer, fourth ed., Taylor and Francis, New York.

Silva, A.M.T., C.G. Silva, G. Dražić and J.L. Faria. 2009. Ce-doped TiO_2 for photocatalytic degradation of chlorophenol. Catal. Today. 144: 13–18.

Vera, M.L., G. Leyva and M.I. Litter. 2017. Simple TiO_2 coatings by sol-gel techniques combined with commercial TiO_2 particles for use in heterogeneous photocatalysis. J. Nanosci. Nanotechnol. 17: 4946–4954.

Xie, J., D. Jiang, M. Chen, D. Li, J. Zhu, X. Lü and C. Yan. 2010. Preparation and characterization of monodisperse Ce-doped TiO_2 microspheres with visible light photocatalytic activity. Colloids Surf. A Physicochem. Eng. Asp. 372: 107–114.

Xu, Y., H. Chen, Z. Zeng and B. Lei. 2006. Investigation on mechanism of photocatalytic activity enhancement of nanometer cerium-doped titania. Appl. Surf. Sci. 252: 8565–8570.

Yanagisawa, K. and J. Ovenstone. 1999. Crystallization of anatase from amorphous titania using the hydrothermal technique: effects of starting material and temperature. J. Phys. Chem. B. 103: 7781–7787.

Zacarías, S.M., M.L. Satuf, M.C. Vaccari and O.M. Alfano. 2012. Efficiency evaluation of different TiO_2 coatings on the photocatalytic inactivation of airborne bacterial spores. Ind. Eng. Chem. Res. 51: 13599–13608.

Zhang, X.W. and L.C. Lei. 2008. One step preparation of visible-light responsive Fe–TiO_2 coating photocatalysts by MOCVD. Mater. Lett. 62: 895–897.

Zhou, G., H. Sun, S. Wang, H. Ming Ang and M.O. Tadé. 2011. Titanate supported cobalt catalysts for photochemical oxidation of phenol under visible light irradiations. Sep. Purif. Technol. 80: 626–634.

Chapter 14

Advances in Nanostructured TiO$_2$ Coatings for Industrial Applications

María Laura Vera,* Hernán Darío Traid, Anabela Natalia Dwojak, Mario Roberto Rosenberger and *Carlos Enrique Schvezov*

1. Introduction

Titanium is a very important element for several industrial applications, being one of the nine most abundant elements in the Earth's crust (0.63 wt.%). The most important mineral sources are ilmenite (FeTiO$_3$) and rutile (TiO$_2$), with increasing annual production of 6,790 and 790 metric tons, respectively, in 2011, with a contained TiO$_2$ known reserve of 750,000 metric tons, mostly ilmenite (MAM 2021). Approximately 95% of titanium is consumed in the form of titanium dioxide (TiO$_2$), mostly as a white pigment in paints, paper, plastics, printing inks, pharmaceuticals, cosmetics, textiles, food industry (food coloring E171), in all cases due to its very low toxicity (Gázquez et al. 2014, USGS 2021). TiO$_2$ is one of the most widely used nanomaterials in our daily lives (Liu et al. 2014a) in such a way that its level of consumption is considered a "quality-of-life" product (Khataee and Mansoori 2012). Moreover, the actual research capabilities have permitted new technological applications of TiO$_2$ due to its semiconducting and other relevant properties at the nanoscale level.

In the present chapter, a broad overview of the industrial applications of TiO$_2$ in energy, construction, air and water purification, biomedical devices and food packaging will be presented. Moreover, the problems to be overcome as well as new questions intrinsically associated with the nanoscale nature of the new applications are mentioned.

2. Characteristics of TiO$_2$

TiO$_2$, usually known as titania, is an n-type semiconductor. It functions as a photoactive material when irradiated with a photon with energy equal to or greater than that of the semiconductor bandgap (E_g), which induces the promotion of an electron from the Valence Band (VB) to the Conduction Band (CB), e$_{CB}^-$, creating positive holes in the VB (h$_{VB}^+$). The charge carriers, h$_{VB}^+$ and e$_{CB}^-$, that successfully migrate to the surface of the semiconductor without recombining can be involved in oxidation and reduction reactions, respectively, generating highly oxidizing or reducing free radicals. The holes can form hydroxyl radicals (HO$^•$), while e$_{CB}^-$ can produce superoxide radicals (O$_2^{•-}$) that,

Instituto de Materiales de Misiones, CONICET, Universidad Nacional de Misiones, Facultad de Ciencias Exactas, Químicas y Naturales, Félix de Azara 1552, 3300 Posadas, Prov. de Misiones, Argentina.
Emails: traidhernan@gmail.com; anabelanataliadwojak@gmail.com; rrmario@gmail.com; schvezov@gmail.com
* Corresponding author: lauravera@fceqyn.unam.edu.ar, veramalau@gmail.com

in combination with H^+, produce hydroperoxyl radicals ($HO_2^{\bullet}$). Hydrogen peroxide is also formed, which leads to the formation of $HO^{\bullet}$ (Eqs. (1)–(11)) (Bahnemann et al. 2007, Litter 2017).

$$TiO_2 + h\nu \rightarrow h_{VB}^+ + e_{CB}^- \tag{1}$$

$$h_{VB}^+ + e_{CB}^- \rightarrow TiO_2 \tag{2}$$

$$e_{CB}^- + A \rightarrow A^{\bullet -} \tag{3}$$

$$h_{VB}^+ + D_{ads} \rightarrow D_{ads}^{\bullet +} \tag{4}$$

$$h_{VB}^+ + OH_{surf}^- (H_2O_{ads}) \rightarrow HO_{surf}^{\bullet} (+ H^+) \tag{5}$$

$$e_{CB}^- + O_2 \rightarrow O_2^{\bullet -} \tag{6}$$

$$2\ HO_{surf}^{\bullet} \rightarrow H_2O_2 \tag{7}$$

$$H_2O_2 + h_{VB}^+ \rightarrow HO_2^{\bullet} + H^+ \tag{8}$$

$$O_2^{\bullet -} + H^+ \rightarrow HO_2^{\bullet} \tag{9}$$

$$2\ HO_2^{\bullet} \rightarrow O_2 + H_2O_2 \tag{10}$$

$$H_2O_2 + e_{CB}^- \rightarrow HO^{\bullet} + OH^- \tag{11}$$

Based on its good properties, TiO_2 is the main semiconductor used as a photocatalyst in heterogeneous photocatalysis processes and consequently, one of the most studied for disinfection and detoxification of air, water and wastewater. As excellent oxidizers, the photogenerated holes can mineralize organic pollutants directly. In addition, $HO^{\bullet}$, which has a strong oxidizing ability, also induces the degradation of organic compounds (R) to less or non-toxic substances like CO_2 and H_2O (Eq. (12)), and the transformation of some toxic species (X) into less harmful ones (Eq. (13)).

$$R + h_{VB}^+/HO^{\bullet} \rightarrow intermediates \rightarrow CO_2 + H_2O \tag{12}$$

$$X^{n+} + h_{VB}^+/HO^{\bullet} \rightarrow intermediates \rightarrow X^{(n+1)+} \tag{13}$$

On the other hand, the reduction of highly toxic metal or metalloid ions (M) could take place directly by photogenerated electrons (Eq. (14)) or indirectly by pathways mediated by free radicals ($R^{\bullet}$) (Eq. (15)) (Litter 1999, 2017).

$$M^{n+} + e_{CB}^- \rightarrow M^{(n-1)} \tag{14}$$

$$R^{\bullet} + M^{n+} \rightarrow R_{ox} + M^{(n-1)+} \tag{15}$$

Moreover, in regards to water splitting, photogenerated electrons can be captured by H^+ in water to generate hydrogen (Eq. (16)), whereas holes will oxidize H_2O to form O_2 (Eq. (17)) (Kang et al. 2019).

$$4\ H^+ + 4\ e_{CB}^- \rightarrow 2\ H_2 \tag{16}$$

$$2\ H_2O + 4\ h_{VB}^+ \rightarrow O_2 + 4\ H^+ \tag{17}$$

TiO_2 is a multifaceted material depending mainly on its polymorphous structure. At low pressure, there are three crystalline phases of TiO_2: rutile (tetragonal), anatase (tetragonal) and brookite (orthorhombic). Rutile is the stable phase; however, anatase and brookite exist in a metastable form as small (nanocrystalline) particles in synthetic and natural samples (Diebold 2003). Through a polymorphic transformation, anatase and brookite are converted to rutile; these transformations are irreversible (Murray 1990).

Anatase is the most photoactive phase of TiO_2, followed by rutile; however, the wide bandgap value of both phases (3.0 and 3.2 eV, respectively) results in the activation of these materials under

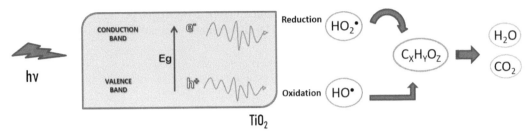

Figure 14.1. Scheme of the removal of organic pollutants on the surface of TiO_2 by photoinduced charge carriers (e_{CB}^{-}/h_{VB}^{+}).

UVA irradiation (Fig. 14.1). Many research efforts are devoted to producing TiO_2 photoactive in the visible part of the spectrum to employ solar light.

Another property of TiO_2 that is fundamental to these increasing applications is its unique photoinduced wettability. TiO_2 surfaces are reversible between superhydrophilicity and hydrophobicity under the alternation of UV irradiation and long-term storage in the dark. The switchable wettability can be observed on both anatase and rutile TiO_2 surfaces in the form of either single crystals or polycrystals, independent of their photocatalytic activities (Liu et al. 2014a). For the metal oxides (TiO_2, SnO_2, ZnO, WO_3 and V_2O_5), the resultant photoinduced behavior between superhydrophilicity and superhydrophobicity is mainly attributed to structural changes at metal oxide surfaces (Liu et al. 2014a).

Ninety percent of the production of titanium in the world is used for the manufacture of TiO_2, which is mainly used for the paint industry and only 10% is used for the manufacture of metal, with 50% of metallic titanium for the manufacture of Ti-6Al-4V alloy (Lütjering and Williams 2007).

The high mechanical resistance of titanium alloys makes them good as structural material. In addition, the high reactivity of titanium with oxygen passivates the titanium, forming a thin and compact layer of titanium oxide on the surface that gives it resistance to corrosion (Lütjering and Williams 2007). This natural layer formed by oxidation in air is very thin, about 5 to 10 nm and can have structural defects, generally microcracks that decrease its protective capacity. A strategy to improve the resistance is to increase the thickness of the titanium oxide layer, mainly by TiO_2 coating fabrication (Lütjering and Williams 2007).

There are many methods of synthesis for the fabrication of nanoscale TiO_2 particles. Some of them are sol-gel, hydrothermal methods, sonochemical methods, microwave-assisted methods, water-assisted techniques, evaporation-induced self-assembly processes and flame spray pyrolysis, among others (Noman et al. 2018, and references therein). There are many reports for production at the laboratory scale but there are some for large-scale production, which generally is protected by patents.

On the other hand, there are numerous methods to produce TiO_2 coatings, such as deposition or diffusion or a combination of both. Some of them are thermal oxidation, anodic oxidation, magnetron sputtering, cathodic arc deposition, plasma immersion ion implantation, ion beam-enhanced deposition, sol-gel process, chemical vapor deposition, physical vapor deposition, electrodeposition, electrophoresis, laser ablation, plasma spray, sputtering, etc. In addition, to produce coatings or compound films of TiO_2 from a suspension nanostructure, deposition techniques are used, e.g., dip-coating, spin-coating and direct immersion of the substrate in the TiO_2 suspension (Vera et al. 2017b, Vera et al. 2020).

The choice of the method and the corresponding synthesis parameters depend on the surface requirements according to the type of application. There is no general theory for the process selection and the performance of a given coating must be proven under conditions as close as possible to the final application (Schvezov et al. 2010).

3. Innovations on the Nanoscale TiO₂ for Industrial Applications

3.1 Energy field

Research on TiO_2 nanomaterials and compounds has given rise to potential energy applications for Li-ion batteries, solar cells and photocatalytic conversion of CO_2 in CH_4, and H_2 production. The application of TiO_2 in these technologies is based on two of the main characteristics of titania: the semiconducting property of the crystalline structures and the versatility of structures, morphologies, geometries of pure TiO_2, doped or compounds obtained with different processes.

3.1.1 Li-ion batteries

Since the first lithium-cathode material commercialized in 1991 (Mizushima et al. 1980), the cell capacity increased by 0.12 Ah yr^{-1}[1] (Brodd 2002, Sauer et al. 2015). At present, the practical specific capacity of the different Li-based cathodes varies between 110 and 280 mAh g^{-1}[2] with the perspective and potential for innovations in cathode technology (Armand et al. 2020).

After the replacement of lithium metal by carbonaceous compounds, graphite became the industrially dominating anode material. However, in the last decade, other materials have been investigated as active anodes. One of them is TiO_2, which is regarded as one of the ideal candidates due to its nontoxicity, low cost, small volume expansion and safe lithiation potential (potential range at that Li$^+$ storage occurs) (Myung et al. 2011, Tang et al. 2014). The use of TiO_2 must overcome other less good properties like its intrinsically low electronic conductivity and low lithium-ion diffusion coefficient, which limits its practical application. One way of overcoming these limitations is to produce hierarchical structures (Ren et al. 2013) and, in this way, shorten the distances for both, electronic and lithium-ion transport and also enhance the stress-strain relief in repetitive lithium-ion insertion/extraction reactions (Li et al. 2012a, Xu et al. 2012). With this approach, chrysanthemum-like TiO_2 nanostructures were produced with surface-folded nanorods showing a capacity of up to 309.3 mAh g^{-1} at a current rate of 0.5 C (1 C = 170 mA g^{-1}) for the first cycle, with a high coulombic efficiency of 93.4% and good cycling stability (Wang et al. 2015). This unique structure simultaneously combines the advantages of the porous structure and the hierarchical structure conferring a large surface-to-volume ratio. This can increase the lithium-ion flux at the electrode-electrolyte interaction, provide a shorter pathway for lithium-ion transport and allow the easy diffusion of the electrolyte inside the TiO_2, leading to an increase in the reactive sites and the electrolyte-electrode interface. In this way, the electrochemical reaction is easier, and the stress-strain release in repetitive lithiation and delithiation reaction is enhanced, achieving a long cycle life.

Another innovative strategy consists of doping. Doping TiO_2 with niobium at high dopant contents (21 at.%), a specific capacity of 58 mAh g^{-1} at 10 C rate is achieved compared to 28, 50, and 40 mAh g^{-1} for pure TiO_2, 9 and 33% Nb-doped, respectively. Nb modifies the electronic structure and creates additional sites for Li$^+$ reaction improving the rate capability and cycling stability. Doping with 2 wt.% vanadium (V^{5+}) increases the reversible specific capacity and rate capability compared with the undoped case. This improvement was attributed to the substitution of the Ti^{4+} ions by V^{5+} ions in the TiO_2 lattice and the creation of more Ti^{4+} vacancies in the lattice, which contributes to the electrical conductivity of the doped sample. Vanadium doping also influences the crystallinity and reduces the particle size, which provides a larger active surface area than that of undoped TiO_2 (Anh et al. 2013).

Copper-doped TiO_2 bronze nanowires have been produced with around 2.0 at.% Cu^{2+}, which leads to a slightly expanded lattice network and a modified electronic structure. The bandgap is reduced from 2.94 to 2.55 eV, resulting in an enhanced electronic conductivity. The doped nanowires

---

[1] Ah yr^{-1}: ampere hour per year.

[2] mAh g^{-1}: milliampere hour per gram.

show a specific capacity of 186.8 mAh g⁻¹ at the 10 C rate with a capacity retention of 64.3% after 2,000 cycles and a specific capacity of 150 mAh g⁻¹ at the 60 C rate.

A new approach to improve the capability, cycling stability and rate capacity is the encapsulation of Mo-doped TiO_2 anatase in nitrogen-doped amorphous carbon with improvements in the initial discharge and charge capacity. The good performance was attributed to the synergistic effect of doping with aliovalent ions and carbon coating (Xia et al. 2019).

Rutile TiO_2 mesocrystals/reduced graphene oxide nanosheets hybrids were produced using a hydrothermal route, obtaining a material composed of ultra-tiny rod-like subunits with the same oriented direction and closely wrapped by the nanosheets of reduced graphene oxide. As an anode material, it exhibited a large capacity of over 150 mAh g⁻¹ at 20 C after 1000 cycles. This high performance was attributed to the intrinsic characteristics of rutile mesocrystals (Lan et al. 2015).

An important line of research is based on the use of aligned nanotubes, amorphous or with the anatase phase obtained by heat treatment. The storage capacity of lithium-ion in amorphous TiO_2 nanotubes is significantly larger than that of the anatase-phase nanotubes. This indicates that the amorphous TiO_2 nanoparticulate anodes can accommodate more Li per TiO_2 molecule than the anatase anode (Zhang et al. 2021). The morphology and size of the nanotubes may have effects on the performance as anode.

Improvements in the specific capacity of the TiO_2 nanotubes can also be achieved by doping the nanotubes and producing TiO_2 composites (Yan et al. 2015, Zhang et al. 2021). Doping is done by different methods using elements such as C (Kim et al. 2014), N (Zhang et al. 2017a), Sn (Weibel et al. 2006 and Fresno et al. 2009) and Nb (Salian et al. 2018). The TiO_2-based composites employ high-conductivity materials like carbon nanomaterial and graphene (Huang et al. 2017, Shen et al. 2011) and metal materials like tin (Yan et al. 2015).

3.1.2 Solar energy harvesting

Solar energy emerges as an alternative to contribute to the challenge of human development for producing usable energy with minimum deterioration of our ecosystems. TiO_2 is an emerging material for a wide range of energy-conversion applications and, at the nanoscale, it has gained attention during the last three decades as an alternative for building photovoltaic solar cells and photoelectrodes for the generation of hydrogen fuel. However, TiO_2 suffers from an inefficient utilization of solar light due to its wide bandgap (such as, 3.2 eV and 3.0 eV for anatase and rutile phases, respectively), its optical response being limited to the UV component (4%) of the solar power spectrum (Gong et al. 2012). One of the lines of development of solar cells using titania is the Dye-Sensitized Solar Cells (DSSC) consisting of a sensitizing dye, a transparent conducting substrate (F-doped tin oxide), a nanometer-sized TiO_2 film, iodide electrolyte and a counter electrode (Pt or carbon). The dye molecules absorb light that excites electrons into the CB of the TiO_2 film. The oxidized dye is restored by electron donation from reducing ions in the electrolyte. These donated electrons are regenerated by the reduction in the electrolyte (Bai et al. 2014, Grätzel 2005, Nowotny 2012). The use of amorphous TiO_2 nanoparticles increases the energy conversion efficiency of the DSSC to 8.71% with the N719 dye under 100 mW cm⁻² simulated light (Aboulouard et al. 2020, Lee and Kang 2010). With TiO_2 nanoparticles synthesized by Flame Spray Pyrolysis (FSP) and hydrothermal sol-gel methods (HT), the photovoltaic performances were η = 2.44, 3.94 and 7.67% with FSP powder, HT powder and an electrode made with commercial titania, respectively under standard conditions, AM 1.5 G,[3] 100 mW cm⁻².

A different approach for building solar cells has been patented, consisting of an array of TiO_2 nanotubes, each nanotube including an outer layer coaxial with an inner layer, where the inner layer comprises p-type TiO_2 and the outer layer comprises n-type TiO_2. An interface between the inner layer and the outer layer defines a p-n junction. It is claimed that the concept takes advantage

---

[3] AM 1.5 G: Air Mass 1.5 Global standard spectrum.

of several light-absorbing materials with a different bandgap, combining them in a sandwich configuration such that each layer absorbs part of the solar spectrum. Theoretical limits of tandem cells including two layers of active material could reach 55%. This multijunction approach allows avoiding the problem of lattice matching requirements for different layers. It also increases the theoretical limit of light to electricity conversion efficiency from 42% for tandem cells to 68% for infinity-layer cells (Qiu et al. 2014).

The mesoscale perovskite-type of solar cells has recently experienced great advances and achieved a power conversion efficiency higher than 19%, as in the case of a triple cation mixed-halide perovskite solar cell with binary trivalent metals incorporated on the TiO_2 electron transport layer (Thambidurai et al. 2019). In order to improve the absorption of light, the film of down-conversion nanocrystals made of europium-doped sodium yttrium fluoride was deposited on the non-conducting side of the conducting glass. The down-conversion nanocrystal layer could absorb high-energy UV photons and convert them to visible light. With this innovation, the average power conversion efficiency was 19.99%, with peaks of 20.17% (Jia 2019).

3.2 Fuel production

The production of fuel, e.g., H_2, methane and methanol, using titania nanostructures and compounds has been studied intensively focusing on different alternatives.

3.2.1 H₂ production

The most direct process of fuel production consists of water electrolysis using photovoltaic electricity (the PV pathway). This pathway involves two stages and two separate devices: solar cells for the generation of photovoltaic electricity and the use of the electricity to produce hydrogen from water with an electrolyzer (Nowotny 2012).

The second process is the direct photocatalysis of hydrogen-containing molecules. In this case, some of the research results are listed in Table 14.1 as examples of the different materials employed.

In this case, it should be noted that more research efforts are needed to support TiO_2 photocatalysts for highly efficient photocatalytic applications (Reddy et al. 2020).

Another line of research for H_2 production is based on bacterial processes in which the addition of different nanoparticles such as TiO_2 improves the efficiency of the H_2 yield (Patel 2018).

Table 14.1. Some achievements in hydrogen production.

Reference	Results
(Meng et al. 2020)	The (Ni_3N–Au_c–TiO_2) composite with an ultralow amount of Au_c (0.00025 wt.%) shows photocatalytic hydrogen production rates that are even higher than that of Pt–TiO_2.
(Mandari et al. 2018)	A novel $Cu_3(PO_4)_2/TiO_2$ nanoparticle (TNP)/CuO nanocomposite had an excellent natural solar-light-driven photocatalytic H_2 production performance, 151 times higher than that of TNP under the optimal conditions.
(Liu et al. 2014b)	TiO_2 nanotube arrays sensitized with silver sulfide (Ag_2S) nanoparticles exhibited remarkable capability to absorb visible light and showed a significant enhancement in the photocatalytic efficiency of hydrogen generation.
(Li et al. 2015)	The sub-10 nm rutile nanoparticles exhibit the state-of-the-art activity among TiO_2-based semiconductors for visible-light-driven water splitting, and the concept of ultrasmall nanoparticles with abundant defects may be extended to the design of other robust semiconductor photocatalysts.
(Karnahl et al. 2014)	Copper photosensitizer-TiO_2 composites exhibit strong absorption in the visible region.
(Haldorai et al. 2014)	TiO_2 nanoparticles decorated with reduced graphene oxide composites exhibit enhanced photocatalytic activity for H_2 production with a rate of 203 mmol h^{-1}, which was higher than that of TiO_2 (35 mmol h^{-1}) and P25 (Degussa-Evonik P25™) (51 mmol h^{-1}).

3.2.2 *Photocatalytic CO₂ reduction*

CO_2 reduction for fuel production pursues two objectives, the reduction of CO_2 contamination and the production of fuel as methane or other organic compounds. However, many authors suggest that it is necessary to increase the conversion efficiency by several orders of magnitude for a practical application (Neatu 2014). In addition, the current state of art is still far from having identified an optimal photocatalyst for CO_2 reduction that may be applied commercially. Therefore, more efforts are necessary for a highly efficient visible-light photocatalytic CO_2 conversion requiring future research (Wang et al. 2019).

Some of the research results on the wide range of photocatalysts based or incorporating nanostructures of TiO_2 are listed in Table 14.2.

Table 14.2. Experimental results of photocatalytic CO_2 reduction.

Reference	Results
(Pathak et al. 2005)	Nanoscale Ag coated TiO_2 nano-particles embedded in Nafion membrane stable films improved the efficiency of photoconversion of CO_2. The primary product is methanol.
(Zhang et al. 2009)	Different crystal phases and sizes of low-dimensional nano-TiO_2 and Pt-metal supported photocatalysts show the best CH_4 yield with the Pt/TiO_2 nanotube photocatalysts, which increases with the increase of the UV irradiation time.
(Vijayan et al. 2010)	Titania nanotubes, 8 to 12 nm in diameter and 50–300 nm in lengths calcined between 200–800°C enhance the photocatalytic efficiency, at 400°C for methane production and 600°C for acetaldehyde oxidation.
(Truong et al. 2012)	TiO_2 catalysts showed excellent visible light absorption and photocatalytic activity for CO_2 reduction to CH_3OH. Bicrystalline anatase-brookite composite afforded the maximum CH_3OH yield.
(Li et al. 2012b)	The CdS (or Bi_2S_3)/TiO_2 nanotube heterostructures enhance the Svisible light absorbance and the photocatalytic performance with the largest yield of methanol on a Bi_2S_3 modified photocatalyst.
(Yin et al. 2018)	Optimized H-TiO_{2-x} (200) shows high activity for CH_4 formation at a rate of 16.2 μmol g^{-1} h^{-1} and a selectivity of 79% under full solar irradiation.
(Liu et al. 2017)	A heterogeneous photo electrocatalyst prepared by covalently binding to a Ru (II) metal-organic complex (Ru-Py) containing exposed pyridyl on the periodic TiO_2 nanotube arrays exhibited excellent CO_2 reduction activity in an aqueous solution.
(Liu et al. 2016)	The basic phenomenon of CO_2 reduction to CH_4 on Pd modified TiO_2 under UV irradiation could be enhanced by Pd or RuO_2 co-doped TiO_2.
(Hwang et al. 2019)	A mixed disordered anatase/ordered rutile (A_d/R_o) TiO_2 provided the highest CH_4 production (3.983 μmol/(g.h)), which is higher than those of metal (W, Ru, Ag and Pt)-doped P25 for CO_2 reduction under visible light.
(Li et al. 2017)	A brookite TiO_2/g-C_3N_4 nanocomposite, in which g-C_3N_4 nanodots with a mean size of ~ 2.8 nm is decorated on brookite TiO_2 quasi-nanocube surfaces showed a pronounced CH_4 generation activity and selectivity.
(Xiong et al. 2015)	TiO_2 nanoparticles modified by Pt^{2+} ions and Pt nanoparticles (Pt^{2+}-Pt^0/TiO_2) enhanced the yields of H_2 and CH_4 but showed no significant effect on the production of CO.

3.3 *Construction field*

TiO_2 coatings are suitable for protecting the surface of various construction materials such as glass, concrete, marble, wood and metals like steel, to treat water staining, moss, algae as well as soot and oil stains, conferring more hydro- and oleo-phobic properties in order to prolong maintenance and reduce cleaning (NFM 2014).

At present, companies like Evonik Degussa, Dupont, Schott, 3M and Corning produce easy-to-clean and anti-fouling coatings for a variety of markets (NFM 2014). The industry is now focusing on self-cleaning coatings using photocatalytic TiO_2, mainly as paints or coatings on glass (e.g., TiO_2-coated tiles and TiO_2-coated glass are commercially available). The photocatalyst layer applied on the glass or tiles is stable, but the organic substrates tend to be degraded by photocatalytic reactions (Chen and Poon 2009). Moreover, paints are photocatalytically effective for about 5 years; after that, as a long-term solution, the product must be reapplied to keep its effectiveness (Graziani et al. 2013). Research on the behavior of photocatalysts under aging, weathering and reactivation conditions is needed to improve the performance of the products and extend the durability of coatings (Graziani et al. 2014).

Under UV irradiation from fluorescence and LED sources, TiO$_2$ has a strong oxidation power and superhydrophilicity. TiO$_2$ attached to glass or tiles can also be used as an antimicrobial to kill bacteria, viruses and fungi or to oxidize/remove foul smells from stains in toilets. Superhydrophilic properties allow dirt and stains to be easily washed away with water or by rainfall on coated exterior surfaces. In other words, TiO$_2$ nanoparticulate self-cleaning coatings greatly benefit building maintenance, especially for skyscrapers (NFM 2014).

Photocatalytic coatings are also used to improve indoor air quality by reducing the amount of Volatile Organic Compounds (VOCs) and other toxic chemicals to which people can be exposed in residences, hotels, restaurants, commercial business facilities, university laboratories and hospitals (NFM 2014). On the other hand, anti-fouling properties are used to extend the life of structures, mainly in marine atmospheres, where protection by corrosion is required.

Research and development efforts in the construction field are oriented to the present challenges of photocatalytic TiO$_2$ coatings, such as the need for efficiency improvements, laboratory to industrial scaling up, and commercial product production (Gopalan et al. 2020).

3.3.1 Self-cleaning coatings

The superhydrophobic property of lotus leaves inspired the concept of "self-cleaning" materials. One of them is TiO$_2$ based-coatings. Besides, under UV radiation, the TiO$_2$ surface becomes highly hydrophilic, which is maintained for a few days even in the dark. As explained, under UV radiation in ambient conditions, TiO$_2$ produces very strong oxidative radicals able to decompose organic contaminants. These findings have opened novel functions of TiO$_2$ coated materials, i.e., building materials with self-cleaning and antibacterial function without using any chemical products, but using only sunlight and rainwater. In this way, self-cleaning coatings reduce costs by reducing the use of chemical detergents and the consumption of energy from high-pressure water jets (NFM 2014).

In order to preserve the façades of buildings with architectural and cultural heritage and to prolong the life of new buildings in polluted urban sites, many developments have been made in the synthesis of TiO$_2$ suspensions to coat buildings. Nowadays the main research is focused on *in situ* durability tests and the possibility to enhance visible light activation of TiO$_2$.

In this regard, rough and porous limestones are being coated with water and alcohol-based colloidal suspensions of TiO$_2$ nanoparticles and, in some cases, were synthesized by sol-gel and hydrothermal process and sprayed with different loads on the stone surface (Calia et al. 2017). In another development, a fluoropolymer host matrix of TiO$_2$ was employed against water penetration into the stones, providing photocatalytic ability (Colangiuli et al. 2019). TiO$_2$ coating has also been applied by spraying TiO$_2$ suspensions on historic stones and architectural surfaces with the results of evident self-cleaning photoinduced effects. This process enhanced the durability of stone surfaces against UV aging, improved the resistance to Relative Humidity (RH)/temperature and abrasion effect and reduced the accumulation of dirt on stone surfaces left in open air for 6 months, without altering the original features (Aldoasri et al. 2017). TiO$_2$ coatings were able to preserve the surface color after exposure of up to 8 months. After this period, the protecting effect was lost and the self-cleaning efficiency was reduced to negligible final rates due to both, partial titania loss and deactivation phenomena (Lettieri et al. 2019).

The resistance of TiO$_2$ coatings applied on limestone to a peeling and a water impact test, reported failures, maintaining however good self-cleaning performance due to the retention of nanoparticles below the surface (Calia et al. 2016).

A new ceramic material, obtained using Packaging Glass Waste (PGW), was developed through the deposition by air-brushing of a nanostructured coating based on titania-silica sol-gel suspension followed by thermal treatment. The results indicate that the applied coatings are transparent and have a good scratch resistance and photocatalytic activity (Taurino et al. 2016).

It was also reported that humidity, in outdoor or indoor ambients, promotes the hydration and improves the photocatalytic activity of clean porous and non-porous TiO_2 coatings synthesized by sol-gel on silicon, but humidity has less influence when a stain is already covering the coating. Hydrated TiO_2 coatings promoted the availability of ·OH and present better stain removal performance (Cedillo Gonzalez et al. 2020).

The remarkable self-cleaning effects of TiO_2 are trying to be combined with its photoactive properties in order to promote visible light absorption by employing a metal-to-metal charge transfer process for hetero-bimetallic Ti (IV)-O-Ce (III) assemblies on the pores of mesoporous silica (Hashimoto 2007). Three different types of materials based on TiO_2, which can function even under room light conditions, were investigated: (1) a TiO_2 film deposited with Cu, which shows a remarkable antibacterial effect, even under very weak UV; (2) a layered TiO_2/WO_3 heterogeneous film, which becomes highly hydrophilic even under fluorescent light; there, the photogenerated holes produced in WO_3 by UV light are transferred to the TiO_2 side and then used for hydrophilic conversion; (3) a nitrogen-doped TiO_2 film, which can be highly hydrophilic by absorbing only visible light (Irie et al. 2004).

Nano-TiO_2 surfaces coated with a 10–20 nm layer of TiO_2 are commercially available, mainly in the Japanese and European markets. For example, Bio Shield Inc. markets a product (NuTiO™) that is a liquid-based self-cleaning coating that contains a nano-TiO_2 suspension. Bio Shield Inc. claims that NuTiO™ can be coated onto any surface to provide a self-cleaning coating that will last up to 10 years before reapplication. The product is claimed to photocatalyze naturally present water and oxygen in the air into hydroxyl and superoxide anions that oxidize and decompose toxic organic chemicals and bacteria (Varner et al. 2010). Similar products are marketed by Green Millennium; for example, TPXsol85™ is a mixed solution of two of their products, PTAsol™ and TOsol™. TPXsol™ is a neutral solution that can be applied to various materials including metals and resins by spraying, dipping or applying by brush to the surface. The final material presents photocatalytic effects and hardness (GM 2021).

Besides, the advantage of the application of superhydrophilic TiO_2 coatings goes further: they have been used in evaporative cooling systems for buildings. Using this technology, water is continuously sprinkled onto the surfaces of buildings that have been coated with TiO_2. With solar irradiation, the surface becomes highly hydrophilic due to the coated TiO_2, which minimizes the amount of water consumption to form a water film. The building is not cooled by the water itself but by the latent heat flux when water evaporates. This application of photocatalytic building materials can result in a significant reduction of electricity consumed for air (Chen and Poon 2009).

3.3.2 Anti-fogging coatings

Anti-fogging coatings are used to improve the attraction between water and a surface by overcoming the surface tension, avoiding the formation of water droplets and permitting the spread over the surface. Thus, anti-fogging coatings prevent fogging on the surface by inhibiting the condensation of water droplets on the surface.

Anti-fogging coatings formulated with anti-fog agents, e.g., TiO_2 chemically bonded into the polymer matrix of the coating, are commercially available, especially for construction application in humid climates or ambients, e.g., architectural buildings, structural and decorative glass, including windows, walls, panels, doors, skylights, dooms and outdoor enclosures, bathroom mirrors and shower doors (FSICT 2021).

In addition, nano-TiO_2 glass, leather, metal, stone and wood coatings are available (De Cie GmbH 2021). These anti-fogging coatings incorporate nanoscale TiO_2 and hybrid polymers by a sol-gel process (NFM 2014).

There are active research and development efforts to improve the performance of these coatings, such as the following. Synthesis of superhydrophilic compound coatings of TiO_2 and SiO_2 on different substrates, combining their photocatalytic and superhydrophobic properties, respectively. A high-rate deposition by arc-discharge of anatase TiO_2/SiO_2 nanocomposite coatings on polymer substrates with durable superhydrophilic and photocatalytic properties promoting antifogging (Chemin et al. 2018). The production of SiO_2/SiO_2-TiO_2 bilayers on polycarbonate with photoinduced superhydrophilicity (contact angle, 5° after UV irradiation), with enhanced hardness and transmittance of the polycarbonate substrate (Shahnooshi et al. 2019). Pure SiO_2 nanowires, TiO_2 nanoparticles and mixed SiO_2-TiO_2 films were deposited directly onto glass substrates by flame spray pyrolysis of organometallic solutions and stabilized by in-situ flame annealing (Tricoli et al. 2009).

3.3.3 Anti-microbial coatings

Biodeterioration is responsible for a large part of the weathering and loss of aesthetic beauty of stones and concrete used on building surfaces and, in the long term, it can compromise the integrity of the substrate (Graziani et al. 2013). Biocidal products are traditionally used to prevent microbial growth but they may be ineffective in the long term, and their potential release to the environment by leaching must be seriously considered (Chen and Poon 2009, Goffredo et al. 2017). The antibacterial property of TiO_2 is employed for protection in a wide variety of situations such as to prevent bacterial colonization on the surface of monuments or medical buildings. The antimicrobial effect is based on its ability to generate active carriers resulting in the formation of effective chemical species such as peroxides (hydrogen peroxide) and other reactive oxygen species. Hospitals and buildings in Asia and recently in Europe have been coated with nano-TiO_2 coatings to protect them against deadly infections and environmental pollution damage. Commercial Gens Nano™ TiO_2 coatings are available, which transform the treated surfaces into antibacterial, antifungal, mold-free surfaces while purifying the air (GENS 2019).

Since the reactivity of TiO_2 is restricted to the wavelength in the UV region, doping with Ag alone or in combination with Fe^{3+} or Sr may become an efficient and economic approach to widen the photocatalytic spectrum. In such a way, the antibacterial activity could be successfully applied to prevent stone biodeterioration without chromatic changes due to UV radiation (La Russa et al. 2014).

3.3.4 Antifouling coatings

Marine fouling, with an estimated annual cost to the shipping industry of over US$ 200 billion, also induces the degradation of underwater archaeological sites (NFM 2014). The common cleaning procedure, consisting of the manual removal of fouling, which requires a continuous maintenance, can be substituted by the inhibition of biological colonization providing long-time protection against biofouling. Even if TiO_2 nanocoatings are not able to fully prevent microalgal biofouling, they become the only alternative, considering that most of the used antifouling paints, especially for ship hulls and based on organo-tin, show a considerable toxicity level.

TiO_2 and Ag-doped TiO_2 applied on stony surfaces have led to a reduction of the biofouling activity (Ruffolo et al. 2013). In addition, superhydrophobic silicone/nanorod-like TiO_2-SiO_2 core-shell composites, applied on structures by dispersion, exhibited promissory fouling release self-cleaning in trials in natural seawater for 6 months in a tropical area (Selim et al. 2019). A TiO_2/fluorinated acrylic nanocomposite paint may also be an alternative to a toxin-free and eco-friendly marine antifouling paint (Zhang et al. 2016).

Antifouling of nanorod-TiO_2 and SiO_2 dispersed in acrylic-based paints for wood coatings was also proposed, with the advantage of the combined effect of hydrophobicity of SiO_2 and photocatalysis of TiO_2 (Kartini et al. 2018).

Moreover, antifouling nanocoatings are being applied in heat exchangers of industries such as brewing, dairy and food processing. In these cases, fouling is due to the adhesion of both organic and inorganic materials on steel walls, which results in reduced thermal transfer efficiency, increased energy costs and contamination problems. By applying antifouling nanocoatings, it is possible to reduce downtime and cut cleaning costs (NFM 2014).

3.3.5 Anticorrosion coatings

Corrosion is a significant problem in industrial equipment, installations and plants caused by environmental factors, such as oxygen and water in mining (ore processing, surface and underground mining and drilling), utilities (seals, accessories and bearings), defense, agriculture (tillage and planting), construction (drill bits, grinder hammer tips and other hardware), shipping, energy (wind power at sea) and transportation (brakes, valve trains, bearings and gears) (NFM 2014). In such cases, nanocoating protection must be developed to be used on steel.

In such a direction, the performance of nanotitania coatings doped with anions of nitrogen, sulfur and chlorine on the surface of 316L stainless steel deposited by a sol-gel process and dip-coating technique was investigated. The results of electrochemical impedance spectroscopy in 0.5 M NaCl showed that the N-modified TiO_2 nanocoating had the highest corrosion resistance due to the addition of nitrogen that improves the compact structure and enhances the hydrophobic property (Yun et al. 2007).

In the petroleum industry, nanocomposite coatings play a strategic role to protect steel oil transportation tanker trucks. TiO_2 nanoparticles were fabricated by a simple template-free sol-gel method and blended with poly-dimethylamino siloxane (PDMAS) to fabricate PDMAS/TiO_2 nanocomposite acting as a multifunctional modifier for polyamine-cured epoxy coating. The results show a superior anticorrosion, mechanical, chemical, electrical, thermal and UV resistance of the PDMAS/TiO_2 epoxy hybrid nanocomposite coatings (Fadl et al. 2020).

Chromium-doped TiO_2 coatings have been found to provide an active photogenerated cathodic protection of 316L stainless steel under illumination and during a long-time dark immersion in 2 M NaCl solution at 60°C, associated with the self-healing property of chromium ions (Li and Fu 2013).

3.3.6 UV-resistant coatings

UV-Titan™ is a commercially available ultrafine, surface treated and transparent TiO_2; its use in automotive coatings, wood lacquers and plastics provides UV protection. As a UV screen, it gives long-lasting protection against the darkening of wood, maintaining the natural look of the wood. UV-Titan™ can also be applied to metals, influencing the flip/flop effect of the metallic pigment (NFM 2014).

In the presence of cerium ammonium nitrate, TiO_2 can be turned from a white powder into a TiO_2/Ce xerogel via a facile bottom-up fabrication process. This transparent TiO_2/Ce xerogel can decrease the surface deterioration induced by UV light and preserve the natural appearance of wood by suppression of free radical generation on wood surfaces upon UV irradiation. In this way, Guo et al. research' would permit the expansion of the applicability of the protective effect of TiO_2 to coatings for natural engineering materials, which will become increasingly important in bioeconomy (Guo et al. 2017).

3.3.7 Air purification

Air purification is also promoted by developing or tuning new construction materials with TiO_2, taking advantage of its photocatalytic capacity to oxidize compounds, such as VOCs and NO_x. These compounds are common pollutants found in relatively higher concentrations in urban areas, mainly in highly trafficked canyon streets or road tunnels. VOCs are reported as carcinogenic, mutagenic or teratogenic compounds, and their combination with NO_x in the presence of sunlight induces the production of tropospheric ozone and photochemical smog (Boonen and Beeldens 2014). The combined effects of ambient (outdoor) and household air pollution cause an estimated deaths of seven million people every year. Moreover, 9 out of 10 people breathe air with pollutant levels that exceed the World Health Organization (WHO) guideline limits.

Since 1990, when the heterogeneous photocatalysis based on TiO_2 to treat trichloroethylene (a VOC) in the gas phase was studied (Dibble and Raupp 1990) up to nowadays, numerous studies have been carried out to address the air pollutant abatement. Usually, the variables assessed in laboratory or bench-scale experiments are radiation source (lamps type, power), photocatalyst (type, doped or not, concentration), pollutant (nature, concentration, flow rate) and relative humidity. The degradation mechanisms and byproducts in photocatalytic experiments are quite well-known (Chen and Poon 2009, Farhanian et al. 2013).

Real-cases bring new challenges. To avoid pollutant diffusion, the distance between the source of the pollutants and the photocatalyst must be the shortest possible. In this sense, in the last decade, TiO_2 has been incorporated in roads onto the top layer of the concrete pavement, and the reaction product is adsorbed on the surfaces and subsequently washed away by rain. About the top 10 mm can maintain the activity even after surface wear for traffic or weathering (Boonen and Beeldens 2014). Using recycled glass, 2 to 5 wt.% of TiO_2 has been added for NO_x removal, with better results using transparent glass compared with a green one, and good resistance performance under a harsh abrasion test (Guo et al. 2012, Guo and Poon 2013).

Photocatalysis applied to indoor environments is usually limited by the low radiation intensity available in such ambients, although just a few photons with proper energy can induce photocatalysis to purify the air because the amounts of pollutants are typically small. This is not the case for tunnels, where bright lamps must be used to significantly reduce the pollutant concentration. Tunnels with photocatalytic materials on roofs, walls and roads showed a decrease in NO_x concentration (Calia et al. 2016).

To improve the efficiency of TiO_2, their doping with heteroatoms has been extensively studied to reduce its bandgap in order to use visible light for air purification (Boonen and Beeldens 2014). Several studies with promising results have been made to improve the indoor air quality, but mainly at the bench-scale (Mamaghani et al. 2018 and references therein). Some scaled-up systems with four parallel experiments in the same conditions were used to test the mineralization of model pollutants in continuous flow (Farhanian et al. 2013).

3.4 Water and wastewater treatment

Water contamination and wastewater without proper treatment can enter the environment leading to serious health problems. Many chemical compounds cannot be removed from water by traditional water treatments. Therefore, present research efforts are oriented to the development of new or improved processes to remove pollutants from water.

The most studied photochemical advanced oxidation/reduction process for water decontamination is heterogeneous photocatalysis with TiO_2 (as described earlier). The overall process produces the degradation of organic compounds such as herbicides, pesticides and emerging contaminants to less or non-toxic substances like CO_2 and H_2O. Many papers and reviews report photocatalytic degradation of different organic compounds (Chen et al. 2020, Guo et al. 2019 and references therein in both). Even highly toxic metal ions (Cr(VI), Hg(II), As(III), U(VI)) can be transformed into less harmful species (Cr(III), Hg^0, As(V), U(IV) respectively) (Lee and Park 2013, Litter 2017 and references therein). Redox transformation involved in each case depends on water or wastewater conditions such as pH, aeration, origin salt and the presence of electron donors, among others (Litter 1999, 2017 and references therein in both).

However, as it was mentioned earlier, the photocatalytic efficiency of TiO_2 is low since only UV radiation is absorbed by the semiconductor due to its wide bandgap (3–3.2 eV). Tailoring the bandgap of TiO_2 to extend the absorption to the visible range is one of the main proposed solutions. With this objective, the bandgap has been modified by doping the TiO_2 crystals (Chang and Liu 2014, Hu et al. 2017) with non-metallic elements such as N (Madhavi and Kondaiah 2018), F, C (Miao et al. 2016), S (Basavarajappa et al. 2020), or with transition metals such as Cr (Wang et al. 2012), Co, V, Fe (Moradi et al. 2018), Cu (Bensouici et al. 2017), Mn (Sudrajat et al. 2020),

Mo, Ni (Shaban et al. 2019), W, Ru (González et al. 2019). The addition of these dopants has two main effects: (1) to change the lattice parameter, which decreases the value of the bandgap below 3 eV, and (2) to create energy trap states in the bandgap permitting the absorption of visible light. These traps also increase the lifetime of the photoelectrons reducing their recombination with holes (Pelaez et al. 2012).

Another line of research consists in depositing nanoparticles of noble metals such as Ag (Liu et al. 2012, Nanaji et al. 2019), Au (Zhang et al. 2017b), Pt, Pd (Jahdi et al. 2020, Maicu et al. 2011) or dye molecules (Rani et al. 2014, Zyoud et al. 2011) that, instead of modifying the TiO_2 bandgap, absorb visible light producing electrons that can be transferred to TiO_2. Both approaches are also investigated in solar cell applications, as it is described earlier.

In addition, the photocatalytic efficiency can also be improved by increasing the surface area of the material. This can be done by producing nanostructured TiO_2 semiconductors in the form of nanoparticles, nanofibers, nanotubes, nanofilms, nanoporous, nanowires or nanorods. Figure 14.2 illustrates some of such nanostructures.

It has been shown that the photocatalytic performance of semiconductors is closely related to their structural and morphological characteristics at the nanoscale, including the size, dimensions, pore structure and volume, specific surface area, exposed surface facets and crystalline phase content (Li et al. 2016). For instance, the anatase phase of TiO_2 is recognized as the most photoactive (Fujishima and Zhang 2006), although the rutile content contributes to the photocatalytic efficiency favoring the charge carrier separation (Hurum et al. 2003). The optimum content of rutile is considered to be around 40% (Hurum et al. 2003, Luttrell et al. 2014, Su et al. 2011, Traid 2017, 2018a).

A very well-known commercial TiO_2 photocatalyst is P25™, with particles of around 30 nm and a photocatalyst area of 50 $m^2 g^{-1}$, composed of 70–80% anatase and 30–20% rutile (Ohtani et al. 2010). This material is widely used because of its relatively high levels of activity in many photocatalytic reaction systems. At present, it is not easy to find a photocatalyst with an activity higher than that of P25™, and, for this reason, P25™ has been used as a *de-facto* standard titania photocatalyst. Photocatalytic reactions of P25™ have been reported in more than 50 thousand papers according to an Internet browser (GS 2021). However, the use of TiO_2 nanoparticles needs a very expensive separation process and recovery of the photocatalyst, which can be eliminated if TiO_2 is supported on a fixed substrate.

There are numerous reports of the synthesis of nanostructured TiO_2 photocatalysts supported on different substrates (glass, polymers, metals like titanium and steel), used in the treatment of specific pollutants in the aqueous phase by heterogeneous photocatalysis. Some of them are listed in Table 14.3.

Among the processes to produce robust photocatalytic coatings, anodic oxidation is one of the most simple and inexpensive (Diamanti et al. 2015, Dikici et al. 2014, Traid 2018a), since it requires neither complex equipment nor high energy consumption, it can be done at an ambient

(a) (b) (c)

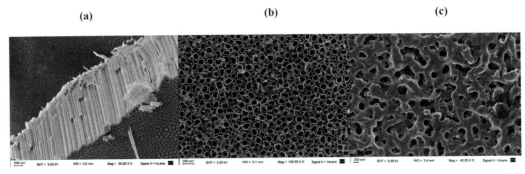

Figure 14.2. SEM images of nanostructured TiO_2 coatings obtained by anodic oxidation on titanium substrates: (a) TiO_2 nanotubes (side view); (b) TiO_2 nanotubes (front view); (c) nanopores of TiO_2.

Table 14.3. Some achievements in the production of supported TiO_2 photocatalysts.

Photocatalyst	Pollutant	Results	Reference
Nanoporous anodic TiO_2 coatings	Cr(VI)/EDTA	The maximum Cr(VI) transformation achieved by the most active coating was 89.1%.	(Traid et al. 2017)
Nanotubular TiO_2 coatings prepared by anodic oxidation	Cr(VI)/EDTA	The most active coating yielded 98% of Cr(VI) transformation.	(Vera et al. 2018b)
Transparent glass sponges coated with TiO_2	Cr(VI)/EDTA	Total transformation of Cr(VI) was achieved by the sponge synthesized at 45 ppi (porous per inch) coated with TiO_2.	(Löffler et al. 2020)
TiO_2 thin film was obtained using a dip-coating method	4-tert-octylphenol	88% of the 4-tert-octylphenol degraded.	(Wu et al. 2012)
TiO_2 coatings deposited by atmospheric plasma spraying on austenitic stainless steel coupons	Methylene blue dye in an aqueous solution	Partial degradation of the pollutants was observed, in addition to a significant discoloration effect.	(Bordes et al. 2015)
TiO_2-P25 nanoparticles were immobilized on polyethylene disks by the ironing method	*p*-nitrophenol	Maximum removal percent (82.6%) achieved at the optimum operational conditions.	(Behnajady et al. 2018)
Recycled polymeric disposals as TiO_2 supports	Wastewater of textile Industry	Mineralization percentages reached were 39 and 42% for the samples collected from two factories.	(Mekkawi et al. 2020)
TiO_2 coatings obtained by sol-gel with the addition of commercial P25 on glass plates	4-chlorophenol degradation and Cr(VI) reduction in the presence of EDTA	High efficiency in 4-CP oxidation and Cr(VI) reduction achieved with the composite coatings.	(Vera et al. 2017b)
TiO_2 films deposited by cathodic arc (CA films) on glass substrates	Cr(VI)/EDTA	Removal of 50% of Cr(VI).	(Kleiman et al. 2011)

temperature, atmospheric pressure, in a relatively short time and with excellent adhesion of the TiO_2 to the substrate. Moreover, there are little or no restrictions on substrate geometries (Kurtz et al. 2018, Traid 2018a), and it is possible to obtain TiO_2 coatings from different surface titanium substrates (Dwojak et al. 2019).

Some of the nanostructures produced by anodic oxidation are shown in Fig. 14.2, such as nanoporous TiO_2 on Ti grade 2 alloy using post-spark conditions ($V \geq 100$ V) during oxidation in a sulfuric acid electrolyte (Dwojak et al. 2017, Kurtz et al. 2018, Traid et al. 2018a, 2018b) and nanotubular coatings up to 100 nm long on Ti grade 2 alloy, using HF-based electrolytes under different oxidation parameters (Vera et al. 2018b). The process is very versatile since different electrolytes can be used such as those based on NH_4F in different solvents like glycerol and ethylene glycol; this process on Ti grade 2 alloy produces nanotubular coatings of several microns in length (Dwojak et al. 2018, 2019, 2021, Kurtz et al. 2018, Vera et al. 2018b).

The nanotubes produced by anodic oxidation are generally amorphous; therefore, it is necessary to heat them to crystallize to the anatase or rutile phases (Vera et al. 2018a), which are necessary for photocatalysis.

The advantage of the anodic oxidation to produce TiO_2 films on non-flat surfaces of metallic Ti allowed to construct a low-cost cylindrical photoreactor that produced a 95% of transformation of Cr(VI) in the presence of ethylenediaminetetraacetic acid (EDTA) after 5 hours of irradiation with a simple household fluorescent cylindrical lamp that partly emits UV radiation (GE, F8T5 BLB) (Traid 2018a).

In addition, the doping of TiO_2 is possible by anodic oxidation with the addition of different ions to the electrolyte. This allowed to obtain porous TiO_2 doped with Ag and Fe by the addition of $Ag_2(SO)_4$ or $Fe_2(SO_4)_3$ to the electrolyte (Kim et al. 2016, Liu et al. 2012, Traid 2018a). Doping with N was also possible by adding urea (Mazierski et al. 2016), ammonium hydroxide (Yuan et al. 2013) or hydrazine (Xu et al. 2010).

3.5 *Biomedical devices*

Titanium and titanium oxides are materials with good biocompatibility and have extended use in medical devices. In particular, three types of medical devices that use TiO_2 will be considered: mechanical prosthetic heart valves, dental implants and hip stem implants, which are fixed or mobile devices, replacing hard or soft tissues and working in different environments and body fluid and hard tissues (Ratner et al. 2004). The ability and performance of these implants in humans are mainly related to the material properties.

3.5.1 *Mechanical prosthetic heart valves*

A mechanical prosthetic heart valve is built with rigid parts that must move correctly in order to replace aortic or mitral natural heart valves when they fail. Nowadays, pyrolytic carbon and titanium alloys are the main materials used in the structural parts of these prostheses (Butany et al. 2003, Mohammadi and Mequanint 2011). The biocompatibility and especially the hemocompatibility of titanium was associated with the naturally present TiO_2 (Eisenbarth et al. 2002, Huang and Yang 2003, Leng et al. 2006, Zhang et al. 1998). This natural layer is very thin and can have structural defects; therefore, to enhance the performance, TiO_2 coatings should be deposited (Lütjering and Williams 2007). To prevent blood clotting and platelet adhesion, a smooth surface is needed (Schvezov et al. 2010, 2017) and the coatings must be smooth with low roughness.

Studies of blood clotting time and platelet adhesion were conducted on TiO_2 coatings produced by ion-beam-assisted deposition technique and Pyrolytic Carbon (PC). The results showed a longer coagulation time and less platelet adhesion on TiO_2 than PC. In addition, *in vivo* implanting showed that the thrombogenicity of the PC was eight times higher than for the TiO_2 coatings, showing a much better hemocompatibility of TiO_2 than that of PC (Zhang et al. 1998). Also, according to other studies (Leng et al. 2002, 2006), TiO_2 films synthesized by plasma immersion ion implantation and deposition or metal vacuum arc source deposition have better blood compatibility than PC. Similar hemocompatibility studies of platelet adhesion and activation, clotting time, thrombin and prethrombin times, protein adsorption and *in vivo* experiments on dogs with TiO_2 coatings synthesized by plasma immersion ion deposition also showed a better hemocompatibility of TiO_2 than PC (Huang and Yang 2003).

TiO_2 coatings are synthesized by three different techniques: anodic, thermal and sol-gel oxidation, on two different substrates, CP titanium and the Ti1.5Al25V alloy, indicated a better cytocompatibility on the coatings made by sol-gel due to the formation of V_2O_5 on anodic and thermal oxidation techniques, which were not observed in the sol-gel coated samples (Eisenbarth et al. 2002).

In addition, a crystalline TiO_2 structure reduced platelet adhesion for amorphous films was found (Huang and Yang 2003). Moreover, the Fermi energy level and the semiconductivity of the TiO_2 coatings have been suggested as suitable for improving the hemocompatibility of heart valves (Zhang et al. 1998), making the charge transfer more difficult from the protein to the surface, inhibiting the decomposition of fibrinogen.

In corrosion studies performed with simulated body fluid of TiO_2 coatings on CP titanium and Ti-6Al-4V synthesized by anodic, thermal and sol-gel oxidation techniques, it was found that, regardless of the technique used, the best corrosion resistance was of the coating with 100 nm thickness (Velten et al. 2001).

In addition, mechanical properties must be evaluated on coatings for medical devices. In the duplex TiO_2/TiN film, the role of TiO_2 is to improve blood compatibility while the TiN film improves the mechanical properties. In addition, the fatigue and the wear resistances of the artificial heart valve modified by TiO$_2$/TiN duplex film are significantly improved compared with bare titanium alloy (Leng et al. 2002, 2006).

TiO$_2$ coatings were synthesized by the sol-gel dip-coating technique on Ti-6Al-4V alloy and were heat-treated at 500ºC for 60 minutes. The tribological behavior of the coating was analyzed using an alternative equipment, with an AISI-52100 steel sphere as counterphase and a load ranging from 1 to 3 N. The results showed that the coated material is more resistant than the bare Ti-6Al-4V alloy (Zhang et al. 2006).

New developments in mechanical prosthetic heart valves have been performed using titanium as a structural material coated with the natural oxide or improved by some of the recently cited physicochemical methods, and the prototypes have been proved *in vitro* and *in vivo* tests as described below.

In 1994, Lapeyre et al. made *in vitro* testing their trileaflet valve prototype, made of a titanium ring coated with natural TiO$_2$; 250 million cycles of test and preclinical trials on calves were performed (Lapeyre et al. 1994). In addition, more recent tests of the now called Triflo™ model (Bruecker and Li 2020, Lapeyre 2010, Vennemann et al. 2016) show a good combination of material and design.

In vivo tests on calves of two models of trileaflet valves made of a titanium ring were made. The results showed a good performance of the design and the materials employed (Gregoric et al. 2004). In addition, a new valve made of titanium housing, which was implanted in sheep, showed no clotting, hemolysis or embolic events (Wium et al. 2020).

In the development of a three-leaf mechanical heart valve (Amerio et al. 2006, Amerio and Schvezov 2019), TiO$_2$ coatings were synthesized on the Ti-6Al-4V alloy (Rosenberger et al. 2005) using the sol-gel (Favilla et al. 2008, 2009) and the anodic oxidation techniques (Vera 2013). The thickness, morphology and crystal structure of the obtained coatings were determined and evaluated for each synthesis condition (Kociubczyk et al. 2015, Schvezov et al. 2010, 2017, Vera et al. 2013, 2014, 2015c, 2017a, c). In addition, the resistance to corrosion (Vera et al. 2015a), wear (Vera et al. 2015d) and erosion (Rosenberger et al. 2013), as well as the hemocompatibility (Schuster et al. 2012, 2015, Vera et al. 2015b) of the coatings were determined, evaluated and employed in order to optimize the material properties for valve construction.

3.5.2 Dental and hip implants

Dental and hip stem implants have different biocompatibility requirements, to induce a good integration with the bone that is in contact, and to achieve strong anchoring to distribute the body load to the bone (Brunette et al. 2001). Strong osseointegration was achieved with the rough surface where the bone-forming cells adhered preferentially to the connective-forming cells (Buser 1997). The TiO$_2$ present in titanium and titanium alloys prostheses contributes to enhance the biocompatibility due to it is biological inertness and has a similar Young module as that of bone (Brunette et al. 2001). Several strategies were used to increase osseointegration, by adding nanostructured TiO$_2$ to control roughness and cell interaction and adding other metal nanoparticles to reduce the bacterial adherence to implants (Allaker 2010). Some examples are described next.

Ti grade 2 screws were coated with 30 nm TiO$_2$ particles by standard dip-coating and modified laser techniques. The laser process consists of irradiating the substrate submerged in an aqueous TiO$_2$ suspension with a Nd-YAG laser (1064 nm) and 1000 mJ during 10 ns. This technique gives a higher quality thin coating film, with improved surface roughness (Azzawi et al. 2018). The samples were implanted in the tibia of rabbits for 4 weeks and then extirpated and mechanically and histologically analyzed. The torque of the laser-treated was 40% higher than those of the dip-coated samples. The histological study showed a better structure, surface roughness values, bone integration and bond strength at the bone-implant interface of modified laser-coated samples (Azzawi et al. 2018).

Tantalum on TiO$_2$ nanotubes by plasma spraying to obtain Ta/TiO$_2$ nanotube composite coatings was tested for cytocompatibility *in vitro* with mouse embryonic osteoblast precursor cells. The results revealed a coating with good surface characteristics, including roughness and hydrophilicity. The Ta/TiO$_2$ nanotube composite coating presented good adhesion, differentiation, mineralization and osteogenesis-related gene expression *in vitro*, all these favorable properties to enhance cytocompatibility (Wang et al. 2020).

Microporous TiO$_2$ doped with silicon (Si-TiO$_2$) coating was made by microarc oxidation on titanium substrate. The surface morphology, roughness and phase composition of the Si-TiO$_2$ coating were similar to those of the Si-free doped. It was found that Si-TiO$_2$ coating can promote osteoblast adhesion, spreading and proliferation (Zhao et al. 2020).

The porous oxide obtained by thermal oxidation of Ti-6Al-4V, obtained between 670 and 750°C improved the bioactivity and osseointegration of implants. Below 900°C, the oxidation products were mainly TiO$_2$ and less Ti$_3$O$_5$, Ti$_2$O$_3$ and TiO (Ma et al. 2020).

The behavior of TiO$_2$ nanotubes of diameters from 15 to 100 nm showed that cell adhesion and spreading are enhanced for nanotube diameters of ~ 15–30 nm, and a strong decay in cell activity is observed for diameters > 50 nm (Bauer et al. 2009). In another study, the relation between nanoscale topography and growth factors on the control of cell behavior of mesenchymal stem cells was observed. The synergic relation between them could be used to design coatings for prostheses (Park et al. 2011).

In the case of synthetized TiO$_2$ powder by the sol-gel route using titanium isopropoxide, precipitating after a 60°C treatment and dried at 150°C, amorphous-TiO$_2$ (275°C) and tetragonal TiO$_2$ (500°C) structures were produced by thermal treatment. A simulated body fluid treatment after 30 days produced rod-like amorphous calcium and/or phosphorous deposits only on the crystalline surfaces. Cytocompatibility tests showed that none of the oxides had a negative or toxic effect on cell viability, and the amorphous TiO$_2$ was more cytocompatible than the crystalline surface (Pradhan et al. 2016).

All these routes could be considered to increase the roughness and osseointegration of bone implanted prostheses. Some of them are simple and not expensive, and they could expand the currently existing patents, e.g., titanium material implants (Oshida 2000), methods for the preparation of implants made of titanium or alloys thereof (Hansson 1998) and methods for obtaining a surface of titanium-based metal implant to be inserted into bone tissues (García Saban et al. 2012), dental implants (Memmolo et al. 2014), and others.

3.6 Food field

3.6.1 Food additive and food films packaging

In the food industry, TiO$_2$ is used as an additive to provide a whitening effect. Its use in food was approved by the Food and Drug Administration (FDA) in 1966 and included in the Codex Alimentarius of the Food and Agriculture Organization (FAO/WHO) in 1969. In addition, nowadays, taking advantage of their properties, it has been used to functionalize food-packaging materials.

Polylactic acid (PLA)/functionalized TiO$_2$ composites have outstanding mechanical, thermal, photocatalytic and antimicrobial properties (Farhoodi et al. 2012, Zhang et al. 2015) and are being investigated for use in many fields, including biomedical, drug delivery applications (Buzarovska et al. 2018, Song et al. 2006, 2008), wound healing dressing materials (Shebi et al. 2019) and food preserving packaging applications to improve the shelf life of fruits and vegetables (Chi et al. 2019, Farhoodi et al. 2012, Fonseca et al. 2015, Li et al. 2018, Segura Gonzalez et al. 2018). For a full review, see Kaseem et al. (2019).

In the case of polypropylene, the addition of TiO$_2$ nanoparticles in composites improves antibacterial, mechanical, UV protection, thermal and hydrophilicity properties of the material without sacrificing other desired characteristics like breathability, handle, fineness, stiffness and low cost (Chaudhari et al. 2013).

In starch bioplastic-based composites, the decomposition temperature of the starch bioplastic was increased by mixing with TiO$_2$ nanoparticles (21 nm). The results showed better compatible morphologies in composite bioplastics, which was associated with a decrease in the number of voids, holes and cracks. Starch bioplastic and bioplastic composites with better thermal, mechanical and chemical properties have been considered suitable for the food and pharmaceutical industry (Amin et al. 2019).

TiO_2 food additive is labeled as E171 in Europe and INS171 in the USA. Since its approval in 1966, its use has been reevaluated several times with the last reevaluation in 2016.

The amount of TiO_2 consumed by the human population varies among countries and has been estimated at around 0.2–0.7 mg/kg bw/d (mg per kg of body weight per day) in the USA and 1 mg TiO_2/kg bw/d (Bachler et al. 2015, Weir et al. 2012) in the UK and Germany. In addition, its consumption varies among the age groups; for instance, it is 0.5 (upper limit 1.1 mg/kg bw/d) for elderly adults and 1.4 mg/kg bw/d (upper limit 3.2 mg/kg bw/d) for children in the Netherlands (Sprong et al. 2015) and Germany (Bachler et al. 2015). It is mainly consumed with TiO_2 added to breath-refreshing micro sweets, savory sauces, dressings, salads and savory-based sandwich spreads, chocolate products, chewing gums, soft drinks, cheese and biscuits (EFSA ANS 2016).

Since the early 1960s, TiO_2 is considered safe for use in food. Since that time, some authors questioned this issue. In the recent reevaluation of TiO_2 (E171) as a food additive (EFSA ANS 2016), the European Food Safety Authority (EFSA) Panel on Food Additives and Nutrient Sources added to Food estimated that the absorption of orally administered TiO_2 particles, including micro- and nanosized (less than 3.2% by mass) fractions, was negligible, reaching at most 0.02–0.1% of the administered dose, and concluded that the use of TiO_2 as a food additive does not raise a genotoxic concern and, therefore, no adverse effects resulting from the eventual accumulation of the absorbed particles was expected. The lowest value found in literature for the no-observed adverse effect levels was 2250 mg TiO_2/kg bw/d (Ropers et al. 2017). However, a year later, Bettini et al. (2017) proved that orally administered food-grade TiO_2 containing nanoscale particles impaired immune homeostasis and induced carcinogenesis in rats. Based on this publication, the French ANSES (Agency for Food, Environmental and Occupational Health) published their opinion on TiO_2 nanoparticles (ANSSAET 2017), in which the necessity of conducting thorough research on the possible dangers connected with the usage of E171 was underlined. France is the first country to ban the use of the E171 food additive because of the possible harmful effects on humans and a lack of scientific data to confirm its safety. The restrictions became effective in 2020 (AD17A 2019). The decision has generated a big controversy (Boutillier et al. 2020).

Some studies revealed that after inhalation or oral exposure, nanoparticles accumulate among other places in the lungs, alimentary tract, liver, heart, spleen, kidneys and cardiac muscle. In addition, they disturb glucose and lipid homeostasis in mice and rats. The risk to human health has been poorly explored (Musial et al. 2020). In the case of TiO_2 nanoparticles, scientific databases inform that they can induce inflammation due to oxidative stress. A regular supply of TiO_2 nanoparticles at small doses can affect the intestinal mucosa, the brain, the heart and other internal organs, which can lead to an increased risk of developing many diseases, tumors or progress of exprocesses (Grissa et al. 2015, Proquin et al. 2016). The mechanism behind the nanotoxicity of nanoparticles has not been discovered yet. Many studies attribute it to oxidative stress, thus nanotoxicity is still an important area for future exploration (Baranowska-Wojcik et al. 2020). Recently, the EFSA Panel on Food Additives and Flavorings has concluded that, based on all the evidence available, a concern for genotoxicity could not be ruled out, and given the many uncertainties, the E171 can no longer be considered safe when used as a food additive (Younes et al. 2021). Therefore, there is a major need for many more studies in this field.

4. Conclusions and Future Outlook

TiO_2 is an emerging material for a wide range of applications and the nanoscale application has gained attention during the last decades as an interesting alternative to improve their performance.

For example, the TiO_2 intrinsically low electronic conductivity, low lithium-ion diffusion coefficient and wide bandgap limit its direct practical use in the energy industry and in photocatalytic-based applications (e.g., construction and water treatment). However, nanoscale innovations such as TiO_2 nanostructures, nanotubes, nanowires, etc. open the possibilities of application in these industries and others. In addition, many ways of doping with innumerable elements have improved the TiO_2 optical response, allowing more efficient utilization of solar light beyond the UV component. In solar cells, that increases conversion efficiency by several orders of magnitude for practical application.

In the construction field, in addition to the enhancement of visible light activation of TiO_2, chemical stability and mechanical resistance of TiO_2 nanostructures and nanocoatings are also being studied in real conditions in order to evaluate durability, life expectancy and reactivation conditions. In this regard, research and development efforts in the construction field are oriented to the present challenges of photocatalytic TiO_2 coatings and are the need for efficiency improvement, laboratory to industrial scaling up and commercial product production.

In biomedical devices, new developments are performed using titanium as a structural material and TiO_2 coatings as a modified interface with human tissues. In this field and in food industry innovation, titanium materials are specially regulated and there are many necessary types of research focused on nanobiotoxicology.

The given improvements have confirmed that nano-TiO_2 plays a significant role in many applications due to its interesting properties in compatibility with current technologies in energy, construction, water treatment, medical and food conservation fields.

Moreover, it has been shown that, for each specific application, TiO_2 properties have been intensively investigated to overcome obstacles in order to achieve practical application and to scale up technologies.

Acknowledgments

This work was supported by Agencia Nacional de Promoción Científica y Tecnológica (ANPCyT) from Argentina under PICT 2017-2133 and PICT 2017-2494 grants.

References

Aboulouard, A., B. Gultekin, M. Can, M. Erol, A. Jouaiti, B. Elhadadi et al. 2020. Dye sensitized solar cells based on titanium dioxide nanoparticles synthesized by flame spray pyrolysis and hydrothermal sol-gel methods: a comparative study on photovoltaic performances. J Mater Res Technol 9: 1569–1577.

AD17A. 2019. Arrêté du 17 avril 2019 Portant Suspension de la Mise sur le Marché des Denrées Contenant l'Additif E 171 (dioxyde de titane-TiO_2). Available online: https://www.legifrance.gouv.fr/affichTexte.do?cidTexte= JORFTEXT000038410047&categorieLien=id (accessed on 24 May 2021).

Aldoasri, M.A., S.S. Darwish, M.A. Adam, N.A. Elmarzugi and S.M. Ahmed. 2017. Protecting of marble stone facades of historic buildings using multifunctional TiO_2 nanocoatings. Sustainabil. Basel 9(11): 2002.

Allaker, R.P. 2010. The use of nanoparticles to control oral biofilm formation. J. Dent. Res. 89(11): 1175–86.

Amerio, O.N., M.R. Rosenberger, P.C. Favilla, M.A. Alterach and C.E. Schvezov. 2006. Prosthetic heart trileaflet valve associated to the last generation of bio- and haemo-compatible materials. Rev. Arg. Cirugía Cardiovascular 4: 70–76.

Amerio, O.N. and C.E. Schvezov. 2019. Trileaflet mechanical prosthetic heart valve. U.S. Patent # 10,478,288-B2.

Amin, M.R., M.A. Chowdhury and M.A. Kowser. 2019. Characterization and performance analysis of composite bioplastics synthesized using titanium dioxide nanoparticles with corn starch. Heliyon 5(8): e02009.

Anh, L.T., A.K. Rai, T.V. Thi, J. Gim, S. Kim, E.C. Shin et al. 2013. Improving the electrochemical performance of anatase titanium dioxide by vanadium doping as an anode material for lithium-ion batteries. J. Power Sources 243: 891–898.

ANSSAET. 2017. Agence Nationale de Sécurité Sanitaire de l'Alimentation, de l'Environnement et du Travail (ANSES). Avis Relatif à Une Demande d'Avis Relatif à l'Exposition Alimentaire Aux Nanoparticules de Dioxyde de Titane; ANSES: Paris, France.

Armand, M., P. Axmann, D. Bresser, M. Copley, K. Edstrom, C. Ekberg et al. 2020. Lithium-ion batteries—Current state of the art and anticipated developments. J. Power Sources 15: 1–26.

Azzawi, Z.G.M, T.I. Hamad and S.A. Kadhim. 2017. A modified laser deposition technique for depositing titania nanoparticles on commercially pure titanium substrates (mechanical evaluation). Int. J. Innov. Res. Sci. Eng. Technol. 6: 350–357.

Azzawi, Z.G.M., T.I. Hamad, S.A. Kadhim and G.A.H. Naji. 2018. Osseointegration evaluation of laser-deposited titanium dioxide nanoparticles on commercially pure titanium dental implants. J. Mater Sci: Mater. Med. 29: 96: 1–11.

Bachler, G., N. von Goetz and K. Hungerbuhler. 2015. Using physiologically based pharmacokinetic (PBPK) modeling for dietary risk assessment of titanium dioxide (TiO$_2$) nanoparticles. Nanotoxicol. 9: 373–380.

Bahnemann, W., M. Muneer and M.M. Haque. 2007. Titanium dioxide-mediated photocatalysed degradation of few selected organic pollutants in aqueous suspensions. Catal Today 124: 133–148.

Bai, Y., I. Mora-Sero, F. De Angelis, J. Bisquert and P. Wang. 2014. Titanium dioxide nanomaterials for photovoltaic applications. Chem. Rev. 114: 10095–10130.

Baranowska-Wójcik, E., D. Szwajgier, P. Oleszczuk and A. Winiarska-Mieczan. 2020. Effects of titanium dioxide nanoparticles exposure on human health—a review. Biol. Trace Elem. Res. 193: 118–129.

Basavarajappa, P.S., S.B. Patil, N. Ganganagappa, K.R. Reddy, A.V. Raghu and C.V Reddy. 2020. Recent progress in metal-doped TiO$_2$, non-metal doped/codoped TiO$_2$ and TiO$_2$ nanostructured hybrids for enhanced photocatalysis. Int. J. Hydrogen Energy 45(13): 7764–7778.

Bauer, S., J. Park, J. Faltenbacher, S. Berger, K. von der Mark and P. Schmuki. 2009. Size selective behavior of mesenchymal stem cells on ZrO$_2$ and TiO$_2$ nanotube arrays. Integr. Biol. 1(8-9): 525–532.

Behnajady, M.A., H. Dadkhah and H. Eskandarloo. 2018. Horizontally rotating disc recirculated photoreactor with TiO$_2$-P25 nanoparticles immobilized onto a HDPE plate for photocatalytic removal of p-nitrophenol. Environ. Technol. 39(8): 1061–1070.

Bensouici, F., M. Bououdina, A.A. Dakhel, R. Tala-Ighil, M. Tounane, A. Iratni and W. Cai. 2017. Optical, structural and photocatalysis properties of Cu-doped TiO$_2$ thin films. Appl. Surf. Sci. 395: 110–116.

Bettini, S., E. Boutet-Robinet, C. Cartier, C. Coméra, E. Gaultier, J. Dupuy et al. 2017. Food-grade TiO$_2$ impairs intestinal and systemic immune homeostasis, initiates preneoplastic lesions and promotes aberrant crypt development in the rat colon. Sci. Rep. 7: 1–13.

Boonen, E. and A. Beeldens. 2014. Recent photocatalytic applications for air purification in Belgium. Coat 4(3): 553–573.

Bordes, M.C., M. Vicent, R. Moreno, J. García-Montaño, A. Serra and E. Sánchez. 2015. Application of plasma-sprayed TiO$_2$ coatings for industrial (tannery) wastewater treatment. Ceramics Int. 41(10): 14468–14474.

Boutillier, S., S. Fourmentin and B. Laperche. 2020. Food additives and the future of health: An analysis of the ongoing controversy on titanium dioxide. Futures 122: 102598.

Brodd, R.J. 2002. Comments on the history of lithium-ion batteries, in: Meeting Abstracts, the Electrochemical Society, ECS. Available online: https://www.electrochem. org/dl/ma/201/pdfs/0259.pdf.

Bruecker, C. and Q. Li. 2020. Possible early generation of physiological helical flow could benefit the triflotrileaflet heart valve prosthesis compared to bileaflet valves. Bioeng. 7(158): 1–16.

Brunette, D.M., P. Tengvall, M. Textor and P. Thomsen. 2001. Titanium in Medicine: Material Science, Surface Science, Engineering, Biological Responses and Medical Applications. Springer Verlag, Switzerland.

Butany, J., M.S. Ahluwalia, C. Munroe, C. Fayet, C. Ahn, P. Blit et al. 2003. Mechanical heart valve prostheses: Identification and evaluation. Cardiovasc. Pathol. 12: 1–22.

Buser, D., R. Mericske-Stern, J.P. Bernard, A. Behneke, N. Behneke, H.P. Hirt et al. 1997. Long-term evaluation of non-submerged ITI implants. Part 1: 8-year life table analysis of a prospective multi-center study with 2 359 implants. Clin. Oral Implants Res. 8: 161–172.

Buzarovska, A., S. Dinescu, L. Chitoiu and M. Costache. 2018. Porous poly(L-lactic acid) nanocomposite scaffolds with functionalized TiO$_2$ nanoparticles: Properties, cytocompatibility and drug release capability. J. Mater Sci. 53: 11151–11166.

Calia, A., M. Lettieri and M. Masieri. 2016. Durability assessment of nanostructured TiO$_2$ coatings applied on limestones to enhance building surface with self-cleaning ability. Build Environ. 110: 1–10.

Calia, A., M. Lettieri, M. Masieri, S. Pal, A. Licciulli and V. Arima. 2017. Limestones coated with photocatalytic TiO$_2$ to enhance building surface with self-cleaning and depolluting abilities. J. Clean Prod. 165: 1036–1047.

Cedillo-González, E.I., J.M. Hernández-López, J.J. Ruiz-Valdés, C. Barbieri and C. Siligardi. 2020. Self-cleaning TiO$_2$ coatings for building materials: The influence of morphology and humidity in the stain removal performance. Constr. Build Mater. 237: 117692.

Chaudhari, S., T. Shaikh and P. Pandy. 2013. A Review on polymer TiO$_2$ nanocomposites. Int. J. Eng. Res. Appl. 3: 1386–1391.

Chang, S.M. and W.S. Liu. 2014. The roles of surface-doped metal ions (V, Mn, Fe, Cu, Ce, and W) in the interfacial behavior of TiO_2 photocatalysts. Appl. Catal. B Environ. 156: 466–475.

Chemin, J.B., S. Bulou, K. Baba, C. Fontaine, T. Sindzingre, N.D. Boscher et al. 2018. Transparent anti-fogging and self-cleaning TiO_2/SiO_2 thin films on polymer substrates using atmospheric plasma. Sci. Rep-Uk 8(1): 1–8.

Chen, D., Y. Cheng, N. Zhou, P. Chen, Y. Wang, K. Li et al. 2020. Photocatalytic degradation of organic pollutants using TiO_2-based photocatalysts: A review. J. Cleaner Prod. 268: 121725.

Chen, J. and C.S. Poon. 2009. Photocatalytic construction and building materials: from fundamentals to applications. Build Environ. 44(9): 1899–1906.

Chi, H., S. Song, M. Luo, G. Zhang, W. Li, L. Li et al. 2019. Effect of PLA nanocomposite films containing bergamot essential oil, TiO_2 nanoparticles, and Ag nanoparticles on shelf life of mangoes. Sci. Hortic. 249: 192–198.

Colangiuli, D., M. Lettieri, M. Masieri and A. Calia. 2019. Field study in an urban environment of simultaneous self-cleaning and hydrophobic nanosized TiO_2-based coatings on stone for the protection of building surface. Sci. Total Environ. 650: 2919–2930.

De Cie GmbH. 2021. (20 April 2021). https://www.decie.de.

Diamanti, M.V., M. Ormellese and M. Pedeferri. 2015. Application-wise nanostructuring of anodic films on titanium: a review. J. Exp. Nanosci. 10(17): 1285–1308.

Dibble, L.A. and G.B. Raupp. 1990. Kinetics of the gas-solid heterogeneous photocatalytic oxidation of trichloroethylene by near UV illuminated titanium dioxide. Catal. Lett. 4(4): 345–354.

Diebold, U. 2003. The surface science of titanium dioxide. Surf. Sci. Report 48: 53–299.

Dikici, T., M. Erol, M. Toparli and E. Celik. 2014. Characterization and photocatalytic properties of nanoporous titanium dioxide layer fabricated on pure titanium substrates by the anodic oxidation process. Ceramics Int. 40(1): 1587–1591.

Dwojak, A.N., M.L. Vera, H.D. Traid, C.E. Schvezov and M.I. Litter. 2017. Evaluación de la eficiencia fotocatalítica de recubrimientos anódicos nanoporosos de TiO_2. Proc. 6° Encuentro de Jóvenes Investigadores en Ciencia y Tecnología de Materiales. Argentina. T06.12.

Dwojak, A.N., M.L. Vera, H.D. Traid, M.I. Litter and C.E. Schvezov. 2018. Evaluación de la morfología de recubrimientos anódicos nanotubulares de TiO_2 de tercera generación. Proc. Jornadas Científico Tecnológicas 45° Aniversario de la Universidad Nacional de Misiones, Argentina.

Dwojak, A.N., M.L. Vera, H.D. Traid, M.I. Litter and C.E. Schvezov. 2019. Fabricación de recubrimientos nanotubulares de TiO_2 mediante oxidación anódica en NH_4-glicerol-agua. Revista SAM 1: 52–57.

Dwojak, A.N., M.L. Vera, H.D. Traid, M.F. Maydana, M.I. Litter and C.E. Schvezov. 2021. Influence of anodizing variables on Cr(VI) photocatalytic reduction using TiO_2 nanotubes obtained by anodic oxidation. Environ. Nanotech. Monitor Manag. 16: 100537.

EFSA ANS. 2016. European Food Safety Authority. Panel on Food Additives and Nutrient Sources added to Food. Scientific opinion on the re-evaluation of titanium dioxide (E 171) as a food additive. EFSA J. 14(83): e04545.

Eisenbarth, E. et al. 2002. Interactions between cells and titanium surfaces. Biomol. Eng. 19: 243–249.

Fadl, A.M., M.I. Abdou, M.A. Hamza and S.A. Sadeek. 2020. Corrosion-inhibiting, self-healing, mechanical-resistant, chemically and UV stable PDMAS/TiO_2 epoxy hybrid nanocomposite coating for steel petroleum tanker trucks. Prog. Org. Coat 146: 105715.

Farhanian, D., F. Haghighat, C.S. Lee and N. Lakdawala. 2013. Impact of design parameters on the performance of ultraviolet photocatalytic oxidation air cleaner. Build Environ. 66: 148–157.

Farhoodi, M., M. Daddashi, M.A. Mohammad, A. Mousavi and Z. Djomeh. 2012. Influence of TiO_2 nanoparticle filler on the properties of PET and PLA nano composites. Polym 36: 745–755.

Favilla, P.C., M.A. Alterach, M.R. Rosenberger, A.E. Ares, C.E. Schvezov and O.N. Amerio. 2008. Properties of titanium oxide films deposited by the sol-gel. Proc TMS 2008. USA.

Favilla, P.C. 2009. Resistencia al desgaste y adherencia de recubrimientos de óxido de titanio sobre Ti6Al4V. Master Thesis, Universidad Nacional de General San Martin, Buenos Aires, Argentina.

Fonseca, C., A. Ochoa, M.T. Ulloa, E. Alvarez, D. Canales and P.A. Zapata. 2015. Poly(lactic acid)/TiO_2 nanocomposites as alternative biocidal and antifungal materials. Mater Sci. Eng. C 57: 314–320.

Fresno, F., D. Tudela, J.M. Coronado and J. Soria. 2009. Synthesis of $Ti_{1-x}Sn_xO_2$ nanosized photocatalysts in reverse microemulsions. Catal. Today 143: 230–236.

FSICT. 2021. FSI Coating Technologies. (15 March 2021). https://fsicti.com/

Fujishima, A. and X. Zhang. 2006. Titanium dioxide photocatalysis: present situation and future approaches. ComptesRendusChimie 9(5-6): 750–760.

García Saban, F.J., J.C. García Saban and M.A. Garcia Saban. 2012. Method for obtaining a surface of titanium-based metal implant to be inserted into bone tissue. U.S. Patent # 2013-0022,787-A1.

Gázquez, M.J., J.P. Bolívar, R. Garcia-Tenorio and F. Vaca. 2014. A review of the production cycle of titanium dioxide pigment. Mater Sci. Appl. 5: 441–458.

GENS. 2019. Green Hearth Nano Science Inc. https://www.gens.ca/gens-nano-self-santizing-coating-for-building-interiors.html.

GM. 2021. Green Millenium. (12 March 2021). http://www.greenmillennium.com/

Goffredo, G.B., S. Accoroni, C. Totti, T. Romagnoli, L. Valentini and P. Munafò. 2017. Titanium dioxide based nanotreatments to inhibit microalgal fouling on building stone surfaces. Build Environ. 112: 209–222.

Gong, D., W. Chye, J. Ho, Y. Tang, Q. Tay, Y. Lai et al. 2012. Silver decorated titanate/titania nanostructures for efficient solar driven photocatalysis. J. Solid State Chem. 189: 117–122.

Gopalan, A.I., J.C. Lee, G. Saianand, K.P. Lee, P. Sonar, R. Dharmarajan et al. 2020. Recent progress in the abatement of hazardous pollutants using photocatalytic TiO$_2$-based building materials. Nanomater. Basel 10(9): 1854.

González, A.S., J.C. Solis-Cortazar, C.A. Pineda-Arellano, E. Ramírez-Morales, A. Monteros and S. Silva-Martínez. 2019. Synthesis of ruthenium-doped TiO$_2$ nanotube arrays for the photocatalytic degradation of terasil blue dye. J. Nanosci. Nanotechnol. 19(8): 5211–5219.

Grätzel, M. 2005. Mesoscopic solar cells ruthenium for electricity and hydrogen production from sunlight. Chem. Lett. 34: 8–13.

Graziani, L., E. Quagliarini, A. Osimani, L. Aquilanti, F. Clementi, C. Yéprémian et al. 2013. Evaluation of inhibitory effect of TiO$_2$ nanocoatings against microalgal growth on clay brick façades under weak UV exposure conditions. Build Environ. 64: 38–45.

Graziani, L., E. Quagliarini, F. Bondioli and M. D'Orazio. 2014. Durability of self-cleaning TiO$_2$ coatings on fired clay brick façades: Effects of UV exposure and wet & dry cycles. Build Environ. 71: 193–203.

Gregoric, I. et al. 2004. Preclinical assessment of a trileaflet mechanical valve in the mitral position in a calf model. Ann. Thorac Surg. 77(1): 196–202.

Grissa, I., J. Elghoul, l. Ezzi, S. Chakroun, E. Kerkeni, M. Hassine et al. 2015. Anemia and genotoxicity induced by sub-chronic intragastric treatment of rats with titanium dioxide nanoparticles. Mutat. Res. Genet. Toxicol. Environ. Mutagen 794: 25–31.

Guo, M.Z., T.C. Ling and C.S. Poon. 2012. TiO$_2$-based self-compacting glass mortar: Comparison of photocatalytic nitrogen oxide removal and bacteria inactivation. Build Environ. 53: 1–6.

Guo, M.Z. and C.S. Poon. 2013. Photocatalytic NO removal of concrete surface layers intermixed with TiO$_2$. Build Environ. 70: 102–109.

Guo, H., D. Klose, Y. Hou, G. Jeschke and I. Burgert. 2017. Highly efficient UV protection of the biomaterial wood by a transparent TiO$_2$/Ce xerogel. ACS Appl. Mater Inter. 9(44): 39040–39047.

Guo, Q., Ch. Zhou, Z. Ma and X. Yang. 2019. Fundamentals of TiO$_2$ photocatalysis: concepts, mechanisms, and challenges. Adv. Mater 31: 1901997.

GS. 2021. Google Scholar (20 April 2021). http://www.scholar.google.com.

Haldorai, Y., A. Rengaraj, C.H. Kwak, Y.S. Huh and Y.K. Han. 2014. Fabrication of nano TiO$_2$ graphene composite: Reusable photocatalyst for hydrogen production, degradation of organic and inorganic pollutants. Synthetic Metals 198: 10–18.

Hansson, S. 1998. Method for the preparation of implants made of titanium or alloys thereof. U.S. Patent # 5,667,385-A.

Hashimoto, K. 2007. TiO$_2$ Photocatalysis towards novel building materials. Proc. Int. RILEM Symposium on Photocatalysis, Environment and Construction Materials. Italy. 3–8.

Hu, X., X. Hu, C. Tang, S. Wen, X. Wu, J. Long and L. Zhou. 2017. Mechanisms underlying degradation pathways of microcystin-LR with doped TiO$_2$ photocatalysis. Chem. Eng. J. 330: 355–371.

Huang, H., J. Yu, Y. Gan, Y. Xia, C. Liang, J. Zhang et al. 2017. Hybrid nanoarchitecture of TiO$_2$ nanotubes and graphene sheet for advanced lithium ion batteries. Mater Res. Bull 96: 425–430.

Huang, N. and P. Yang. 2003. Hemocompatibility of titanium oxide films. Biomater. 23: 2177–2187.

Hurum, D.C., A.G. Agrios, K.A. Gray, T. Rajh and M.C. Thurnauer. 2003. Explaining the enhanced photocatalytic activity of Degussa P25 mixed-phase TiO$_2$ using EPR. J. Phys. Chem. B 107(19): 4545–4549.

Hwang, H.M., S. Oh, J.H. Shim, Y.M. Kim, A. Kim, D. Kim et al. 2019. Phase-selective disordered anatase/ordered rutile interface system for visible-light-driven, metal-free CO$_2$ reduction. ACS Appl. Mater Interfaces 11: 35693–35701.

Irie, H., K. Sunada and K. Hashimoto. 2004. Recent developments in TiO$_2$ photocatalysis: novel applications to interior ecology materials and energy saving systems. Electrochemistry 72(12): 807–812.

Jahdi, M., S.B. Mishra, E.N. Nxumalo, S.D. Mhlanga and A.K. Mishra. 2020. Smart pathways for the photocatalytic degradation of sulfamethoxazole drug using F-Pd co-doped TiO$_2$ nanocomposites. Appl. Catal. B 267: 118716.

Jia, J., J. Dong, J. Lin, Z. Lan, L. Fan and J. Wu. 2019. Improved photovoltaic performance of perovskite solar cells by utilizing down-conversion NaYF$_4$:Eu^{3+} nanophosphors. J. Mater Chem. C 7: 937–942.

Kang, X., S. Liu, Z. Dai, Y. He, X. Song and Z. Tan. 2019. Titanium dioxide: From engineering to applications. Catal. 9(191): 1–32.

Karnahl, M., E. Meja, N. Rockstroh, S. Tschierlei, S.P. Luo, K. Grabow et al. 2014. Photocatalytic hydrogen production with copper photosensitizer-titanium dioxide composites. Chem. Cat. Chem. 6: 82–86.

Kartini, I., I.Y. Khairani, K. Triyana and S. Wahyuni. 2018. Nanostructured titanium dioxide for functional coatings. pp. 445–468. *In*: Yang, D. (ed.). Titanium Dioxide—Material for a Sustainable Environment. Intech Open, London, UK.

Kaseem, M., K. Hamad and Z. Ur Rehman. 2019. Review of recent advances in polylactic acid/TiO_2 composites. Mater 12: 3659–3675.

Khataee, A. and G.A. Mansoori. 2012. Nanostructured Materials Titanium Dioxide Properties, Preparation and Applications. World Scientific Publishing Co. Pte. Ltd., Singapore.

Kim, H.S., S.H. Yu, Y.E. Sung and S.H. Kang. 2014. Carbon treated self- ordered TiO_2 nanotube arrays with enhanced lithium-ion intercalation performance. J. Alloys Compd. 597: 275–281.

Kim, Y.S., K.R. Shin, G.W. Kim, Y.G. Ko and D.H. Shin. 2016. Photocatalytic activity of TiO_2 film containing Fe_2O_3 via plasma electrolytic oxidation. Surf. Eng. 32(6): 443–447.

Kleiman, A., A. Márquez, M.L. Vera, J.M. Meichtry and M.I. Litter. 2011. Photocatalytic activity of TiO_2 thin films deposited by cathodic arc. Appl. Catal. B 101(3-4): 676–681.

Kociubczyk, A.I., M.L. Vera, C.E. Schvezov, E. Heredia and A.E. Ares. 2015. TiO_2 coatings in alkaline electrolytes using anodic oxidation technique. Procedia Mater Sci. 8: 65–72.

Kurtz, A.E., H.D. Traid, M.L. Vera and M.I. Litter. 2018. Influencia de la geometría del sustrato en estructuras nanotubulares de TiO_2 obtenidas por oxidación anódica. Proc. 18° Congreso Internacional de Metalurgia y Materiales SAM-CONAMET, Argentina. 1150–1152.

La Russa, M.F., A. Macchia, S.A. Ruffolo, F. De Leo, M. Barberio, P. Barone et al. 2014. Testing the antibacterial activity of doped TiO_2 for preventing biodeterioration of cultural heritage building materials. Int. Biodeter. Biodegr. 96: 87–96.

Lan, T., H. Qiu, F. Xie, J. Yang and M. Wei. 2015. Rutile TiO_2 mesocrystals/reduced graphene oxide with high-rate and long-term performance for lithium-ion batteries. Sci. Rep. 5(1): 1–6.

Lapeyre, D.M. et al. 1994. *In vivo* evaluation of a trileaflet mechanical heart valve. ASAIO J. 40(3): M707–M713.

Lapeyre, D. 2010. Mechanical prosthetic heart valve. U.S. Patent # 20,100,131,056-A1.

Lee, S.Y. and S.J. Park. 2013. TiO_2 photocatalyst for water treatment applications. J. Ind. Eng. Chem. 19(6): 1761–1769.

Lee, Y. and M. Kang. 2010. The optical properties of nanoporous structured titanium dioxide and the photovoltaic efficiency on DSSC. Mater Chem. Phys. 122: 284–289.

Leng, Y.X., N. Huang, P. Yang, J.Y. Chen, H. Sun, J. Wang et al. 2002. Influence of oxygen pressure on the properties and biocompatibility of titanium oxide fabricated by metal plasma ion implantation and deposition. Thin Solid Films 420-421: 408–413.

Leng, Y.X., J.Y. Chen, P. Yang, J. Wang, A.S. Zhao, G.J. Wan et al. 2006. The microstructure and mechanical properties of TiN and TiO_2/TiN duplex films synthesized by plasma immersion ion implantation and deposition on artificial heart valve. Surf. Coat Technol. 201(3-4): 1012–1016.

Lettieri, M., D. Colangiuli, M. Masieri and A. Calia. 2019. Field performances of nanosized TiO_2 coated limestone for a self-cleaning building surface in an urban environment. Build Environ. 147: 506–516.

Li, J.M., W. Wan, F. Zhu, Q. Li, H.H. Zhou, J.J. Li et al. 2012a. Nanotube-based hierarchical titanate microspheres: an improved anode structure for Li-ion batteries. Chem. Commun. 48: 389–391.

Li, S. and J. Fu. 2013. Improvement in corrosion protection properties of TiO_2 coatings by chromium doping. Corros Sci. 68: 101–110.

Li, K., B. Peng, J. Jin, L. Zan and T. Peng. 2017. Carbon nitride nanodots decorated brookite TiO_2 quasi nanocubes for enhanced activity and selectivity of visible-light-driven CO_2 reduction. Appl. Catal. B 203: 910–916.

Li, L., J. Yan, T. Wang, Z.J. Zhao, J. Zhang, J. Gong and N. Guan. 2015. Sub-10 nm rutile titanium dioxide nanoparticles for efficient visible-light-driven photocatalytic hydrogen production. Nat. Commun. 6: 5881.

Li, X., H. Liu, D. Luo, J. Li, Y. Huang, H. Li et al. 2012b. Adsorption of CO_2 on heterostructure CdS(Bi_2S_3)/TiO_2 nanotube photocatalysts and their photocatalytic activities in the reduction of CO_2 to methanol under visible light irradiation. Chem. Eng. J. 180: 151–158.

Li, X., J. Yu and M. Jaroniec. 2016. Hierarchical photocatalysts. Chem. Soc. Rev. 45(9): 2603–2636.

Li, W., L. Li, H. Zhang, M. Yuan and Y. Qin. 2018. Evaluation of PLA nanocomposite films on physicochemical and microbiological properties of refrigerated cottage cheese. J. Food Process Pres. 42: e13362.

Litter, M.I. 1999. Heterogeneous photocatalysis; transition metal ions in photocatalytic systems. Appl. Catal. B 23: 89–114.

Litter, M.I. 2017. Last advances on TiO_2-photocatalytic removal of chromium, uranium and arsenic. Curr. Opin. Green Sustain Chem. 6: 150–158.

Liu, H., A.Q. Dao and C. Fu. 2016. Activities of combined TiO_2 semiconductor nanocatalysts under solar light on the reduction of CO_2. J. Nanosci. Nanotechnol. 16: 3437–3446.

Liu, J., H. Shi, Q. Shen, C. Guo and G. Zhao. 2017. Efficiently photoelectrocatalyze CO_2 to methanol using Ru(II)-pyridyl complex covalently bonded on TiO_2 nanotube arrays. Appl. Catal. B Environ. 210: 368–378.

Liu, K., M. Cao, A. Fujishima and L. Jiang. 2014a. Bio-inspired titanium dioxide materials with special wettability and their applications. Chem. Rev. 114: 10044–10094.

Liu, R., P. Wang, X. Wang, H. Yu and J. Yu. 2012. UV- and visible-light photocatalytic activity of simultaneously deposited and doped Ag/Ag (I)-TiO_2 photocatalyst. J. Phys Chem. C 116(33): 17721–17728.

Liu, X., Z. Liu, J. Lu, X. Wu and W. Chu. 2014b. Silver sulfide nanoparticles sensitized titanium dioxide nanotube arrays synthesized by in situ sulfurization for photocatalytic hydrogen production. J. Coll. Interface Sci. 413: 17–23.

Löffler, F.B., F.J. Altermann, E.C. Bucharsky, K.G. Schell, M.L. Vera, H.D. Traid and M.I. Litter. 2020. Morphological characterization and photocatalytic efficiency measurements of pure silica transparent open-cell sponges coated with TiO_2. Inter J. Appl. Ceramic Technol. 17(4): 1930–1939.

Lütjering, G. and J. Williams. 2007. Titanium. Springer, Switzerland.

Luttrell, T., S. Halpegamage, J. Tao, A. Kramer, E. Sutter and M. Batzill. 2014. Why is anatase a better photocatalyst than rutile?-Model studies on epitaxial TiO_2 films. Sci. Rep. 4(1): 1–8.

Ma, K., R. Zhang, J. Sun and C. Liu. 2020. Oxidation mechanism of biomedical titanium alloy surface and experiment. Inter. J. Corr. 2020: 1678615.

Madhavi, V. and P. Kondaiah. 2018. Influence of silver nanoparticles on titanium oxide and nitrogen doped titanium oxide thin films for sun light photocatalysis. Appl. Surf. Sci. 436: 708–719.

Maicu, M., M.C. Hidalgo, G. Colón and J.A. Navío. 2011. Comparative study of the photodeposition of Pt, Au and Pd on pre-sulphated TiO_2 for the photocatalytic decomposition of phenol. J. Photochem. Photobiol. A Chem. 217(2-3): 275–283.

MAM. 2021. Metalpedia Asian Metal. (20 March 2021). http://metalpedia.asianmetal.com/metal/titanium.

Mamaghani, A.H., F. Haghighat and C.S. Lee. 2018. Photocatalytic degradation of VOCs on various commercial titanium dioxides: Impact of operating parameters on removal efficiency and by-products generation. Build Environ. 138: 275–282.

Mandari, K.K., J.Y. Do, S.V.P. Vattikuti, A.K.R. Police and M. Kang. 2018. Solar light response with noble metal-free highly active copper(II) phosphate/titanium dioxide nanoparticle/copper(II) oxide nanocomposites for photocatalytic hydrogen production. J. Alloys Compd. 750: 292–303.

Mazierski, P., M. Nischk, M. Gołkowska, W. Lisowski, M. Gazda, M.J. Winiarski and A. Zaleska-Medynska. 2016. Photocatalytic activity of nitrogen doped TiO_2 nanotubes prepared by anodic oxidation: The effect of applied voltage, anodization time and amount of nitrogen dopant. Appl. Catal. B Environ. 196: 77–88.

Mekkawi, D.M., N.A. Abdelwahab, W.A. Mohamed, N.A. Taha and M.S.A. Abdel-Mottaleb. 2020. Solar photocatalytic treatment of industrial wastewater utilizing recycled polymeric disposals as TiO_2 supports. J. Cleaner Prod. 249: 119430.

Memmolo, M., J. Hall and E. De Haller. 2014. Dental Implant. U.S. Patent # 2018-03553268-A1.

Meng, X., W. Kuang, W. Qi, Z. Cheng, T. Thomas, S. Liu et al. 2020. Ultra low loading of Au clusters on nickel nitride efficiently boosts photocatalytic hydrogen production with titanium dioxide. Chem. Cat. Chem. 12: 2752–2759.

Miao, R., Z. Luo, W. Zhong, S.Y. Chen, T. Jiang, B. Dutta and S.L. Suib. 2016. Mesoporous TiO_2 modified with carbon quantum dots as a high-performance visible light photocatalyst. Appl. Catal. B Environ. 189: 26–38.

Mizushima, K., P.C. Jones, P.J. Wiseman and J.B. Goodenough. 1980. $LixCoO_2$ (0 < x < 1): a new cathode material for batteries of high energy density. Mater Res. Bull 15: 783–789.

Mohammadi, H and K. Mequanint. 2011. Prosthetic aortic heart valves: modeling and design. Med. Eng. Phys. 33(2): 131–147.

Moradi, V., M.B. Jun, A. Blackburn and R.A. Herring. 2018. Significant improvement in visible light photocatalytic activity of Fe doped TiO_2 using an acid treatment process. Appl. Surf. Sci. 427: 791–799.

Murray, J.L. 1990. Phase diagrams of binary titanium alloys. Monograph series on alloy phase diagrams. ASM International, USA, 211–227.

Musial, J., R. Krakowiak, D.T. Mlynarczyk, T. Goslinski and B.J. Stanisz. 2020. Titanium dioxide nanoparticles in food and personal care products—What do we know about their safety? Nanomater. 10(1110): 1–23.

Myung, S.T., N. Takahashi, S. Komaba, C.S. Yoon, Y.K. Sun, K. Amine et al. 2011. Nanostructured TiO_2 and its application in lithium ion storage. Adv. Funct. Mater 21: 3231–3241.

Nanaji, K., R.K.S.K. Janardhana, T.N. Rao and S. Anandan. 2019. Energy level matching for efficient charge transfer in Ag doped-Ag modified TiO_2 for enhanced visible light photocatalytic activity. J. Alloys Compd. 794: 662–671.

Neatu, S., J.A. Maciá-Agulló and H. Garcia. 2014. Solar light photocatalytic CO_2 reduction: general considerations and selected bench-mark photocatalysts. Int. J. Mol. Sci. 15: 5246–5262.

NFM. Nanocoatings. Future Markets. 2014. Construction & Exterior Protection. Edition 3.

Noman, M.T., M.A. Ashraf and A. Ali. 2018. Synthesis and applications of nano-TiO_2: a review. Environ. Sci. Poll Res. 26: 3262–3291.

Nowotny, J. 2012. Oxide Semiconductors for Solar Energy Conversion Titanium Dioxide. CRC Press. Boca Raton, USA.

Ohtani, B., O.O. Prieto-Mahaney, D. Li and R. Abe. 2010. What is Degussa (Evonik) P25? Crystalline composition analysis, reconstruction from isolated pure particles and photocatalytic activity test. J. Photochem. Photobiol. A Chem. 216(2-3): 179–182.

Oshida, Y. 2000. Titanium material implants. U.S. Patent # 6,183,255-B1.

Park, J., S. Bauer, A. Pittrof, M.S. Killian, P. Schmuki and K. von der Mark. 2011. Synergistic control of mesenchymal stem cell differentiation by nanoscale surface geometry and immobilized growth factors on TiO_2 nanotubes. Small 8(1): 98–107.

Patel, S.K.S., J.K. Lee and V.C. Kalia. 2018. Nanoparticles in biological hydrogen production: an overview. Indian J Microbiol. 58: 8–18.

Pathak, P., M.J. Meziani, L. Castillo and Y.P. Sun. 2005. Metal-coated nanoscale TiO_2 catalysts for enhanced CO_2 photoreduction. Green Chem. 7: 667–670.

Pelaez, M., N.T. Nolan, S.C. Pillai, M.K. Seery, P. Falaras, A.G. Kontos and D.D. Dionysiou. 2012. A review on the visible light active titanium dioxide photocatalysts for environmental applications. Appl. Catal. B Environ. 125: 331–349.

Pradhan, D., A.W. Wren, S.T. Misture and N.P. Mellott. 2016. Investigating the structure and biocompatibility of niobium and titanium oxides as coatings for orthopedic metallic implants. Mater Sci. Eng. C 58: 918–926.

Proquin, H., C. Rodríguez-Ibarra, C.G.J. Moonen, I.M. Urrutia Ortega, J.J. Briedé, T.M. de Kok et al. 2016. Titanium dioxide food additive (E171) induces ROS formation and genotoxicity: Contribution of micro and nano-sized fractions. Mutagen 32: 139–149.

Qiu, X., M.P. Paranthaman, M. Chi, I.N. Ivanov and Z. Zhang. 2014. Array of titanium dioxide nanostructures for solar energy utilization. U.S. Patent # 8,920,767-B2.

Rani, M., S.J. Abbas and S.K. Tripathi. 2014. Influence of annealing temperature and organic dyes as sensitizers on sol-gel derived TiO_2 films. Mater Sci. Eng. B 187: 75–82.

Ratner, B.D., A.S. Hoffman, F.J. Schoen and J.E. Lemons. 2004. Biomaterials Science: An Introduction to Materials in Medicine, Academic Press, London.

Reddy, K.R., M.S. Jyothi, A.V. Raghu, V. Sadhu, S. Naveen and T.M. Aminabhavi. 2020. Nanocarbons-supported and polymers-supported titanium dioxide nanostructures as efficient photocatalysts for remediation of contaminated wastewater and hydrogen production. pp. 139–169. *In*: Inamuddin, Asiri A. and Lichtfouse E. (eds.). Nanophotocatalysis and Environmental Applications. Environmental Chemistry for a Sustainable World. Springer, Cham. Switzerland.

Ren, Z., Y. Guo, C.H. Liu and P.X. Gao. 2013. Hierarchically nanostructured materials for sustainable environmental applications. Front Chem. 1: 1–22.

Ropers, M.H., H. Terrisse, M. Mercier-Bonin and B. Humbert. 2017. In Application of Titanium Dioxide. Chapter Titanium Dioxide as Food Additive. INTECH. Rijeka. Edited by Magdalena Janus. Croatia.

Rosenberger, M., P. Favilla, M. Alterach, O. Amerio and C. Schvezov. 2005. Modelización, diseño y construcción de un prototipo de prótesis de válvula cardiaca. Rev. CENIC Ciencias Biológicas 36: 1–9.

Rosenberger, M.R., L.A. Guerrero, M.L. Vera and C.E. Schvezov. 2013. Erosión de materiales para aplicaciones en prótesis de válvulas cardíacas. Ann AFA 24(2): 71–76.

Ruffolo, S.A., A. Macchia, M.F. La Russa, L. Mazza, C. Urzi, F. De Leo et al. 2013. Marine antifouling for underwater archaeological sites: TiO_2 and Ag-Doped TiO_2. Int. J. Photoenergy 251647: 1–6.

Salian, G.D., B.M. Koo, C. Lefevre, T. Cottineau, C. Lebouin, A.T. Tesfaye, P. Knauth, V. Keller and T. Djenizian. 2018. Niobium alloying of self-organized TiO_2 nanotubes as an anode for lithium-ion microbatteries. Adv. Mater Technol. 3: 1700274.

Sauer, I.L., J.F. Escobar, M.F.P. da Silva, C.G. Meza, C. Centurion and J. Goldemberg. 2015. Bolivia and Paraguay: a beacon for sustainable electric mobility? Renew. Sustain Energy Rev. 51: 910–925.

Schuster, J.M., C.E. Schvezov and M.R. Rosenberger. 2012. Evaluación de la hemocompatibilidad del óxido de titanio y del carbón pirolítico. Ann. AFA 24: 77–82.

Schuster, J.M., C.E. Schvezov and M.R. Rosenberger. 2015. Analysis of the results of surface free energy measurement of Ti6Al4V by different methods. Procedia Mater Sci. 8: 732–741.

Schvezov, C.E., M.A. Alterach, M.L. Vera, M.R. Rosenberger and A.E. Ares. 2010. Characteristics of haemocompatible TiO_2 nano-films produced by the sol-gel and anodic oxidation techniques. JOM 84–87.

Schvezov, C.E., M.L. Vera, J.M. Schuster and M.R. Rosenberger. 2017. Production and characterization of TiO_2 nanofilms for hemocompatible and photocatalytic applications. JOM 69(10): 2038–2044.

Segura Gonzalez, E.A., D. Olmos, M.A. Lorente, I. Velaz and J. Gonzalez-Benito. 2018. Preparation and characterization of polymer composite materials based on PLA/TiO₂ for antibacterial packaging. Polym. 10: e1365.

Selim, M.S., S.A. El-Safty, A.M. Azzam, M.A. Shenashen, M.A. El-Sockary and O.M. Abo Elenien. 2019. Superhydrophobic silicone/TiO₂-SiO₂ nanorod-like composites for marine fouling release coatings. Chem. Sel. 4: 3395–3407.

Shaban, M., A.M. Ahmed, N. Shehata, M.A. Betiha and A.M. Rabie. 2019. Ni-doped and Ni/Cr co-doped TiO₂ nanotubes for enhancement of photocatalytic degradation of methylene blue. J. Coll Interface Sci. 555: 31–41.

Shahnooshi, M., A. Eshaghi and A.A. Aghaei. 2019. Transparent anti-fogging and anti-scratch SiO₂/SiO₂-TiO₂ thin film on polycarbonate substrate. Mater Res. Express 6(8): 086447.

Shebi, A. and S. Lisa. 2019. Evaluation of biocompatibility and bactericidal activity of hierarchically porous PLA-TiO₂ nanocomposite films fabricated by breath-figure method. Mater Chem. Phys 230: 308–318.

Shen, L., X. Zhang, H. Li, C. Yuan and G. Cao. 2011. Design and tailoring of a three-dimensional TiO₂-graphene-carbon nanotube nanocomposite for fast lithium storage. J. Phys Chem. Lett. 2: 3096–3101.

Song, M., C. Pan, J.Y. Li, X.M. Wang and Z.Z. Gu 2006. Electrochemical study on synergistic effect of the blending of nano TiO₂ and PLA polymer on the interaction of antitumor drug with DNA. Electroanalysis 18: 1995–2000.

Song, M., C. Pan, C. Chen, J.Y. Li, X.M. Wang and Z.Z. Gu. 2008. The application of new nanocomposites: Enhancement effect of polylactide nanofibers/nano-TiO₂ blends on biorecognition of anticancer drug daunorubicin. Appl. Surf. Sci. 255: 610–612.

Sprong, C., M. Bakker, M. Niekerk and M. Vennemann. 2015. Exposure assessment of the food additive titanium dioxide (E 171) based on use levels provided by the industry (Internet). Available from: http://www.rivm.nl/dsresou.

Su, R., R. Bechstein, L. Sø, R.T. Vang, M. Sillassen, B. Esbjornsson and F. Besenbacher. 2011. How the anatase-to-rutile ratio influences the photoreactivity of TiO₂. J. Phys Chem. C 115(49): 24287–24292.

Sudrajat, H., S. Babel, A.T. Ta and T.K. Nguyen. 2020. Mn-doped TiO₂ photocatalysts: Role, chemical identity, and local structure of dopant. J. Phys. Chem. Solids 144: 109517.

Tang, Y.X., Y.Y. Zhang, J.Y. Deng, J.Q. Wei, H.L. Tam, B.K. Chandran et al. 2014. Mechanical force driven growth of elongated bending TiO₂-based nanotubular materials for ultrafast rechargeable lithium ion batteries. Adv. Mater 26: 6111–6118.

Taurino, R., L. Barbieri and F. Bondioli. 2016. Surface properties of new green building material after TiO₂-SiO₂ coatings deposition. Ceramic Int 42(4): 4866–4874.

Thambidurai, M., S. Foo, K.M. Salim, P.C. Harikesh, A. Bruno, N.F. Jamaludin et al. 2019. Improved photovoltaic performance of triple-cation mixed-halide perovskite solar cells with binary trivalent metals incorporated into the titanium dioxide electron transport layer. J. Mater Chem. C 7: 5028–5036.

Traid, H.D., M.L. Vera, A.E. Ares and M.I. Litter. 2017. Advances on the synthesis of porous TiO₂ coatings by anodic spark oxidation. Photocatalytic reduction of Cr(VI). Mater Chem. Phys 191: 106–113.

Traid, H.D. 2018a. Síntesis de recubrimientos porosos y nanotubulares de TiO₂ anódico aplicados a fotocatálisis heterogénea. Ph.D. Thesis, Universidad Nacional de Misiones, Misiones, Argentina.

Traid, H.D., A.N. Dwojak, M.L. Vera, A.E. Ares and M.I. Litter. 2018b. Porous titanium dioxide coatings synthesized by anodic oxidation. Matéria 23(2): e-12060.

Tricoli, A., M. Righettoni and S.E. Pratsinis. 2009. Anti-fogging nanofibrous SiO₂ and nanostructured SiO₂−TiO₂ films made by rapid flame deposition and in situ annealing. Langmuir 25(21): 12578–12584.

Truong, Q.D., T.H. Le, J.Y. Liu, C.C. Chung and Y.C. Ling. 2012. Synthesis of TiO₂ nanoparticles using novel titanium oxalate complex towards visible light-driven photocatalytic reduction of CO₂ to CH₃OH. Appl. Catal. A Gen 28: 437–438.

USGS. 2021. United States Geological Survey. Titanium Statics Information (20 February 2021). https://www.usgs.gov/centers/nmic/titanium-statistics-and-information.

Varner, K.E., K. Rindfusz, A. Gaglione and E. Viveiros. 2010. Nano Titanium Dioxide Environmental Matters: State of the Science Literature Review. U.S. Environmental Protection Agency, Washington, DC, EPA/600/R-10/089.

Velten, D., V. Biehl, F. Aubertin, B. Valeske, W. Possart and J. Breme. 2001. Preparation of TiO₂ layers on cp-Ti and Ti6Al4V by thermal and anodic oxidation and by sol-gel coating techniques and their characterization. J. Biomed. Mater 59: 18–28.

Vennemann, B.M., T. Rösgen, T.P. Carrel and D. Obrist. 2016. Time-resolved micro PIV in the pivoting area of the triflo mechanical heart valve. Cardiovasc Eng. Technol. 7: 210–222.

Vera, M.L. 2013. Obtención y caracterización de películas hemocompatibles de TiO₂. Ph.D. Thesis, Universidad Nacional de San Martín, Buenos Aires, Argentina.

Vera, M.L., M.A. Alterach, M.R. Rosenberger, D.G. Lamas, C.E. Schvezov and A.E. Ares. 2014. Characterization of TiO₂ nanofilms obtained by sol-gel and anodic oxidation. Nanomater. Nanotechnol. 4: 1–10.

Vera, M.L., E. Linardi, L. Lanzani, C. Mendez, C.E. Schvezov and A.E. Ares. 2015a. Corrosion resistance of titanium dioxide anodic coating on Ti-6Al-4V. Mater Corr. 66(10): 1140–1149.

Vera, M.L., J. Schuster, M.R. Rosenberger, H. Bernard, C.E. Schvezov and A.E. Ares. 2015b. Evaluation of the haemocompatibility of TiO$_2$ coatings obtained by anodic oxidation of Ti-6Al-4V. Procedia Mater Sci. 8: 366–374.

Vera, M.L., M.R. Rosenberger, C.E. Schvezov and A.E. Ares. 2015c. Fabrication of TiO$_2$ crystalline coatings by combining anodic oxidation and heat treatments. Int. J. Biomater. Article ID 395657, 1–9.

Vera, M.L., M.R. Rosenberger, C.E. Schvezov and A.E. Ares. 2015d. Wear resistance of anodic titanium dioxide nano-coatings produced on Ti-6Al-4V alloy. Nanomater. Nanotechnol. 5(1): 1–7.

Vera, M.L., Á. Colaccio, M.R. Rosenberger, C.E. Schvezov and A.E. Ares. 2017a. Influence of the electrolyte concentration on the smooth TiO$_2$ anodic coatings on Ti-6Al-4V. Coat 7(3): 39: 1–11.

Vera, M.L., G. Leyva and M.I. Litter. 2017b. Simple TiO$_2$ coatings by sol-gel techniques combined with commercial TiO$_2$ particles for use in heterogeneous photocatalysis. J. Nanosci. Nanotechnol. 17(7): 4946–4954.

Vera, M.L., M.C. Avalos, M.R. Rosenberger, R.E. Bolmaro, C.E. Schvezov and A.E. Ares. 2017c. Evaluation of the influence of texture and microstructure of titanium substrates on TiO$_2$ anodic coatings at 60 V. Mater Charact. 131: 348–358.

Vera, M.L., E.R. Henrikson, H.D. Traid, A.E. Ares and M.I. Litter. 2018a. Influence of thermal treatments in nanotubular anodic coatings of TiO$_2$. Matèria 23(2): e-12126.

Vera, M.L., H.D. Traid, E.R. Henrikson, A.E. Ares and M.I. Litter. 2018b. Heterogeneous photocatalytic Cr(VI) reduction with short and long nanotubular TiO$_2$ coatings prepared by anodic oxidation. Mater Res. Bull 97: 150–157.

Vera, M.L., H.D. Traid and M.I. Litter. 2020. TiO$_2$ coatings prepared by sol-gel and electrochemical methodologies. pp. 39–73. *In*: Almeida, R.M. et al. (eds.). Sol-Gel Derived Optical and Photonic Materials. Woodhead Publishing, Elsevier, New York, NY, USA.

Vijayan, B., N.M. Dimitrijevic, T. Rajhand and K. Gray. 2010. Effect of calcination temperature on the photocatalytic reduction and oxidation processes of hydrothermally synthesized titania nanotubes. J. Phys Chem. C 114: 12994–13002.

Wang, C., H. Shi and Y. Li. 2012. Synthesis and characterization of natural zeolite supported Cr-doped TiO$_2$ photocatalysts. Appl. Surf. Sci. 258(10): 4328–4333.

Wang, C., Z. Sun, Y. Zheng and Y.H. Hu. 2019. Recent progress in visible light photocatalytic conversion of carbon dioxide. J. Mater Chem. 7: 865–887.

Wang, F., C. Li, S. Zhang and H. Liu. 2020. Tantalum coated on titanium dioxide nanotubes by plasma spraying enhances cytocompatibility for dental implants. Surf. Coat Technol. 382: 125161.

Wang, L., Z. Nie, C. Cao, Y. Zhu and S. Khalid. 2015. Chrysanthemum-like TiO$_2$ nanostructures with exceptional reversible capacity and high coulombic efficiency for lithium storage. J. Mater Chem. A 3: 6402–6407.

Weibel, A., R. Bouchet, S.L.P. Savin, A.V. Chadwick, P.E. Lippens, M. Womes et al. 2006. Local atomic and electronic structure in nanocrystalline Sn-doped anatase TiO$_2$. Chem. Phys Chem. 7: 2377–2383.

Weir, A., P. Westerhoff, L. Fabricius, K. Hristovski and N. von Goetz. 2012. Titanium dioxide nanoparticles in food and personal care products. Environ. Sci. Technol. 46: 2242–2250.

Wium, E., C.J. Jordaan, L. Botes and F.E. Smit. 2020. Alternative mechanical heart valves for the developing world. Asian Cardiovasc. Thorac Ann. 28(7): 431–443.

Wu, Y., H. Yuan, X. Jiang, G. Wei, C. Li and W. Dong. 2012. Photocatalytic degradation of 4-tert-octylphenol in a spiral photoreactor system. J. Environ. Sci. 24(9): 1679–1685.

Xia, Y., Ch. Rong, X. Yang, F. Lu and X. Kuang. 2019. Encapsulating Mo-doped TiO$_2$ anatase in N-doped amorphous carbon with excellent Lithium storage performances. Front Mater 6(1): 1–12.

Xiong, Z., H. Wang, N. Xu, H. Li, B. Fang, Y. Zhao et al. 2015. Photocatalytic reduction of CO$_2$ on Pt^{2+}-Pt^0/TiO$_2$ nanoparticles under UV/Vis light irradiation: A combination of Pt^{2+} doping and Pt nanoparticles deposition. Int. J. Hydrogen Energy 40: 10049–10062.

Xu, D.S., J.M. Li, Y. Xiang and J. Li. 2012. From titanates to TiO$_2$ nanostructures: Controllable synthesis, growth mechanism, and applications. Sci. China Chem. 55: 2334–2345.

Xu, J., Y. Ao, M. Chen and D. Fu. 2010. Photoelectrochemical property and photocatalytic activity of N-doped TiO$_2$ nanotube arrays. Appl. Surf. Sci. 256(13): 4397–4401.

Yan, X., Z. Wang, M. He, Z. Hou, T. Xia, G. Liu et al. 2015. TiO$_2$ nanomaterials as anode materials for lithium-ion rechargeable batteries. Energy Technol. 3: 801–814.

Yin, G., X. Huang, T. Chen, W. Zhao, Q. Bi, J. Xu et al. 2018. Hydrogenated blue titania for efficient solar to chemical conversions: preparation, characterization, and reaction mechanism of CO$_2$ reduction. ACS Catal. 8: 1009–1017.

Younes, M., G. Aquilina, L. Castle, K.-H. Engel, P. Fowler, M.J. Frutos Fernandez et al. 2021. Safety assessment of titanium dioxide (E171) as a food additive. EFSA J. 19(5): 6585.

Yuan, B., Y. Wang, H. Bian, T. Shen, Y. Wu and Z. Chen. 2013. Nitrogen doped TiO_2 nanotube arrays with high photoelectrochemical activity for photocatalytic applications. Appl. Surf. Sci. 280: 523–529.

Yun, H., J. Li, H.B. Chen and C.J. Lin. 2007. A study on the N-, S- and Cl-modified nano-TiO_2 coatings for corrosion protection of stainless steel. Electrochim Acta 52(24): 6679–6685.

Zhang, J., M. Pan, C. Luo, X. Chen, J. Kong and T. Zhou. 2016. A novel composite paint (TiO_2/fluorinated acrylic nanocomposite) for antifouling application in marine environments. J. Environ. Chem. Eng. 4(2): 2545–2555.

Zhang, F. et al. 1998. Artificial heart valves: improved hemocompatibility by titanium oxide coatings prepared by ion beam assisted deposition. Surf. Coat Technol. 103-104: 146–150.

Zhang, H., J. Huang, L. Yang, R. Chen, W. Zou, X. Lin et al. 2015. Preparation, characterization and properties of PLA/TiO_2 nanocomposites based on a novel vane extruder. RSC Adv. 5: 4639–4647.

Zhang, M., K. Yin, Z.D. Hood, Z. Bi, C.A. Bridges, S. Dai et al. 2017a. *In situ* TEM observation of the electrochemical lithiation of N-doped anatase TiO_2 nanotubes as anodes for lithium-ion batteries. J. Mater Chem. A 5: 20651–20657.

Zhang, M.M., J.Y. Chen, H. Li and C.R. Wang. 2021. Recent progress in Li-ion batteries with TiO_2 nanotube anodes grown by electrochemical anodization. Rare Met. 40: 249–271.

Zhang, Q.H., W.D. Han, Y.J. Hong and J.-G. Yu. 2009. Photocatalytic reduction of CO_2 with H_2O on Pt-loaded TiO_2 catalyst. Catal. Today 148: 335–340.

Zhang, Y., H. Hu, M. Chang, D. Chen, M. Zhang, L. Wu and X. Li. 2017b. Non-uniform doping outperforms uniform doping for enhancing the photocatalytic efficiency of Au-doped TiO_2 nanotubes in organic dye degradation. Ceramics Int. 43(12): 9053–9059.

Zhang, W., C. Wang and W. Liu. 2006. Characterization and tribological investigation of sol-gel ceramic films on Ti-6Al-4V. Wear 260(4-5): 379–386.

Zhao, Q., X. Li, S. Guo, N. Wang, W. Liu, L. Shi et al. 2020. Osteogenic activity of a titanium surface modified with silicon-doped titanium dioxide. Mater Sci. Eng. C 110: 110682.

Zyoud, A., N. Zaatar, I. Saadeddin, M.H. Helal, G. Campet, M. Hakim and H.S. Hilal. 2011. Alternative natural dyes in water purification: anthocyanin as TiO_2-sensitizer in methyl orange photo-degradation. Solid State Sci. 13(6): 1268–1275.

Chapter 15

Nanostructured Films by Physical Vapor Deposition for Photocatalytic Applications

Ariel Kleiman, Magalí Xaubet and *Adriana Márquez**

1. Introduction

Physical Vapor Deposition (PVD) describes a variety of methods for depositing thin films that involve the physical ejection of material in the vapor phase, like atoms or molecules, and the condensation and nucleation of these particles onto a substrate. By introducing chemically reactive gases into the vapor, new compounds can be formed. PVD processes allow the deposition in mono-layered, multi-layered and multi-graduated coating systems, as well as special alloy composition and structures (Mattox 2010). The processes are environmentally friendly (Navinšek et al. 1999). PVD techniques, such as sputtering and cathodic arc, have been widely used to deposit thin coatings to improve corrosion resistance and mechanical and tribological properties of surfaces for applications in the metal-mechanical, automotive and aerospace industries (Mehran et al. 2017, Fottovati et al. 2019, Deng et al. 2020, Krella 2020). The fabrication of wear-resistant and biocompatible coatings for use in the medical industry is increasing (Geyao et al. 2020, Hussein et al. 2020). Photocatalytic films deposited by PVD for environmental applications are in full development (Kavaliunas et al. 2020, Kleiman et al. 2020, Ma et al. 2021, Velardi et al. 2021).

The heterogeneous photocatalytic process is a surface-based mechanism that is initiated by the absorption of a photon by a semiconductor with energy equal to or greater than the material bandgap, thus producing electron-hole pairs. The photogenerated electrons and holes become trapped in surface sites, where the reduction of electrons or the oxidation of holes by donating species adsorbed on the surface of the photocatalyst can occur (Linsebigler et al. 1995). Among the materials proposed as photocatalysts, titanium dioxide and zinc oxide have proved to present the highest photocatalytic activity under UV irradiation (Hernández-Alonso et al. 2009, Wang et al. 2020). Furthermore, doping with transition metals or nitrogen has made both oxides photocatalytically active under visible light (Delegan et al. 2014, Abdel-wahab et al. 2016, Ong et al. 2018, Nasirian et al. 2018).

After an early stage in which bulk photocatalysts were used, research on photocatalytic materials had focused on particles in aqueous suspensions, which exhibited a larger surface area and, hence,

Universidad de Buenos Aires, Facultad de Ciencias Exactas y Naturales, Departamento de Física - Universidad de Buenos Aires, Consejo Nacional de Investigaciones Científicas y Técnicas, Instituto de Física del Plasma (INFIP), Facultad de Ciencias Exactas y Naturales, Intendente Güiraldes 2160, Pabellón 1, Ciudad Universitaria, 1428 Buenos Aires, Argentina.
Emails: kleiman@df.uba.ar; xmagali@df.uba.ar
* Corresponding author: amarquez@df.uba.ar

a better photocatalytic performance (Fox and Dulay 1993). When working with aqueous TiO_2 suspensions under UV illumination, a good photocatalytic efficiency was observed for the inactivation of microorganisms and the transformation of organic and inorganic compounds, processes with significant impact in water treatment (Hashimoto et al. 2005, Akpan and Hameed 2009). However, the use of suspended particles presents a great difficulty in the final step of separating the powder from the solution, and it is not suitable for continuous flow systems. Therefore, the immobilization of the photocatalyst, either as a supported powder or as a thin film on a substrate, has emerged as an alternative method. In addition, immobilized photocatalysts have also attracted attention for their application in gas phase photocatalysis and superhydrophilic self-cleaning surfaces (Zhang et al. 2012, Schneider et al. 2014).

Photocatalytic films have been produced by a great variety of techniques, sol-gel and dip-coating being the most widespread methods (Varshney et al. 2016). Most researches have focused on the control of the processes in order to optimize the exposed surface area of the photocatalysts, seeking films with columnar structure, nanotubes or mesoporous films. One of the main advantages of PVD processes compared with conventional methods is that the films present better adhesion to the substrate, which makes them suitable for reuse (Lin et al. 2006, Kleiman et al. 2011). In the late 90s, the number of published works on photocatalytic films obtained by PVD methods began to grow rapidly. Within the large number of publications in the early years, extensive studies can be found on the characteristics of the films obtained, including analysis of structural, morphological, mechanical and optical properties, and discussions on the potential efficiency of films in different applications based on the observed properties. However, it was not until several years later that the photocatalytic activity of PVD films began to be studied directly, as was the case with films synthesized by other methods. Despite their relevance for photocatalytic applications, the durability properties of photocatalysts obtained by any technique have rarely been studied (Snyder et al. 2013, Patel et al. 2014, Kleiman et al. 2020). Experimental reactors for the treatment of wastewaters combining TiO_2 deposited by PVD as a photocatalyst coating with a Dielectric Barrier Discharge (DBD) at atmospheric pressure have been proposed. The DBD-photocatalyst hybrid process has demonstrated a higher rate of decomposition of oil hydrocarbons, phenols and synthetic surfactants, compared with the rate found using just the DBD (Grinevich et al. 2011).

This chapter covers an overview of the results achieved with nanostructured photocatalytic films obtained by PVD. The basic principles of the PVD techniques, the devices used for film deposition and the structural characteristics of the films are described. The literature on the photocatalytic performance of TiO_2 and ZnO films prepared by the most used PVD techniques is reviewed. Finally, the relevant characteristics of photocatalytic films deposited by PVD and the challenges to be addressed for using these films in practical applications are described.

2. Physical Vapor Deposition Techniques

The deposition of thin films using PVD processes has been widely investigated in order to improve the physical and chemical properties of surfaces of different materials, such as mechanical and tribological characteristics, corrosion resistance, optical and electronic features and aesthetic appearance (Martin 2009). Nowadays, PVD technologies are employed in industrial applications, namely for microelectronics, cuttings tools and decorative purposes (Baptista et al. 2018a, Abegunde et al. 2019, Deng et al. 2020).

PVD processes are atomistic deposition processes described in terms of three basic steps:

1) The creation of vapor phase species by the vaporization of solid or liquid sources that provide the base elements of the thin film,

2) The transport of the species from the source to the substrate through a vacuum or low-pressure gaseous environment,

3) The condensation and nucleation of the species on the substrate.

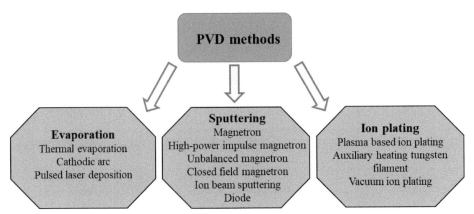

Figure 15.1. Basic PVD processes mentioning the most employed methods.

Films of elements or alloys are obtained in vacuum or non-reactive gaseous environments, while compounds can be obtained by introducing a reactive gas into the chamber, which chemically reacts with the vapor phase species. Mono-layered, multi-layered and gradient coatings can be easily deposited with PVD methods (Valkov et al. 2018, Hovsepian and Ehiasarian 2019). The thickness of the films deposited with PVD ranges from a few nanometers to some microns, with typical deposition rates between 1 and 100 nanometers per minute.

Evaporation, sputtering and ion plating are considered the three basic PVD processes (Martin 2009). From these basic processes, the evolution of the technology and the adaptation of the systems to the market requirements have led to multiple variants of PVD methods. Figure 15.1 contains a list of some of the most widely used PVD methods linked with the different basic processes.

2.1 Evaporation

In the evaporation process, the vapor species are produced by heating a source material by mechanisms such as thermal evaporation, arc discharge or laser beam.

2.1.1 Thermal evaporation

In thermal vaporization, the surface or the volume of the source material is heated at a temperature at which the vapor pressure is high, resulting in evaporation or sublimation of the material (Ohering 2002). The composition of the vaporized species depends on the relative vapor pressures of the components of the source material. As the deposition is generally carried out in vacuum, with a gas pressure range from 10^{-3} Pa to 10^{-7} Pa, the evaporated atoms follow a line-of-sight trajectory to the substrate with little or no collision with gas molecules. The typical distance from the source to the substrate is 10–50 cm. The most used heating techniques are resistive heating, e-beams and Radio Frequency (RF) inductive heating (Mattox 2010). Electron-beam heating is produced by the conversion of the high kinetic energy of the electrons into thermal energy on the surface of the evaporant material (Singh and Wolfe 2005).

2.1.2 Cathodic arc deposition

A cathodic arc is a low voltage, high current discharge between two electrodes in a vacuum chamber in which the cathode material vaporizes (Anders 2009). The current is concentrated in non-stationary spot-locations on the cathode surface, where high plasma densities are reached. The highly ionized plasma expanding from the cathode spots contains metallic ions with multiple charge states and kinetic energy in the range 18–150 eV (Boxman et al. 1996a). Together with the plasma, some macroparticles (MPs) are emitted from the cathode spot. When a substrate intercepts the plasma jet, the plasma condenses on the surface. Cathodic Arc Deposition (CAD) is characterized by high rates of deposition and by producing dense, well-adherent films. A diagram of the cathodic arc discharge

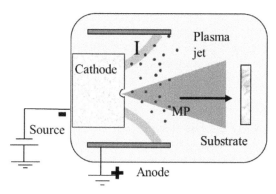

Figure 15.2. Schematic diagram of a continuous cathodic arc discharge.

is shown in Fig. 15.2. The discharge can be pulsed or continuous. In pulsed devices, a capacitor bank, which is discharged through a resistor and an inductor connected in series, is triggered repeatedly. The advantage of pulsed arcs is that they do not require cathode cooling, but the steady-state coating rate is relatively lower than the rate of continuous arcs (Sanders and Anders 2000).

The substrate can be biased negatively with respect to the plasma potential with continuous or pulsed voltage. When high voltages pulses of 1–10 kV are applied directly to the substrate, ions are implanted on the surface during the pulse, whereas in the time intervals between the pulses, the deposition of a film occurs. Then, the process, named Plasma-Based Ion Implantation and Deposition (PBII&D), combines the implantation and deposition of the ion species (Anders 2000, Manova et al. 2010).

The presence of MPs, which produce protuberances and depressions on the surface, may deteriorate some properties of the coatings and be inconvenient for some applications. Different filter designs have been developed to separate the MPs from the ion flux. In general terms, magnetic filters avoid a direct line of sight between the cathode and the substrate and guide the plasma flux to the substrate by means of electromagnetic fields (Schultrich 2018). Descriptions of the filters and their functional principles are provided in several reviews (Boxman et al. 1996b, Martin and Bendavid 2001, Aksenov et al. 2003). Two configurations of MP filters are schematized in Fig. 15.3. The most popular filter design is the toroidal duct (Fig. 15.3a) with its variations. The plasma beam is guided through the toroidal duct by a magnetic field parallel to its wall. Aksenov et al. (1982) designed the first magnetic island filter (Fig. 15.3b), an arrangement that is less complex and more compact than the toroidal duct. The magnetic island filter consists of a straight duct with an external solenoid and an electromagnet enclosed in a housing made of a non-magnetic material located inside the duct on its axis. The solenoid field is excited in the opposite axial direction to the field generated by the internal electromagnet, and the cross-section of the housing is sufficient to block

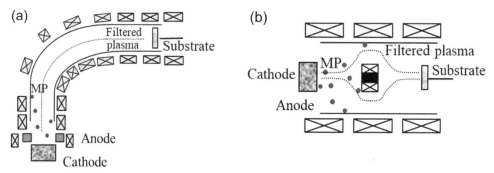

Figure 15.3. Schemes of macroparticle filter configurations: (a) toroidal duct filter and (b) magnetic island filter.

the line of sight between the cathode and the substrate. The magnetic field strength is similar to that used in toroidal filters. Based on the magnetic island filter scheme, several variants were developed, showing an efficiency higher than 25% in the plasma transport and reducing the number of MPs by approximately 98% (Bolt et al. 1999, Kleiman et al. 2008).

2.1.3 Pulsed laser deposition

Pulsed Laser Deposition (PLD) arises from the interaction of a high-power pulsed laser with the surface of a solid target where the beam is focused. The energy absorbed from the beam by a small area of the target surface ablates the material in the form of a plasma plume. The evaporated material absorbs energy from the laser beam producing the expansion of the plume in which all species expand away from the target surface with an identical angular distribution but with different velocity distributions. The vapor material condenses onto a substrate located in front of the target at a relatively short distance (Lowndes et al. 1996, Rao 2013). A schematic diagram of the PLD setup is shown in Fig. 15.4. Plume propagation has been extensively studied using optical absorption and emission spectroscopy combined with ion probe measurements (Ojeda et al. 2018, Irimiciuc et al. 2021). One of the most remarkable characteristics of PLD is the preservation of the stoichiometry of the target material in the ablation process as a consequence of the fast and intense heating of the surface by the laser beam. Ceramics, superconducting or ferroelectric oxides can be deposited by PLD with the original chemical composition.

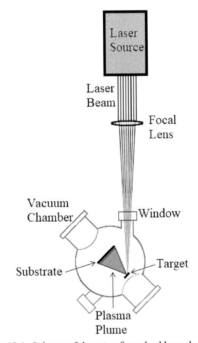

Figure 15.4. Scheme of the setup for pulsed laser deposition.

2.2 Sputtering

In sputtering processes, some material is emitted from a target as a result of the bombardment of the surface with accelerated positive ions, which transmit sufficient energy and momentum to cause the ejection of one or more atoms. Typical plasmas for sputtering applications are obtained with Direct Current (DC) or RF glow discharges of moderate voltage (hundreds to a few thousand volts) between two electrodes in presence of an inert gas such as Ar or Kr, at pressures between 0.1 and 10 Pa. For a reactive sputtering process, a chemically reactive gas as oxygen or nitrogen is also introduced into

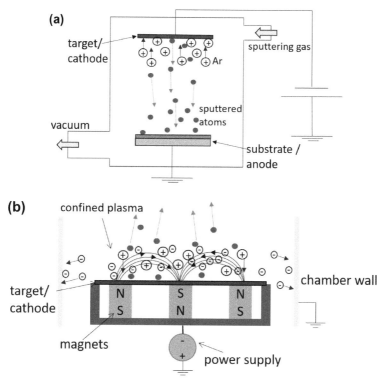

Figure 15.5. Schemes of sputtering systems: (a) DC sputtering with two electrodes and (b) target holder with permanent magnets employed in magnetron sputtering configurations.

the chamber. One of the most common sputtering systems is the two-electrode configuration, known as diode (Fig. 15.5a), where the target acts as cathode placed facing the substrate. The plasmas of these discharges have a low degree of ionization, with characteristic values between 10^{-2} and 10^{-4}. Other configurations, as confocal multitargets or triode with the introduction of a third electrode to increase the ionization, are also employed (Rossnagel et al. 1998).

Magnetron-based sputtering is one of the most widespread technologies in PVD processes (Baptista et al. 2018a). Magnetron sputtering systems are basically a diode with permanent magnets or coils in the target holder to magnetically enhance the ion sputtering, as outlined in Fig. 15.5b. The magnetic field confines the secondary electrons produced by the ion bombardment of the target close to the target surface, increasing the probability of ionizing the working gas. As ions are much heavier than electrons, the confining magnetic field does not significantly affect their trajectories nor the sputter bombardment process. The increased ionization rate allows operating the discharges at lower pressures with higher deposition rates and larger deposition areas (Baptista et al. 2018b).

The most recent variant of magnetron sputtering is the high-power impulse magnetron sputtering (HiPIMS). A high-power pulse supply connected to the target provides it with a large energy impulse over a very short period (Anders 2014). The voltage pulse is around 1 kV with widths in the order of 100 ms and frequencies up to 500 Hz. The high energy pulse leads to high-density plasma with a high ionization degree of the sputtered target material without overheating the target (Anders and Yang 2018).

2.3 Ion plating

Ion plating combines the deposition of vaporized material with the bombardment of the depositing film by atomic-sized energetic particles to modify the properties of the film. The depositing material can be vaporized by either evaporation, sputtering or cathodic arc. The bombarding particles are

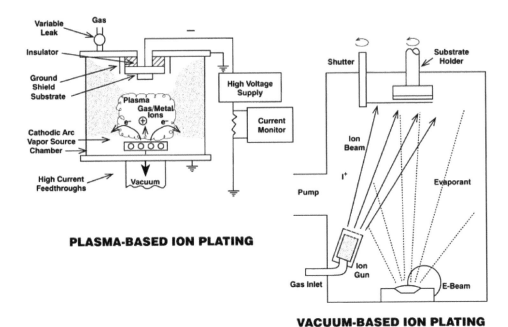

Figure 15.6. Plasma-based ion plating system using a cathodic arc vaporization source with bombardment from the plasma and vacuum-based ion plating system using a thermal evaporation source and an 'ion gun' for bombardment. Reprinted from (Mattox 2000) with permission from Elsevier.

generally energetic ions of an inert or reactive gas and high-energy neutrals formed by charge exchange collisions between thermal neutrals and high-energy ions, but ions can also be formed from the condensing film material (Martin 2009). In plasma-based ion plating, the ions can be extracted from a glow discharge generated by biasing the substrate to high negative potentials (in the order of kilovolts) in a gas environment at a pressure between 0.1 and 1 Pa. The substrate is bombarded by high-energy gas ions, which sputter off the material present on the surface, carrying out a surface cleaning until the film material covers the surface. The pressure range of the glow discharge enhances the scattering of the vapor particles in all directions through gas collisions contributing to the growth of a more uniform film. The continuous bombardment of the depositing material modifies the microstructure and the properties of the film. The drawbacks of this method can be the decrease of the deposition rate due to the sputtering off of the deposited material and the heating of the substrate induced by the intense ion bombardment. Alternatives to avoid these effects are to use an auxiliary heating tungsten filament to provide the electrons necessary to support the gaseous discharge or a vacuum-based ion plating system that adds to the evaporation source a separate ion gun in a vacuum environment. Schemes of ion plating configurations are depicted in Fig. 15.6 (Mattox 2000).

2.4 Microstructure of PVD films

The microstructure of films deposited by PVD depends on the element composition, angle, flux and kinetic energy of species incident at the growing surface, as well as the temperature and the material of the substrate. These factors directly influence the mobility of the incident species (adatoms) on the substrate. Surface and bulk diffusion of adatoms are not the only physical processes involved in nucleation and growth of the film, but also additional processes such as resputtering, reflected neutrals, desorption, shadowing and ion implantation may occur. The diffusion and desorption of the adatoms are also related to the characteristic diffusion and sublimation activation energies whose magnitudes scale with the melting temperature (T_m) of the condensate. Therefore, the energy of the incident species along with the substrate temperature (T_s) are decisive for the dominance of one or more of the physical processes during the nucleation and growth of the film (Petrov et al. 2003). The energy of the incident species ranges from a few tenths of electron-volts for thermal evaporation to tens for sputtering, and up to hundreds for arc evaporation; even higher energies are achieved with ion bombardment. The adhesion of the film to the substrate is determined by the type of interface developed between the film and the substrate, which depends on the substrate morphology, cleanliness and interfacial reactions at the surface. Deposition starts as three-dimensional nuclei formed at favored sites on the substrate surface. These initial nuclei grow laterally and in thickness until they coalesce and re-nucleate to form a continuous film. The average thickness at which a continuous film begins to form depends on the substrate temperature and the deposition rate (Pashley et al. 1964, Bunshah 1977).

Numerous investigations have addressed the influence of different PVD process parameters, such as the bias voltage, gas type, flow, pressure, substrate temperature and pretreatments, on the microstructure and morphology of the films and the correlation with their properties (Gjevori et al. 2009, Gupta et al. 2021, Herrera-Jimenez et al. 2021). The results showed that the effect of the deposition parameters on the structural morphology of metal, semiconductor and ceramic films presents similar features in a first approximation. Phenomenological descriptions of the relationship between the deposition variables and the microstructure characteristics obtained in PVD films have been represented through Structure Zone Diagrams (SZDs). Movchan and Demchishin (1969) proposed the first SZD from observations of thick evaporated coatings (0.3 to 2 mm) of pure metals (Ti, Ni, W) and oxides (ZrO_2 and Al_2O_3). Three structural zones as a function of the homologous temperature, $T_h = T_s/T_m$, were identified. Zone 1, where $T_h < 0.3$ for pure metals and $T_h < 0.26$ for oxides, is the low-temperature region. The low adatom mobility and shadowing effects lead to the formation of tapered crystallites with domed tops and separated by voided boundaries several nanometers wide, leaving a high density of lattice imperfections and pores at grain boundaries. Zone 2, for 0.26–$0.3 < T_h < 0.45$, is characterized by a denser columnar grain structure in which the grain size grows as temperature increases. The development of this structure is governed by surface and grain boundary diffusion. In Zone 3, $T_h > 0.45$, bulk diffusion and recrystallization dominate the polyhedral structure evolution with equiaxed grains that enlarge as temperature augments.

Movchan and Demchishin's microstructural representation was later modified by Thorton (1974). Thorton's diagram adds the influence of gas pressure during the deposition process in the microstructure evolution, which is relevant in sputtering or ion plating processes. This model introduces a transition zone, Zone T, between Zones 1 and 2, consisting of tightly packed fibrous grains with weak grain boundaries. Zone T is not noticeable in pure metals but becomes evident in complex alloys, compounds or deposits produced at higher gas pressures. More recently, Anders (2010) proposed an expanded SZD to encompass the description of the microstructure obtained in processes involving high-energy ions. The parameters represented in the axes of Thorton's model were replaced by parameters directly related to the film growth process, as shown in Fig. 15.7. The linear homologous temperature was substituted by a generalized temperature T^* (on a logarithmic scale), that takes into account the substrate temperature plus a temperature shift caused by the potential energy of particles reaching on the surface. The pressure gas was replaced by a normalized energy flux E^* (on a logarithmic scale), describing displacement and heating effects caused by the kinetic energy of the bombarding particles, normalized by a characteristic energy of the material, such as the cohesive energy. The vertical axis of the diagram incorporates the representation of an effective film thickness t^*, considering thickness reduction effects by densification and sputtering comprising even the ion etching process. Although the SZD is a qualitative representation of the morphology evolution and the boundary temperature values between the different zones are indicatives, the shift of the zone boundaries to lower temperatures becomes apparent as E^* increases in the deposition process.

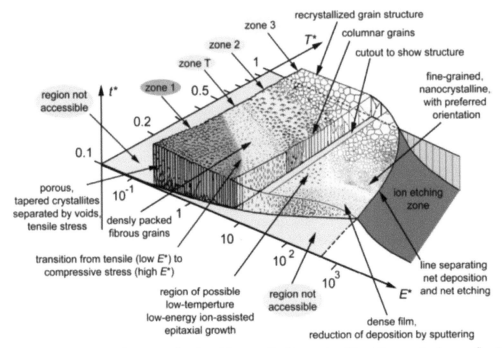

Figure 15.7. Structure zone diagram as a function of the generalized temperature T^* and the normalized energy flux E^*; t^* represents the effective thickness. The numbers on the axes are only for orientation, the actual values depending on the material, and many other conditions. Reprinted from (Anders 2010) with permission from Elsevier.

3. Photocatalytic Films by PVD

A wide variety of materials has been analyzed for their use as photocatalysts in the last 50 years. From the outset, TiO_2 has been by far the most studied material due to its relatively high photocatalytic activity, excellent functionality, good chemical stability and low toxicity (Lazar et al. 2010, Nakata and Fujishima 2012). Anatase is generally considered the most photocatalytically active phase of TiO_2, although anatase/rutile mixtures with low content of the latter have demonstrated the best photocatalytic performance (Jang et al. 2001, Hurum et al. 2003, Loeb et al. 2019). Among all the other materials tested as photocatalysts, ZnO has attracted the most attention and has been established as the first alternative to titanium dioxide (Hernández-Alonso et al. 2009).

Regarding PVD techniques, the production of TiO_2 and ZnO films has been mainly addressed by sputtering, CAD and PLD. In turn, different variants, with different configurations and/or types of discharge, have been used for each technique. Table 15.1 shows ranges of the values typically used for different parameters of the PVD processes. A metal (Ti or Zn) cathode and oxygen ambient is used in CAD. Sometimes, a small flow of argon is also used in order to provide arc stability (Elzwawi et al. 2012). DC or pulsed arcs are used, and the plasma can be filtered or unfiltered. Coatings obtained with filtered arcs exhibit fewer or no macroparticles, while films deposited with unfiltered arcs have a more uniform thickness over larger areas (Takikawa et al. 1999, Huang et al. 2013). The main difference in the deposition parameters between TiO_2 and ZnO is that the arc current is somewhat lower for the latter. Production of films by sputtering has been approached by using metal or oxide targets and mostly with DC or RF power sources. Argon and oxygen are used as sputtering and reactive gases, respectively. When using oxide targets, the proportion of oxygen is generally lower and in many cases, only argon is employed. Both types of targets are used also in PLD. The most used laser sources are KrF excimer (wavelength 248 nm) and different harmonics of Nd:YAG (1064 nm, 532 nm and 266 nm). Normally, the pulses are in the range of the nanoseconds; however, femtosecond pulses were also used (Kumi-Barimah et al. 2020). Ambient

Table 15.1. Typical value ranges of process parameters for TiO_2 and ZnO deposition by PVD techniques.

Technique		Cathode/ target	Arc current/ Power/Laser fluence (pulse, repetition rate)	Pressure (Pa)	Ambient gas	Cathode/ target-to-substrate (cm)	Deposition rate (nm/min)
CAD	Unfiltered	Ti/Zn	50–150 A (TiO_2) 30–80 A (ZnO)	0.1–8	O_2 + Ar (0–10%)	20–40	20–150 (TiO_2) 1–20 (ZnO)
	Filtered			0.01–2		15–25*	
Sputtering		Ti/Zn	30–500 W	0.2–8	Ar + O_2 (1–50%)	4–25	1–30
		TiO_2/ZnO			Ar + O_2 (0–30%)		
PLD		Ti/Zn	1–7 J/cm² (1–25 ns) (2–10 Hz)	2–70	O_2	2–8	2–5
		TiO_2/ZnO		10^{-5}–100	Vacuum – O_2		

* From the filter duct exit

gas is generally oxygen, although the use of oxide targets allows performing the process under high vacuum conditions. Many different materials have been employed as substrates; regarding photocatalytic films, glass and silicon are the most used. Substrates are usually heated or biased in order to modify or improve the film characteristics.

PVD techniques are characterized by high growth rates; values reported for the three techniques are summarized in Table 15.1. Cathodic arcs achieve the highest rates, followed by sputtering. The deposition rate depends on many factors, with the parameters included in Table 15.1 having a strong influence. The growth rate decreases with the increasing substrate-to-target distance (Arakelova et al. 2016) and with the increasing pressure (Dreesen et al. 2009, Guillén et al. 2021). For sputtered films, it has been found that the deposition rate also decreases when the oxygen percentage is increased for a fixed total pressure (Wicaksana et al. 1992). On the other hand, films grow faster with the higher the laser fluence (Kawasaki et al. 2013), sputtering power (Ahumada-Lazo et al. 2014) or arc current (Weng et al. 2011). It is worth noting that CAD deposition rates for ZnO are generally lower than for TiO_2, as the arc currents used are also lower, as mentioned above. Deposition rate decreasing with increasing temperature has been reported for TiO_2 grown by CAD (Çetinörgü-Goldenberg et al. 2013) as well as for ZnO grown by CAD (Elzwawi et al. 2012) and sputtering (Bensmaine and Benyoucef 2014). This behavior was not found for TiO_2 films prepared by sputtering, with the rates being very similar for all temperatures between RT and 500°C (Meng et al. 1993b). It has been also found that deposition rate can be enhanced by using targets with conical protrusions instead of flat ones (Arakelova et al. 2016).

TiO_2 and ZnO films prepared by PVD are highly transparent. In the visible range, film transmittance generally exhibits interference effects due to uniformity and low roughness, with the mean transmittance value being typically above 75%. A sharp fall is observed in the UV region due to absorption (Kleiman et al. 2007, Tsoutsouva et al. 2011).

3.1 Titanium dioxide

3.1.1 Crystalline structure

The crystalline structure of TiO_2 films obtained by PVD techniques strongly depends on the substrate temperature during the growth process and on the energy of the particles impinging on the substrate. The diagram presented in Fig. 15.8 (Löbl et al. 1994) shows qualitatively the conditions for the occurrence of amorphous, anatase or rutile TiO_2 films. At Room Temperature (RT) and low energy, amorphous TiO_2 is obtained. The crystalline phases can be formed at higher substrate temperature and/or ion energy, with the rutile structure requiring higher values than anatase. The energy of the ions depends on the emission mechanism and electrical parameters, as well as the pressure of the gas

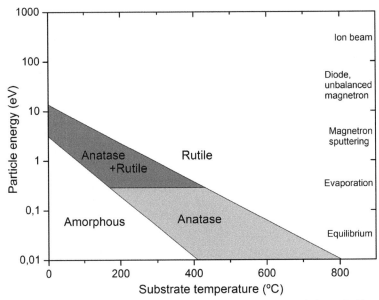

Figure 15.8. Phase diagram for obtaining different TiO$_2$ structures with PVD techniques (Löbl et al. 1994).

and the distance between the cathode or target and the substrate, and can be increased by biasing the substrate. The crystalline structure of the substrate may also influence that of the film.

Table 15.2 summarizes results extracted from works in which the crystalline structure was studied for different deposition temperatures, fixing other parameters of the process. In general terms, when no bias is applied, CAD with the substrate at RT leads to amorphous films; at 300–400°C, the films grow in the anatase phase and, at a higher temperature, anatase/rutile mixtures or pure rutile are obtained. However, anatase films grown at RT on silicon substrates have also been

Table 15.2. Crystalline phases as a function of the substrate temperature during deposition with unbiased substrates.

Technique	Substrate	Conditions	Temperature (°C)/Structure	Reference
Unfiltered CAD	Glass	$I = 120$ A $p = 1$–3 Pa $t = \sim 300$ nm	RT / α 200 / α 300 / A or α 400 / A	(Kleiman et al. 2007)
Unfiltered CAD	Stainless steel	$I = 120$ A $p = 2$–5 Pa $t = \sim 300$ nm	RT / α 400 / A+R 500 / A+R 560 / R	(Franco Arias et al. 2012)
DC Sputtering	Si	$I = 0.09$ A $p = 1.3$ Pa (10% O$_2$) $t = 400$–450 nm	RT / α 200 / A 300 / A+R 400 / A	(Wicaksana et al. 1992)
DC Sputtering	Glass	$I = 0.35$ A $p = 0.8$ Pa (7.5% O$_2$) $t = 730$–790 nm	RT / A (101) 300 / A (101) + (004) 400 / A (101) + (004) 500 / A (004)	(Meng et al. 1993b)
PLD (TiO$_2$ target)	SiO$_2$	Fluence = 2 J/cm^2 $p = 10$ Pa (O$_2$) $t = \sim 300$ nm	100 / A (101) 300 / A (101) 400 / A (101) + (004) 500 / A (101) + (004)	(György et al. 2005)

α: amorphous; A: anatase; R: rutile
I: arc or sputtering current; p: working pressure; t: thickness

reported (Bendavid et al. 1999). In the case of films prepared by sputtering or PLD, the anatase phase was also achieved at RT for certain deposition parameters. It has been reported that preferential plane orientation in anatase films changed with the temperature, either for sputtering using Ti targets (Meng et al. 1993b) or PLD using TiO_2 targets (György et al. 2005), as shown in Table 15.2. In addition, films prepared by PLD from an anatase target underwent a complete transformation to rutile for substrate temperature of 700°C, while for lower temperatures anatase/rutile mixtures were obtained (Kumi-Barimah et al. 2020).

By biasing the substrate with negative voltages, crystalline phases can be obtained at a lower temperature. Results extracted from works in which the crystalline structure was analyzed as a function of the bias voltage are summarized in Table 15.3. Films deposited by CAD on silicon substrates grew in the rutile phase when applying negative voltages of 100 V with the substrate at 300°C (Zhang et al. 1998) or even at RT (Bendavid et al. 1999). Films prepared by sputtering on silicon were amorphous without polarization, while anatase was obtained when applying a bias of –50 V (Sekhar et al. 2011) or –100 V (Kondaiah et al. 2012). When using glass substrates, polarization is not as effective as when using conductive substrates. CAD films grown at RT resulted amorphous for all bias values up to –400 V (Bendavid et al. 2000). However, films grown by sputtering on glass substrates exhibited a higher amount of rutile as the bias voltage increased (Nezar et al. 2017), as shown in Table 15.3. On the other hand, using the PBII&D technique, for 3 kV pulses (30 ms, 3 kHz) the films grew amorphous, while with voltages in the range 5–10 kV, films in the rutile phase were obtained, with the temperature being lower than 100°C (Manova et al. 2009).

Besides substrate temperature and polarization, another important parameter in PVD techniques is the gas pressure. High pressure leads to the reduction of the energy of the particles on their way to the substrate due to collisions with gas particles. On the other hand, low oxygen pressure may reduce the chances of chemical reactions, prevent the formation of oxides or lead to the formation of sub-stoichiometric compounds (Simionescu et al. 2019, Guillén et al. 2021). All films prepared at different sputtering pressures in the range 0.2–2 Pa exhibited the anatase phase, but the intensity

Table 15.3. Crystalline phases as a function of the bias voltage applied to the substrate.

Technique	Substrate	Conditions	Bias voltage (V) / Structure	Reference
Filtered CAD	Si	$I = 100$ A $p = 0.2$ Pa $T = 300$°C	0 / α –100 / R –200 / R –300 / R –400 / R	(Zhang et al. 1998)
Filtered CAD	Si	$I = 120$ A $p = 0.35$ Pa RT	0 / A –50 / α –100 / R –200 / R –300 / R –400 / R	(Bendavid et al. 1999)
RF Sputtering	Si	Power = 200 W $p = 5$ Pa (1% O_2) RT	0 / α –100 / A –150 / A	(Kondaiah et al. 2012)
DC Sputtering	Si	Power = 200 W $p = 0.5$ Pa (12% O_2) RT	0 / α –50 / A –100 / A –150 / A	(Sekhar et al. 2011)
RF Sputtering	Glass	Power = 250 W $p = 2.66$ Pa (25% O_2) RT	0 / A –75 / A (70%) + R (30%) –100 / R (55%) + A (45%)	(Nezar et al. 2017)

α: amorphous; A: anatase; R: rutile

I: arc current; p: working pressure; T: temperature

of the diffraction peaks decreased with the increasing pressure, the film deposited at the highest pressure being almost amorphous (Meng and dos Santos 1993a). In similar pressure ranges, the crystalline structure has also been reported to change from anatase to amorphous (Wicaksana et al. 1992) or from rutile to anatase (Gu et al. 2003) by increasing the pressure. The latter was also observed for sputtering (Pihosh et al. 2009) and PLD (Iwulska and Śliwiński 2011) in films prepared with a TiO_2 target.

The crystalline structure of the substrate has also shown to affect the crystalline structure of the films. The structure of Si and the orientation of the exposed plane have been reported to favor the formation of crystalline phases and textures (Mändl et al. 2000). It has also been found that depositing a thin layer of Ti between the substrate and the TiO_2 film promotes the formation of rutile at a lower temperature, either by sputtering (Krishna et al. 2011) or cathodic arc (Franco Arias et al. 2017).

Other parameters that influence the structure have also been analyzed. It was reported that increasing the laser fluence in PLD increases the proportion of rutile in anatase/rutile mixtures obtained at 400°C (Yamaki et al. 2002). Besides, the film thickness effect on the crystalline phase has also been studied for sputtered films prepared with a TiO_2 target; when the thickness was 30 nm, the film was amorphous, while for films 60 nm and thicker, the anatase phase was observed (Pihosh et al. 2009).

As with films obtained by chemical methods, phase transformations can also be reached by annealing the samples after deposition. Originally amorphous films have crystallized in the anatase phase, anatase/rutile mixtures or pure rutile when annealed at 400°C (Takikawa et al. 1999), 500°C (Kleiman et al. 2007) or 800°C (Leng et al. 2004), respectively. These temperature values are in agreement with the diagram in Fig. 15.8 in the absence of energetic particles. However, the outcome after the heat treatment depends not only on the annealing temperature but also on the characteristics of the samples and, in turn, on the deposition conditions. Films obtained by PLD at RT, annealed at 400°C, resulted in different anatase proportions when different pressures were used for deposition (Brunella et al. 2007). On the other hand, amorphous films deposited by CAD with thickness lower than 1 μm crystallized in the anatase phase when annealed at 400°C, while thicker films also presented small amounts of rutile (Kleiman et al. 2020).

3.1.2 Morphology/microstructure

PVD techniques lead to dense, uniform films with columnar structure as can be seen in Fig. 15.9 (Henkel et al. 2018), where cross-section images from films with different thicknesses grown by sputtering are depicted. The same microstructure composed of columnar grains has been also reported for films obtained by CAD (Kleiman et al. 2014) and PLD (Brunella et al. 2007). The density of the films has been found to be similar to that of bulk TiO_2 (Kleiman et al. 2014, Simionescu et al. 2019). This can be *a priori* a drawback in comparison with porous films, which present a larger specific surface area. However, a considerable crack formation was observed after heat treatment in films prepared by sputtering with thickness greater than 400 nm, thus enlarging the exposed surface (Henkel et al. 2018). As seen in Fig. 15.9, the cracking was more significant the thicker the film was. A similar feature was found for films grown by CAD, the cauliflower-shaped structures due to the presence of cracks being noticeable for films with thickness greater than 700 nm (Kleiman et al. 2020). Besides the effect of the annealing, crack formation can be tuned through the O_2/Ar ratio in sputtered films (Vahl et al. 2019a). It was also found that the higher the sputtering pressure, the greater the film porosity, using either Ti (Meng and dos Santos 1993a) or TiO_2 (Eufinger et al. 2007) targets. Attempts to increase the reactive surface area have also been made by decorating the surface of sputtered films with TiO_2 nanoparticles with different non-PVD methods (Vahl et al. 2019b).

Grain size is typically measured from imaging techniques such as Atomic Force Microscopy (AFM), while crystallite size is determined by peak broadening in X-Ray Diffraction (XRD) patterns using Scherrer's formula (Langford and Wilson 1978). Amorphous deposited films crystallized

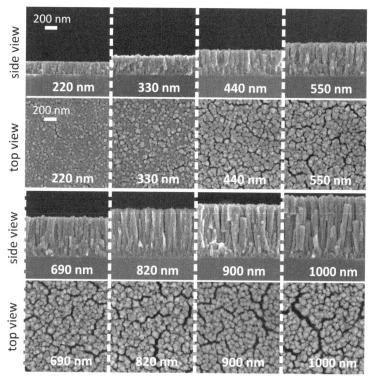

Figure 15.9. Scanning electron microscopy images of sputtered TiO$_2$ films of different thicknesses deposited at RT and annealed at 650°C (Henkel et al. 2018).

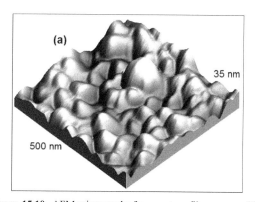
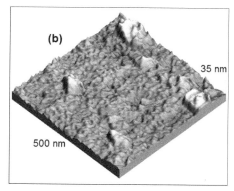

Figure 15.10. AFM micrographs from anatase films prepared by CAD (a) at RT, annealed at 400°C, (b) with the substrate at 400°C.

during post-deposition annealing generally exhibit larger grains than those crystallized *in situ* during deposition (Löbl et al. 1994). Figure 15.10 shows AFM images of the surface of films prepared by CAD at different temperatures. Anatase films deposited at 400°C exhibited a grain size of 15–30 nm and a surface roughness of 2.4 nm, while for films grown at RT, post-annealed at 400°C, the grain size was 50–80 nm and the roughness was 5 nm (Kleiman et al. 2007). Increasing grain size with increasing annealing temperature was observed in sputtered films prepared with a TiO$_2$ target as well, the size being 14.5 nm for the as-deposited film at RT and 31.9 nm after annealing at 550°C (Doghmane et al. 2020). On the other hand, the crystallite size was found to decrease with increasing argon flow rate in sputtered samples (Zhang et al. 2004), and with increasing laser fluence in PLD films (Yamaki et al. 2002).

3.2 Zinc oxide

Although diverse polymorphs have been synthesized or predicted to be stable under certain conditions, hexagonal wurtzite is the most stable crystal structure and by far the most commonly observed for ZnO (Sponza et al. 2015). Films of ZnO prepared by PVD techniques are usually highly textured. The preferred orientation depends on the deposition parameters, the (002) plane being the most observed orientation since it presents the lowest surface energy (Fujimura et al. 1993). Furthermore, this orientation has proved to optimize the photocatalytic activity (Jang et al. 2006).

Films prepared by PLD grew crystalline at all temperatures between RT and 900°C; the preferred orientation changed from the plane (101) to (002) with increasing temperature (Angappane et al. 2009). When the films were already (002) oriented even at low temperatures, the rise in temperature enhanced the intensity of this diffraction peak (Tsoutsouva et al. 2011). Similar results were also reported for sputtered films; while the (002) peak intensity increased with temperature, its width decreased, indicating a larger crystallite size (Bensmaine and Benyoucef 2014). For PLD films, the grain size was reported to increase from 50 nm at RT to 100 nm at 750°C, while the surface roughness augmented from 2 to 4 nm (Angappane et al. 2009). A similar change in roughness was observed for CAD films with increasing temperature; but in contrast, the preferred orientation changed from the plane (002) to (103) as the temperature increased (Xu et al. 2001). Post-deposition annealing of the samples has also demonstrated to improve crystallinity by increasing crystallite size. For a film deposited at 400°C by PLD, the crystallite size was 18 nm; after annealing at different temperatures up to 700°C, the crystallite increased monotonically up to 43 nm (Zhao et al. 2006). Similarly, in films prepared by CAD, the crystallite size increased after annealing at 800°C; the preferred (002) orientation did not change (Kara et al. 2014).

It was also studied that the structure and texture vary with pressure. For films deposited by CAD at RT, the relative intensity of the (002) peak grew with increasing pressure in the range 0.4–2 Pa and decreased slightly with a further increase in the pressure (Huang et al. 2013). For PLD films, the (002) peak sharpened with increasing oxygen pressure up to 50 Pa (Tsoutsouva et al. 2011). Other parameters have also been reported to influence film structure. In CAD films deposited with negative bias voltages, the crystallite size and the intensity of the (002) peak increased with voltages up to 400 V, then decreased as the voltage increased further (Zhao et al. 2011). The texture in the (002) direction was enhanced by increasing the laser fluence to 1.6 J/cm^2, while the film turned amorphous for higher fluences (Haider et al. 2021). For PLD also, the relative height of the (002) diffraction peak increased when the mean kinetic energy of the plasma augmented from 60 to 80 eV and decreased when the energy was 100 eV (Guerrero de León et al. 2020). In addition, for sputtered films, the texture with the preferred (002) orientation was enhanced the most at a certain target-to-substrate distance (12 cm) and was weakened when moving the substrate away from or toward the target (Arakelova et al. 2016). The authors attributed this effect to the deposition rate, the influence of which on the film texture has been studied earlier (Fujimura et al. 1993).

3.3 Doped films

TiO_2 and, to a lesser extent, ZnO have been established as the most widely studied photocatalysts due to their high photoactivity under UV radiation. However, the low efficiency under visible light due to the wide bandgap (3.2 eV and 3.37 eV, respectively) is their main drawback, as only approximately 4% of the sunlight can be harnessed. Therefore, several investigations have been undertaken in order to expand the range in the UV-visible spectrum where they are active and to increase the fraction of sunlight capable of initiating photocatalytic processes. The first approach to extend the photocatalytic activity of TiO_2 towards the range of visible light was doping with nitrogen (Asahi et al. 2001), which became the most widely used dopant in TiO_2 (Nasirian et al. 2018) and ZnO (Samadi et al. 2016) films. Many other dopants including non-metals such as boron, sulfur or carbon and rare earth and transition metals have also been studied for both oxides (Samadi et al. 2016, Fang et al. 2017, Ong et al. 2018). Despite reducing the bandgap, doping with transition metals has been reported to turn anatase thermally unstable (Pelaez et al. 2012). In contrast, N-doped TiO_2 has achieved high chemical stability while reducing the bandgap as well (Bakar and Ribeiro 2016).

3.3.1 Nitrogen doping

Regarding films prepared by PVD methods, nitrogen doping has been addressed mainly through *in situ* doping during the film growth, simultaneously injecting O_2 and N_2 as reactive gases when metal targets or cathodes are used or just N_2 for oxide targets, in addition to, eventually, non-reactive argon. The use of NH_3 as an additional nitrogen source (Xu et al. 2006) or air as a source of both oxygen and nitrogen (Chan and Lu 2008) has been also studied. Another reported method is the use of N-doped TiO_2 targets (Hu et al. 2012). On the other hand, post-deposition doping has been performed by ion implantation in nitrogen ambient (Esparza et al. 2017).

The deposition rate tended to increase with the injection of small N_2 flow rates as long as the total pressure was not increased. For sputtering, using a Ti target, when Ar and O_2 flows were kept constant, increasing N_2 flow rate led to lower deposition rates, as the total pressure was higher (Liu et al. 2013). For a total constant pressure, the deposition rate was slightly higher when N_2 was employed along with O_2 and Ar (Javid et al. 2019). Using a TiO_2 target, and keeping the total pressure fixed, the deposition rate increased with the N_2/Ar ratio up to 5% N_2 and then decreased for increasing N_2 percentage (Delegan et al. 2014). In the case of CAD, the deposition rate increased when the N_2 percentage was increased up to 30%; for higher N_2 amounts, the behavior was different for different substrate temperatures (Çetinörgü-Goldenberg et al. 2013). The growth rate also increased for an increasing N_2/Ar ratio when sputtering a ZnO target (Wang et al. 2007).

In general, the nitrogen content in the samples increased with the partial pressure of N_2 during the film growth. Films prepared by unfiltered CAD reached nearly 5 at.% N for a N_2/O_2 flow ratio of 1, with transition to a semi metallic film, which is no longer transparent, for a flow ratio above 2 (Manova et al. 2017). A similar behavior was observed for films grown by PBII&D, except that films became semi metallic for a lower N_2 percentage (Asenova et al. 2014). With a filtered arc, films achieved 3% N for a N_2 partial pressure of 77% (Çetinörgü-Goldenberg et al. 2013). For PLD films prepared from a Ti target, the nitrogen content in the sample reached 8% for a N_2 percentage of around 90% in the N_2/O_2 gas mixture (Farkas et al. 2018). When sputtering a TiO_2 target, the N content increased with a N_2/Ar ratio up to 6% N for 15% N_2, and then slightly decreased to 5% for 25% N_2 (Delegan et al. 2014). When doping was carried out by post-deposition implantation of N ions, the incorporated nitrogen was concentrated near the film surface. The nitrogen content was observed down to a depth of around 200 nm, with a maximum nitrogen concentration of 20 at.% at 50 nm depth and 3–6 at.% at the surface (Esparza et al. 2017).

In most studies, the amount of nitrogen incorporated was small; therefore, the structural characteristics of N-doped films generally corresponded well to pure TiO_2 or ZnO (Chiu et al. 2007). However, N doping can affect the growth of crystalline phases. CAD samples obtained with a N_2 flow higher than 20% grew amorphous, when anatase was expected according to results for undoped samples (Çetinörgü-Goldenberg et al. 2013). Similar results have been reported for PLD films, which presented the anatase phase using small N_2 flow rates (up to 20% N_2) and were amorphous when the N_2 percentage was 50% or higher, even though the total pressure was constant (Farkas et al. 2018). Films prepared by sputtering with a TiO_2 target and a substrate temperature of 470°C grew in the anatase phase when no nitrogen was used. By injecting a small N_2 flux, the rutile phase was observed, the N content in the film being 0.2–0.3%. Increasing the N_2 percentage reverted the film structure back to anatase for samples with 1.4–3.6% N. Films with 6.2% N exhibited anatase diffraction peaks together with additional weak peaks that were associated with $TiO_{2-x}N_x$ (Delegan et al. 2014). Crystallization in the rutile phase for low N content could be related to the fact that the growth rate was maximal under those deposition conditions. X-ray patterns resembling that of TiN were obtained when sputtering a Ti target in environments with a high proportion of nitrogen, either with a 5/1 N_2/O_2 ratio (Grigorov et al. 2011) or using air (Chan and Lu 2008).

Grain size and surface roughness of N-doped films were reported in the same range as for the undoped samples. However, both magnitudes were found to decrease with increasing N_2 partial pressure for either CAD (Çetinörgü-Goldenberg et al. 2013) or sputtering (Chiu et al. 2007, Nejand et al. 2010, Liu et al. 2013). Films produced by PLD using a N:TiO_2 target showed larger grains and higher roughness with increasing temperature from RT to 400°C and, at the same time, the film turned from amorphous to anatase (Hu et al. 2012).

Regarding N-doped ZnO, films grown with N_2 percentages up to 25% showed ZnO structure with preferential (002) orientation. Samples prepared with higher N_2 percentages exhibited another diffraction peak that was not identified (Wang et al. 2007), but might be associated with zinc nitride (Tu et al. 2006). The crystallite size was found to increase with the N_2 percentage, reaching a maximum for 25% N_2, then decreasing as the amount of nitrogen was further increased. A similar relation between the crystallite size and the N_2 percentage was observed for films prepared by CAD. In this case, ZnO structure with (002) orientation was observed for all the studied N_2 flow rates (Tuzemen et al. 2014).

The incorporation of nitrogen generally led to narrowing the bandgap and thus shifting the absorption edge towards longer wavelengths. At the same time, with the increase in nitrogen content, a decrease in the transparency of the films was usually observed. For sputtered N:TiO_2 films, the bandgap changed from 3.2 to 2.3 eV when the N content increased from 0 to 3.4%; however, the bandgap increased to 2.4–2.5 eV for N contents around 6%, and titanium oxynitride was observed by XRD (Delegan et al. 2014). For CAD films, the bandgap was reduced from 3.3 (TiO_2) to 2.7 eV (around 5% N), while for samples prepared by PBII&D the bandgap changed from 3.2 (TiO_2) to 2.8 eV (2–3% N) (Manova et al. 2017). Similar behavior was observed for PLD films, the bandgap decreasing linearly with the increasing nitrogen content, reaching 2.8 eV for 8% N (Farkas et al. 2018). For N:ZnO films deposited by sputtering, the bandgap was narrowed from 2.1 (10% N_2) to 1.5 eV (50% N_2) while staying almost the same as the N_2 flow rate was further increased (Wang et al. 2007).

3.3.2 Metal doping and co-doping

Many different metals and strategies have been reported for doping TiO_2 and ZnO with the aim of enhancing the photocatalytic activity and/or extending it to the visible range. As with N-doped films, low dopant contents led to structures that resemble those of undoped oxides.

Fe-doped TiO_2 films were deposited by sputtering using a Ti target with Fe pieces of different sizes (Fe/Ti area ratio: 0.64 and 2.5). While pure TiO_2 was anatase, Fe-doped films resulted amorphous; for the highest Fe amount, XRD revealed the presence of Fe_3O_4. The film transmittance in the visible range decreased slightly due to doping, and the absorption edge was shifted to the visible spectrum (Carneiro et al. 2005). Another approach was the use of a Fe-doped rutile target with 15 at.% Fe for PLD. With the substrate at RT, the film was also amorphous; at 300°C, anatase was observed, and higher temperatures led to mixed samples with increasing rutile content. The Fe concentration increased from 0.5 at RT to 0.8 at.% at 700°C, while the bandgap decreased from 3.46 to 3.23 eV (Meng et al. 2019).

Cr doping on sputtered TiO_2 films was carried out by metal ion implantation during the final step of deposition. Cr^{3+} ions were implanted at different fluences between 1×10^{15} and 2×10^{17} atoms/cm^2; the ion energy was 60 keV. For fluences up to 1×10^{16} atoms/cm^2, the films presented the anatase phase; for higher fluences, the TiO_2 was amorphous and the presence of chromium oxide was observed. The bandgap was 3.4 eV for pure TiO_2 and was reduced to a minimum value of 3.2 eV for the lowest Cr fluence (Zhao et al. 2017).

ZnO films were doped with Ni by co-sputtering ZnO and Ni targets. The Ni content varied between 3.5 and 7 wt.% by changing the Ni sputter power. Small amounts of Ni enhanced the orientation in the (002) direction, while the greatest Ni content degraded the film crystallinity. The transmittance (80% for pure ZnO) greatly decreased with increasing Ni content. The bandgap was narrowed from 3.2 to 1.4 eV (Abdel-wahab et al. 2016). Co-sputtering of ZnO and metal targets was also used for the deposition of Ag-doped ZnO films. Ag doping reduced the crystallite size and degraded the crystalline quality; Ag peaks were observed by XRD. For films deposited at RT and 50°C, no ZnO peaks were detected; at 100°C, ZnO structure with (002) orientation was observed. The bandgap was reduced from 3.1 to 2.7 eV with increasing substrate temperature (Rashid et al. 2014).

Ti doping on ZnO was also studied. Films were deposited by dual cathode (Zn and Ti) CAD. The titanium arc current was varied in order to obtain samples with different Ti content between 0.64 and 0.83%. The films presented a ZnO structure with a preferred (002) orientation; the crystallite size and crystalline quality decreased with the increasing Ti content. The mean transmittance was around 90%, and the bandgap fell within the range of 3.21–3.22 eV for all the films (Hsu et al. 2016). In another work, following a similar methodology, the films were deposited amorphous at RT and were crystallized in the (002) direction of ZnO with annealing at 500°C. The bandgap was reduced from 3.42 (as-deposited film) to 3.25 eV by annealing (Wu et al. 2011).

Another strategy was the deposition of ZnO/CdO films by PLD through simultaneous ablation of Cd and Zn targets. XRD analysis showed diffraction peaks corresponding to both oxides. The bandgap was 2.83 eV for bare CdO and 2.52 eV for ZnO/CdO. Both samples were highly transparent, the mean transmittance being approximately 90% (Mostafa and Mwafy 2020).

Co-doping with nitrogen and metals was also studied. Pure, N-doped, V-doped and V-N co-doped TiO_2 films were deposited by sputtering using a TiO_2 target with and without partial covering with small pieces of metallic vanadium. All the samples, which were annealed at 500°C, exhibited anatase/rutile mixtures, the atomic percentages for the dopants being 2% N and 1% V. The bandgap was reduced from 3.1 eV (TiO_2) to 3.0 eV (N-doped), 2.8 eV (V-doped) and 2.5 eV (V–N co-doped) (Patel et al. 2014). In another work, pure, N-doped and Ta-N co-doped TiO_2 samples were prepared by sputtering using a Ti target with an amounted pellet of 99% TiO_2, 1% Ta_2O_5 in an $Ar/O_2/N_2$ gas mixture. All the films grew in the anatase phase at 400°C. The bandgap was narrowed for the doped films; the values being 3.20 eV for TiO_2, 3.16 eV for $TiO_{1.985}N_{0.015}$ and $Ti_{0.995}Ta_{0.005}O_{1.987}N_{0.013}$, and 3.07 eV for $Ti_{0.994}Ta_{0.006}O_{1.977}N_{0.023}$ (Obata et al. 2007).

4. Photocatalytic Performance of PVD Films

The photocatalytic activity is generally evaluated through the efficiency in the degradation of organic dyes; among them, Methylene Blue (MB) is the most frequently used (Ansari et al. 2016, Ong et al. 2018). In a few works, the photocatalytic performance of films prepared by PVD was directly assessed in the degradation of phenolic compounds and other actual pollutants, which is important in view of their application in water treatment. For instance, the photocatalytic efficiency of $N:TiO_2$ films prepared by sputtering was assessed for the degradation of chlortetracycline (CTC), which is an antibiotic widely used in the agricultural industry and arises as an emerging contaminant that could seep into the drinkable water supply (Delegan et al. 2014). Also, the photocatalytic activity of TiO_2 grown by CAD was evaluated by the efficiency to reduce Cr (VI) in the presence of ethylenediaminetetraacetic acid (EDTA) (Kleiman et al. 2011, 2020). Cr (VI) is a toxic and carcinogenic pollutant present in wastewaters from different industries. The Cr (VI)/EDTA system has been found to be an excellent model system to test TiO_2 photocatalysts (Litter 2017). Different photocatalysts prepared by PVD techniques and the pollutants employed to test their photocatalytic performance are listed in Table 15.4.

The assessment of the photocatalytic activity depends not only on the photocatalyst characteristics but also on the design and parameters of the photocatalytic experiments. Therefore, the comparison between the degradation rates reported for different experiments is generally not straightforward. When testing sputtered Ag:ZnO films in the degradation of 2-chlorophenol under visible light, the photocatalytic activity was found higher for lower pollutant concentration (Rashid et al. 2014). The same behavior was observed for pure ZnO under UV irradiation, for which the photocatalytic activity also increased with UV intensity, and was found higher at pH 3 than at pH 5 (Rashid et al. 2015). In contrast, also for sputtered ZnO under UV, the photocatalytic activity

Table 15.4. Photocatalytic films deposited with different PVD techniques for the degradation of different types of pollutants.

Technique	Photocatalyst	Pollutant	Reference
Sputtering	Ni:ZnO	Methyl green	(Abdel-wahab et al. 2016)
Sputtering	ZnO	Orange G	(Ahumada-Lazo et al. 2014)
Sputtering	N:TiO$_2$	Methylene blue	(Asahi et al. 2001)
PLD	TiO$_2$	Stearic acid	(Brunella et al. 2007)
Sputtering	Fe:TiO$_2$	Rhodamine B	(Carneiro et al. 2005)
Arc Ion Plating	TiO$_2$	Methylene blue	(Chang et al. 2006)
Sputtering	N:TiO$_2$	Methylene blue	(Chiu et al. 2007)
Sputtering	N:TiO$_2$	CTC	(Delegan et al. 2014)
Sputtering	TiO$_2$	Methylene blue	(Ghori et al. 2018)
Sputtering	TiO$_2$	Methylene blue	(Henkel et al. 2018)
PLD	N:TiO$_2$	Methyl orange	(Hu et al. 2012)
Sputtering	TiO$_2$	Rhodamine B	(Hui et al. 2017)
Sputtering	doped TiO$_2$	Oxalic acid	(Kavaliunas et al. 2020)
Sputtering	N:TiO$_2$	2-propanol	(Kitano et al. 2006)
CAD	TiO$_2$	Cr (VI)	(Kleiman et al. 2011)
CAD	TiO$_2$	Cr (VI)	(Kleiman et al. 2020)
Sputtering	Mg:ZnO	Methylene blue	(Kuru and Narsat 2019)
Sputtering	TiO$_2$/ZnO	Rhodamine B	(Ma et al. 2021)
PLD	ZnO/CdO	4-nitrophenol	(Mostafa and Mwafy 2020)
Sputtering	N:TiO$_2$	Methylene blue	(Nakano et al. 2007)
Sputtering	N:TiO$_2$/ZnO	Methylene blue	(Nejand et al. 2010)
Sputtering	Ta-N:TiO$_2$	Oleic acid	(Obata et al. 2007)
Sputtering	V-N:TiO$_2$	Various pollutants	(Patel et al. 2014)
Sputtering	TiO$_2$	Methylene blue	(Pihosh et al. 2009)
Sputtering	Ag:ZnO	2-chlorophenol	(Rashid et al. 2014)
Sputtering	ZnO	2-chlorophenol	(Rashid et al. 2015)
Sputtering	ZnO	Methylene blue	(Rivera et al. 2019)
CAD	N:TiO$_2$	Methylene blue	(Thorat et al. 2019)
PLD	doped TiO$_2$	Methylene blue	(Velardi et al. 2021)
PLD	TiO$_2$	Methylene blue	(Yamaki et al. 2002)
Sputtering	TiO$_2$	Methyl orange	(Zhang et al. 2004)
Sputtering	Cr:TiO$_2$	Methylene blue	(Zhao et al. 2017)

in the degradation of orange G was higher for pH 7 than under acid (pH 4) or basic (pH 10) medium (Ahumada-Lazo et al. 2014).

4.1 *TiO$_2$ and ZnO–UV photocatalysis*

When sputtered ZnO and Mg:ZnO films, annealed at different temperatures up to 600°C, were tested in the degradation of MB, the best performance found for an intermediate temperature (400°C), especially for the doped film, was associated with the larger crystallite size and better crystallinity (Kuru and Narsat 2019).

For TiO$_2$ films prepared by arc ion plating, the photocatalytic activity in the degradation of MB was higher with the increase of the O$_2$ partial pressure during deposition, which led to a reduction in the bandgap. The authors also found a correlation between the degradation rate and the proportion of anatase in the obtained anatase/rutile mixtures, as the efficiency was higher when the amount of anatase was higher (Chang et al. 2006). For TiO_2 films deposited by PLD with different laser fluences, the best performance in the degradation of MB was achieved with the film prepared with the lowest laser fluence; this result was also attributed to the greater proportion of anatase phase and to the fact that the grains were less densely packed (Yamaki et al. 2002). On the other hand, despite having lower grain size and roughness, anatase films deposited by CAD at 400°C performed very similarly in the reduction of Cr (VI) to films deposited at RT and annealed at 400°C (Kleiman et al. 2011).

Photocatalytic activity is generally improved with increasing film thickness, but when the thickness exceeds the depth at which substantial photon or reagent depletion takes place, no further enhancement is obtained (Camera-Roda and Santarelli 2007). Thus, the thickness at which the degradation rate is not significantly enhanced anymore can be considered as the optimal thickness. For TiO_2 films deposited by sputtering, the degradation rate for MB increased with the thickness up to approximately 400–500 nm for films with cracks on the surface and to 250–350 nm for uncracked films (Henkel et al. 2018). The efficiency was not significantly improved for thicker films, thus indicating that the optimal thickness was reached. For sputtered TiO_2 films also, the degradation rate for rhodamine B reached a limit value when the thickness was 260 nm (Hui et al. 2017). For TiO_2 films prepared by arc ion plating, the optimal thickness for MB degradation was approximately 1000 nm (Chang et al. 2006). The enhancement of the photocatalytic activity by film thickness was also observed for TiO_2 films grown by CAD; in this case, the photocatalytic reduction of Cr (VI) in the presence of EDTA was the model pollutant used. Complete Cr (VI) reduction was observed after 240 and 300 minutes for samples with thicknesses 1100 and 1000 nm, respectively. In this case, as the optimal thickness was not achieved, the efficiency could be further improved by further increasing the thickness (Kleiman et al. 2020). For ZnO films prepared by sputtering with thicknesses up to approximately 400 nm, the photocatalytic activity increased with the film thickness for the degradation of MB (Rivera et al. 2019), orange G (Ahumada-Lazo et al. 2014), and 2-chlorophenol (Rashid et al. 2015), and no limit value was found in that range.

As mentioned above, results obtained in different experiments cannot be easily compared. However, in a few cases, samples prepared by different methods were tested as photocatalysts under the same experimental conditions. Sputtered TiO_2 films exhibited a much better performance in the degradation of MB under UV light than spin-coated TiO_2 films, which had a porous structure composed of spherical TiO_2 nanoparticles (Ghori et al. 2018). TiO_2 films prepared by PLD were also more efficient in the degradation of stearic acid than anodized samples (Brunella et al. 2007). In contrast, in the reduction of Cr (VI) in the presence of EDTA, the photocatalytic activity of TiO_2 deposited by CAD was lower than that of TiO_2 P25 films with similar thickness prepared by dip-coating (Kleiman et al. 2011). However, the possibility of increasing the thickness allowed CAD films to outperform dip-coated samples, as the latter exhibited poor adhesion, and TiO_2 partially detached in contact with water when the thickness was above 480 nm (Kleiman et al. 2020).

4.2 Doped films–visible light photocatalysis

While nitrogen doping has proved to enhance the photocatalytic activity of TiO_2 under visible light, different behaviors of the materials have been reported under UV irradiation, which could depend on the properties of the films as well as the spectrum of the light source. Although with a relatively low efficiency, N-doped TiO_2 films prepared by PLD presented higher photoactivity for the degradation of MB than pure TiO_2 under UV (Velardi et al. 2021). Higher UV activity by N doping for MB degradation was also observed for ZnO/TiO_2 bilayers prepared by sputtering (Nejand et al. 2010). In contrast, under UV also, N-doped and undoped sputtered films exhibited similar activity (Asahi et al. 2001), while other N-doped films presented lower photocatalytic efficiency than undoped samples (Kitano et al. 2006). In addition, the MB degradation rate for sputtered TiO_2 films under UV irradiation was significantly enhanced by N doping under a small N_2 flow rate during deposition, but the performance worsened when the flow increased (Chiu et al. 2007). Under UV-visible light, increased MB degradation rate by N doping was observed for CAD films; N:TiO$_2$ with 8 at.% of N achieved 89% degradation after 180 minutes, while the best performance for pure TiO$_2$ was 71% (Thorat et al. 2019).

Under visible light, sputtered N: TiO$_2$ films showed significant photocatalytic activity for decomposition of MB, while the activity of pure TiO$_2$ was almost negligible (Asahi et al. 2001, Nakano et al. 2007). N:TiO$_2$ films deposited by PLD were tested in the decomposition of methyl orange under visible light. The efficiency of amorphous films deposited at RT and 200°C was rather low, while an anatase film grown at 400°C exhibited a good photocatalytic activity, almost achieving complete decomposition after 4 hours (Hu et al. 2012). Sputtered N:TiO$_2$ films were tested in the electrophotocatalytic degradation of CTC under solar irradiation. The efficiency was found to increase with the nitrogen content up to 3.4% N and to decrease with further increasing N percentage (Delegan et al. 2014). The most efficient sample was the one exhibiting the lowest bandgap (2.25 eV), which was also the one that presented the highest content of anatase among the doped films. Oxidative degradation of 2-propanol was studied for sputtered N:TiO$_2$ samples under different light sources. Under UV light, the highest activity was observed for pure TiO$_2$, and the activity decreased with the increasing N content. Under visible light, the efficiency grew with increasing nitrogen content (up to 6%) and then decreased. The best N:TiO$_2$ sample reached complete 2-propanol oxidation after 8 hours under sunlight irradiation, while pure TiO$_2$ failed to achieve 50% yield (Kitano et al. 2006). A similar behavior was observed for TiO$_2$ grown by sputtering and Cr-doped by ion implantation. The photocatalytic activity was studied through the degradation of MB under UV and visible irradiation. Under UV, the best performance was obtained for pure TiO$_2$ and the efficiency tended to decrease with the Cr content. Under visible light, the efficiency grew with the increasing Cr content up to a certain value and then decreased. There was no clear correlation between efficiency and bandgap (Zhao et al. 2017). Photocatalytic activity for the degradation of MB, 4-nitrophenol and 4-chlorophenol under visible light was also shown by pure, N-doped, V-doped and V-N co-doped TiO$_2$, prepared by sputtering. For the three pollutants, the best performance was achieved by V-N co-doping, followed by V-doped, N-doped and pure TiO$_2$ films (Patel et al. 2014).

Pure, N-doped and Ta-N co-doped TiO$_2$ films prepared by sputtering were tested for the photocatalytic decomposition of oleic acid under UV and visible light in order to assess their potential application as self-cleaning surfaces. The activity was evaluated by the decrease in water contact angles having oleic acid applied to the film surface. All the films showed a similar performance under UV, with pure TiO$_2$ showing slightly higher efficiency. Under visible light, TiO$_2$ showed low activity, while co-doped films exhibited the best performance, becoming the best option as a self-cleaning surface (Obata et al. 2007).

Ag:ZnO films prepared by sputtering had photocatalytic activity superior to that of pure ZnO under visible light. Doped samples achieved 54% efficiency after 3 hours irradiation for the degradation of 2-chlorophenol, while only 21% was degraded with pure ZnO. Increasing the substrate temperature from RT to 100°C also improved the photocatalytic activity, as the crystallinity was enhanced and the bandgap was narrowed (Rashid et al. 2014). Ni-doped ZnO films prepared by co-sputtering ZnO and Ni targets presented visible light activity for the degradation of methyl green. The efficiency was higher with a higher Ni content and a narrower bandgap (Abdel-wahab et al. 2016). Degradation of 4-nitrophenol under UV-visible light by CdO films deposited by PLD was outperformed by ZnO/CdO prepared by simultaneous ablation of Cd and Zn targets, due to the narrowing of the bandgap (Mostafa and Mwafy 2020).

Improvement of the photocatalytic activity under visible light by doping was also observed in bacterial (*Escherichia coli*) inactivation. Nitrogen doping by ion implantation enhanced the antibacterial effect of TiO$_2$ films deposited by CAD (Esparza et al. 2017). Cu-doped TiO$_2$ films deposited by HiPIMS achieved also complete inactivation after 10 minutes irradiation, while complete inactivation using pure TiO$_2$ took 40 minutes in the best case (Rtimi et al. 2013).

4.3 Reuse of photocatalysts

The possibility of reusing photocatalysts has great importance for their use in water treatment or other practical applications. However, the efficiency of the photocatalysts when reused has not been extensively studied, and only a few reports can be found in literature. Some of the results are listed in Table 15.5, where the efficiency is calculated as a percentage of the yield in the first use of the photocatalyst. It can be seen that, in most reports, the performance decreased very slightly after successive tests. In some cases, the photocatalytic activity was reduced just for the second run and then remained practically the same. Not included in the table, Cu-doped TiO_2 films deposited by HiPIMS showed the same efficiency up to the fourth use in bacterial inactivation, and slightly lower performance on the eighth run (Rtimi et al. 2013).

It is worth mentioning that, in the case of Cr (VI) reduction by CAD-TiO_2, for which Table 15.5 shows a significant diminution in the photocatalytic performance after reuse, the efficiency was calculated from the ratio between the rate constants instead of the percentage yield. Furthermore, the rate constants only accounted for the effect of the film, since the contribution of the homogeneous Cr (VI) photochemical reduction, calculated from experiments without TiO_2, was subtracted. A similar behavior was also observed for P-25 films prepared by dip-coating (Kleiman et al. 2020). This decrease in the rate constant after reuse was associated with the deactivation caused by Cr (III) deposition on the surface during the photocatalytic test; however, Cr (III) can be removed from the surface using an EDTA or citric acid solution (Djellabi et al. 2016), allowing the reuse of the photocatalyst.

Table 15.5. Efficiency of photocatalysts when reused.

Film	Technique	Pollutant	Light	Use #	Efficiency (% of 1st use)	Reference
V-N:TiO_2	Sputtering	Methylene blue	Visible	2–4	95	(Patel et al. 2014)
ZnO	Sputtering	2-chlorophenol	UV	2–3	93	(Rashid et al. 2015)
Ag:ZnO	Sputtering	2-chlorophenol	Visible	2–4	95–84	(Rashid et al. 2014)
ZnO	Sputtering	Orange G	UV	2–5	> 90*	(Ahumada-Lazo et al. 2014)
ZnO/CdO	PLD	4-nitrophenol	UV-visible	2–6	98–92	(Mostafa and Mwafy 2020)
TiO_2	CAD	Cr (VI)	UV	2–4	65–32**	(Kleiman et al. 2020)

* Did not decrease monotonically at successive runs
** Calculated from rate constant instead of percent yield

5. Final Remarks

PVD techniques lead to the deposition of dense, uniform films with good adhesion to the substrate. These characteristics make PVD films very attractive for optical coatings and different industrial applications. However, the compactness of PVD films may be inconvenient for photocatalytic applications, where porous films or large exposed surface areas are required for a good photocatalytic performance. Nonetheless, the porosity in PVD films can be tuned to a certain extent by the ambient pressure during deposition, and the appearance of cracks in the surface, which is more pronounced the thicker the film, enlarges the exposed surface area and can be also tailored by gas pressure and heat treatment.

The photocatalytic activity of the films is also enhanced by increasing the film thickness and narrowing the bandgap. Other factors that improve the photocatalytic efficiency are the crystalline quality of ZnO and the proportion of anatase in TiO_2, although small amounts of rutile have proved to enhance the photoactivity as well. PVD techniques allow increasing the film thickness to reach the optimal thickness at which the maximum efficiency is achieved without detriment to the film adhesion and transparency. The crystalline structure of the films can be easily controlled by the PVD process parameters (temperature, bias, pressure, power supply) and/or by post-deposition annealing. The bandgap is significantly reduced by doping the films, which can be done directly during deposition by injecting gaseous nitrogen into the deposition chamber and/or by using doped or alloyed targets.

In addition, ZnO and TiO_2 films prepared by PVD are highly transparent. Despite transparency is reduced by doping, the transmittance of films remains relatively high for low amounts of dopants, which are nevertheless effective in narrowing the bandgap. The high transparency of PVD films makes them suitable photocatalysts for water treatment reactors, where the irradiation may come from outside, passing through a transparent support and the film.

PVD techniques are environmentally friendly, which is a great advantage for large-scale production. Nowadays, systems capable to produce PVD coatings on an industrial scale for large surfaces are available. In these devices, uniform coatings of great extension are achieved by using multiple plasma sources and/or rotating or displacing the substrate. The existing industrial equipment can easily be adapted for the production of photocatalytic films, making it possible to tackle the development of large self-cleaning surfaces and hybrid reactors for water treatment.

Acknowledgments

This work was supported by grants of Universidad de Buenos Aires (PID 20020190100292BA), CONICET (PUE 22920180100050CO and PIP 11220170100711CO), and ANPCyT (PICT-2017-2869).

References

Abdel-wahab, M.Sh., A. Jilani, I.S. Yahia and A.A. Al-Ghamdi. 2016. Enhanced the photocatalytic activity of Ni-doped ZnO thin films: Morphological, optical and XPS analysis. Superlattices Microstruct. 94: 108–118.

Abegunde, O.O., E.T. Akinlabi, O.P. Oladijo, S. Akinlabi and A.U. Ude. 2019. Overview of thin film deposition techniques. AIMS Mater. Sci. 6: 174–199.

Ahumada-Lazo, R., L.M. Torres-Martínez, M.A. Ruíz-Gómez, O.E. Vega-Becerra and M.Z. Figueroa-Torres. 2014. Photocatalytic efficiency of reusable ZnO thin films deposited by sputtering technique. Appl. Surf. Sci. 322: 35–40.

Akpan, U.G. and B.H. Hameed. 2009. Parameters affecting the photocatalytic degradation of dyes using TiO_2-based photocatalysts: a review. J. Hazard. Mater. 170: 520–529.

Aksenov, I.I., V.A. Belous, V.G. Padalka and V.M. Khoroshikh. 1982. Swedish Patent # 82188-8.

Aksenov, I.I., V.E. Strel'nitskij, V.V. Vasilyev and D.Y. Zaleskij. 2003. Efficiency of magnetic plasma filters. Surf. Coat. Technol. 163: 118–127.

Anders, A. (ed.). 2000. Handbook of Plasma Immersion Ion Implantation and Deposition. Wiley, New York, USA.

Anders, A. 2009. Cathodic Arcs: from Fractal Spots to Energetic Condensation. Springer Science & Business Media, New York, USA.

Anders, A. 2010. A structure zone diagram including plasma-based deposition and ion etching. Thin Solid Films 518: 4087–4090.

Anders, A. 2014. A review comparing cathodic arcs and high power impulse magnetron sputtering (HiPIMS). Surf. Coat. Technol. 257: 308–325.

Anders, A. and Y. Yang. 2018. Plasma studies of a linear magnetron operating in the range from DC to HiPIMS. J. Appl. Phys. 123: 043302.

Angappane, S., N.R. Selvi and G.U. Kulkarni. 2009. ZnO(101) films by pulsed reactive crossed-beam laser ablation. Bull. Mater. Sci. 32: 253–258.

Ansari, S.A., M.M. Khan, M.O. Ansari and M.H. Cho. 2016. Nitrogen-doped titanium dioxide (N-doped TiO_2) for visible light photocatalysis. New J. Chem. 40: 3000–3009.

Arakelova, E., A. Khachatryan, A. Kteyan, K. Avjyan and S. Grigoryan. 2016. ZnO film deposition by DC magnetron sputtering: Effect of target configuration on the film properties. Thin Solid Films 612: 407–413.

Asahi, R., T. Morikawa, T. Ohwaki, K. Aoki and Y. Taga. 2001. Visible-light photocatalysis in nitrogen-doped titanium oxides. Science 293: 269–271.

Asenova, I., D. Manova and S. Mändl. 2014. Incorporation of nitrogen into TiO_2 thin films during PVD processes. J. Phys. Conf. Ser. 559: 012008.

Bakar, S.A. and C. Ribeiro. 2016. Nitrogen-doped titanium dioxide: An overview of material design and dimensionality effect over modern applications. J. Photochem. Photobiol. C 27: 1–29.

Baptista, A., F. Silva, J. Porteiro, J. Míguez and G. Pinto. 2018a. Sputtering physical vapour deposition (PVD) coatings: A critical review on process improvement and market trend demands. Coatings 8: 402.

Baptista, A., F.J.G. Silva, J. Porteiro, J.L. Míguez, G. Pinto and L. Fernandes. 2018b. On the physical vapour deposition (PVD): evolution of magnetron sputtering processes for industrial applications. Procedia Manuf. 17: 746–757.

Bendavid, A., P.J. Martin, Å. Jamting and H. Takikawa. 1999. Structural and optical properties of titanium oxide thin films deposited by filtered arc deposition. Thin Solid Films 355–356: 6–11.

Bendavid, A., P.J. Martin and H. Takikawa. 2000. Deposition and modification of titanium dioxide thin films by filtered arc deposition. Thin Solid Films 360: 241–249.

Bensmaine, S. and B. Benyoucef. 2014. Effect of the temperature on ZnO thin films deposited by r.f. magnetron. Phys. Procedia 55: 144–149.

Brunella, M.F., M.V. Diamanti, M.P. Pedeferri, F.D. Fonzo, C.S. Casari and A.L. Bassi. 2007. Photocatalytic behavior of different titanium dioxide layers. Thin Solid Films 515: 6309–6313.

Bolt, H., F. Koch, J.L. Rodet, D. Karpov and S. Menzel. 1999. Al_2O_3 coatings deposited by filtered vacuum arc—characterization of high temperature properties. Surf. Coat. Technol. 116: 956–962.

Boxman, R.L., D.M. Sanders and P.J. Martin (eds.). 1996a. Handbook of Vacuum Arc Science & Technology: Fundamentals and Applications. William Andrew, Norwich, NY, USA.

Boxman, R.L., V. Zhitomirsky, B. Alterkop, E. Gidalevich, I. Beilis, M. Keidar et al. 1996b. Recent progress in filtered vacuum arc deposition. Surf. Coat. Technol. 86: 243–253.

Bunshah, R.F. 1977. 3.1 mechanical properties of PVD films. Vacuum 27: 353–362.

Camera-Roda, G. and F. Santarelli. 2007. Optimization of the thickness of a photocatalytic film on the basis of the effectiveness factor. Catal. Today 129: 161–168.

Carneiro, J.O., V. Teixeira, A. Portinha, L. Dupák, A. Magalhães and P. Coutinho. 2005. Study of the deposition parameters and Fe-dopant effect in the photocatalytic activity of TiO_2 films prepared by dc reactive magnetron sputtering. Vacuum 78: 37–46.

Çetinörgü-Goldenberg, E., L. Burstein, I. Chayun-Zucker, R. Avni and R.L. Boxman. 2013. Structural and optical characteristics of filtered vacuum arc deposited N:TiO_x thin films. Thin Solid Films 537: 28–35.

Chan, M.H. and F.H. Lu. 2008. Preparation of titanium oxynitride thin films by reactive sputtering using air/Ar mixtures. Surf. Coat. Technol. 203: 614–618.

Chang, J.T., C.W. Su and J.L. He. 2006. Photocatalytic TiO_2 film prepared using arc ion plating. Surf. Coat. Technol. 200: 3027–3024.

Chiu, S.M., Z.S. Chen, K.Y. Yang, Y.L. Hsu and D. Gan. 2007. Photocatalytic activity of doped TiO_2 coatings prepared by sputtering deposition. J. Mater. Process. Technol. 192-193: 60–67.

Deng, Y., W. Chen, B. Li, C. Wang, T. Kuang and Y. Li. 2020. Physical vapor deposition technology for coated cutting tools: A review. Ceram. Int. 46: 18373–18390.

Delegan, N., R. Daghrir, P. Drogui and M.A. El Khakani. 2014. Bandgap tailoring of in-situ nitrogen-doped TiO_2 sputtered films intended for electrophotocatalytic applications under solar light. J. Appl. Phys. 116: 153510.

Djellabi, R., F.M. Ghorab, S. Nouacer, A. Smara and O. Khireddine. 2016. Cr(VI) photocatalytic reduction under sunlight followed by Cr(III) extraction from TiO_2 surface. Mat. Letters 176: 106–109.

Doghmane, H.E., T. Touam, A. Chelouche, F. Challali and B. Bordji. 2020. Investigation of the influences of post-thermal annealing on physical properties of TiO_2 thin films deposited by RF sputtering. Semiconductors 54: 268–273.

Dreesen, L., F. Cecchet and S. Lucas. 2009. DC magnetron sputtering deposition of titanium oxide nanoparticles: Influence of temperature, pressure and deposition time on the deposited layer morphology, the wetting and optical surface properties. Plasma Process. Polym. 6: S849–S854.

Elzwawi, S., H.S. Kim, R. Heinhold, M. Lynam, G. Turner and J.G. Partridge. 2012. Device quality ZnO grown using a filtered cathodic vacuum arc. Physica B 407: 2903–2906.

Esparza, J., G. García Fuentes, R. Bueno, R. Rodríguez, J.A. García, A.I. Vitas et al. 2017. Antibacterial response of titanium oxide coatings doped by nitrogen plasma immersion ion implantation. Surf. Coat. Technol. 314: 67–71.

Eufinger, K., D. Poelman, H. Poelman, R. De Gryse and G.B. Marin. 2007. Photocatalytic activity of dc magnetron sputter deposited amorphous TiO_2 thin films. Appl. Surf. Sci. 254: 148–152.

Fang, W., M. Xing and J. Zhang. 2017. Modifications on reduced titanium dioxide photocatalysts: a review. J. Photochem. Photobiol. C 32: 21–39.

Farkas, B., P. Heszler, J. Budai, A. Oszkó, M. Ottosson and Zs. Geretovszky. 2018. Optical, compositional and structural properties of pulsed laser deposited nitrogen-doped Titanium-dioxide. Appl. Surf. Sci. 433: 149–154.

Fotovvati, B., N. Namdari and A. Dehghanghadikolaei. 2019. On coating techniques for surface protection: A review. J. Manuf. Mater. Process. 3: 28.

Fox, M.A. and M.T. Dulay. 1993. Heterogeneous photocatalysis. Chem. Rev. 93: 341–357.

Franco Arias, L., A. Kleiman, E. Heredia and A. Márquez. 2012. Rutile titanium dioxide films deposited with a vacuum arc at different temperatures. J. Phys. Conf. Ser. 370: 012027.

Franco Arias, L.M., A. Kleiman, D. Vega, M. Fazio, E. Halac and A. Márquez. 2017. Enhancement of rutile phase formation in TiO₂ films deposited on stainless steel substrates with a vacuum arc. Thin Solid Films 638: 269–276.

Fujimura, N., T. Nishihara, S. Goto, J. Xu and T. Ito. 1993. Control of preferred orientation for ZnOₓ films: control of self-texture. J. Cryst. Growth 130: 269–279.

Geyao, L., D. Yang, C. Wanglin and W. Chengyong. 2020. Development and application of physical vapor deposited coatings for medical devices: A review. Procedia CIRP 89: 250–262.

Ghori, M.Z., S. Veziroglu, B. Henkel, A. Vahl, O. Polonskyi, T. Strunskus et al. 2018. A comparative study of photocatalysis on highly active columnar TiO₂ nanostructures in-air and in-solution. Sol. Energy Mater. Sol. Cells 178: 170–178.

Gjevori, A., K. Nonnenmacher, B. Ziberi, D. Hirsch, J.W. Gerlach, D. Manova and S. Mändl. 2009. Investigation of nucleation and phase formation of photocatalytically active TiO₂ films by MePBIID. Nucl. Instrum. Meth. B 267: 1658–1661.

Grigorov, K.G., I.C. Oliveira, H.S. Maciel, M. Massi, M.S. Oliveira Jr., J. Amorim et al. 2011. Optical and morphological properties of N-doped TiO₂ thin films. Surf. Sci. 605: 775–782.

Grinevich, V.I., E.Y. Kvitkova, N.A. Plastinina and V.V. Rybkin. 2011. Application of dielectric barrier discharge for waste water purification. Plasma Chem. Plasma Process. 31: 573–583.

Gupta, G., R.K. Tyagi, S.K. Rajput, P. Saxena, A. Vashisth and S. Mehndiratta. 2021. PVD based thin film deposition methods and characterization/property of different compositional coatings—A critical analysis. Mater. Today: Proc. 38: 259–264.

Gu, G.R., Y.A. Li, Y.C. Tao, Z. He, J.J. Li, H. Yin et al. 2003. Investigation on the structure of TiO₂ films sputtered on alloy substrates. Vacuum 71: 487–490.

Guerrero de León, J.A., A. Pérez-Centeno, G. Gómez-Rosas, E. Camps, J.S. Arias-Cerón, M.A. Santana-Aranda et al. 2020. ZnO thin films grown at different plasma energies by the laser ablation of metallic Zn with a 532 nm wavelength. Mater. Res. Express 7: 016423.

Guillén, E., M. Krause, I. Heras, G. Rincón-Llorente and R. Escobar-Galindo. 2021. Tailoring crystalline structure of titanium oxide films for optical applications using non-biased filtered cathodic vacuum arc deposition at room temperature. Coatings 11: 233.

György, E., G. Socol, E. Axente, I.N. Mihailescu, C. Ducu and S. Ciuca. 2005. Anatase phase TiO₂ thin films obtained by pulsed laser deposition for gas sensing applications. Appl. Surf. Sci. 247: 429–433.

Haider, A.J., A.A. Jabbar and G.A. Ali. 2021. A review of pure and doped ZnO nanostructure production and its optical properties using pulsed laser deposition technique. J. Phys. Conf. Ser. 1795: 012015.

Hashimoto, K., H. Irie and A. Fujishima. 2005. TiO₂ photocatalysis: a historical overview and future prospects. Jpn. J. Appl. Phys. 44: 8269.

Henkel, B., A. Vahl, O.C. Aktas, T. Strunskus and F. Faupel. 2018. Self-organized nanocrack networks: a pathway to enlarge catalytic surface area in sputtered ceramic thin films, showcased for photocatalytic TiO₂. Nanotechnology 29: 035703.

Hernández-Alonso, M.D., F. Fresno, S. Suárez and J.M. Coronado. 2009. Development of alternative photocatalysts to TiO₂: Challenges and opportunities. Energy Environ. Sci. 2: 1231–1257.

Herrera-Jimenez, E.J., E. Bousser, T. Schmitt, J.E. Klemberg-Sapieha and L. Martinu. 2021. Effect of plasma interface treatment on the microstructure, residual stress profile, and mechanical properties of PVD TiN coatings on Ti-6Al-4V substrates. Surf. Coat. Technol. 413: 127058.

Hovsepian, P.E. and A.P. Ehiasarian. 2019. Six strategies to produce application tailored nanoscale multilayer structured PVD coatings by conventional and High Power Impulse Magnetron Sputtering (HIPIMS). Thin Solid Films 688: 137409.

Hu, J., H. Tang, X. Lin, Z. Luo, H. Cao, Q. Li et al. 2012. Doped titanium dioxide films prepared by pulsed laser deposition method. Int. J. Photoenergy 2012: 758539.

Huang, Z., P. Luo, W. Chen, S. Pan and D. Chen. 2013. Hemocompatibility of ZnO thin films prepared by filtered cathodic vacuum arc deposition. Vacuum 89: 220–224.

Hui, W., S. Guodong, Z. Xiaoshu, Z. Wei, H. Lin and Y. Ying. 2017. *In-situ* synthesis of TiO₂ rutile/anatase heterostructure by DC magnetron sputtering at room temperature and thickness effect of outermost rutile layer on photocatalysis. J. Environ. Sci. 60: 33–42.

Hurum, D.C., A.G. Agrios, K.A. Gray, T. Rajh and M.C. Thurnauer. 2003. Explaining the enhanced photocatalytic activity of Degussa P25 mixed-phase TiO₂ using EPR. J. Phys. Chem. B 107: 4545–4549.

Hussein, M.A., N.K. Ankah, A.M. Kumar, M.A. Azeem, S. Saravanan, A.A. Sorour et al. 2020. Mechanical, biocorrosion, and antibacterial properties of nanocrystalline TiN coating for orthopedic applications. Ceram. Int. 46: 18573–18583.

Hsu, S.F., M.H. Weng, J.H. Chou, C.H. Fang and R.Y. Yang. 2016. Effect of the Ti-target arc current on the properties of Ti-doped ZnO thin films prepared by dual-target cathodic arc plasma deposition. Ceram. Int. 42: 14438–14442.

Irimiciuc, S.A., S. Chertopalov, J. Lancok and V. Craciun. 2021. Langmuir probe technique for plasma characterization during pulsed laser deposition process. Coatings 11: 762.

Iwulska, A. and G. Śliwiński. 2011. Preparation of porous TiO_2 films by means of pulsed laser deposition for photocatalytic applications. Photonics Lett. Pol. 3: 98–100.

Jang, H.D., S.K. Kim and S.J. Kim. 2001. Effect of particle size and phase composition of titanium dioxide nanoparticles on the photocatalytic properties. J. Nanopart. Res. 3: 141–147.

Jang, E.S., J.H. Won, S.J. Hwang and J.H. Choy. 2006. Fine tuning of the face orientation of ZnO crystals to optimize their photocatalytic activity. Adv. Mater. 18: 3309–3312.

Javid, A., M. Kumar, M. Ashraf, J.H. Lee and J.G. Han. 2019. Photocatalytic antibacterial study of N-doped TiO_2 thin films synthesized by ICP assisted plasma sputtering method. Physica E 106: 187–193.

Kara, K., E.S. Tuzemen and R. Esen. 2014. Annealing effects of ZnO thin films on p-Si(100) substrate deposited by PFCVAD. Turk. J. Phys. 38: 238–244.

Kavaliunas, V., E. Krugly, M. Sriubas, H. Mimura, G. Laukaitis and Y. Hatanaka. 2020. Influence of Mg, Cu, and Ni dopants on amorphous TiO_2 thin films photocatalytic activity. Materials 13: 886.

Kawasaki, H., D. Taniyama, T. Ohshima, T. Ihara, Y. Yagyu and Y. Suda. 2013. Titanium oxide thin film preparation by pulsed laser deposition method using a powder target. Trans. Mater. Res. Soc. Japan 38: 69–72.

Kitano, M., K. Funatsu, M. Matsuoka, M. Ueshima and M. Anpo. 2006. Preparation of nitrogen-substituted TiO_2 thin film photocatalysts by the radio frequency magnetron sputtering deposition method and their photocatalytic reactivity under visible light irradiation. J. Phys. Chem. B 110: 25266–25272.

Kleiman, A., A. Márquez and D.G. Lamas. 2007. Anatase TiO_2 films obtained by cathodic arc deposition. Surf. Coat. Technol. 201: 6358–6362.

Kleiman, A., A. Márquez and R.L. Boxman. 2008. Performance of a magnetic island macroparticle filter in a titanium vacuum arc. Plasma Sources Sci. Technol. 17: 015008.

Kleiman, A., A. Márquez, M.L. Vera, J.M. Meichtry and M.I. Litter. 2011. Photocatalytic activity of TiO_2 thin films deposited by cathodic arc. Appl. Catal. B 101: 676–681.

Kleiman, A., D.G. Lamas, A.F. Craievich and A. Márquez. 2014. X-ray reflectivity analysis of titanium dioxide thin films grown by cathodic arc deposition. J. Nanosci. Nanotechnol. 14: 3902–3909.

Kleiman, A., J.M. Meichtry, D. Vega, M.I. Litter and A. Márquez. 2020. Photocatalytic activity of TiO_2 films prepared by cathodic arc deposition: Dependence on thickness and reuse of the photocatalysts. Surf. Coat. Technol. 382: 125154.

Kondaiah, P., M.C. Sekhar, S.J. Chandra, R. Martins, S. Uthanna and E. Elangovan. 2012. Influence of substrate bias voltage on the physical, electrical and dielectric properties of rf magnetron sputtered TiO_2 films. IOP Conf. Ser. Mater. Sci. Eng. 30: 012005.

Krella, A. 2020. Resistance of PVD coatings to erosive and wear processes: a review. Coatings 10: 921.

Krishna, D.S.R., Y. Sun and Z. Chen. 2011. Magnetron sputtered TiO_2 films on a stainless steel substrate: Selective rutile phase. Thin Solid Films 519: 4860–4864.

Kumi-Barimah, E., R. Penhale-Jones, A. Salimian, H. Upadhyaya, A. Hasnath and G. Jose. 2020. Phase evolution, morphological, optical and electrical properties of femtosecond pulsed laser deposited TiO_2 thin films. Sci. Rep. 10: 10144.

Kuru, M. and H. Narsat. 2019. The effect of heat treatment temperature and Mg doping on structural and photocatalytic activity of ZnO thin films fabricated by RF magnetron co-sputtering technique. J. Mater. Sci.: Mater. Electron. 30: 18484–18495.

Langford, J.I. and A.J.C. Wilson. 1978. Scherrer after sixty years: A survey and some new results in the determination of crystallite size. J. Appl. Crystallogr. 11: 102–113.

Lazar, M.A., J.K. Tadvani, W.S. Tung, L. Lopez and W.A. Daoud. 2010. Nanostructured thin films as functional coatings. IOP Conf. Ser. Mater. Sci. Eng. 12: 012017.

Leng, Y.X., J.Y. Chen, H. Sun, P. Yang, G.J. Wan, J. Wang et al. 2004. Properties of titanium oxide synthesized by pulsed metal vacuum arc deposition. Surf. Coat. Technol. 176: 141–147.

Lin, C.K., T.J. Yang, Y.C. Feng, T.T. Tsung and C.Y. Su. 2006. Characterization of electrophoretically deposited nanocrystalline titanium dioxide films. Surf. Coat. Technol. 200: 3184–3189.

Linsebigler, A.L., G. Lu and J.T. Yates Jr. 1995. Photocatalysis on TiO_2 surfaces: principles, mechanisms, and selected results. Chem. Rev. 95: 735–758.

Litter, M.I. 2017. Last advances on TiO$_2$-photocatalytic removal of chromium, uranium and arsenic. Current Opinion in Green Sustain. Chem. 6: 150–158.

Liu, H., T. Yao, W. Ding, H. Wang, D. Ju and W. Chai. 2013. Study on the optical property and surface morphology of N doped TiO$_2$ film deposited with different N$_2$ flow rates by DCPMS. J. Environ. Sci. 25(Suppl.): S54–S58.

Löbl, P., M. Huppertz and D. Mergel. 1994. Nucleation and growth in TiO$_2$ films prepared by sputtering and evaporation. Thin Solid Films 251: 72–79.

Loeb, S.K., P.J.J. Alvarez, J.A. Brame, E.L. Cates, W. Choi, J. Crittenden et al. 2019. The technology horizon for photocatalytic water treatment: sunrise or sunset? Environ. Sci. Technol. 53: 2937–2947.

Lowndes, D.H., D.B. Geohegan, A.A. Puretzky, D.P. Norton and C.M. Rouleau. 1996. Synthesis of novel thin-film materials by pulsed laser deposition. Science 273: 898–903.

Ma, H., B. Hao, W. Song, J. Guo, M. Li and L. Zhang. 2021. A high-efficiency TiO$_2$/ZnO nano-film with surface oxygen vacancies for dye degradation. Materials 14: 3299.

Mändl, S., G. Thorwarth and B. Rauschenbach. 2000. Textured titanium oxide thin films produced by vacuum arc deposition. Surf. Coat. Technol. 133–134: 283–288.

Manova, D., A. Gjevori, F. Haberkorn, J. Lutz, S. Dimitrov, J.W. Gerlach et al. 2009. Formation of hydrophilic and photocatalytically active TiO$_2$ thin films by plasma based ion implantation and deposition. Phys. Status Solidi A 206: 71–77.

Manova, D., J.W. Gerlach and S. Mändl. 2010. Thin film deposition using energetic ions. Materials 3: 4109–4141.

Manova, D., L. Franco Arias, A. Hofele, I. Alani, A. Kleiman, I. Asenova et al. 2017. Nitrogen incorporation during PVD deposition of TiO$_2$:N thin films. Surf. Coat. Technol. 312: 61–65.

Martin, P.J. and A. Bendavid. 2001. Review of the filtered vacuum arc process and materials deposition. Thin Solid Films 394: 1–14.

Martin, P.M. (ed.). 2009. Handbook of Deposition Technologies for Films and Coatings: Science, Applications and Technology. William Andrew, Norwich, NY, USA.

Mattox, D.M. 2000. Ion plating—past, present and future. Surf. Coat. Tech. 133: 517–521.

Mattox, D.M. 2010. Handbook of Physical Vapor Deposition (PVD) Processing. William Andrew, Norwich, NY, USA.

Mehran, Q.M., M.A. Fazal, A.R. Bushroa and S. Rubaiee. 2017. A critical review on physical vapor deposition coatings applied on different engine components. Crit. Rev. Solid State Mater. Sci. 43: 158–175.

Meng, L.J. and M.P. dos Santos. 1993a. Investigations of titanium oxide films deposited by d.c. reactive magnetron sputtering in different sputtering pressures. Thin Solid Films 226: 22–29.

Meng, L.J., M. Andritschky and M.P. dos Santos. 1993b. The effect of substrate temperature on the properties of d.c. reactive magnetron sputtered titanium oxide films. Thin Solid Films 223: 242–247.

Meng, L., Z. Wang, L. Yang, W. Ren, W. Liu, Z. Zhang et al. 2019. A detailed study on the Fe-doped TiO$_2$ thin films induced by pulsed laser deposition route. Appl. Surf. Sci. 474: 211–217.

Mostafa, A.M. and E.A. Mwafy. 2020. Synthesis of ZnO/CdO thin film for catalytic degradation of 4-nitrophenol. J. Mol. Struct. 1221: 128872.

Movchan, B.A. and A.V. Demshishin. 1969. Structure and properties of thick condensates of nickel, titanium, tungsten, aluminum oxides, and zirconium dioxide in vacuum. Fiz. Metal. Metalloved. 28: 653–660.

Nakano, Y., T. Morikawa, T. Ohwaki and Y. Taga. 2007. Origin of visible-light sensitivity in N-doped TiO$_2$ films. Chem. Phys. 339: 20–26.

Nakata, K. and A. Fujishima. 2012. TiO$_2$ photocatalysis: design and applications, J. Photochem. Photobiol. C 13: 169–189.

Nasirian, M., Y.P. Lin, C.F. Bustillo-Lecompte and M. Mehrvar. 2018. Enhancement of photocatalytic activity of titanium dioxide using non-metal doping methods under visible light: a review. Int. J. Environ. Sci. Technol. 15: 2009–2032.

Navinšek, B., P. Panjan and I. Milošev. 1999. PVD coatings as an environmentally clean alternative to electroplating and electroless processes. Surf. Coat. Technol. 116: 476–487.

Nejand, B.A., S. Samhadi and V. Ahmadi. 2010. Optical and photocatalytic characteristics of nitrogen doped TiO$_2$ thin film deposited by magnetron sputtering. Sci. Iran. Trans. F: Nanotechnol. 17: 102–107.

Nezar, S., S. Sali, M. Faiz, M. Mekki, N.A. Laoufi, N. Saoula et al. 2017. Properties of TiO$_2$ thin films deposited by rf reactive magnetron sputtering on biased substrates. Appl. Surf. Sci. 395: 172–179.

Obata, K., H. Irie and K. Hashimoto. 2007. Enhanced photocatalytic activities of Ta, N co-doped TiO$_2$ thin films under visible light. Chem. Phys. 339: 124–132.

Ohering, M. 2002. Materials Science of Thin Films Deposition and Structure. Academic Press, San Diego, CA, USA.

Ojeda-G-P, A., M. Döbeli and T. Lippert. 2018. Influence of plume properties on thin film composition in pulsed laser deposition. Adv. Mater. Interfaces 5: 1701062.

Ong, C.B., L.Y. Ng and A.W. Mohammad. 2018. A review of ZnO nanoparticles as solar photocatalysts: synthesis, mechanisms and applications. Renew. Sustain. Energy Rev. 81: 536–551.

Pashley, D.W., M.J. Stowell, M.H. Jacobs and T.J. Law. 1964. The growth and structure of gold and silver deposits formed by evaporation inside an electron microscope. Philos. Mag.: J. Theor. Exp. Appl. Phys. 10: 127–158.

Patel, N., R. Jaiswal, T. Warang, G. Scarduelli, A. Dashora, B.L. Ahuja et al., 2014. Efficient photocatalytic degradation of organic water pollutants using V-N-codoped TiO_2 thin films. Appl. Catal. B 150–151: 74–81.

Pelaez, M., N.T. Nolan, S.C. Pillai, M.K. Seery, P. Falaras, A.G. Kontos et al. 2012. A review on the visible light active titanium dioxide photocatalysts for environmental applications. Appl. Catal. B 125: 331–349.

Petrov, I., P.B. Barna, L. Hultman and J.E. Greene. 2003. Microstructural evolution during film growth. J. Vac. Sci. Technol. A. 21: S117–S128.

Pihosh, Y., M. Goto, A. Kasahara and M. Tosa. 2009. Photocatalytic property of TiO_2 thin films sputtered-deposited on unheated substrates. Appl. Surf. Sci. 256: 937–942.

Rao, M.C. 2013. Pulsed laser deposition—ablation mechanism and applications. Int. J. Mod. Phys. Conf. Ser. 22: 355–360.

Rashid, J., M.A. Barakat, N. Salah and S.S. Habib. 2014. Ag/ZnO nanoparticles thin films as visible light photocatalyst. RSC Adv. 4: 56892–56899.

Rashid, J., M.A. Barakat, N. Salah and S.S. Habib. 2015. ZnO-nanoparticles thin films synthesized by RF sputtering for photocatalytic degradation of 2-chlorophenol in synthetic wastewater. J. Ind. Eng. Chem. 23: 134–139.

Rivera, Z.R., A.M. Alvarez and M.L. Olvera Amador. 2019. Effect of Thickness on Photocatalytic Properties of ZnO thin films Deposited by RF Magnetron Sputtering. pp. 1–6. 16th International Conference on Electrical Engineering, Computing Science and Automatic Control (CCE), 11–13 September 2019, Mexico City, Mexico.

Rossnagel, S.M., R. Powell and A. Ulman (eds.). 1998. PVD for Microelectronics: Sputter Deposition to Semiconductor Manufacturing. Elsevier, Amsterdam, the Netherlands.

Rtimi, S., O. Baghriche, C. Pulgarin, J.C. Lavanchy and J. Kiwi. 2013. Growth of TiO_2/Cu films by HiPIMS for accelerated bacterial loss of viability. Surf. Coat. Technol. 232: 804–813.

Samadi, M., M. Zirak, A. Naseri, E. Khorashadizade and A.Z. Moshfegh. 2016. Recent progress on doped ZnO nanostructures for visible-light photocatalysis. Thin Solid Films 605: 2–19.

Sanders, D.M. and A. Anders. 2000. Review of cathodic arc deposition technology at the start of the new millennium. Surf. Coat. Tech. 133: 78–90.

Schneider, J., M. Matsuoka, M. Takeuchi, J. Zhang, Y. Horiuchi, M. Anpo et al. 2014. Understanding TiO_2 photocatalysis: mechanisms and materials. Chem. Rev. 114: 9919–9986.

Schultrich, B. 2018. Vacuum arc with particle filtering. pp. 493–526. *In*: Tetrahedrally Bonded Amorphous Carbon Films I. Springer, Berlin, Heidelberg, Germany.

Sekhar, M.C., P. Kondaiah, S.J. Chandra, G.M. Rao and S. Uthanna. 2011. Effect of substrate bias voltage on the structure, electric and dielectric properties of TiO_2 thin films by dc magnetron sputtering. Appl. Surf. Sci. 258: 1789–1796.

Simionescu, O.G., C. Romanitan, O. Tutunaru, V. Ion, O. Buiu and A. Avram. 2019. RF magnetron sputtering deposition of TiO_2 thin films in a small continuous oxygen flow rate. Coatings 9: 442.

Singh, J. and D. Wolfe. 2005. Review nano and macro-structured component fabrication by electron beam-physical vapor deposition (EB-PVD). J. Mater Sci. 40: 1–26.

Snyder, A., Z. Bo, R. Moon, J.C. Rochet and L. Stanciu. 2013. Reusable photocatalytic titanium dioxide–cellulose nanofiber films. J. Colloid Interface Sci. 399: 92–98.

Sponza, L., J. Goniakowski and C. Noguera. 2015. Structural, electronic, and spectral properties of six ZnO bulk polymorphs. Phys. Rev. B 91: 075126.

Takikawa, H., T. Matsui, T. Sakakibara, A. Bendavid and P.J. Martin. 1999. Properties of titanium oxide film prepared by reactive cathodic vacuum arc deposition. Thin Solid Films 348: 145–151.

Thorat, N., R. Varma, R. Mundotia, A. Kale, P. Sarawade, U. Mhatre et al. 2019. Photocatalytic activity of nanostructured TiO_2 and N-TiO_2 thin films deposited onto glass using CA-PVD technique. AIP Conf. Proc. 2115: 030318.

Thornton, J.A. 1974. Influence of apparatus geometry and deposition conditions on the structure and topography of thick sputtered coatings. J. Vac. Sci. Technol. 11: 666–670.

Tsoutsouva, M.G., C.N. Panagopoulos, D. Papadimitriou, I. Fasaki and M. Kompitsas. 2011. ZnO thin films prepared by pulsed laser deposition. Mater. Sci. Eng. B 176: 480–483.

Tu, M.L., Y.K. Su and C.Y. Ma. 2006. Nitrogen-doped p-type ZnO films prepared from nitrogen gas radio-frequency magnetron sputtering. J. Appl. Phys. 100: 53705.

Tuzemen, E.S., K. Kara, S. Elagoz, D.K. Takci, I. Altuntas and R. Esen. 2014. Structural and electrical properties of nitrogen-doped ZnO thin films. Appl. Surf. Sci. 318: 157–163.

Vahl, A., J. Dittmann, J. Jetter, S. Veziroglu, S. Shree, N. Ababii et al. 2019a. The impact of O_2/Ar ratio on morphology and functional properties in reactive sputtering of metal oxide thin films. Nanotechnology 30: 235603.

Vahl, A., S. Veziroglu, B. Henkel, T. Strunskus, O. Polonskyi, O.C. Aktas et al. 2019b. Pathways to tailor photocatalytic performance of TiO_2 thin films deposited by reactive magnetron sputtering. Materials 12: 2840.

Valkov, S., M.P. Nikolova, E. Yankov, T. Hikov, R. Bezdushnyi, D. Dechev et al. 2018. Comparison of the phase composition and nanohardness of gradient TiN/TiO_2 coatings on Ti5Al4V alloy deposited by different PVD methods. IOP Conf. Ser.: Mater. Sci. Eng. 416: 012034.

Varshney, G., S.R. Kanel, D.M. Kempisty, V. Varshney, A. Agrawal, E. Sahle-Demessie et al. 2016. Nanoscale TiO_2 films and their application in remediation of organic pollutants. Coord. Chem. Rev. 306: 43–64.

Velardi, L., L. Scrimieri, L. Maruccio, V. Nassisi, A. Serra, D. Manno et al. 2021. Synthesis and doping of TiO_2 thin films via a new type of laser plasma source. Vacuum 184: 109890.

Wang, J., V. Sallet, F. Jomard, A.M. Botelho do Rego, E. Elamurugu, R. Martins et al. 2007. Influence of the reactive N_2 gas flow on the properties of rf-sputtered ZnO thin films. Thin Solid Films 515: 8780–8784.

Wang, Y.H., K.H. Rahman, C.C. Wu and K.C. Chen. 2020. A review on the pathways of the improved structural characteristics and photocatalytic performance of titanium dioxide (TiO_2) thin films fabricated by the magnetron-sputtering technique. Catalysts 10: 598.

Weng, M.H., C.T. Pan, R.Y. Yang and C.C. Huang. 2011. Structure, optical and electrical properties of ZnO thin films on the flexible substrate by cathodic vacuum arc technology with different arc currents. Ceram. Int. 37: 3077–3082.

Wicaksana, D., A. Kobayashi and A. Kinbara. 1992. Process effects on structural properties of TiO_2 thin films by reactive sputtering. J. Vac. Sci. Technol. A 10: 1479–1482.

Wu, C.S., B.T. Lin and R.Y. Yang. 2011. Structural and optical properties of Ti-doped ZnO thin films prepared by the cathodic vacuum arc technique with different annealing processes. Thin Solid Films 519: 5106–5109.

Xu, X.L., S.P. Lau, J.S. Chen, Z. Sun, B.K. Tay and J.W. Chai. 2001. Dependence of electrical and optical properties of ZnO films on substrate temperature. Mater. Sci. Semicond. Process. 4: 617–620.

Xu, P., L. Mi and P.N. Wang. 2006. Improved optical response for N-doped anatase TiO_2 films prepared by pulsed laser deposition in $N_2/NH_3/O_2$ mixture. J. Cryst. Growth 289: 433–439.

Yamaki, T., T. Sumita, S. Yamamoto and A. Miyashita. 2002. Preparation of epitaxial TiO_2 films by PLD for photocatalyst applications. J. Cryst. Growth 237–239: 574–579.

Zhang, F., X. Wang, C. Li, H. Wang, L. Chen and X. Liu. 1998. Rutile-type titanium oxide films synthesized by filtered arc deposition. Surf. Coat. Technol. 110: 136–139.

Zhang, W., Y. Li, S. Zhu and F. Wang. 2004. Influence of argon flow rate on TiO_2 photocatalyst film deposited by dc reactive on magnetron sputtering. Surf. Coat. Technol. 182: 192–198.

Zhang, L., R. Dillert, D. Bahnemann and M. Vormoor. 2012. Photo-induced hydrophilicity and self-cleaning: models and reality. Energy Environ. Sci. 5: 7491–7507.

Zhao, J., L. Hu, Z. Wang, J. Sun and Z. Wang. 2006. ZnO thin films on Si(111) grown by pulsed laser deposition from metallic Zn target. Appl. Surf. Sci. 253: 841–845.

Zhao, Y., S. Hou, L. Fang, Y. Wu, W. Li, G. Sheng et al. 2011. The effect of negative substrate bias on the strain prosperities of ZnO films deposited by PFCVAD. Adv. Mater. Res. 287–290: 2373–2380.

Zhao, Y.X., S. Han, Y.H. Lin, C.H. Hu, L.Y. Hua, C.T. Lee et al. 2017. Photocatalytic properties of TiO_2 films prepared by bipolar pulsed magnetron sputtering. Surf. Coat. Technol. 320: 630–634.

Chapter 16

Nanomaterial-Based Fluorescent Development of Latent Fingerprints

Meng Wang,[1,]* *Qing Bao*[2] and *Chuanbin Mao*[3,]*

1. Introduction

1.1 Fingerprint

1.1.1 Papillary ridge skin

Skin is an essential and the largest organ of the human body. The epidermis on areas of the finger and palm appears as a series of papillary ridges and depressed furrows, and the sweat pores are intensively located on the papillary ridges (Wilshire 1996). These volar areas of the skin possessing complex forms and patterns are known to display papillary (or friction) ridge skin. The morphology of the papillary ridge skin is a direct reflection of its function. The papillary ridges facilitate the hands to grip surfaces firmly, and the sweat pores allow the perspiration to excrete.

The papillary ridge patterns on fingers are extremely specific and are different not only from one person to another but also from one finger to another. The papillary ridge patterns are also quite stable, and are topologically invariant from the birth of a person (Thomas 1978). Furthermore, these patterns can be easily deposited on various substrates to form a fingerprint, clearly reflecting the papillary ridge patterns. Therefore, fingerprints with these unique characteristics have already become a well-established aid to personal identification in forensic sciences.

1.1.2 Papillary ridge features

As introduced by Ashbaugh in 1999, the papillary ridge features of fingerprints can be classified and recorded by three levels of detail (Champod et al. 2004). Level 1 feature is the overall pattern printed by the flow of papillary ridges. There are four basic types of level 1 features in fingerprints, namely, arch, loop, whorl and compound (Knowles 1978). Level 2 feature means major ridge path deviations, also named minutiae or Galton characteristics. There are seven basic types of level 2 features in fingerprints, namely, termination, bifurcation (or fork), enclosure (or lake), dot (or island), short independent ridge, hook (or spur) and crossover (Knowles 1978). Some common features such as wrinkles and creases, as well as some occasional features such as scars, cuts, calluses, warts, blisters and subsidiary (or incipient) ridges, are also referred to as the level 2 traits (Ashbaugh 1991, 1992). Level 3 feature refers to intrinsic or innate ridge formations. In general, the

---

[1] Department of Trace Examination, National Police University of China, Shenyang, Liaoning 110035, China.
[2] School of Materials Science and Engineering, Zhejiang University, Hangzhou, Zhejiang 310058, China.
[3] Department of Chemistry & Biochemistry, University of Oklahoma, Norman, Oklahoma 73019, USA.
Email: baoqing0410@163.com
* Corresponding authors: mengwang@alum.imr.ac.cn; maophage@gmail.com

papillary ridge shapes and the sweat pore features are referred to as the level 3 features (Champod et al. 2004).

1.1.3 Fingerprint types

When a finger touches the surface of the substrate, a fingerprint can be formed via a material transfer process or impression effect. A fingerprint is the visualized mirror image of the corresponding papillary ridge patterns. At crime scenes, there are two main categories of fingerprints that can usually be found: patent (or visible) fingerprints and latent (or invisible) fingerprints (Champod et al. 2004).

Visible fingerprint refers to the fingerprint visible to naked eyes without any specific treatment. Visible fingerprints can be subdivided into positive fingerprints, negative fingerprints and plastic (or molded) fingerprints. A positive fingerprint means the fingerprint ridges are stained with a colored material, including blood, paint, ink or mud. A negative fingerprint is formed when the fingerprint ridges remove the adhering surface substance such as soot or dust. A plastic fingerprint is the three-dimensional impression caused when the finger presses into a malleable or pliable substrate such as putty, wax, clay, tacky paint, melted plastic or heavy grease.

A latent fingerprint is almost invisible to naked eyes. In general, a latent fingerprint is formed by the transfer of secretions from the fingers onto the touched substrates. Latent fingerprints are the most common and important form of fingerprints at crime scenes.

1.1.4 Fingerprint residues

In practice, the chemical constituents in fingerprint residues are quite complex and labile (Amoros and Puit 2014, Archer et al. 2005, Croxton et al. 2010). Numerous substances may be present in the fingerprint deposit, including gland secretions, epidermis exuviations and extrinsic attachments. Moreover, these gland secretions tend to be altered via volatilization, oxidation and bacterial degradation processes.

There are three major secretory glands (i.e., eccrine, sebaceous and apocrine glands) in the dermis, which are responsible for the secretion of sweat (Lee and Gaensslen 2001). The eccrine glands are extensively distributed all over the body, especially in the finger, palm and sole regions. The sebaceous glands are mainly present in regions containing hair follicles, especially in the scalp, face, neck and back. The apocrine glands are typically localized in the axillary, beast, inguinal and genital regions. The main components of secretions from the above three secretory glands are summarized in Table 16.1 (Cadd et al. 2015). It should be noted that the composition of gland secretions in sweat is approximately 99% water. The secretions from eccrine glands are usually hydrophilic, while the ones from sebaceous glands are typically hydrophobic.

Among the three major secretory glands, the eccrine gland is the only cutaneous appendage for the papillary ridge skin. In other words, sebaceous and apocrine glands cannot be found in the finger, palm and sole regions. However, sebaceous and apocrine secretions can be easily transferred from other parts of the body (e.g., hair and face) onto fingers and palms by simple touch actions. In most cases, the fingerprint residues contain the secretions of only the eccrine and sebaceous gland, of which nearly 99% of residues are composed of water. The components of eccrine and sebaceous gland secretions detected in fingerprint residues are summarized in Table 16.2 (Girod et al. 2012).

Table 16.1. The main constituents of gland secretions.

Source	Inorganic constituents	Organic constituents
Eccrine glands	Water, chlorides, metal ions, ammonia, sulfate, phosphate	Amino acids, lactic acids, sugars, urea, creatinine, choline, uric acid
Sebaceous glands	—	Fatty acids, glycerides, wax esters, sterol esters, sterols, hydrocarbons, squalene, alcohols
Apocrine glands	Water, iron	Proteins, carbohydrates, cholesterol

Table 16.2. The constituents of gland secretions in fingerprint residues (excluding water).

Source	Constituents	Quantitative Data (mass or concentration)
Eccrine glands	Proteins and polypeptides	384 µg
	Lactic acids	9~10 µg or 154 mmol L^{-1}
	Chloride	1~15 µg
	Urea	0.4~1.8 µg
	Sodium	0.2~6.9 µg or 54.0 mmol L^{-1}
	Potassium	0.2~5 µg or 39.9 mmol L^{-1}
	Amino acids	0.2~1 µg
	Total ammonia	0.2~0.3 µg or 5.13 mmol L^{-1}
	Phenol	0.06~0.25 µg
	Calcium	0.03~0.3 µg or 5.49 mmol L^{-1}
	Sulfide	0.02~0.2 µg
	Magnesium	1.67 mmol L^{-1}
	Uric acid	150 µmol L^{-1}
Sebaceous glands	Free fatty acids	37.6%
	Wax esters and diglycerides	25%
	Monoglycerides, triglycerides and cholesterol esters	21%
	Squalene	14.6% or 28~5311 ng
	Cholesterol	3.8% or 1032 ng

Another substance in fingerprint residues is epidermis exuviations, which may contain an important constituent named deoxyribonucleic acid (DNA). It is no surprise that a large amount of DNA is typically present in blood fingerprint residues. The epidermal cells can be easily shed from the skin surface when rubbing the skin or touching the substrate (Freinkel and Woodley 2001). Therefore, there is a certain degree of probability that the DNA can also be deposited in common fingerprint residues (McRoberts 2011). As important biological evidence, the DNA molecules in fingerprint residues will greatly improve the utility value of fingerprints in forensic sciences.

In addition to gland secretions and epidermis exuviations, fingerprint residues may contain various extrinsic attachments, such as cosmetics, dust, blood, food residues and bacteria spores. More importantly, some special substances such as explosives, drugs and hazardous goods may also be deposited in fingerprint residues. As special trace evidence, the extrinsic attachments in fingerprint residues will further expand the application of fingerprints in forensic sciences.

1.1.5 Substrate types

As introduced by Champod et al. (2004), the substrates bearing fingerprints can be classified into three categories, namely, porous, nonporous and semiporous substrates.

The substrate whose surface is inclined to absorb the fingerprint residuals is grouped as a porous substrate. Typical examples of porous substrates are paper, cardboard, cotton fabric, untreated wood, cellulose, etc. Fingerprints deposited on porous substrates are somewhat durable because most hydrophilic fingerprint residuals are absorbed and protected in the internal layer of the substrate. The substrate whose surface does not absorb any of the fingerprint residuals is grouped as a nonporous substrate. Typical examples of a nonporous substrate are glass, metal, rubber, polyethylene plastic, glazed ceramics, glossy paint, etc. Fingerprints deposited on nonporous substrates are somewhat fragile because nearly all the fingerprint residuals are exposed on the outermost surface of the substrate. The substrate whose surface possesses intermediate characteristics between the porous substrate and nonporous substrate is generally categorized as a semiporous substrate. Typical examples of semiporous substrates are painted surfaces, waxed surfaces, wallpaper, wall paints, varnished wood, polymer banknotes, cellophane, glossy magazine covers, etc.

1.2 Latent fingerprint development

Latent fingerprint refers to the fingerprint almost invisible to the naked eye, which is the most frequent form of fingerprint at crime scenes (Lee and Gaensslen 2001). In practice, however, latent fingerprint also is the most intractable and problematic: it is present but invisible. How to make latent fingerprints visible is an important precondition for fingerprint analysis and identification. Thus, lots of methods for latent fingerprint development have emerged and developed (Ramotowski 2013). The main principle underlying latent fingerprint development is to create enough contrasts between the papillary ridges and the corresponding substrates (Su 2016). Till now, latent fingerprint development has gradually grown into a comprehensive technique, which involves the appropriate application of traditional (e.g., optical, physical and chemical) methods as well as emerging (e.g., biological, nanomaterial, electrochemical and instrumental) methods (Wang et al. 2018a, Wei et al. 2016, Xu et al. 2015).

1.2.1 Traditional developing method

Due to the variety of substrate surfaces as well as the difference in fingerprint residuals, the method for latent fingerprint development is even multifarious. In most previously reported literature, the categorization of the traditional developing method somewhat lacked rigor. Table 16.3 gives a general categorization of different traditional methods for latent fingerprint development. Therein, powder dusting, cyanoacrylate ester fuming, 1,8-diazafluoren-9-one staining and amino black staining are the most commonly used developing method at crime scenes.

Table 16.3. Traditional methods for latent fingerprint development.

Category	Types		Examples
	General classification	**Specific classification**	
Irradiation	Ultraviolet irradiation	Reflectance mode	Ultraviolet scattering
		Fluorescence mode	Ultraviolet-to-visible fluorescence
	Visible irradiation	Reflectance mode	Visible scattering
		Fluorescence mode	Visible-to-visible fluorescence
Solids	Powder dusting	Nonfluorescent powder	Metal flakes, metal oxides, magnetic powders, graphite, pollen, starch
		Fluorescent powder	Inorganic phosphors, organic fluorescent dyes
	Flame fuming	Combustion product	Rosin, camphor, pine tar, nitrocellulose plastics, soft resinous pine
Liquids	Solution staining	Physical absorption	Ag-physical developers, amino black, gentian violet, Nile red, Nile blue, Sudan black, oil red O
		Chemical reaction	Silver nitrate, ninhydrin, 1,2-indanedione, 1,8-diazafluoren-9-one, tetramethyl benzidine
	Suspension staining	Aqueous solution	Small particle reagent, multimetal deposition (in an aqueous solvent)
		Nonaqueous solution	Multimetal deposition (in organic solvent)
Vapors	Atmospheric fuming	Volatile organics	Cyanoacrylate esters, p-dimethylaminocinnimaldehyde
		Volatile inorganics	Iodine, osmium tetroxide, ruthenium tetroxide
	Vacuum treatment	Metallic materials	Zinc and gold (vacuum metal deposition)
		Other materials	Cyanoacrylate esters (vacuum fuming)

1.2.2 *New developing method*

With the development of emerging technology, rapid advances in nanomaterial, biology, electrochemistry and instrumental analysis have led to a mushrooming of new methods for latent fingerprint development, including nanomaterial- and electrochemistry-based developing methods as well as a spectroscopy imaging method (Becue 2016, Hazarika and Russell 2012, Lesniewski 2016). Table 16.4 gives a general categorization of different new methods for latent fingerprint development. Among them, a nanomaterial-based developing method has now become the fastest-growing area in latent fingerprint development, owing to its good performance, easy operation, high efficiency and wide applicability (Wang et al. 2017a).

Table 16.4. New methods for latent fingerprint development.

Category	Types	Examples
Nanomaterial-based method	Physical strategy	Physical adsorption, hydrophobic effect
	Chemical strategy	Chemical bonding
	Biological strategy	Antibody combining, aptamer recognition
Electrochemical method	Electrochemical deposition	Anodic deposition, cathodic deposition, potentiodynamic deposition
	Electrochemiluminescence	$Ru(bpy)_3^{2+}$/TPrA, RUB/TPrA, enzyme-linked immunoassay
	Scanning electrochemical probe	Scanning electrochemical microscopy, scanning Kelvin probe microscopy
Spectroscopy method	Mass Spectrometry (MS) imaging	Desorption electrospray ionization MS, surface assisted laser desorption/ionization time-of-flight MS, matrix assisted desorption ionization MS, time-of-flight secondary ionization MS, laser desorption ionization MS, desorption electro-flow focusing ionization MS
	Vibrational spectroscopy imaging	Fourier transform infrared spectroscopy, Raman spectroscopy

bpy: bipyridyl; TPrA: tripropylamine; RUB: rubrene

1.3 *Evaluation of fingerprint development*

For a long time, most studies have focused on the application of new materials and the improvement of new strategies for latent fingerprint development. However, how to accurately evaluate the effects of fingerprint development remains unknown. Moreover, some evaluating descriptions used in literature were somewhat inappropriate, inconsistent and even ambiguous. In 2017, Wang et al. first put forward four quality metrics to evaluate the effects of fingerprint development, namely, contrast, sensitivity, selectivity and toxicity (Wang et al. 2017a). Up to now, however, only a few pieces of literature have reported on evaluating these metrics for fingerprint development.

1.3.1 *Contrast*

The concept of contrast in fingerprint development means the contrast between the fingerprint (developing signal) and the substrate (background noise). There are three effective methods to improve the contrast in fingerprint development: (i) to enhance the developing signal intensity; (ii) to adjust the developing signal color; and (iii) to reduce the background noise (Wang et al. 2017a).

In the powder dusting method, for example, pathways to enhance the developing contrast mainly depend on the color and the fluorescent property of the substrate. As for substrates with a single color and weak fluorescence, they are utilizing non-fluorescent powder, and simultaneously adjusting the original color of the powder is an effective way. As for substrates with multiple colors and weak fluorescence, utilizing fluorescent powder or further increasing the fluorescent intensity of the powder is an effective way. As for substrates with a single color and strong fluorescence, they are using non-fluorescent or fluorescent powder, and simultaneously adjusting the original or fluorescent color of the powder is an effective way. As for substrates with multiple colors and strong fluorescence, they are utilizing special fluorescent powder (i.e., upconversion luminescent powder) and simultaneously reducing the background fluorescence is an effective way.

In 2008, Humphreys et al. first proposed a technique to qualitatively evaluate the effectiveness of fingerprint development methods by analyzing the contrast between the fingerprint and the substrates. In this method, a fiber-optic spectrophotometer is attached to a microscope with axial illumination to measure the intensity counts of the fingerprint and substrates (Humphreys et al. 2008). The contrast of different fingerprints was quantitatively expressed as a relative contrast index. In 2011, Vanderwee et al. further investigated and tested the credibility and repeatability of the above evaluative technique by using three different microspectrophotometers (Vanderwee 2011). From the results, the above method can construct a repeatable and objective measurement on each instrument even though it cannot produce absolute or universal values. In 2013, Matuszewski and Szafalowicz reported another method to evaluate the contrast in a fingerprint. In this method, the scanner-acquired fingerprint image is quantified using the histogram function of Adobe Photoshop software (Matuszewski and Szafalowicz 2013). However, this evaluative method was only applicable in dealing with the cases of dark fingerprints on white substrates or white fingerprints on dark substrates.

The above research work could only be qualified for the contrast of inked fingerprints and fingerprints developed by the non-fluorescent method but do not involve fingerprints developed by the fluorescent method. To overcome this deficiency, in 2021, Wang et al. first proposed a method based on spectral analysis to evaluate the effectiveness of fingerprints development by fluorescently imaging with the aid of a fluorescence spectrophotometer (Wang et al. 2021). As shown in Fig. 16.1, a fingerprint was developed using fluorescent nanoparticles, and the fluorescence spectrum was collected through a fluorescence spectrophotometer. The developed contrast was quantified using two indexes. The first index was the intensity Index (I) which was defined as the integrated fluorescence intensities of the fingerprint divided by the background. The second index was the Chroma index (C), which was defined as the color difference between the signal and the background in the chromaticity graph. Finally, the contrast was determined by the product of the common logarithm of the intensity index and the chroma index.

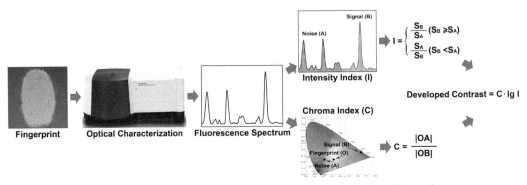

Figure 16.1. Scheme for evaluating the contrast of fluorescently developed fingerprints.

1.3.2 Sensitivity

The sensitivity in the area of fingerprint development is defined as the visibility and clearness of the papillary ridge features. There are three efficient strategies to increase the developing sensitivity: (i) to assure the quality of developing materials with small size and proper shape; (ii) to improve the operating skills; and (iii) to optimize the developing conditions (Wang et al. 2017a).

In the powder dusting method, for example, the micromorphology and size of the powders have a great effect on the developing sensitivity. On one hand, the powders with strong adsorption properties are inclined to conglutinate the minutiae of the papillary ridge features, resulting in a decreased developing sensitivity. In general, the powders with lamellar shape will possess an excessively strong adsorption property, while the ones with spherical, cubic, polyhedral or rod-like shapes will possess an appropriate adsorption property. On the other hand, powders with big particle sizes are liable to cover the sweat pore features, resulting in a decreased developing sensitivity. It was reported that the nanosized developing powders could clearly reveal more sweat pore features (Wang et al. 2015a). As shown in Fig. 16.2, latent fingerprints deposited on glass were developed by powders with various morphologies and different sizes. When developed by bronze powders (Fig. 16.2a), the minutiae of the papillary ridges were seriously conglutinated, and the sweat pore features were thickly covered (Fig. 16.2a), due to their lamellar shape and big particle size; when developed by micronsized $NaYF_4$:Yb,Er powders (Fig. 16.2b), the minutiae of papillary ridges were obtained, but the sweat pore features were almost covered (Fig. 16.2b) due to their rod-like shape and big particle size; when developed by nanosized $NaYF_4$:Yb,Er powders (Fig. 16.2c), the minutiae of papillary ridges and the sweat pore features were clearly observed (Fig. 16.2c), due to their cubic shape and small particle size.

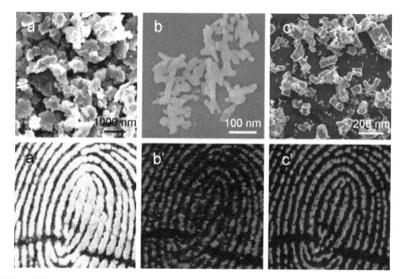

Figure 16.2. Using different powders to develop latent fingerprints: (a) bronze powders, (b) micron-sized $NaYF_4$:Yb,Er powders, and (c) nano-sized $NaYF_4$:Yb,Er powders.

1.3.3 Selectivity

The concept of selectivity in the area of fingerprint development is defined as the specificity of developing materials in adhering or reacting only with the papillary ridges in the fingerprint but not with the furrows on the substrate. There are three efficient strategies to improve the developing selectivity: (i) to control the character of developing materials with proper tackiness and targeted property; (ii) to promote the operating skills; and (iii) to optimize the developing conditions (Wang et al. 2017a).

In the powder dusting method, for example, the tackiness of the powders and the roughness of the substrates have a great effect on the developing selectivity. In general, the stickier the powder, or the rougher the substrate, the stronger adhesion between the powder and the substrate it will be, and further, the lower developing selectivity it will achieve. As for latent fingerprints deposited on the same glass and then developed by two types of powders with different tackiness, a high developing selectivity can be achieved when using $NaYF_4$:Yb,Er nanopowders with appropriate tackiness (Fig. 16.3a), while a low developing selectivity can be achieved when using green fluorescent powders with strong tackiness (Fig. 16.3b). As for latent fingerprints deposited on two types of

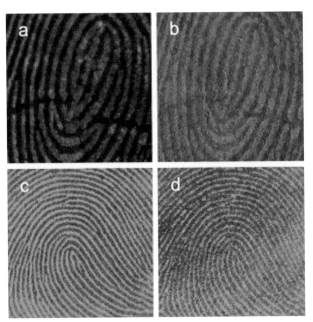

Figure 16.3. Using different powders to develop latent fingerprints on a variety of substrates: (a) glass, $NaYF_4$:Yb,Er nanopowders; (b) glass, green fluorescent powders; (c) printing paper, magnetic powders; (d) filter paper, magnetic powders.

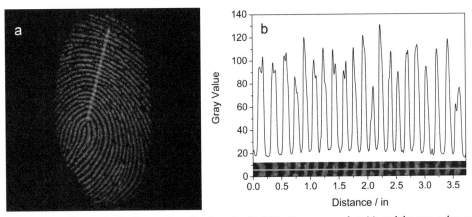

Figure 16.4. Latent fingerprint development on glass by using $NaYbF_4$: Tm nanopowders (a), and the gray value curve of the selected area in the developed fingerprint image.

paper with different roughness and then developed by the same magnetic powders, the printing paper has a smooth surface, resulting in a high developing selectivity (Fig. 16.3c), while the filter paper has a rough surface, showing a low developing selectivity (Fig. 16.3d).

In 2020, Wang et al. first proposed a method based on gray analysis to evaluate the selectivity of fluorescently developed fingerprints with the aid of Image J and Origin software. As shown in Fig. 16.4, first, the gray value curve of a selected area in the developed fingerprint image was obtained from Image J; then, the gray value curve was analyzed by means of Origin; finally, the developing selectivity was derived by the ratio of the integrated peaks (corresponding to papillary ridges) to the valleys (corresponding to furrows) in the gray value curve.

1.3.4 Toxicity

The toxicity in the area of fingerprint development is defined as not only the direct toxicity of the developing materials and the corresponding equipment to human health, but also the potential damage to DNA in the fingerprint residues. There are two efficient strategies to reduce the developing toxicity: (i) to reduce the toxicity of developing materials; and (ii) to reduce the toxicity of the developing methods (Wang et al. 2017a).

In the powder dusting method, for example, the most flagrant threat to human health is flying dust. Besides, the ultraviolet illumination used in a fingerprint development may cause damage not only to the eyes and skin of the operator but also to the DNA in the fingerprint residues. The macroscopic toxicity to human health can be alleviated by safeguarding procedures and improving methods. However, the microscopic toxicity to touch DNA detection remains a challenge because the multiple detections of fingerprints and touch DNA seem to be incompatible. Specifically, the DNA in the fingerprint residues is likely to damage the process of latent fingerprint development, and the touch DNA extraction also reduces the effectiveness of latent fingerprint development. In 2016, Wang first reported the development of latent fingerprints by using $LaPO_4$:Ce,Tb luminescent nanopowders and the detection of touch DNA in developed fingerprints (Wang 2016b). In 2019, Peng et al. also reported the use of Ce^{3+} and La^{3+} ions doped binuclear luminescent nanocomposites for latent fingerprint development without damage to touch DNA.

2. Nanomaterials in Latent Fingerprint Development

In the past 20 years, the application of various nanomaterials, especially fluorescent nanomaterials, in latent fingerprint development has attracted considerable attention, leading to the establishment of nanomaterial-based fluorescent development of latent fingerprints (Wang et al. 2019b). Nanomaterials that are typically used in fingerprint development are quantum dots (QDs), metal and metal oxide nanoparticles, rare earth (RE) ions doped down- or up-conversion luminescent nanoparticles and carbon dots (Wang et al. 2019b). A timeline of typical nanomaterials used for latent fingerprint development is presented in Fig. 16.5. The first use of fluorescent nanomaterials in latent fingerprint development was reported by Menzel et al. in 2000 (Menzel et al. 2000b). Compared to traditional methods, the use of novel fluorescent nanomaterials for latent fingerprint development has several outstanding advantages, including high contrast due to their strong fluorescence emission and unique optical properties, high sensitivity due to their small particle size and tunable morphology and high selectivity due to their suitable tackiness and flexible surface functionalization (Wang et al. 2017a).

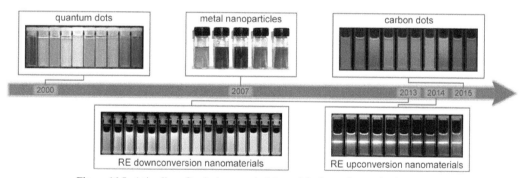

Figure 16.5. A timeline of typical nanomaterials used for latent fingerprint development.

2.1 Quantum dots

QDs refer to fluorescent semiconductor nanoparticles composed of elements from groups II–VI (e.g., ZnS, CdS, CdSe and CdTe) or III–V (e.g., InP, InAs and GaAs) in the periodic table, whose spatial dimension is smaller than or close to twice the exciton Bohr radius of the corresponding bulk semiconductor material. The prominent advantages offered by QDs are size-tunable fluorescence, broad absorption spectrum, narrow emission spectrum, high brightness, good photo-stability and flexible surface functionalization.

2.1.1 *Ordinary quantum dots*

In 2000, the use of a suspension of CdS-DSS QDs (CdS QDs capped with dioctyl sulfosuccinate molecules) for developing the latent fingerprints on aluminum foil, painted metal and sticky-side substances was first reported (Menzel et al. 2000b). In the same year, Menzel et al. further reported the similar use of CdS/PAMAM (CdS QDs modified with polyamidoamine dendrimers) nanocomposite (NC) powders for latent fingerprint development (Menzel et al. 2000a). Although they are the pioneers of fingerprint development using fluorescent nanomaterials, their developing results need to be improved. Since then, a variety of ordinary QDs, including CdS, ZnS, CdSe and CdTe, as well as QDs with one semiconductor (e.g., CdSe) as a core and another one (e.g., ZnS) as a shell, have been used for developing latent fingerprints.

Table 16.5 summarizes the details of latent fingerprint development based on a variety of ordinary QDs.

The ordinary QDs can emit bright fluorescence with good photostability, ensuring an improved developing signal. The usual QDs also possess a small particle size (1~10 nm) and spherical morphology, resulting in high developing sensitivity. The excitation of standard QDs usually requires NUV or blue light, leading to minor or negligible damage to touch DNA in the fingerprint residuals. However, the chemical stability of bare QDs is quite weak, which makes them suitable to be oxidized in air and further produce fluorescence quenching. To improve their chemical stability,

Table 16.5. Summary of latent fingerprint development based on ordinary QDs.

Year	Nanomaterials	λ_{ex} (nm)	Strategy/consuming time	Reference
2000	CdS-DSS	NUV	Physical adsorption/medium	Menzel et al. 2000b
2000	CdS/PAMAM	NUV	Physical adsorption/medium	Menzel et al. 2000a
2008	CdS/PAMAM	365	Physical adsorption/14 h	Jin et al. 2008
2009	CdS/Chitosan	450	Physical adsorption/short	Dilag et al. 2009
2009	CdSe-TGA	380	Physical adsorption/15 min	/Wang et al. 2009
2009	CdTe-TGA	300~400	Physical adsorption/30 min	Becue et al. 2009
2010	CdTe-TGA	365	Physical adsorption/15 min	Liu et al. 2010
2011	CdS/PPH	450	Physical adsorption/short	Algarra et al. 2011
2011	CdTe/MMT	365	Physical adsorption/short	Gao et al. 2011a
2011	CdTe-Hydrazine	365	Physical adsorption/0.5~1 h	Gao et al. 2011b
2013	CdS-p(DMA)	350	Physical adsorption/short	Dilag et al. 2013
2013	CdS-p(DMA-co-MMA)	350	Physical adsorption/short	Dilag et al. 2013
2013	CdS-p(DMA-co-Sty)	350	Physical adsorption/short	Dilag et al. 2013
2013	CdSe/PPH	450	Physical adsorption/short	Algarra et al. 2013
2013	ZnS:Cu-MPA	300~400	Physical adsorption/30 min	Moret et al. 2013
2013	CdTe-MSA	300~400	Physical adsorption/1~2 s	Cai et al. 2013
2014	CdSe/PPH	450	Physical adsorption/short	Algarra et al. 2014
2014	ZnS:Mn-(*N*-L-Cys)	302	Physical adsorption/5 s	Xu et al. 2014
2014	CdSe-Tween 20	365/440	Hydrophobic effect/> 15 min	Wang et al. 2014b
2014	CdTe-LBA	365	Aptamer recognition/30 min	Wang et al. 2014a
2017	CdTe-(*N*-L-Cys)	UV	Physical adsorption/5 s	Li et al. 2017b
2017	CdTe-MPA	UV	Physical adsorption/< 10 s	Singh et al. 2017

DSS: dioctyl sulfosuccinate; PAMAM: polyamidoamine; TGA: thioglycolic acid; PPH: porous phosphate heterostructure; MMT: montmorillonite; DMA: dimethylacrylamide; DMA-co-MMA: dimethylacrylamide-co-methylmethacrylate; DMA-co-Sty: dimethylacrylamide-co-styrene; MPA: 3-mercaptopropionic acid; MSA: mercaptosuccinic acid; *N*-L-Cys: *N*-acetyl-L-cysteine; LBA: lysozyme binding aptamer; NUV: near ultraviolet; UV: ultraviolet.

bare QDs are often modified with stable substances (e.g., PPH and MMT) to form NC powders or capped with special ligand molecules (e.g., TGA and MPA) to form colloidal solutions. In addition, the potential toxicity of cadmium-containing QDs should not be ignored.

2.1.2 Core-shell quantum dots

To reduce the intrinsic toxicity, increase the chemical stability and enhance the fluorescent intensity of QDs, bare QDs are often coated with a layer of silica. In 2012, Gao et al. reported the use of core-shell $CdTe@SiO_2$ (CdTe QDs coated with silica) NC powders for developing the latent fingerprints on different substrates (Gao et al. 2012). However, some substances such as black rubber and paper still emitted strong background fluorescence under 365 nm excitation.

To adjust the fluorescent color of QDs, bare QDs are often modified with other types of QDs. In 2015, Wu et al. first prepared silica-coated ZnCdS:Cu QDs and then coated the QDs with a shell of ZnCdS QDs. They finally conjugated polyallylamine molecules (PAA) onto the resultant QDs, forming new core-shell-structured QDs termed $ZnCdS:Cu@SiO_2/ZnCdS-PAA$. Then they used the core-shell-structured QDs to develop the latent fingerprints on glass and paper, simultaneously detecting the trinitrotoluene (TNT) components in the fingerprint residuals (Wu et al. 2015). Under 365 nm excitation, the fingerprints without TNT components could exhibit green emission, while the ones with TNT components could become red fluorescent.

To enrich the magnetic property of QDs, bare QDs can also be modified with magnetic materials such as Fe_3O_4 nanoparticles. In 2019, Wang et al. reported the use of core-shell $Fe_3O_4@SiO_2/CdTe$-GSH (Fe_3O_4 nanoparticles coated with silica, then modified with glutathione molecules capped CdTe QDs) NC powders for developing the latent fingerprints on glass, black plastics, ceramics, banknotes, paper and leather. (Wang 2019a). Unfortunately, some substances such as banknotes and paper could emit strong background fluorescence under UV excitation.

Table 16.6 summarizes the details of latent fingerprint development based on a variety of core-shell QDs.

Compared with ordinary QDs, core-shell QDs possess low toxicity, good stability and improved fluorescent property, which can further promote the results of latent fingerprint development. However, the corresponding preparation process of core-shell QDs is more complex and time-consuming. Furthermore, when using ordinary QDs or core-shell QDs, the background fluorescence interference derived from UV excitation cannot be avoided.

Table 16.6. Summary of latent fingerprint development based on core-shell QDs.

Year	Nanomaterials	λ_{ex} (nm)	Strategy/consuming time	Reference
2012	$CdTe@SiO_2$	365	Physical adsorption/short	Gao et al. 2012
2014	$SiO_2/CdTe@SiO_2$	352	Physical adsorption/2 h	Dong et al. 2014
2015	$CdSe/ZnS@SiO_2$	354	Physical adsorption/short	Ryu et al. 2015
2015	$CdTe@SiO_2/Ag$	365	Physical adsorption/short	Niu et al. 2015
2015	$ZnCdS: Cu@SiO_2/ZnCdS-PAA$	365	Chemical bonding/30 s	Wu et al. 2015
2019	$Fe_3O_4@SiO_2/CdTe$-GSH	UV	Physical adsorption/short	Wang et al. 2019a

PAA: polyacrylic acid; GSH: glutathione.

2.2 Metal and metal oxide nanoparticles

Metal and metal oxide nanoparticles (NPs) refer to metal and metal oxide whose particle size is reduced to the nanoscale. In general, Au and Fe_3O_4 NPs are most widely used in latent fingerprint development. The prominent advantages offered by metal and metal oxide NPs are a suitable size and flexible surface functionalization.

2.2.1 Metal nanoparticles

In 2007, Leggett et al. first reported the use of Au-Ab (nano-gold conjugated with antibody) NP suspension combined with organic fluorescent dyes for developing the smokers' latent fingerprints on glass, simultaneously detecting the metabolites of nicotine in fingerprint residuals (Leggett et al. 2007). Under 470 and 560 nm light excitation, the developed fingerprints could exhibit high contrast, sensitivity and selectivity. Since then, a variety of metal and metal oxide NPs modified with targeting molecules including antibodies and aptamers have been used for developing the latent fingerprints.

Table 16.7 summarizes the details of latent fingerprint developments based on a variety of metal NPs.

As metal NPs are typically non-fluorescent, they are often used by combining with various organic fluorescent dyes to improve the developing contrast. However, the surface modification and the targeting molecule conjugation of these NPs are very complicated and tedious.

Table 16.7. Summary of latent fingerprint developments based on metal NPs.

Year	Nanomaterials	λ_{ex} (nm)	Strategy/consuming time	Reference
2007	Au-Ab, Alexa Fluor dyes	470/560	Antibody combining/1 h	Leggett et al. 2007
2011	Au-Ab, Fluorescent Red	590	Antibody combining/45 min	Spindler et al. 2011
2013	Au-Aptamer	—	Aptamer recognition/14.5 h	Li et al. 2013
2015	Au-Ab	—	Chemical bonding/2.5 h	Peng et al. 2015
2016	Ag-Aptamer	UV	Aptamer recognition/2 h	Ran et al. 2016

Ab: antibody.

2.2.2 Metal oxide nanoparticles

In 2008, Hazarika et al. reported the similar use of Fe_3O_4-Ab (nano-Fe_3O_4 conjugated with antibody) NP suspension combined with organic fluorescent dyes for developing latent fingerprints on glass, simultaneously detecting the drug metabolites in fingerprint residuals (Hazarika et al. 2008). Under 470 nm excitation, the developed fingerprints could also exhibit high contrast, sensitivity and selectivity.

Besides, some nonmetallic oxide NPs, typically organic fluorescent dyes loaded SiO_2 NCs, can also be used for latent fingerprint development. In 2008, Theaker et al. reported the use of SiO_2/Rhodamine 6 G (nano-SiO_2 modified with Rhodamine 6 G) NC powders and suspensions for developing latent fingerprints on glass (Theaker et al. 2008). Under 415 nm excitation, the developed fingerprints could emit strong fluorescence; however, the strong background interference could not be avoided, and the developing sensitivity and selectivity were not high enough.

Table 16.8 summarizes the details of latent fingerprint developments based on a variety of metal oxide NPs.

Table 16.8. Summary of latent fingerprint developments based on metal oxide NPs.

Year	Nanomaterials	λ_{ex} (nm)	Strategy/consuming time	Reference
2008	Fe_3O_4-Ab, Alexa Fluor dyes	470/560	Antibody combining/medium	Hazarika et al. 2008
2008	ZnO:Li-Tween 20	350	Hydrophobic effect/2~5 min	Choi et al. 2008
2008	SiO_2/Rhodamine 6 G	415	Physical adsorption/short	Theaker et al. 2008
2009	Fe_3O_4-Ab, Alexa Fluor dyes	470	Antibody combining/15 min	Hazarika et al. 2009
2010	Fe_3O_4-Ab, Alexa Fluor dyes	470/560	Antibody combining/medium	Hazarika et al. 2010
2013	Fe_3O_4@Ag	—	Physical adsorption/short	Zhang et al. 2013
2016	Fe_3O_4/PCDA	254	Physical adsorption/short	Lee et al. 2016
2017	SiO_2/MB	—	Physical adsorption/short	Zhang et al. 2017
2017	ZrO_2/CuO	254	Physical adsorption/short	Renuka et al. 2017

Ab: antibody; PCDA: 10,12-pentacosadiynoic acid; MB: Methylene Blue.

Similar to the case of metal NPs, most metal oxide NPs are non-fluorescent. They are often used by combining with various organic fluorescent dyes to improve the developing contrast, leading to a very complicated and tedious operating process.

2.3 Rare earth-doped down conversion nanomaterials

RE-doped down conversion nanomaterials (DCNMs) refer to luminescent NMs with RE ions acting as activators, which can convert shorter wavelength excitations (e.g., UV light) into longer wavelength emissions (e.g., visible light). Inorganic DCNMs and organic-inorganic DC nanocomplexes are two major categories of RE-doped DCNMs. The prominent advantages offered by RE-doped DCNMs are strong luminescence, suitable size, controllable morphology, excellent photo- and chemical stability, easy synthesis and flexible surface functionalization.

2.3.1 Inorganic downconversion nanomaterials

In 2013, Saif first reported the use of $Y_2Zr_2O_7$:Eu/SiO_2 and $Y_2Zr_2O_7$:Tb/SiO_2 (Eu^{3+} or Tb^{3+} ions doped $Y_2Zr_2O_7$ DCNMs modified with silica) NC powders for developing the latent fingerprints on glass, aluminum foil, plastic bags and compact disks (Saif 2013). However, the developing contrast on the plastic bags and compact disks was relatively low due to the strong background fluorescence interference. In addition, the developing sensitivity and selectivity on all substrates were not high enough. Since then, a variety of RE-doped inorganic DCNMs and organic-inorganic DC nanocomplexes have been used for developing latent fingerprints.

In 2015, Wang et al. reported the similar use of YVO_4:Eu and $LaPO_4$:Ce,Tb DCNM powders for developing latent fingerprints on glass, aluminum foil, plastic bags and compact disks (Wang et al. 2015b). It should be noted that the fingerprint development on all substrates exhibited high contrast, sensitivity and selectivity, due to their high luminescence, suitable size and morphology and appropriate tackiness.

In 2019, Babu et al. reported the use of TiO_2:Eu DCNM powders for developing latent fingerprints on aluminum foil, plastic sheets, magazine covers, scissors, soft drink cans and bar codes (Babu et al. 2019). Under 254 nm light, however, the strong background fluorescence interference could not be avoided. Therefore, when conducting latent fingerprint developments under short-wavelength UV light, the strong luminescence of the RE-doped inorganic DCNMs should be ensured to get a strong enough developing signal.

Table 16.9 summarizes the details of latent fingerprint developments based on a variety of RE-doped inorganic DCNMs.

Due to the strong and stable luminescence, suitable size and morphology and appropriate tackiness, latent fingerprints developed by using RE-doped inorganic DCNMs can usually obtain high contrast, sensitivity and selectivity. In addition, the synthesis of these DCNMs is relatively easy and mild. However, the excitation of these DCNMs usually requires short-wavelength UV light, which may lead to the background fluorescence interferences, as well as the potential damage to touch DNA in fingerprint residuals.

2.3.2 Organic-inorganic downconversion nanocomplexes

In 2018, Peng et al. first reported the use of $Eu_{0.5}Tb_{0.5}(AA)_3$Phen and Tb $(AA)_3$Phen (RE^{3+} ions coordinated with acrylic acid and 1,10-phenanthroline) nanocomplex powders for developing latent fingerprints on plastic sheets, aluminum alloys, ceramic tiles, painted wood, envelopes and leather (Peng et al. 2018b). The fingerprint development on all substrates exhibited high contrast, sensitivity and selectivity.

In 2021, Zhu et al. reported the use of Eu $(PTA)_3$Phen (RE^{3+} ions coordinated with terephthalic acid and 1,10-phenanthroline) nanocomplex suspension for developing latent fingerprints on glass, painted wood, ceramic tiles, marble, laminate floor and metal sheets (Zhu et al. 2021). The fingerprint development on all substrates exhibited high contrast, sensitivity and selectivity.

Table 16.9. Summary of latent fingerprint development based on RE inorganic DCNMs.

Year	Nanomaterials	λ_{ex} (nm)	Strategy/consuming time	Reference
2013	$Y_2Zr_2O_7$:Eu/SiO_2	254	Physical adsorption/short	Saif 2013
2013	$Y_2Zr_2O_7$:Tb/SiO_2	254	Physical adsorption/short	Saif 2013
2014	$Sr_4Al_{14}O_{25}$:Eu,Dy	365	Physical adsorption/short	Sharma et al. 2014
2015	YVO_4:Eu-PEI	254	Physical adsorption/short	Wang et al. 2015b
2015	$LaPO_4$:Ce,Tb-EG	254	Physical adsorption/short	Wang et al. 2015b
2015	$Y_2Ti_2O_7$:Eu/SiO_2	254	Physical adsorption/short	Saif et al. 2015
2016	$YAlO_3$:Tm	365	Physical adsorption/short	Darshan et al. 2016a
2016	$YAlO_3$:Sm	365	Physical adsorption/short	Darshan et al. 2016b
2016	$YAlO_3$:Nd	365	Physical adsorption/short	Darshan et al. 2016c
2016	$YAlO_3$:Tb	254	Physical adsorption/short	Darshan et al. 2016d
2016	$La_2Ti_2O_7$:Eu/SiO_2	254	Physical adsorption/short	Saif et al. 2016
2017	$LaVO_4$:Eu-PAA	300	Physical adsorption/short	Chen et al. 2017a
2017	$LaVO_4$:Dy-PAA	300	Physical adsorption/short	Chen et al. 2017a
2017	$Y_4Zr_3O_{12}$:Eu	254	Physical adsorption/short	Park and Yang 2017
2017	$BaTiO_3$:Dy	254	Physical adsorption/short	Dhanalakshmi et al. 2017
2017	SiO_2@$SrTiO_3$:Eu,Li	254	Physical adsorption/short	Sandhyarani et al. 2017
2017	Zn_2TiO_4:Sm	254	Physical adsorption/short	Girish et al. 2017
2017	Mg_2SiO_4:Eu	254	Physical adsorption/short	Naik et al. 2017
2017	Mg_2SiO_4:Tb	254	Physical adsorption/short	Naik et al. 2017
2017	$CdSiO_3$:Eu	254	Physical adsorption/short	Basavaraj et al. 2017a
2017	$CdSiO_3$:Tb	254	Physical adsorption/short	Basavaraj et al. 2017a
2017	$CdSiO_3$:Dy-Starch	254	Physical adsorption/short	Basavaraj et al. 2017b
2017	$La_2(MoO_4)_3$:Eu-CA	254	Physical adsorption/short	Li et al. 2017a
2017	ZrO_2:Dy-CTAB	254	Physical adsorption/short	Yadav et al. 2017
2018	$CaZrO_3$:Eu	254	Physical adsorption/short	Navami et al. 2018
2018	SiO_2@$ZnAl_2O_4$:Eu	254	Physical adsorption/short	Komahal et al. 2018
2018	$BaTiO_3$:Nd-CTAB	254	Physical adsorption/short	Dhanalakshmi et al. 2018a
2018	$La_2Ti_2O_7$:Eu	254	Physical adsorption/short	Park et al. 2018a
2018	$BaTiO_3$:Eu@SiO_2	254	Physical adsorption/short	Muniswamy et al. 2018
2018	$Gd_2Ti_2O_7$:Eu	365	Physical adsorption/short	Park et al. 2018b
2018	$BaTiO_3$:Eu	254	Physical adsorption/short	Dhanalakshmi et al. 2018b
2018	Y_2O_3:Eu	254	Physical adsorption/short	Marappa et al. 2018
2018	Y_2O_3:Eu	254	Physical adsorption/short	Venkatachalaiah et al. 2018
2018	CeO_2:Eu-EGGG	254	Physical adsorption/short	Deepthi et al. 2018
2018	MoO_3:Eu	254	Physical adsorption/short	Yogananda et al. 2018
2018	SiO_2@LaOF:Eu	254	Physical adsorption/short	Suresh et al. 2018
2018	Fe_3O_4/$La_2(MoO_4)_3$:Eu	254	Physical adsorption/short	Yu et al. 2018
2019	TiO_2:Eu	254	Physical adsorption/short	Babu et al. 2019

PEI: polyethyleneimine; EG: Ethylene Glycol; PAA: polyacrylic acid; CA: Citric Acid; CTAB: cetrimonium bromide; EGGG: epigallocatechin gallate.

Table 16.10. Summary of latent fingerprint developments based on RE-doped organic-inorganic DC nanocomplexes.

Year	Nanomaterials	λ_{ex} (nm)	Strategy/consuming time	Reference
2018	$Eu_{0.5}Tb_{0.5}(AA)_3Phen$	312	Physical adsorption/short	Peng et al. 2018b
2018	$Tb(AA)_3Phen$	312	Physical adsorption/short	Peng et al. 2018b
2019	$Ce_xLa_{1-x}(SSA)_3Phen$	312	Physical adsorption/short	Peng et al. 2019
2021	$Eu(PTA)_3Phen$	254	Chemical bonding/1~11 s	Zhu et al. 2021

AA: acrylic acid; Phen: 1,10-phenanthroline; PTA: terephthalic acid.

Table 16.10 summarizes the details of latent fingerprint developments based on a variety of RE-doped organic-inorganic DC nanocomplexes.

RE-doped organic-inorganic DC nanocomplexes also have similar advantages as RE-doped inorganic DCNMs, including strong luminescence, suitable morphology, appropriate tackiness and easy synthesis. Therefore, latent fingerprints developed by using RE-doped organic-inorganic DC nanocomplexes can also obtain high contrast, sensitivity and selectivity. Nevertheless, the background fluorescence interference and the potential damage to touch DNA still cannot be avoided due to the usage of short-wavelength UV light.

2.4 Rare earth-doped up conversion nanomaterials

On the other hand, RE-doped DCNMs, RE-doped up conversion NMs (UCNMs) refer to luminescent NMs with RE ions acting as activators. These kinds of nanomaterials can absorb longer wavelength light (e.g., near-infrared light) and then emit at shorter wavelengths (e.g., visible light) through a two-photon or multiphoton mechanism. Except for some prominent advantages such as strong luminescence, suitable size, controllable morphology, excellent photo- and chemical stability and flexible surface functionalization, RE-doped UCNMs also possess one competitive advantage, which is the near-infrared (NIR) responsive luminescence. Under NIR excitation, RE-doped UCNMs can emit strong visible luminescence, while common substrates cannot emit any fluorescence, thus the background fluorescence interference is skillfully avoided.

2.4.1 Ordinary up conversion nanomaterials

In 2014, Wang et al. first reported the use of $NaYF_4$:Yb,Er-LBA ($NaYF_4$:Yb,Er UCNMs conjugated with lysozyme binding aptamer) suspensions for developing latent fingerprints on glass, marble and coins (Wang et al. 2014a). Under 980 nm NIR excitation, the developed fingerprints could exhibit high contrast, sensitivity and selectivity. Unfortunately, all the fingerprints could not be captured, maybe due to the restriction of the photographic technique or equipment. Since then, a variety of ordinary and core-shell UCNMs have been used for developing latent fingerprints.

In 2015, Wang et al. reported the use of $NaYF_4$:Yb,Er-OA ($NaYF_4$:Yb,Er UCNMs capped with oleic acid molecules) UCNM powders for developing latent fingerprints on different substrates (Wang et al. 2015c). The substrates bearing latent fingerprints were divided into three groups, including those with a single color and weak fluorescence (i.e., glass, white ceramic tile and black marble), with multiple colors and weak fluorescence (i.e., multicolor marbles), and with single (or multiple) color and strong fluorescence (i.e., note papers, Chinese paper money and plastic plates). Under 980 nm NIR excitation, the developed fingerprints on all substrates could exhibit high contrast, sensitivity and selectivity. All the fingerprints could also be clearly photographed. In addition, the background fluorescence interferences could be completely avoided even on substrates with strong fluorescence properties.

In 2020, Wang et al. reported the use of $NaYbF_4$:Tm-OA ($NaYbF_4$:Tm UCNMs capped with oleic acid molecules) UCNM powders for dual-mode development of latent fingerprints on different substrates (Wang et al. 2020). Excited with 980 nm NIR light, the developed fingerprint could emit both strong visible light (475 nm) and ultra-strong NIR light (800 nm). Under NIR-to-visible mode,

Table 16.11. Summary of latent fingerprint developments based on RE-doped ordinary UCNMs.

Year	Nanomaterials	λ_{ex} (nm)	Strategy/consuming time	Reference
2014	NaYF$_4$:Yb,Er-LBA	980	Aptamer recognition/30 min	Wang et al. 2014a
2015	NaYF$_4$:Yb,Er-OA	980	Physical adsorption/short	Wang et al. 2015c
2015	NaYF$_4$:Yb,Er-OA	980	Physical adsorption/short	Wang et al. 2015a
2015	NaYF$_4$:Yb,Ce,Er-PEI	980	Physical adsorption/short	Xie et al. 2015
2016	NaYF$_4$:Yb,Gd,Er-OA	980	Physical adsorption/short	Li et al. 2016c
2016	NaYF$_4$:Yb,Er-SDS	980	Hydrophobic effect/10 s	Wang 2016a
2017	NaYbF$_4$:Ho-OA	980	Physical adsorption/short	Du et al. 2017
2017	Eu(DBM)$_3$Phen/NaGdF$_4$:Yb, Er	254, 980	Physical adsorption/short	Shahi et al. 2017
2020	NaYbF$_4$:Tm-OA	980	Physical adsorption/short	Wang et al. 2020

LBA: Lysozyme Binding Aptamer; OA: Oleic Acid; PEI: polyethyleneimine; SDS: Sodium Dodecyl Sulfate; DBM: dibenzoyl methane; Phen: 1,10-phenanthroline.

800 nm emission light and 980 nm excitation light were blocked from entering the camera by using a shortwave-pass filter, resulting in an image of a fingerprint with visible luminescence. Under NIR-to-NIR mode, only 800 nm emission light could enter the camera by using a bandpass filter, resulting in an image of a fingerprint with NIR luminescence. Thus, this dual-mode developing method could achieve double fluorescent images from single NIR excitation, which would be beneficial to contrast and supplement each other. The developed fingerprints on all substrates could exhibit high contrast, sensitivity and selectivity.

Table 16.11 summarizes the details of latent fingerprint development based on a variety of RE-doped ordinary UCNMs.

RE-doped ordinary UCNMs have similar advantages as RE-doped DCNMs, including strong and stable luminescence, suitable size and morphology and appropriate tackiness. Therefore, latent fingerprints developed by using RE-doped ordinary UCNMs can also obtain high contrast, sensitivity and selectivity. More importantly, the excitation of RE-doped ordinary UCNMs usually requires NIR light, which can fundamentally avoid the background fluorescence interferences. However, the synthetic conditions for these UCNMs are relatively severe, probably involving high temperature, high pressure, long time, water-free and oxygen-free systems.

2.4.2 Core-shell upconversion nanomaterials

To enrich the luminescent property of ordinary UCNMs, bare UCNMs can be coated with other kinds of NMs, meanwhile, other kinds of NMs can also be coated with UCNMs to form the core-shell or core-shell-shell luminescent NMs. In 2016, Li et al. reported the use of NaYbF$_4$:Tm@NaYF$_4$:Yb@NaNdF$_4$:Yb (NaYbF$_4$:Tm UCNMs coated with NaYF$_4$:Yb and further coated with NaNdF$_4$:Yb) NC suspensions for dual-mode development of latent and blood fingerprints on soft-drink labels (Li et al. 2016b). When excited with 808 nm NIR light, the developed fingerprint could give both a 696 nm DC emission and a 980 nm UC emission. However, the selectivity of blood fingerprint development needs to be improved.

In 2017, Zhou et al. reported the use of Cu$_{2-x}$S@SiO$_2$@Y$_2$O$_3$:Yb, Er (Cu$_{2-x}$S semiconductor plasmon NPs coated with silica and further coated with Y$_2$O$_3$:Yb, Er UCNMs) NC powders for developing the latent fingerprints on paper (Zhou et al. 2017). When excited with 980 nm NIR light, the UC luminescence from Cu$_{2-x}$S@SiO$_2$@Y$_2$O$_3$: Yb, Er NCs could be enhanced about 30-fold compared with that from SiO$_2$@Y$_2$O$_3$: Yb, Er NCs, due to the Cu$_{2-x}$S-induced local surface plasmon resonance effect.

Table 16.12 summarizes the details of latent fingerprint development based on a variety of RE doped core-shell UCNMs.

Table 16.12. Summary of latent fingerprint developments based on RE-doped core-shell UCNMs.

Year	Nanomaterials	λ_{ex} (nm)	Strategy/consuming time	Reference
2016	$NaYbF_4$:Tm@$NaYF_4$:Yb@$NaNdF_4$:Yb	808	Hydrophobic effect/16 minutes	Li et al. 2016b
2017	$Cu_{2-x}S$@SiO_2@Y_2O_3:Yb, Er	980	Physical adsorption/short	Zhou et al. 2017
2018	$NaYF_4$:Yb,Gd,Er@SiO_2	980	Physical adsorption/short	Wang et al. 2018b

Compared with ordinary UCNMs, core-shell UCNMs usually possess improved luminescent intensity and enriched luminescent properties, which can further promote the contrast of latent fingerprint development. However, the corresponding preparation process of core-shell UCNMs is more complex and time-consuming.

2.5 Carbon dots

Carbon Dots (CDs), as a new class of fluorescent NMs, refer to zero-dimensional fluorescent carbon NMs with particle sizes below 10 nm, which were first discovered during the purification of single-walled carbon nanotubes. The prominent advantages offered by CDs are tunable fluorescence, excellent chemical- and photostability and flexible surface functionalization.

2.5.1 Ordinary carbon dots

In 2015, Fernandes et al. first reported the use of CDs/SiO_2, CDs/TiO_2 and CDs/laponite (CDs modified with silica, titania and laponite) NC powders for latent fingerprint development on glass (Fernandes et al. 2015). When excited with 410 nm UV, 440 nm blue and 470 nm green light, the developed fingerprint could emit blue, green and red fluorescence, respectively. However, the developing results corresponding to CDs/TiO_2 NCs were relatively poor. Since then, a variety of ordinary and core-shell CDs have been used for developing latent fingerprints.

In the same year, Dilag et al. reported the use of CDs/p(DMA) (CDs modified with poly(dimethylacrylamide)) NC suspension for latent fingerprint development on aluminum foil (Dilag et al. 2015). When excited with 350 nm light, the developed fingerprint could emit yellow, green and red fluorescence, using different optical filtering methods. However, the preparation of CDs/p(DMA) NCs was quite complex and tedious.

Table 16.13 summarizes the details of latent fingerprint developments based on a variety of ordinary CDs.

CDs have similar advantages as QDs, including bright fluorescence, small particle size, suitable morphology and flexible surface functionalization. Therefore, latent fingerprints developed by using ordinary CDs can also obtain high contrast, sensitivity and selectivity. More importantly, compared with QDs, the intrinsic toxicity of CDs is negligible. However, to avoid their fluorescence quenching induced by the aggregation effect, pure CDs are often modified with stable substances (e.g., SiO_2, TiO_2 and starch) to form NC powders.

Table 16.13. Summary of latent fingerprint developments based on ordinary CDs.

Year	Nanomaterials	λ_{ex} (nm)	Strategy/consuming time	Reference
2015	CDs/SiO_2	410, 440, 470	Physical adsorption/short	Fernandes et al. 2015
2015	CDs/TiO_2	410, 440, 470	Physical adsorption/short	Fernandes et al. 2015
2015	CDs/laponite	410, 440, 470	Physical adsorption/short	Fernandes et al. 2015
2015	CDs/p(DMA)	350	Hydrophobic effect/5 s	Dilag et al. 2015
2016	CDs/Starch	365	Physical adsorption/short	Li et al. 2016a
2017	CDs	UV	Physical adsorption/30 min	Chen et al. 2017b
2018	CDs/SiO_2	365	Physical adsorption/5 min	Zhu et al. 2018

DMA: dimethylacrylamide.

2.5.2 Core-shell carbon dots

To hinder the fluorescence quenching of CDs, bare CDs are often coated with a layer of silica. In 2017, Zhao et al. reported the use of core-shell CDs@SiO$_2$ (CDs coated with silica) NC powders for developing the latent fingerprints on glass, aluminum foil, black plastic bags, drug packing and leather (Zhao et al. 2017). Similar to other DCNMs, however, the developing contrast on fluorescent substrates was relatively low. In addition, the developing selectivity on the smooth substrate was not high enough.

Table 16.14 summarizes the details of latent fingerprint development based on a variety of core-shell CDs.

Silica or titania coating is an effective way to avoid the fluorescence quenching of CD powders induced by the aggregation effect. However, the corresponding preparation process of core-shell CD powders is more complex and time-consuming.

Table 16.14. Summary of latent fingerprint development based on core-shell CDs.

Year	Nanomaterials	λ_{ex} (nm)	Strategy/consuming time	Reference
2017	CDs@SiO$_2$	365	Physical adsorption/short	Zhao et al. 2017
2017	CDs@TiO$_2$	normal light	Physical adsorption/30 seconds	Amith Yadav et al. 2017
2018	SiO$_2$@CDs	365	Physical adsorption/short	Peng et al. 2018a

3. Strategies in Latent Fingerprint Development

In the NM-based fluorescent development of latent fingerprints, the ultimate purpose is to make NPs selectively combined with fingerprint residuals. In general, there are three basic strategies to achieve the above purpose, including physical, chemical and biological strategies. A timeline of typical strategies used for latent fingerprint development is presented in Fig. 16.6. In general, physical strategy can be divided into physical adsorption and hydrophobic effect approaches, chemical strategy refers to chemical bonding approach and biological strategy can be divided into antibody combination and aptamer recognition approaches. Perhaps, physical adsorption is the oldest and most effective way for latent fingerprint development.

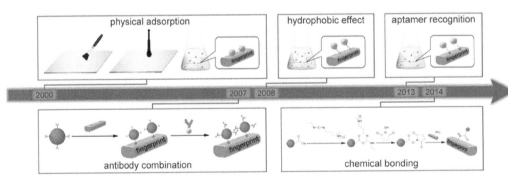

Figure 16.6. A timeline of typical strategies used for latent fingerprint development.

3.1 Physical strategy

In the physical strategy, NPs have selectively adhered to fingerprint residuals via physical adsorption or a hydrophobic effect mechanism. As for the physical adsorption approach, fingerprint development can be conducted through either powder dusting or suspension staining. As for the hydrophobic effect approach, fingerprint development must be conducted in suspension. The powder dusting treatment is always simple and efficient, which is only suitable for developing fresh fingerprints on smooth substrates. At the same time, the suspension staining procedure is more complex and time-consuming, which can be used to deal with aged fingerprints.

3.1.1 Physical adsorption

In 2015, Wang et al. reported the use of $NaYF_4$:Yb, Er UCNM powders for developing latent fingerprints on glass, marble, ceramic tiles, aluminum alloys, aluminum foil, stainless steel, plastic cards, wood floor, painted wood, floor leather and various papers (Wang et al. 2015a). The $NaYF_4$:Yb, Er UCNMs used here were ordinary luminescent powders, which could emit green luminescence under 980 nm NIR excitation. First, the UCNM dry powders are directly deposited onto the fingerprint by a light dusting action. Then, the excess powders are carefully removed by a gentle brushing action. During this process, the NPs can adhere to the aqueous (e.g., sweat) or oily (e.g., grease) components in fingerprint residues via physical adsorption. Finally, the developed fingerprint is excited by 980 nm NIR light to emit green luminescence, revealing the fingerprint image. They also found that the particle size had a great effect on developing sensitivity: when developed by nanosized $NaYF_4$:Yb, Er powders, even the detail of sweat pores could be clearly observed, which was almost impossible to achieve by micron-sized $NaYF_4$:Yb, Er powders. The ordinary powder dusting technique is the simplest and most commonly used procedure for latent fingerprint development, and it possesses a series of advantages such as high contrast, high sensitivity, easy operation and high efficiency. However, the most evident drawback of this procedure is dust blowing, which is dangerous to human health.

One effective way to avoid dust blowing is by modifying ordinary powders with magnetic materials. In 2016, Lee et al. reported the use of Fe_3O_4/DA (magnetic Fe_3O_4 NPs modified with diacetylene) NC powders for developing latent fingerprints on paper and aluminum foil (Lee et al. 2016). The magnetically responsive Fe_3O_4/DA powders used here could display not only stimulus-responsive color but also red fluorescence. First, the Fe_3O_4/DA dry powders are applied to a fingerprint by using a magnet and further adhered to the sebaceous components in fingerprint residues via physical adsorption. During this process, the magnetic powders can be tightly fastened around the magnet, thus preventing the powders from dust blowing. Then, the developed fingerprint is irradiated under 254 nm light for 10~30 seconds to induce the photopolymerization of DA monomers, generating the blue-colored polydiacetylenes (PDAs). Finally, the resultant fingerprint is heat-treated at 80°C for 20 seconds, promoting the conversion from blue-phase PDAs to red ones, meanwhile emitting red fluorescence.

Another effective way to avoid dust blowing is making ordinary powders into suspension form. In 2011, Gao et al. reported the use of CdTe-Hydrazine (CdTe QDs capped with hydrazine molecules) QD suspension for developing latent fingerprints on glass, ceramic tiles, painted polymers, plastic sheets, black rubber and aluminum foil (Gao et al. 2011). First, the pH of CdTe-Hydrazine QD suspension is adjusted to 6.4 before use. Then, the fingerprint is immersed into the above QD suspension for 30~60 minutes. Based on the fact that the isoelectric point of amino acids is around 6.02, when the pH is adjusted higher than 6.4, the amino acids in the fingerprint residuals will possess negative charges. During the staining process, positive charged QDs can be adsorbed onto the surface of negatively charged fingerprint residuals via electrostatic adsorption. After that, the excess powders are carefully rinsed with water. Finally, the developed fingerprint is excited by 365 nm light to emit fluorescence, revealing the fingerprint image. Although the developing method is novel, the staining time is too long in practice.

3.1.2 Hydrophobic effect

In 2016, Wang reported the use of $NaYF_4$:Yb, Er UCNM suspensions for developing latent fingerprints on glass, stainless steel, aluminum alloy, aluminum foil, marble, ceramic tiles, plastic cards, painted wood and Chinese paper money (Wang 2016a). First, the suspension containing UCNMs, surfactants (Sodium Dodecyl Sulfonate, SDS) and water is prepared before use. Here, the hydrophilic sulfonic groups on one end of SDS molecules can cap the surfaces of the NPs, with their hydrophobic alkyl chains outward, resulting in a stable dispersed UCNM suspension. Then, the fingerprint is immersed in the above UCNM suspension for 10 seconds. During the staining process, the hydrophobic alkyl chains could adsorb the oily components in the fingerprint residues via the hydrophobic effect. After that, the excess powders are carefully rinsed with water. Finally, the developed fingerprint is excited by 980 nm NIR light to emit fluorescence, revealing the fingerprint image. In particular, this method can be used to detect fingerprints that are difficult to detect, including aged fingerprints, fingerprints on wet substrates and fingerprints on multicolored substrates with strong fluorescence, high contrast, sensitivity and selectivity. Compared with the physical adsorption-based suspension staining method, the staining time in this method is significantly shortened with the aid of surfactants.

3.2 Chemical strategy

In the chemical strategy, NPs are selectively combined with fingerprint residuals mainly via a chemical bonding mechanism (Moret et al. 2014). Fingerprint development using the chemical bonding method must be conducted in suspension, and the surfaces of NPs are typically capped or modified with active functional groups (e.g., carboxyl group and amino group). In general, chemical-based developing strategies can deal with not only fresh fingerprints but also with some aged fingerprints.

In 2017, Wang et al. reported the use of Zn_2GeO_4:Ga, Mn nanosuspensions for developing the latent fingerprints on poker cards, soft drink cans, knives and desks (Wang et al. 2017b). First, the surfaces of Zn_2GeO_4:Ga,Mn NPs are coated with a layer of silica and further modified with a carboxyl group via the typical Stober-based method (Han et al. 2017). Then, the carboxyl groups on the surfaces of these NPs are activated by the aid of 1-ethyl-3-(3-dimethylaminopropyl)-carbodiimide (EDC) and combined with N-hydroxysuccinimide (NHS). After that, the NPs modified with activated carboxyl groups are made into suspension and then applied to a latent fingerprint for 30 minutes. During the staining process, the activated carboxyl groups on surfaces of NPs can react with the amino groups in the latent fingerprint residuals to form amido bonds under mild conditions. Later, the excess powders are carefully rinsed with water. Finally, the developed fingerprint is excited by UV light to emit fluorescence, revealing the fingerprint image. The developing method is novel. However, the preparation of carboxyl group functionalized Zn_2GeO_4:Ga, Mn NPs is too tedious, involving silica coating, amino group modification and carboxyl group conjugation. In addition, the staining time is too long in practice.

To simplify the tedious procedure of carboxyl group modification, the NMs can be well designed. In 2021, Zhu et al. reported the use of [Eu (PTA)$_3$Phen] (Eu^{3+} ions coordinated with terephthalic acid and 1,10-phenanthroline) nanocomplex suspension for developing latent fingerprints on glass, painted wood, ceramic tiles, marble, plastic cards, laminate floor and metals (Zhu et al. 2021). Here, the reagent of PTA is ingeniously selected as not only the reactant but also the carboxyl modifier. On one hand, the carboxyl groups on one end of PTA molecules can coordinate with the Eu^{3+} ion to form a nanocomplex; on the other hand, the carboxyl groups on the other end of PTA molecules are outward, modifying the surfaces of the nanocomplex carboxyl groups. Thus, the synthesis and surface modification of NMs can be accomplished simultaneously, which is more efficient. Later the latent fingerprint development is carried out with the aid of EDC combined with NHS. More importantly, the whole staining time is reduced to only 3 seconds.

The chemical bonding strategy can be applied in not only developing latent fingerprints but also in detecting special components in fingerprint residuals. In 2015, Wu et al. reported the use of core-shell ZnCdS:Cu@SiO$_2$/ZnCdS-PAA (ZnCdS:Cu QDs coated with silica, then modified with ZnCdS QDs, and finally conjugated with polyallylamine molecules) NC suspension for developing latent fingerprints on glass and paper, simultaneously detecting the TNT components in fingerprint residuals (Wu et al. 2015). The QD NCs with core-shell structure and dual-emitting properties are well designed, being composed of a red-emitting core (i.e., ZnCdS: Cu), and a green-emitting shell (i.e., ZnCdS) and further functionalized with amino groups (i.e., PAA). First, the fingerprint is immersed into the above QD NC suspension for 30 seconds. During the staining process, the QD NCs can be adsorbed onto or linked with the surface of fingerprint residuals via physical adsorption or chemical bonding. When there are no TNT components in the fingerprint residuals, the ZnCdS shell will emit green fluorescence under 365 nm excitation. On the other hand, the TNT components in fingerprint residuals can react with primary amines in PAA molecules to form the Meisenheimer complexes (Wu et al. 2015), causing the quenching of green fluorescence. Consequently, the ZnCdS: Cu core will give red fluorescence under 365 nm excitation. Only through the fluorescence color, can one easily determine whether TNT components are present in the fingerprint residuals.

3.3 Biological strategy

In the biological strategy, NPs are selectively combined with fingerprint residuals via antibody combination or aptamer recognition mechanism. Fingerprint development using either an antibody combination or aptamer recognition method must be conducted in suspension, and the surfaces of NPs are typically capped or modified with antibodies or aptamers, respectively. In general, biological-based developing strategies can develop both fresh and aged fingerprints.

3.3.1 Antibody combination

In 2007, Leggett et al. reported the use of Au-Ab (nano-gold conjugated with anti-cotinine antibody) NP suspension combined with Alexa Fluor-Ab (secondary antibody labeled with organic fluorescent dyes) for developing the smokers' latent fingerprints on glass, simultaneously detecting the nicotine metabolite (cotinine) in fingerprint residuals (Leggett et al. 2007). First, the anti-cotinine antibody conjugated nano-gold is suspended in water and then applied to the latent fingerprint at 37°C for 30 minutes. The excess substances are carefully rinsed with water. During the first staining process, the nano-gold could be combined with the cocaine metabolite in fingerprint residues via the combination between cotinine and anti-cotinine antibody. Then, Alexa Fluor labeled secondary antibody is suspended and applied on latent fingerprint at 37°C for another 30 minutes, and the excess substances are carefully rinsed with water. During the second staining process, the Alexa Fluor could be linked with the nano-gold on fingerprint surfaces via the combination between primary antibody and secondary antibody. Thus, the conjugation between the Alexa Fluor and special fingerprint residuals is accomplished via a two-step staining process. Finally, the developed fingerprint is excited by blue or green light to emit green or red fluorescence, revealing the fingerprint image.

Since then, this antibody combination model has been established as a universal method for the development of latent fingerprints with simultaneous identification of special fingerprint residues. For example, Hazarika et al. in 2008 reported the use of Fe_3O_4-Ab (nano-Fe_3O_4 conjugated with primary antibody) NP suspension combined with Alexa Fluor-Ab (secondary antibody labeled with organic fluorescent dyes) for developing latent fingerprints on glass, simultaneously detecting the tetrahydrocannabinol (THC), methadone, 2-ethylidene-1,5-dimethyl-3,3,-diphenylpyrrolidine (EDDP) and benzoylecgonine in fingerprint residuals (Hazarika et al. 2008); Hazarika et al. in 2009 reported the similar use of Fe_3O_4-Ab NP suspension combined with Alexa Fluor-Ab for developing the smokers' latent fingerprints on glass, simultaneously detecting the nicotine metabolite in fingerprint residuals. However, the preparation of antibody conjugated NPs here requires complicated and tedious steps, which will take 24~48 hours (Hazarika et al. 2009). In addition, the staining condition is not mild, and the staining time is too long, which is not user-friendly in practice.

3.3.2 Aptamer recognition

In 2013, Li et al. reported the use of Au-CBA (nano-gold conjugated with cocaine-binding aptamer) NP suspension for developing latent fingerprints on glass, simultaneously detecting the cocaine components in fingerprint residuals (Li et al. 2013). First, the glass substrate is cleaned by sonication for 15 minutes sequentially with soapy water, acetone, ethanol and deionized water, and further blocked with casein at 37°C for 2 hours and at 4°C for another 12 hours. The CBA conjugated nano-gold is made into suspension and then applied to the latent fingerprint at room temperature for 30 minutes. The excess substances are carefully rinsed with water. During the staining process, the nanogold can be physically adsorbed onto or chemically linked with the surface of fingerprint residuals via physical adsorption or aptamer recognition. Finally, the Localized Surface Plasmon Resonance (LSPR) of the nanogold developed fingerprint is observed using an inverted microscope under white light, and the cocaine components are further quasi-quantitative detected using a Plasmonic Resonance Rayleigh Scattering (PRRS) spectroscopy. When there are no cocaine components in the fingerprint residuals, the nanogold will generate a green scattering signal in the darkfield image. On the hand, the nanogold will be aggregated due to the combination between the cocaine and the CBA, resulting in a red scattering signal. Only through the scattering color of nanogold, it is possible to easily determine whether it contains cocaine components in the fingerprint residuals. However, the preparation of CBA modified nanogold NPs here requires complicated and tedious steps, which will take more than 40 hours, and the treatment before and during the fingerprint development is also time-consuming.

In 2014, Wang et al. reported the similar use of CdTe-LBA (CdTe QDs conjugated with lysozyme-binding aptamer) and $NaYF_4$:Yb, Er-LBA NP suspension for developing latent fingerprints on glass, marble and coins (Wang et al. 2014a). First, LBA conjugated CdTe QDs and $NaYF_4$:Yb, Er UCNMs are separately suspended and then applied on latent fingerprint at room temperature for 30 minutes. During the staining process, the NPs could be combined with the lysozyme components in the fingerprint residues via the combination between lysozyme and LBA. Then, the excess substances are carefully rinsed with water. Finally, the CdTe QDs and $NaYF_4$:Yb, Er UCNMs developed fingerprint are separately excited by 365 nm and 980 nm light-emitting green fluorescence and revealing the fingerprint image. However, this method still encounters similar problems of complicated preparation and long staining time.

4. Summary and Outlook

In this chapter, the recent advances in nanomaterial-based fluorescent development of latent fingerprints from two aspects: one about the developing materials, and the other about the developing strategies are described. Regarding developing materials, this chapter makes emphasis on fingerprint developments based on QDs, metal and metal oxide nanoparticles, rare-earth-doped down- and up-conversion luminescent nanomaterials and fluorescent carbon dots. On the aspect of developing strategies, the focus on fingerprint developments based on physical adsorption, hydrophobic effect, chemical bonding, antibody combination, and aptamer recognition are concentrated here. In addition, the benefits and drawbacks of each nanomaterial and strategy are discussed. It should be noted that there is no ideal or universal nanomaterial or strategy for latent fingerprint development. Therefore, it is necessary to carefully choose the appropriate nanomaterial or suitable strategy according to the actual condition of fingerprints as well as the character of substrates.

At present, the nanomaterials used in latent fingerprint development exhibit the transition tendency as follows: from down conversion NMs to up conversion NMs, from single-performance NMs to multiple-performance NMs, from toxic NMs to non-toxic NMs. The strategies used in latent fingerprint development exhibit the transition tendency as follows: from simple physical methods to complex chemical and biological methods, from non-specific adhesion to target recognition and from qualitative imaging to quantitative detection. As a result, the effect of latent fingerprint development exhibits the transition tendency as follows: from high background interferences to low fluorescent noises, from clear development of papillary ridges to distinct identification of sweat pores, from simple utilization of material evidence to simultaneous detection of biological evidence (i.e., touch DNA).

Acknowledgments

Meng Wang would like to thank the financial support of the National Science Foundation of China (Nos. 21205139, 21802169), and Liaoning BaiQianWan Talents Program in 2020.

References

Algarra, M., J. Jimenez-Jimenez, R. Moreno-Tost, B.B. Campos and J.C.G. Esteves da Silva. 2011. CdS nanocomposites assembled in porous phosphate heterostructures for fingerprint detection. Opt. Mater. 33: 893–898.

Algarra, M., J. Jimenez-Jimenez, M.S. Miranda, B.B. Campos, R. Moreno-Tost, E. Rodriguez-Castellon et al. 2013. Solid luminescent CdSe-thiolated porous phosphate heterostructures. Application in fingermark detection in different surfaces. Surf. Interface Anal. 45: 612–618.

Algarra, M., K. Radotic, A. Kalauzi, D. Mutavdzic, A. Savic, J. Jimenez-Jimenez et al. 2014. Fingerprint detection and using intercalated CdSe nanoparticles on non-porous surfaces. Anal. Chim. Acta 812: 228–235.

Amoros, B.G. and M.D. Puit. 2014. A model study into the effects of light and temperature on the degradation of fingerprint constituents. Sci. Justice 54: 346–350.

Archer, N.E., Y. Charles, J.A. Elliott and S. Jickells. 2005. Changes in the lipid composition of latent fingerprint residue with time after deposition on a surface. Forensic Sci. Int. 154: 224–239.

Ashbaugh, D.R. 1991. Palmar flexion crease identification. J. Forensic Identification. 41: 255–273.

Ashbaugh, D.R. 1992. Incipient ridges and the clarity spectrum. J. Forensic Identification. 42: 106–114.

Babu, K.R.V., C.G. Renuka, R.B. Basavaraj, G.P. Darshan and H. Nagabhushana. 2019. One pot synthesis of $TiO_2:Eu^{3+}$ hierarchical structures as a highly specific luminescent sensing probe for the visualization of latent fingerprints. J. Rare Earth. 37: 134–144.

Basavaraj, R.B., H. Nagabhushana, G.P. Darshan, B.D. Prasad, M. Rahul, S.C. Sharma et al. 2017a. Red and green emitting CTAB assisted $CdSiO_3:Tb^{3+}/Eu^{3+}$ nanopowders as fluorescent labeling agents used in forensic and display applications. Dyes Pigments 147: 364–377.

Basavaraj, R.B., H. Nagabhushana, G.P. Darshan, B.D. Prasad, S.C. Sharma and K.N. Venkatachalaiah. 2017b. Ultrasound assisted rare earth doped wollastonite nanopowders: labeling agent for imaging eccrine latent fingerprints and cheiloscopy applications. J. Ind. Eng. Chem. 51: 90–105.

Becue, A., S. Moret, C. Champod and P. Margot. 2009. Use of quantum dots in aqueous solution to detect blood fingermarks on non-porous surfaces. Forensic Sci. Int. 191: 36–41.

Becue, A. 2016. Emerging fields in fingermark (meta)detection—a critical review. Anal. Methods 8: 7983–8003.

Cadd, S., M. Islam, P. Manson and S. Bleay. 2015. Fingerprint composition and aging: A literature review. Sci. Justice 55: 219–238.

Cai, K.Y., R.Q. Yang, Y.J. Wang, X.J. Yu and J.J. Liu. 2013. Super fast detection of latent fingerprints with water soluble CdTe quantum dots. Forensic Sci. Int. 226: 240–243.

Champod, C., C. Lennard, P. Margot and M. Stoilovic. 2004. Fingerprints and other ridge skin impressions. CRC Press: Florida.

Chen, C.L., Y. Yu, C.G. Li, D. Liu, H. Huang, C. Liang et al. 2017a. Facile synthesis of highly water-soluble lanthanide-doped t-$LaVO_4$ NPs for antifake ink and latent fingermark detection. Small 13: 1702305.

Chen, J., J.S. Wei, P. Zhang, X.Q. Niu, W. Zhao, Z.Y. Zhu et al. 2017b. Red-emissive carbon dots for fingerprints detection by spray method: coffee ring effect and unquenched fluorescence in drying process. ACS Appl. Mater. Interfaces 9: 18429–18433.

Choi, M.J., K.E. McBean, P.H.R. Ng, A.M. McDonagh, P.J. Maynard, C. Lennard et al. 2008. An evaluation of nanostructured zinc oxide as a fluorescent powder for fingerprint detection. J. Mater. Sci. 43: 732–737.

Croxton, R.S., M.G. Baron, D. Butler, T. Kent and V.G. Sears. 2010. Variation in amino acid and lipid composition of latent fingerprints. Forensic Sci. Int. 199: 93–102.

Darshan, G.P., H.B. Premkumar, H. Nagabhushana, S.C. Sharma, S.C. Prashantha, H.P. Nagaswarup et al. 2016a. Blue light emitting ceramic nano-pigments of Tm^{3+} doped $YAlO_3$: applications in latent finger print, anti-counterfeiting and porcelain stoneware. Dyes Pigments 131: 268–281.

Darshan, G.P., H.B. Premkumar, H. Nagabhushana, S.C. Sharma, S.C. Prashanth and B.D. Prasad. 2016b. Effective fingerprint recognition technique using doped yttrium aluminate nano phosphor material. J. Colloid Interf. Sci. 464: 206–218.

Darshan, G.P., H.B. Premkumar, H. Nagabhushana, S.C. Sharma, B.D. Prasad and S.C. Prashantha. 2016c. Neodymium doped yttrium aluminate synthesis and optical properties—a blue light emitting nanophosphor and its use in advanced forensic analysis. Dyes Pigments 134: 227–233.

Darshan, G.P., H.B. Premkumar, H. Nagabhushana, S.C. Sharma, B.D. Prasad, S.C. Prashantha et al. 2016d. Superstructures of doped yttrium aluminates for luminescent and advanced forensic investigations. J. Alloy. Compd. 686: 577–587.

Deepthi, N.H., G.P. Darshan, R.B. Basavaraj, B.D. Prasad and H. Nagabhushana. 2018. Large-scale controlled bio-inspired fabrication of 3D CeO_2:Eu^{3+} hierarchical structures for evaluation of highly sensitive visualization of latent fingerprints. Sensor. Actuat. B-Chem. 255: 3127–3147.

Dhanalakshmi, M., H. Nagabhushana, G.P. Darshan, R.B. Basavaraj and B.D. Prasad. 2017. Sonochemically assisted hollow/solid $BaTiO_3$:Dy^{3+} microspheres and their applications in effective detection of latent fingerprints and lip prints. J. Sci. Adv. Mater. Devices 2: 22–33.

Dhanalakshmi, M., H. Nagabhushana, S.C. Sharma, R.B. Basavaraj, G.P. Darshan and D. Kavyashree. 2018a. Bio-template assisted solvothermal synthesis of broom-like $BaTiO_3$:Nd^{3+} hierarchical architectures for display and forensic applications. Mater. Res. Bull. 102: 235–247.

Dhanalakshmi, M., H. Nagabhushana, G.P. Darshan and B.D. Prasad. 2018b. Ultrasound assisted sonochemically engineered effective red luminescent labeling agent for high resolution visualization of latent fingerprints. Mater. Res. Bull. 98: 250–264.

Dilag, J., H. Kobus and A.V. Ellis. 2009. Cadmium sulfide quantum dot/chitosan nanocomposites for latent fingermark detection. Forensic Sci. Int. 187: 97–102.

Dilag, J., H. Kobus and A.V. Ellis. 2013. CdS/polymer nanocomposites synthesized via surface initiated RAFT polymerization for the fluorescent detection of latent fingermarks. Forensic Sci. Int. 228: 105–114.

Dilag, J., H. Kobus, Y. Yu, C.T. Gibson and A.V. Ellis. 2015. Non-toxic luminescent carbon dot/poly (dimethylacrylamide) nanocomposite reagent for latent fingermark detection synthesized via surface initiated reversible addition fragmentation chain transfer polymerization. Polym. Int. 64: 884–891.

Dong, W.J., Y. Cheng, L. Luo, X.Y. Li, L.N. Wang, C.G. Li et al. 2014. Synthesis and self-assembly of hierarchical SiO_2-QDs@SiO_2 nanostructures and their photoluminescence applications for fingerprint detection and cell imaging. RSC Adv. 4: 45939–45945.

Du, P., P. Zhang, S.H. Kang and J.S. Yu. 2017. Hydrothermal synthesis and application of Ho^{3+}-activated $NaYbF_4$ bifunctional upconverting nanoparticles for *in vitro* cell imaging and latent fingerprint detection. Sensor. Actuat. B-Chem. 252: 584–591.

Fernandes, D., M.J. Krysmann and A. Kelarakis. 2015. Carbon dot based nanopowders and their application for fingerprint recovery. Chem. Commun. 51: 4902–4905.

Freinkel, R.K. and D.T. Woodley. 2001. The Biology of Skin. The Parthenon: New York.

Gao, F., C.F. Lv, J.X. Han, X.Y. Li, Q. Wang, J. Zhang et al. 2011a. CdTe montmorillonite manocomposites: control synthesis, UV radiation-dependent photoluminescence, and enhanced latent fingerprint detection. J. Phys. Chem. C 115: 21574–21583.

Gao, F., J.X. Han, J. Zhang, Q. Li, X.F. Sun, J.C. Zheng et al. 2011b. The synthesis of newly modified CdTe quantum dots and their application for improvement of latent fingerprint detection. Nanotechnology 22: 075705.

Gao, F., J. Han, C.F. Lv, Q. Wang, J. Zhang, Q. Li et al. 2012. Application of core-shell-structured CdTe@SiO_2 quantum dots synthesized via a facile solution method for improving latent fingerprint detection. J. Nanopart. Res. 14: 1191.

Girish, K.M., S.C. Prashantha, R. Naik and H. Nagabhushana. 2017. Zn_2TiO_4: a novel host lattice for Sm^{3+} doped reddish orange light emitting photoluminescent material for thermal and fingerprint sensor. Opt. Mater. 73: 197–205.

Girod, A., R. Ramotowski and C. Weyermann. 2012. Composition of fingermark residue: A qualitative and quantitative review. Forensic Sci. Int. 223: 10–24.

Han, Y., Z. Lu, Z. Teng, J. Liang, Z. Guo, D. Wang, M. Han and W. Yang. 2017. Unraveling the growth mechanism of silica particles in the stöber method: *in situ* seeded growth model. Langmuir 33: 5879–5890.

Hazarika, P., S.M. Jickells, K. Wolff and D.A. Russell. 2008. Imaging of latent fingerprints through the detection of drugs and metabolites. Angew. Chem. Int. Ed. 47: 10167–10170.

Hazarika, P., S.M. Jickells and D.A. Russell. 2009. Rapid detection of drug metabolites in latent fingermarks. Analyst 134: 93–96.

Hazarika, P., S.M. Jickells, K. Wolff and D.A. Russell. 2010. Multiplexed detection of metabolites of narcotic drugs from a single latent fingermark. Anal. Chem. 82: 9150–9154.

Hazarika, P. and D.A. Russell. 2012. Advances in fingerprint analysis. Angew. Chem. Int. Ed. 51: 3524–3531.

Humphreys, J.D., G. Porter and M. Bell. 2008. The quantification of fingerprint quality using a relative contrast index. Forensic Sci. Int. 178: 46–53.

Jin, Y.J., Y.J. Luo, G.P. Li, J. Li, Y.F. Wang, R.Q. Yang et al. 2008. Application of photoluminescent CdS/PAMAM nanocomposites in fingerprint detection. Forensic Sci. Int. 179: 34–38.

Knowles, A.M. 1978. Aspects of physicochemical methods for the detection of latent fingerprints. J. Phys. E: Sci. Instrum. 11: 713–721.

Komahal, F.F., H. Nagabhushana, R.B. Basavaraj, G.P. Darshan, B.D. Prasad, S.C. Sharma et al. 2018. Design of bi-functional composite core-shell $SiO_2@ZnAl_2O_4:Eu^{3+}$ array as a fluorescent sensors for selective and sensitive latent fingerprints visualization protocol. Adv. Powder Technol. 29: 1991–2002.

Lee, H.C. and R.E. Gaensslen. 2001. Advances in Fingerprint Technology, 2nd ed. CRC Press: Florida.

Lee, J., C.W. Lee and J.M. Kim. 2016. A magnetically responsive polydiacetylene precursor for latent fingerprint analysis. ACS Appl. Mater. Interfaces 8: 6245–6251.

Leggett, R., E.E. Lee-Smith, S.M. Jickells and D.A. Russell. 2007. "Intelligent" fingerprinting: simultaneous identification of drug metabolites and individuals by using antibody-functionalized nanoparticles. Angew. Chem. Int. Ed. 46: 4100–4103.

Lesniewski, A. 2016. Hybrid organic–inorganic silica based particles for latent fingermarks development: A review. Synth. Met. 222: 124–131.

Li, K., W.W. Qin, F. Li, X.C. Zhao, B.W. Jiang, K. Wang et al. 2013. Nanoplasmonic imaging of latent fingerprints and identification of cocaine. Angew. Chem. Int. Ed. 52: 11542–11545.

Li, H.R., X.J. Guo, J. Liu and F. Li. 2016a. A synthesis of fluorescent starch based on carbon nanoparticles for fingerprints detection. Opt. Mater. 60: 404–410.

Li, J.C., X.J. Zhu, M. Xue, W. Feng, R.L. Ma and F.Y. Li. 2016b. Nd^{3+}-sensitized upconversion nanostructure as a dual-channel emitting optical probe for near infrared-to-near infrared fingerprint imaging. Inorg. Chem. 55: 10278–10283.

Li, B.Y., X.L. Zhang, L.Y. Zhang, T.T. Wang, L. Li, C.G. Wang et al. 2016c. NIR-responsive $NaYF_4:Yb,Er,Gd$ fluorescent upconversion nanorods for the highly sensitive detection of blood fingerprints. Dyes Pigments 134: 178–185.

Li, F., S.Q. Liu, R.Y. Qi, H.R. Li and T.F. Cui. 2017a. Effective visualization of latent fingerprints with red fluorescent $La_2(MoO_4)_3:Eu^{3+}$ microcrystals. J. Alloy. Compd. 727: 919–924.

Li, Y.Q., C.Y. Xu, C. Shu, X.D. Hou and P. Wu. 2017b. Simultaneous extraction of level 2 and level 3 characteristics from latent fingerprints imaged with quantum dots for improved fingerprint analysis. Chinese Chem. Lett. 28: 1961–1964.

Liu, J.J., Z.X. Shi, Y.C. Yu, R.Q. Yang and S.L. Zuo. 2010. Water-soluble multicolored fluorescent CdTe quantum dots: synthesis and application for fingerprint developing. J. Colloid Interf. Sci. 342: 278–282.

Marappa, B., M.S. Rudresha, R.B. Basavaraj, G.P. Darshan, B.D. Prasad, S.C. Sharma et al. 2018. EGCG assisted $Y_2O_3:Eu^{3+}$ nanopowders with 3D micro-architecture assemblies useful for latent finger print recognition and anti-counterfeiting applications. Sensor. Actuat. B-Chem. 264: 426–439.

Matuszewski, S. and M. Szafalowicz. 2013. A simple computer-assisted quantification of contrast in a fingerprint. J. Forensic Sci. 58: 1310–1313.

McRoberts, A. 2011. The Fingerprint Sourcebook. National Institute of Justice: Washington DC.

Menzel, E.R., M. Takatsu, R.H. Murdock, K. Bouldin and K.H. Cheng. 2000a. Photoluminescent CdS/dendrimer nanocomposites for fingerprint detection. J. Forensic Sci. 45: 770–773.

Menzel, E.R., S.M. Savoy, S.J. Ulvick, K.H. Cheng, R.H. Murdock and M.R. Sudduth. 2000b. Photoluminescent semiconductor nanocrystals for fingerprint detection. J. Forensic Sci. 45: 545–551.

Moret, S., A. Becue and C. Champod. 2013. Cadmium-free quantum dots in aqueous solution: Potential for fingermark detection, synthesis and an application to the detection of fingermarks in blood on non-porous surfaces. Forensic Sci. Int. 224: 101–110.

Moret, S., A. Becue and C. Champod. 2014. Nanoparticles for fingermark detection: an insight into the reaction mechanism. Nanotechnology 25: 425502.

Muniswamy, D., H. Nagabhushana, R.B. Basavaraj, G.P. Darshan and B.D. Prasad. 2018. Surfactant-assisted $BaTiO_3:Eu^{3+}@SiO_2$ core-shell superstructures obtained by ultrasonication method: dormant fingerprint visualization and red component of white light-emitting diode applications. ACS Sustainable Chem. Eng. 6: 5214–5226.

Naik, R., S.C. Prashantha and H. Nagabhushana. 2017. Effect of Li^+ codoping on structural and luminescent properties of $Mg_2SiO_4:RE^{3+}$ (RE = Eu, Tb) nanophosphors for displays and eccrine latent fingerprint detection. Opt. Mater. 72: 295–304.

Navami, D., R.B. Basavaraj, S.C. Sharma, B.D. Prasad and H. Nagabhushana. 2018. Rapid identification of latent fingerprints, security ink and WLED applications of $CaZrO_3:Eu^{3+}$ fluorescent labelling agent fabricated via bio-template assisted combustion route. J. Alloy. Compd. 762: 763–779.

Niu, P.H., B.C. Liu, Y.J. Li, Q. Wang, A. Dong, H.T. Hou et al. 2015. $CdTe@SiO_2/Ag$ nanocomposites as antibacterial fluorescent markers for enhanced latent fingerprint detection. Dyes Pigments 119: 1–11.

Park, J.Y. and H.K. Yang. 2017. Novel red-emitting $Y_4Zr_3O_{12}$:Eu^{3+} nanophosphor for latent fingerprint technology. Dyes Pigments 141: 348–355.

Park, J.Y., S.J. Park, M. Kwak and H.K. Yang. 2018a. Rapid visualization of latent fingerprints with Eu-doped $La_2Ti_2O_7$. J. Lumin. 201: 275–283.

Park, S.J., J.Y. Kim, J.H. Yim, N.Y. Kim, C.H. Lee, S.J. Yang et al. 2018b. The effective fingerprint detection application using $Gd_2Ti_2O_7$:Eu^{3+} nanophosphors. J. Alloy. Compd. 741: 246–255.

Peng, T.H., W.W. Qin, K. Wang, J.Y. Shi, C.H. Fan and D. Li. 2015. Nanoplasmonic imaging of latent fingerprints with explosive RDX residues. Anal. Chem. 87: 9403–9407.

Peng, D., X. Liu, M.J. Huang, D. Wang and R.L. Liu. 2018a. A novel monodisperse SiO_2@C-dot for the rapid and facile identification of latent fingermarks using self-quenching resistant solid-state fluorescence. Dalton Trans. 47: 5823–5830.

Peng, D., X. Wu, X. Liu, M.J. Huang, D. Wang and R.L. Liu. 2018b. Color-tunable binuclear (Eu, Tb) nanocomposite powder for the enhanced development of latent fingerprints based on electrostatic interactions. ACS Appl. Mater. Interfaces 10: 32859–32866.

Peng, D., M.J. Huang, Y.R. Xiao, Y.Y. Zhang, L. Lei and J. Zhu. 2019. Highly-selective recognition of latent fingermarks by La-sensitized Ce nanocomposites via electrostatic binding. Chem. Commun. 55: 10579–10582.

Ramotowski, R.S. 2013. Lee and Gaensslen's Advances in Fingerprint Technology, 3rd ed. CRC Press: Florida.

Ran, X., Z.Z. Wang, Z.J. Zhang, F. Pu, J.S. Ren and X.G. Qu. 2016. Nucleic-acid-programmed Ag-nanoclusters as a generic platform for visualization of latent fingerprints and exogenous substances. Chem. Commun. 52: 557–560.

Renuka, L., K.S. Anantharaju, Y.S. Vidya, H.P. Nagaswarupa, S.C. Prashantha, S.C. Sharma et al. 2017. A simple combustion method for the synthesis of multi-functional ZrO_2/CuO nanocomposites: Excellent performance as sunlight photocatalysts and enhanced latent fingerprint detection. Appl. Catal. B-Environ. 210: 97–115.

Ryu, S.J., H.S. Jung and J.K. Lee. 2015. Latent fingerprint detection using semiconductor quantum dots as a fluorescent inorganic nanomaterial for forensic application. B. Korean Chem. Soc. 36: 2561–2564.

Saif, M. 2013. Synthesis of down conversion, high luminescent nano-phosphor materials based on new developed Ln^{3+}:$Y_2Zr_2O_7$/SiO_2 for latent fingerprint application. J. Lumin. 135: 187–195.

Saif, M., M. Shebl, A.I. Nabeel, R. Shokry, H. Hafez, A. Mbarek et al. 2015. Novel non-toxic and red luminescent sensor based on Eu^{3+}:$Y_2Ti_2O_7$/SiO_2 nano-powder for latent fingerprint detection. Sensor. Actuat. B-Chem. 220: 162–170.

Saif, M., N. Alsayed, A. Mbarek, M. El-Kemary and M.S.A. Abdel-Mottaleb. 2016. Preparation and characterization of new photoluminescent nanopowder based on Eu^{3+}:$La_2Ti_2O_7$ and dispersed into silica matrix for latent fingerprint detection. J. Mol. Struct. 1125: 763–771.

Sandhyarani, A., M.K. Kokila, G.P. Darshan, R.B. Basavaraj, B.D. Prasad, S.C. Sharma et al. 2017. Versatile core-shell SiO_2@$SrTiO_3$:Eu^{3+},Li^+ nanopowders as fluorescent label for the visualization of latent fingerprints and anti-counterfeiting applications. Chem. Eng. J. 327: 1135–1150.

Shahi, P.K., P. Singh, A.K. Singh, S.K. Singh, S.B. Rai and R. Prakash. 2017. A strategy to achieve efficient dual-mode luminescence in lanthanide-based magnetic hybrid nanostructure and its demonstration for the detection of latent fingerprints. J. Colloid Interf. Sci. 491: 199–206.

Sharma, V., A. Das, V. Kumar, O.M. Ntwaeaborwa and H.C. Swart. 2014. Potential of $Sr_4Al_{14}O_{25}$:Eu^{2+},Dy^{3+} inorganic oxide-based nanophosphor in latent fingermark detection. J. Mater. Sci. 49: 2225–2234.

Singh, S., Y.M. Sabri, D. Jampaiah, P.R. Selvakannan, A. Nafady, A.E. Kandjani et al. 2017. Easy, one-step synthesis of CdTe quantum dots via microwave irradiation for fingerprinting application. Mater. Res. Bull. 90: 260–265.

Spindler, X., O. Hofstetter, A.M. McDonagh, C. Roux and C. Lennard. 2011. Enhancement of latent fingermarks on non-porous surfaces using anti-L-amino acid antibodies conjugated to gold nanoparticles. Chem. Commun. 47: 5602–5604.

Su, B. 2016. Recent progress on fingerprint visualization and analysis by imaging ridge residue components. Anal. Bioanal. Chem. 408: 2781–2791.

Suresh, C., H. Nagabhushana, R.B. Basavaraj, G.P. Darshan, D. Kavyashree, B.D. Prasad et al. 2018. SiO_2@LaOF:Eu^{3+} core-shell functional nanomaterials for sensitive visualization of latent fingerprints and WLED applications. J. Colloid Interf. Sci. 518: 200–215.

Theaker, B.J., K.E. Hudson and F.J. Rowell. 2008. Doped hydrophobic silica nano- and micro-particles as novel agents for developing latent fingerprints. Forensic Sci. Int. 174: 26–34.

Thomas, G.L. 1978. The physics of fingerprints and their detection. J. Phys. E: Sci. Instrum. 11: 722–731.

Vanderwee, J., G. Porter, A. Renshaw and M. Bell. 2011. The investigation of a relative contrast index model for fingerprint quantification. Forensic Sci. Int. 204: 74–79.

Venkatachalaiah, K.N., H. Nagabhushana, R.B. Basavaraj, G.P. Darshan, B.D. Prasad and S.C. Sharma. 2018. Flux blended synthesis of novel Y_2O_3:Eu^{3+} sensing arrays for highly sensitive dual mode detection of LFPs on versatile surfaces. J. Rare Earth. 36: 954–964.

Wang, Y.F., R.Q. Yang, Y.J. Wang, Z.X. Shi and J.J. Liu. 2009. Application of CdSe nanoparticle suspension for developing latent fingermarks on the sticky side of adhesives. Forensic Sci. Int. 185: 96–99.

Wang, J., T. Wei, X.Y. Li, B.H. Zhang, J.X. Wang, C. Huang et al. 2014a. Near-infrared-light-mediated imaging of latent fingerprints based on molecular recognition. Angew. Chem. Int. Ed. 53: 1616–1620.

Wang, Y.F., R.Q. Yang, Z.X. Shi, J.J. Liu, K. Zhang and Y.J. Wang. 2014b. The effectiveness of CdSe nanoparticle suspension for developing latent fingermarks. J. Saudi Chem. Soc. 18: 13–18.

Wang, M., M. Li, M.Y. Yang, X.M. Zhang, A.Y. Yu, Y. Zhu et al. 2015a. NIR-induced highly sensitive detection of latent fingermarks by $NaYF_4$:Yb,Er upconversion nanoparticles in a dry powder state. Nano Res. 8: 1800–1810.

Wang, M., M. Li, A.Y. Yu, J. Wu and C.B. Mao. 2015b. Rare earth fluorescent nanomaterials for enhanced development of latent fingerprints. ACS Appl. Mater. Interfaces 7: 28110–28115.

Wang, M., Y. Zhu and C.B. Mao. 2015c. Synthesis of NIR-responsive $NaYF_4$:Yb,Er upconversion fluorescent nanoparticles using an optimized solvothermal method and their applications in enhanced development of latent fingerprints on various smooth substrates. Langmuir 31: 7084–7090.

Wang, M. 2016a. Latent fingermarks light up: facile development of latent fingermarks using NIR-responsive upconversion fluorescent nanocrystals. RSC Adv. 6: 36264–36268.

Wang, M. 2016b. Synthesis of $LaPO_4$:Ce,Tb fluorescent nanopowders and their applications in nondestructive development of latent fingerprints. Spectrosc. Spectr. Anal. 36: 1412–1417.

Wang, M., M. Li, A.Y. Yu, M.Y. Yang and C.B. Mao. 2017a. Fluorescent nanomaterials for the development of latent fingerprints in forensic sciences. Adv. Funct. Mater. 27: 1606243.

Wang, J., Q.Q. Ma, H.Y. Liu, Y.Q. Wang, H.J. Shen, X.X. Hu et al. 2017b. Time-gated imaging of latent fingerprints and specific visualization of protein secretions via molecular recognition. Anal. Chem. 89: 12764–12770.

Wang, Y.Q., J. Wang, Q.Q. Ma, Z.H. Li and Q. Yuan. 2018a. Recent progress in background-free latent fingerprint imaging. Nano Res. 11: 5499–5518.

Wang, L.J., W.H. Gu, Z.B. An and Q.Y. Cai. 2018b. Shape-controllable synthesis of silica coated core/shell upconversion nanomaterials and rapid imaging of latent fingerprints. Sensor. Actuat. B-Chem. 266: 19–25.

Wang, Z.L., X. Jiang, W.B. Liu, G.L. Lu and X.Y. Huang. 2019a. A rapid and operator-safe powder approach for latent fingerprint detection using hydrophilic Fe_3O_4@SiO_2-CdTe nanoparticles. Sci. China Chem. 62: 889–896.

Wang, M., J.S. Ju, Z.X. Zhu, D.P. Shen, M. Li, C.J. Yuan et al. 2019b. Recent progress in nanomaterial-based fluorescent development of latent fingerprints. Sci. Sin. Chim. 49: 1425–1441.

Wang, M., D.P. Shen, Z.X. Zhu, J.S. Ju, J. Wu, Y. Zhu et al. 2020. Dual-mode fluorescent development of latent fingerprints using $NaYbF_4$:Tm upconversion nanomaterials. Mater. Today Adv. 8: 100113.

Wang, M., D.P. Shen, Z.X. Zhu, M. Li, C.J. Yuan, Y. Zhu et al. 2021. Quantifying contrast of latent fingerprints developed by fluorescent nanomaterials based on spectral analysis. Talanta. 231: 122138.

Wei, Q.H., M.Q. Zhang, B. Ogorevc and X.J. Zhang. 2016. Recent advances in the chemical imaging of human fingermarks (a review). Analyst 141: 6172–6189.

Wilshire, B. 1996. Advances in fingerprint detection. Endeavour 20: 12–15.

Wu, P., C.Y. Xu, X.D. Hou, J.J. Xu and H.Y. Chen. 2015. Dual-emitting quantum dot nanohybrid for imaging of latent fingerprints: simultaneous identification of individuals and traffic light-type visualization of TNT. Chem. Sci. 6: 4445–4450.

Xie, H.H., Q. Wen, H. Huang, T.Y. Sun, P.H. Li, Y. Li et al. 2015. Synthesis of bright upconversion submicrocrystals for high-contrast imaging of latent-fingerprints with cyanoacrylate fuming. RSC Adv. 5: 79525–79531.

Xu, C.Y., R.H. Zhou, W.W. He, L. Wu, P. Wu and X.D. Hou. 2014. Fast imaging of eccrine latent fingerprints with nontoxic Mn-doped ZnS QDs. Anal. Chem. 86: 3279–3283.

Xu, L.R., C.Z. Zhang, Y.Y. He and B. Su. 2015. Advances in the development and component recognition of latent fingerprints. Sci. China Chem. 58: 1090–1096.

Yadav, H.J.A., B. Eraiah, H. Nagabhushana, G.P. Darshan, B.D. Prasad. S.C. Sharma et al. 2017. Facile ultrasound route to prepare micro/nano superstructures for multifunctional applications. ACS Sustainable Chem. Eng. 5: 2061–2074.

Yadav, H.J.A., B. Eraiah, R.B. Basavaraj, H. Nagabhushana, G.P. Darshan, S.C. Sharma et al. 2018. Rapid synthesis of C-dot@TiO_2 core-shell composite labeling agent: Probing of complex fingerprints recovery in fresh water. J. Alloy. Compd. 742: 1006–1018.

Yogananda, H.S., R.B. Basavaraj, G.P. Darshan, B.D. Prasad, R. Naik, S.C. Sharma et al. 2018. New design of highly sensitive and selective MoO_3:Eu^{3+} micro-rods: probing of latent fingerprints visualization and anti-counterfeiting applications. J. Colloid Interf. Sci. 528: 443–456.

Yu, A.Y., R.Q. Yang and M. Wang. 2018. Synthesis of $La_2(MoO_4)_3$:Eu magnetically fluorescent nanoparticles and its application for latent fingerprint development. Spectrosc. Spectr. Anal. 38: 144–150.

Zhang, L.Y., X.F. Zhou and T. Chu. 2013. Preparation and evaluation of Fe_3O_4-core@Ag-shell nanoeggs for the development of fingerprints. Sci. China Chem. 56: 551–556.

Zhang, M.Q., Y.Y. Ou, X. Du, X.Y. Li, H.W. Huang, Y.Q. Wen et al. 2017. Systematic study of dye loaded small mesoporous silica nanoparticles for detecting latent fingerprints on various substrates. J. Porous Mater. 24: 13–20.

Zhao, Y.B., Y.J. Ma, D. Song, Y. Liu, Y.P. Luo, S. Lin et al. 2017. New luminescent nanoparticles based on carbon dots/SiO_2 for the detection of latent fingermarks. Anal. Methods 9: 4770–4775.

Zhou, D.L., D.Y. Li, X.Y. Zhou, W. Xu, X. Chen, D.L. Liu et al. 2017. Semiconductor plasmon induced up-conversion enhancement in $mCu_{2-x}S@SiO_2@Y_2O_3$:Yb^{3+}/Er^{3+} core-shell nanocomposites. ACS Appl. Mater. Interfaces 9: 35226–35233.

Zhu, B.Y., G.J. Ren, M.Y. Tang, F. Chai, F.Y. Qu, C.G. Wang et al. 2018. Fluorescent silicon nanoparticles for sensing Hg^{2+} and Ag^+ as well visualization of latent fingerprints. Dyes Pigments 149: 686–695.

Zhu, Z.X., M. Wang, M. Li, C.J. Yuan and J. Wu. 2021. Europium nanocomplex for development of latent fingerprints based on carboxyl activation mechanism. Chinese J. Anal. Chem. 49: 237–245.

Chapter 17

Nanoparticles Assisted Personal Respiratory Protection Equipment
A Novel Industrial Application of Nanomaterials

Sammani Ramanayaka,[1] Jayanta Kumar Biswas,[2] Soumyajit Biswas,[3] Kusalvin Dabare[1,4] and Meththika Vithanage[1,*]

1. Introduction

Airborne transmission of respiratory infectious diseases has been known to be one of the most severe public threats in the world for decades. The dramatic spread within a concise time and fast sickening of these infections can lead society towards a pandemic. For centuries, people have faced strikes of airborne respiratory infections caused by bacteria and viruses. Since the 19th century, the world population has confronted a few avian pandemics such as the Spanish flu, Severe Acute Respiratory Syndrome (SARS), and the Middle East Respiratory Syndrome (MERS), leading to the loss of thousands of precious lives (Piret and Boivin 2021). However, the outbreak of severe acute respiratory syndrome coronavirus 2 (SARS-CoV-2) in 2019 has grasped the attention of the world to airborne transmission of diseases and the protection techniques, such as face masks as Personal Protective Equipment (PPE), social distancing and vaccination. Although it is not the cure, face masks were accepted as the best method to control the rapid spreading of diseases. Later, the World Health Organization (WHO), the specialized agency of the United Nations, which is responsible for international public health, recommended wearing a face mask in public places during the COVID-19 pandemic (UN 2022).

The history behind the utilization of face masks runs from the Middle Ages to the Renaissance (Matuschek et al. 2020). It has been reported that face masks were worn by doctors who treated patients suffering from plagues in Europe and the Spanish flu, which was caused by an A/H1N1 virus in 1918. More than protection from airborne diseases, face masks could protect against an

[1] Ecosphere Resilience Research Center, Faculty of Applied Sciences, University of Sri Jayewardenepura, Nugegoda 10250, Sri Lanka.

[2] Department of Ecological Studies & International Centre for Ecological Engineering, University of Kalyani, Kalyani, West Bengal, India.

[3] Department of Biochemistry & Biophysics, University of Kalyani, Kalyani, West Bengal, India.

[4] College of Chemical Sciences, Institute of Chemistry, Adamantane House, Rajagriya, Sri Lanka.

Emails: sammani@sci.sjp.ac.lk; jkbiswas@klyuniv.ac.in; sbbiswas4@gmail.com; kusalvind@gmail.com

* Corresponding author: meththika@sjp.ac.lk

array of nosocomial threats in hospital environments (Chandrashekar et al. 1997). In the COVID-19 pandemic, face masks reduced the risk factor in both aspects of nosocomial and community spread. Nevertheless, the face mask material that provides the physical barrier to respiratory and mucosalivary droplets strongly influences the protection provided for infected and disinfected individuals (Chua et al. 2020). Furthermore, face masks are significantly different from respirators. Cloth, surgical and other enhanced protection cloth masks are face masks whereas N95 and KN95 are respirators (Rubbo and Abbott 1968). Home-made cloth face masks and enhanced protection cloth face masks are reusable whereas the surgical masks, N95 and KN95 respirators are disposable. Generally, face masks do not tightly fit the wearer, while respirators do. Therefore, complete protection without any leakage can be expected from respirators over face masks. As a whole, most face masks provide some sort of protection against dust, pollen, chemical fumes and pathogens such as bacteria in a size range of 100 μm–750 nm (Chua et al. 2020). However, viruses are usually in the 20–400 nm range, where SARS-CoV-2 has been identified as 60–140 nm in size (Zhu et al. 2020). Nevertheless, masks and respirators with larger membrane pores, for instance, cotton and synthetic fabrics, will not be effective against viruses or bioaerosols as membranes consisting of ultra-small pore sizes. Further, materials modified with water-resistant properties will offer additional protection against pathogen-laden respiratory droplets from coughing and sneezing. Moreover, modified designs and enhanced membranes of masks will reduce discomfort and easy breathability over conventional face masks. The global shortage of face masks, respirators and raw materials during the Covid-19 pandemic, owing to the high demand, grasped the research attention in developing alternatives and introducing modified, highly effective face masks.

To develop a multifunctional high impact face mask, nanomaterials (NMs) have been identified as the best in terms of their unique properties such as self-sanitizing ability, filtration efficiency via ultra-small pore sizes and comfortability. Nanoparticle (NP)-assisted face masks provide the maximum protection from bioaerosols due to the extreme hydrophobicity of nanomembranes and their excellent filtering competence (Abbasinia et al. 2018). Moreover, the next generation of face masks will be personalized 3D printed structures and will encourage the reduction in waste generation. Nevertheless, the involvement of nanoscience and nanotechnology in face mask production will revolutionize the PPE industry. Hence, this chapter focuses on (i) the impact of nanomaterials on bioaerosols, (ii) the filtration capabilities of nanomembranes, and (iii) the reusability and comfort of face masks, (iv) real scale production and facing challenges, and (v) future perspectives.

2. Interactions of Nanomaterials with Bioaerosols

2.1 Nanomaterial-microbe crosstalk: physicochemical profile as a determinant

Nanomaterials and microbes exhibit an amazing diversity in nature, functionalities and complexities (Ingle et al. 2021, Rai and Biswas 2018, Westmeier et al. 2018). The intensity and nature of their interactions depend on functional attributes of nanomaterials, physicochemical variations in microbial surface structures and 'fingerprint' features. These qualities define their toxicity and pathobiological manifestations (Docter et al. 2015, Feliu et al. 2016). Microbes exhibit a vast diversity in biochemical and molecular compositions. While incredible bacterial variations emerge from the molecular composition of the cell wall, fungal variations arise from the hydrophobic cell surface of its spores (Brown et al. 2015). In addition, microbial shapes also influence NM-microbe interactions. However, the sizes of diverse bioparticles (e.g., bacterial and fungal spores) are comparable considering the overall negative surface charge carried by them. The positively charged NMs are frequently more efficient in binding than the negatively charged NMs (Courtney et al. 2016, Gupta et al. 2016). Above all, the size of NMs is the most important factor for nanomaterial-microbial interaction, dominating over all other factors including material, shape or charge.

The majority of studies have focused on the antibacterial activity of NMs rather than a comprehensive insight addressing the diverse aspects of nanomaterial-microbe crosstalk. Bacterial pathogens respond differently to NPs since Gram-positive and Gram-negative bacteria have distinct surface chemistry displaying different interactions. Gram-negative bacteria have two membranes constituted of integrated porins and membrane proteins (Stauber et al. 2018). The glycocalyx present on the surface of the outer membrane of Gram-negative bacteria consists of liposaccharides that are primary targets for specific interactions with carbohydrate-recognizing structures, such as cationic amphiphiles (Lan et al. 2012, Wu et al. 2011). The net negative bacterial surface charge is attributed mainly to dense glycocalyx with many deprotonated carbohydrates that bind cationic supramolecular decoys. Besides, bacterial adhesins present in the pili of most Gram-negative bacteria interact with carbohydrates and facilitate bacterial adhesion to (bio)material surfaces and/or cell membranes. On the other hand, the Gram-positive bacteria lack a second outer membrane or specific receptors. Since microbial cell wall thickness is an essential determinant for the interaction of the NMs Gram-negative bacteria are more susceptible to cell membrane damage compared to Gram-positive bacteria which is attributed to the difference in the peptidoglycan layer (Mollasalehi and Yazdanparast 2013). A thick (30–100 nm) peptidoglycan layer of Gram-positive bacteria compared to Gram-negative bacteria (a few nanometers) prohibits the penetration of NMs into the bacterial cells (Bouwmeester et al. 2009).

2.2 Influence of (patho)physiological and natural environment on NMs-Microbe crosstalk

On entering the complex physiological or ecological system, different biomolecules can quickly be adsorbed onto all known NMs. Such "coronated" NMs emerge as entities distinct from the original or manufactured ones concerning their nature and destiny in those environments (Docter et al. 2015, Feliu et al. 2016). Physiological milieus such as plasma, saliva, intestinal fluids or lung surfactants (lipoprotein complexes) are complex yet predictable in their composition, comprising a characteristic set of polysaccharides, proteins and lipids. While natural environments with their vast array of abiotic and biotic factors show an incredible diversity and complexity, and present heterogeneous, mixed coronas (eco-coronas) due to interactions among NMs bioaerosols and the environment factors (Docter et al. 2015). Westmeier et al. (2018) studied the mechanism of NM-microbe complex formation by assembling diverse biomolecule coronas, including plasma proteins, lung surfactant lipids and environmental humic acids (Westmeier et al. 2018). A high concentration of biomolecules was reported to prevent NM binding to bacteria. The biomolecule coronas assembly coming in the first contact with NMs and microbes has a significant role in the nature of NM-microbe interactions. Biomolecules produced by bacteria, such as flagellin, form coronas on NMs with antimicrobial potential. Such physical shielding imparts bacterial resistance to NMs by reducing NM-bacteria complex formation (Westmeier et al. 2018). Under a pathophysiological environment, different proteins or lipids significantly alter the bacterial resistance and antibiotic action of NMs (Panáček et al. 2018, Westmeier et al. 2018).

2.3 Outcome of nanomaterial-microbe interaction: antimicrobial action and oxidative stress

The chemistry, size and shape of the nanoparticles influence cellular uptake, internalization/subcellular localization and capability to catalyze oxidative reactions (Biswas et al. 2018, Xia et al. 2006). NPs enter microbial cells through endocytosis pathways. Positively charged NPs can enter faster and remain longer in the cell than negatively or neutrally charged NPs (Oh and Park 2014). Such uptake is started by van der Waals forces, electrostatic and steric interactions or interfacial tension effects (Peters et al. 2006). Then, NPs get localized and are found in various locations within the cell (Garcia-Garcia et al. 2005, Xia et al. 2006). Depending on their cellular localization, NPs cause damage to that particular cellular and biochemical entity and, ultimately results in cell death. The best-developed paradigm of nanotoxicity inflicted by NPs is the generation of Reactive Oxygen Species (ROS) and oxidative stress (Fig. 17.1). Like energy-deficient robbers, ROS attack and snatch energy from cell or biochemical molecules to satisfy themselves. These cellular costs

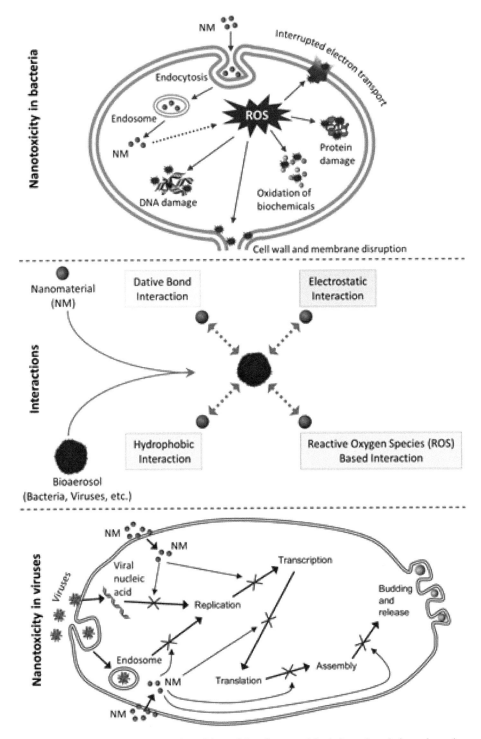

Figure 17.1. Simplified schematic representation of the toxicity of nanoparticles in bacteria and viruses borne by aerosols called bioaerosol.

lead to a vicious cycle of ROS production, DNA damage and signal activation, resulting in cell death pathways (Ingle et al. 2020, Wu et al. 2011). The toxicity of nanoparticles on microbes is discussed below based on the main categories of NMs such as metallic NPs, metal oxide NPs, carbon-based nanomaterials, quantum dots and graphene.

Metal nanoparticles can generate ROS, acting as catalysts in Fenton-type reactions (Risom et al. 2005). Interaction of metal NPs with sulfur-containing proteins present in the cell wall of the pathogen may cause cellular disruption due to electrostatic forces, bringing about changes in the integrity of the lipid bilayer, membrane permeability and alteration in transport processes leading to cell death (Aziz et al. 2019, Ingle et al. 2021). Nanoparticles may enter mitochondria and alter their function, leading to ROS production (Xia et al. 2006). They have been confirmed as effective biocidal agents against a broad spectrum of bacteria (Table 17.1) such as *Escherichia coli, Staphylococcus aureus, Staphylococcus epidermis, Leuconostoc mesenteroides, Bacillus subtilis, Vibrio cholerae, Pseudomonas aeruginosa, Salmonella typhae, Klebsiella mobilis* and *Klebsiella pneumoniae* (Marambio-Jones and Hoek 2010). For example, entering the cell, AgNPs interact with cellular proteins, DNA and other biomolecules, leading to the destruction of the cell (Ingle et al. 2020, Ray et al. 2009). Disruption of signal transduction by dephosphorylation of peptides/proteins can lead to loss of cell viability (Ingle et al. 2021, Shahverdi et al. 2007). AgNP shows virucidal activity but the actual mechanism is yet to be unveiled clearly (Chen and Liang 2020).

Nanoparticles interfere in (m)any of the four stages of the viral life cycle, i.e., binding to a host cell, internalization, replication inside the cell and release (budding) (Fig. 17.1). For example, silver and copper NPs release Ag^+ and Cu^{2+} cations that can damage the viral genome and disintegrate the viral membrane (Singh et al. 2017). NPs show efficacy against coronaviruses (Du et al. 2018). Zinc, silver and copper cations showed antiviral activity against different coronaviruses like the Porcine Epidemic Diarrhea Virus (PEDV), which have characteristics similar to SARS-CoV-2 (Bright et al. 2009). Intracellular zinc ions inhibit replication in various RNA viruses, including influenza virus, respiratory syncytial virus, picornaviruses and coronaviruses such as SARS-CoV, interfering with the RNA polymerase activity.

Metal oxide nanoparticles such as zinc oxide and titanium dioxide nanoparticles have been reported to have contradictory effects on different types of cells (Table 17.1). They were shown to be more toxic to Gram-positive *S. aureus* than Gram-negative *E. coli* or *P. aeruginosa* by Premanathan et al. (2011), while Sinha et al. (2011) observed a more cytotoxic effect on Gram-negative cells (Premanathan et al. 2011, Sinha et al. 2011). Flow cytometry and electron microscopic study with *E. coli* and *S. typhimurium* revealed aggregation and adhering of ZnO and TiO_2 NPs to the cell surface and internalization into the cell cytoplasm (Kumar et al. 2011). ZnO and TiO_2 NPs induce oxidative stress and DNA damage, leading to reduced viability of the bacterial cell (Fig. 17.1). TiO_2, ZnO and SnO_2 exhibit strong photocatalytic properties with biocidal action on pathogenic bacteria, fungi and viruses (Wang et al. 2020, Wiehe et al. 2019, Zhang et al. 2019).

Conversely, TiO_2 NPs are reported to inactivate the H3N2 influenza virus by direct contact and damage the viral membrane in the absence of light, which points to some biocidal mechanism other than photodynamic inhibition (Mazurkova et al. 2010). ZnO NPs inhibit the proliferation of the H1N1 influenza virus by the dissolution of Zn^{2+} ions and SARS-CoV-2 by Zn^{2+}-induced inhibition of RNA polymerase (Ghaffari et al. 2019, Skalny et al. 2020). Hasan et al. (2020) reported that the nanostructured aluminum surfaces emulated from insect wing architecture could be effective as antibacterial and antiviral means. They fabricated nanostructures by random alignment as ridges on surfaces of aluminum alloy by a wet-etching process and reported that Respiratory Syncytial Virus (RSV) and rhinovirus (RV) showed persistency while nonenveloped virus showed susceptibility. They also induced the natural degradation of the enveloped virus (RSV) by disrupting the envelope. Besides, they reduced the transmission of both viruses via surface contact.

The carbon-based nanomaterials are extensively used as antibacterial agents (Table 17.1). The antibacterial activity of the carbon-based NMs involves a combination of both a physical and a chemical process (Hajipour et al. 2012). Carbon-based NMs can cause morphological alterations

Table 17.1. Toxic effects of chief categories of nanomaterials on different microorganisms.

Type of nanomaterials	Microorganisms	Nanotoxic effects	References
Metals Silver (Ag)	*E. coli*	• Bactericidal effect • Cell wall pitting • Accumulation of NPs inside the cell membrane	(Sondi and Salopek-Sondi 2004)
	V. cholerae *S. typhus* *P. aeruginosa*	• Inhibition of bacterial growth • Size-dependent bactericidal effect	(Morones et al. 2005)
	Pseudomonas putida KT2440	• Bactericidal effect • Cell death	(Gajjar et al. 2009)
	Streptococcus mutans	• Size-dependent antibacterial activity • Bactericidal effect	(Espinosa-Cristóbal et al. 2009)
	Nitrosomonas europaea	• Extreme sensitivity to Ag NPs and Ag ions • Smaller sized (20 nm) NP more toxic than larger sized (80 nm) NPs	(Radniecki et al. 2011)
	Enterobacter sp. *Marinobacter* sp. *Bacillus subtilis halophilic bacterium* sp. *EMB4*	• Nanotoxicity more pronounced on Gram –ve than Gram +ve cells • Morphological changes and NP accumulation in the cytoplasm	(Sinha et al. 2011)
	B. subtilis *Agrobacterium tumefaciens*	• High toxicity to all bacterial strains • Damage to the cellular structure at a concentration of 1 g/mL • *Agrobacterium tumefaciens* most susceptible to nanotoxicity	(Wang et al. 2012)
Colloidal Ag	*Pseudomonas aeruginosa* *Micrococcus luteus* *Bacillus subtilis* *Bacillus barbaricus* *Klebsiella pneumoniae*	• Adsorption of NPs on bacterial cells via electrostatic attraction • Size-dependent bactericidal effect; increased with increasing NP adsorption on the cell surface	(Khan et al. 2011)
Gold (Au) Platinum (Pt)	*Gram –ve Salmonella Enteritidis* *Listeria monocytogenes* *Salmonella typhimurium strain TA102*	• Nano Au could not internalize bacterial cells while nano Pt could enter into the cells • NPs aggregated within the flagella	(Sawosz et al. 2010)
Metal oxides ZnO	*Enterobacter* sp. *Marinobacter* sp. *Bacillus subtilis* *Halophilic bacterium* sp. *EMB4* *Staphylococcus aureus* *E. coli* *Nitrosomonas europaea* *P. aeruginosa PA01* *E. coli (K12 substrain DH10B)* *E. coli K88* *Salmonella typhimurium*	• Varied nanotoxicity between Gram –ve and Gram +ve cells, morphological changes, and NP accumulation in the cytoplasm • Concentration-dependent toxicity • Loss of cell viability • Inhibition and suspension of growth • Generation of ROS, induction of oxidative stress, DNA damage leading to cell death (apoptosis)	(Kumar et al. 2011, Premanathan et al. 2011, Sinha et al. 2011)
MgO	*E. coli strain C3000* *B. megaterium strain* *B. subtilis spores*	• Biocidal activity	(Stoimenov et al. 2002)
CeO$_2$	*E. coli wild strain RR1* *B. subtilis* *Shewanella oneidensis*	• Strain and size-dependent inhibition of bacterial growth • Bactericidal to *E. coli* and *B. subtilis*	(Thill et al. 2006)

Table 17.1 contd. ...

...Table 17.1 contd.

Type of nanomaterials	Microorganisms	Nanotoxic effects	References
Al_2O_3	*E. coli strain (NCIM no. 2666)* *Cupriavidus metallidurans CH34* *E. coli MG1655* *B. licheniformis* *Ruminococcus flavefaciens 007C*	• Concentration-dependent toxicity • Bacteriostatic action; delay in initiation of bacterial growth • Strain-specific toxicity • Nanotoxicity influenced by the chemical composition, size, shape, charge of NP • Cell wall rupture and NP internalization	(Pakrashi et al. 2011, Pelletier Dale et al. 2010, Sadiq et al. 2009)
TiO_2	*E. coli* *Ruminococcus flavefaciens 007C* *B. subtilis (ATCC 9372)* *E. coli (ATCC700926)*	• Toxicity dependent on electrostatic interactions between cell and charged NP • pH, ionic strength, and electrolyte composition involved in regulating above interactions • Concentration-dependent antimicrobial activity -changes in cell morphology, cellular shrinkage and fragmentation • Induce oxidative stress and DNA damage leading to reduced cell viability	(Kumar et al. 2011, Park et al. 2012)
Fe_3O_4	*E. coli*	• Concentration-dependent growth inhibition	(Chatterjee et al. 2011)
Hematite		• Morphological changes after NP exposure • Treated cells stiffer and harder compared to control	(Zhang et al. 2012)
Carbon-based Nanomaterials and Quantum Dots Single wall nanotubes (SWNTs) Multi wall nanotubes (MWNTs) Aqueous phase C60 NP Colloidal graphite	*E. coli* *P. aeruginosa* *B. subtilis* *Staphylococcus epidermis*	• SWCNT showed the highest toxicity • Colloidal graphite and MWCNT showed moderate antibacterial activity • *B. subtilis* showed enhanced tolerance to carbon-based nanomaterials toxicity up to 1 hour	(Kang et al. 2009)
MWCNT	*Cupriavidus metallidurans CH34* *E. coli MG1655*	• Bactericidal effect on *E. coli* • *C. metallidurans* highly resistant • Bacterial adsorption on CNT observed • No internalization recorded	(Simon-Deckers et al. 2009)
CdTe quantum dots	*E. coli* *P. aeruginosa (Gram –ve)* *Staphylococcus aureus* *B. subtilis (Gram +ve)* *Halobacterium halobium R1* *P. aeruginosa strain PA01*	• ROS mediated toxicity • *S. aureus* most resistant to toxicity compared to *E. coli* and *P. aeruginosa* • Release of Cd^{2+} upon bacterial degradation of QDs leads to cytotoxicity • Changes in cell morphology and metabolism • Internalization of QDs	(Dumas et al. 2009)
MWCNTs (C60 fullerene, Fullerene soot)	*P. fluorescens*	• MWCNTs and fullerene soot altered lipid, protein and DNA profile of bacteria • C60 was most toxic with extensive alterations • Fullerene soot is the least toxic	(Riding et al. 2012)
Fullerene C60	*B. subtilis* *E. coli* *Agrobacterium tumefaciens*	• Growth inhibition at 10 and 20 g/mL *Agrobacterium tumefaciens* most susceptible to nanotoxicity • Initial delay in growth of *B. subtilis* and *E. coli*	(Wang et al. 2012)

or structural disintegration of the microbial cell membrane, resulting in cell death. Chemically, carbon-based NMs induce oxidative stress by ROS generation leading to loss of cell viability. NPs such as C60 fullerenes, SWNTs, QDs and ultrafine particles produce ROS, primarily on exposure to light (particularly UV light) or transition metals (Sayes et al. 2004). Maternal exposure to titanium oxide (TiO_2) NPs alters the gene expression related to apoptosis and oxidative stress (Shimizu et al. 2009).

Few studies have shown bactericidal activities of carbon nanotubes (CNTs) comprising single-walled CNTs (SWNTs) and multi-walled CNTs (MWCNTs). SWNTs exhibited toxicity to several bacterial such as *Escherichia coli, Pseudomonas aeruginosa, Bacillus subtilis* and *Staphylococcus epidermidis* (Kang et al. 2009). SWNTs cause a loss in cell viability in the case of *E. coli* strain K12 resulting from disruption of cell membrane integrity (Kang et al. 2007). MWCNTs induce oxidative stress by generating ROS and cause alteration in biochemical molecules such as lipids, carbohydrates and DNA, leading to cell membrane disruption and cell death (Riding et al. 2012). C60 fullerene has more toxicity resulting in biochemical aberrations in altered lipid, protein and DNA profile but is mainly restricted to the outer cell components, in contrast to MWCNTs, upsetting the internal cellular machinery (Sinha and Khare 2013).

Carbon Dots (CD) endowed with an extraordinarily high surface-to-volume ratio and high ability to form stable and homogeneous water dispersions have great antiviral potential. Their antiviral action against the pseudorabies virus (PRV) and Porcine Reproductive and Respiratory Syndrome Virus (PRRSV), the DNA and RNA viruses, respectively, involve mechanisms interfering with viral replication by activating the interferon response (Dong et al. 2020). Uniform and stable cationic CD fabricated from curcumin display antiviral properties against coronaviruses and the Porcine Epidemic Diarrhea Virus (PEDV). Du et al. (2018) reported that this type of CD significantly inhibits PEDV viral entry, the synthesis of negative-strand RNA and the budding of these viruses, which offers clues to fighting other coronaviruses such as SARS-CoV-2 (Du et al. 2018).

Quantum dots, the unique class of fluorescing nanoparticles cause cytotoxic effects. For example, the cytotoxicity of cadmium telluride quantum dots on extremely halophilic bacteria *Halobacterium halobium* R1, *P. aeruginosa* and *B. subtilis* emerges from the Cd^{2+} ions released out of the degradation of the CdTe QDs (Dumas et al. 2009, Li et al. 2012). Its use in humans leads to health concerns.

Graphene possesses good antiviral capacity. Being a single layer of carbon atoms, graphene has a remarkably high surface-to-mass ratio. The large surface of graphene provides a very high ligand contact area for the adsorption of negatively charged sulfates. These can interact with, positively charged residues of viruses and inhibit microorganisms (Ziem et al. 2017). Ziem et al. (2017) synthesized Thermally Reduced Graphene Oxide (TRGO) derivative functionalized with dendritic polyglycerol (dPG) and modified the material by sulfation, finding an amplified antiviral activity against African swine fever virus, orthopoxvirus and herpes virus strains, where dPG binds to the heparin-binding domain of surface protein A27 (Ziem et al. 2017). Graphene derivatives have also been exploited as drug delivery systems like reverse transcriptase inhibitors conjugated with graphene quantum dots to treat HIV and hypericin-GO (graphene oxide) against reovirus (Du et al. 2018). Hypericin has been a computationally identified antiviral compound with promising potential against COVID-19 (Smith and Smith 2020).

3. Facemasks Containing Nanoparticles

The COVID-19 outbreak has emphasized the intense necessity of new generation face masks that provide maximum personal protection and low environmental impact, for instance, their disposal. Hence, nanotechnology-based face masks have been recognized as the superlative solution for the timely challenge. Different methodologies such as incorporating nanoparticles in a mask membrane via spray or dip coating, fabricating membranes with nanofibers/nanowires and adsorption of nanoparticles to membrane have been followed to fabricate functionalized face masks (Table 17.2).

Table 17.2. Enhanced performance of nanoparticles assisted face masks.

Fabrication method	Material modification	Nanomaterial type	Active effect	Remarks	References
Spray coating nanoparticle emulsion	Outermost layer	TiO$_2$, Ag(I), and Al$_2$O$_3$ nanoparticles	Antimicrobial effect of Ag(I) and TiO$_2$	100% inhibition of *E. coli* and *S. aureus* in 24 h	(Li et al. 2006)
Dual-channel spray-assisted nanocoating	Nonwoven surgical mask	Copper nanoparticles (CuNPs)	Combined photocatalytic and photothermal properties; Antimicrobial properties of Cu nanoparticles	Increased hydrophobicity, reusability and self-sterilizing ability	(Kumar et al. 2021)
Dip coating by starch capped nanoparticles	Whole face mask	Ag(I)	Antimicrobial effect of Ag(I)	100% inhibition of *E. coli* and *S. aureus* in 24 hours	(Hiragond et al. 2018)
Composite of nanofiber and nanoparticles	Nanoparticles electrospun nanofiber membrane	Polyacrylonitrile nanofiber and Ag(I) nanoparticles	Antimicrobial effect of Ag(I) and ultra-small pores	Washable mask with better antibacterial properties in one cycle loading of AgNPs (2247 ppm/gr nanofiber)	(Kharaghani et al. 2018)
		Polyacrylonitrile nanofiber and CuO nanoparticles	Antimicrobial effect of CuO and ultra-small pores	Excellent microbial activity	(Hashmi et al. 2019)
	Nanocomposite fibrous (polypropylene) membrane	h-BN (Hexagonal boron nitride) nanoparticles	Antibacterial activity	Destroy 99.3% for *E. coli* and 96.1% for *S. aureus*	(Xiong et al. 2021)
Nanowires based filter	Nanowire membrane	TiO$_2$ nanowires	Photocatalytic properties	Reusable filters	(Horváth et al. 2020)
Adsorbed or radiation-induced crosslinked nanoparticles	Graphene oxide-based cotton fabric	Graphene oxide nanoparticles	Antimicrobial effect of graphene oxide	Inactivate > 98% of bacteria in less than 4 hours Washable (inactivation efficiency of bacteria > 90% after 100 washings No skin irritation	(Zhao et al. 2013)

Nanocoating is one of the best approaches in new generation face masks. The existing PPEs, specifically face masks, can be modified with nanocoatings to upgrade the protection against bioaerosols at a comparatively low cost (Kumar et al. 2021, Li et al. 2006). Further, nanocoating provides more than antibacterial properties, such as superb hydrophobicity and self-sanitizing ability (Kumar et al. 2021). Nanoparticles with photocatalytic properties can generate ROS and inactivate viruses via oxidization (Kumar et al. 2021). Additionally, the antibacterial effect has been facilitated by temperature generated from photothermal processes under solar light (Kumar et al. 2021). A team of scientists from the Queensland University of Technology (QUT, Australia) has come forward with a biodegradable, environmentally friendly face mask with a highly breathable nanocellulose membrane. Cellulose fibers from the waste plant material and sugar cane bagasse have been used in the production of nanomembranes (Widdowson 2020). Kharaghani et al. (2018) introduced a nanocomposite consisting of nanofibers with embedded Ag(I) nanoparticles. Polyacrylonitrile (PAN) nanofibers were synthesized using the electrospinning technology and antimicrobial properties were incorporated with Ag(I) nanoparticles (Kharaghani et al. 2018). Similarly, copper oxide (CuO) nanoparticles also provide an excellent antibacterial compound, which can be used in nanoparticle-assisted face mask manufacturing (Hashmi et al. 2019). Moreover, a nonwoven network of TiO_2 nanowires with photocatalytic properties has been studied for the feasibility of use as a nanomembrane. However, a considerably low yield in synthesis confirms the limitations in mass-scale production of TiO_2 nanowires (Horváth et al. 2020). Furthermore, scientists have explored a variety of nanomaterials without limiting them to metal nanoparticles. As a result, graphene oxide nanoparticles adsorbed on cotton fiber filters have been introduced to face mask manufacturing processes (Zhao et al. 2013). Nevertheless, scientists are in continuous research to develop potential nanomembranes that are highly effective, eco-friendly and of low cost. Interestingly, leading suppliers of nanofiber production equipment such as Elmarco have already implemented their Nanospider Antimicroweb technology into the facemasks manufacturing industry (Barhate et al. 2008).

3.1 Membrane filtration efficiencies

Excellent filtration efficiency is the most essential characteristic of a successful membrane. Three major parameters that influence the performance of the filter are pressure drop through the mask, filtration efficiency and lifetime. Nanofibrous membranes among respiratory and breathing air filter manufacturers are more popular due to their high particle capturing efficiency. Additionally, nanofiber membranes demonstrate an increased lifetime with decreasing pressure drop through the mask (Akduman and Kumbasar 2018, Barhate et al. 2008, Wang et al. 2015). Sakariya and Smaldone (2014) reported a filtration efficiency comparison between a nanofiber prototype surgical face mask and the N95 respirator where the nanofiber prototype mask revealed 98.98% of filtration efficiency while N95 showed only 82.68% (Skaria and Smaldone 2014). A research team from the Korean Advanced Institute of Science and Technology (KAIST) has developed an orthogonal nanofibrous structure that enhances the filtering efficiency by minimizing the pressure delivery towards the air filter. Furthermore, it has proven a 94% filtering efficiency and no deformation after 20 hand washes (Office 2020). Interestingly, nylon 6-polyacrylonitrile nanofiber-nets binary (N6–PAN NNB) structured membrane has shown the best filtration efficiency (99.99%) reported so far among nanoparticle assisted face masks (Wang et al. 2015).

Nanomembranes that do not contain any material to inactivate airborne pathogens always achieve high filtration efficiencies (O'Kelly et al. 2021). However, membranes with antibacterial, photocatalytic or photothermal properties also demonstrate a considerably high filtration efficiency and inactivation of airborne pathogens. In nanofibrous filters, four main filtration mechanisms can be followed: surface straining, depth straining, depth filtration and cake filtration, whereas in practice two or more mechanisms are involved at the same time (Barhate et al. 2008). However, the depth filtration mechanism is not applicable in air filtration processes. The surface straining mechanism is useful for long-lasting dust cleaning filters while the depth straining mechanism is governed by particle size or shape (Purchas and Sutherland 2002). Although particles are tiny in size, they will be trapped in the depth of the media. The depth straining mechanism is followed by processes such as microfiltration and nanofiltration (Barhate et al. 2008). In the cake filtration mechanism, particles are deposited as layers in a nanofiber structural matrix, forming a cake. The influence of two or more of these mechanisms in filtering air-borne particles/pathogens enhances filtration efficiency (Barhate et al. 2008). Furthermore, the high surface area to volume ratio of nanoparticles/fibers, lightweight and high intra-fiber porosity of the nanomembranes enhance the efficiency of the facemask as a whole (Akduman and Kumbasar 2018, Barhate et al. 2008, Chuanfang 2012). Different manufacturers of face masks and the filtration efficiencies of their products are listed in Table 17.3.

Table 17.3. Filtering efficiencies of different face mask membranes.

Year	Country	Type of filter	Filtration efficiency	Reference
2014	USA	Nanofiber membrane	> 98.98%	(Skaria and Smaldone 2014)
2015	China	Polysulfone based nanofiber membrane	> 90%	(Li and Gong 2015)
2015	China	Nanofiber nets of binary structured nylon 6–polyacrylonitrile membrane	< 99.99%	(Wang et al. 2015)
2020	Israel	Metal oxide nanoparticle infused cotton polyester	> 98%	(Leichman 2020)
2020	Australia	Cellulose nanofiber membrane	-	(Widdowson 2020)
2020	South Korea	Orthogonal nanofiber membrane	> 94%	(Office 2020)

3.2 Face fit

Face masks, one of the most important types of PPEs, must be fitted perfectly to the user's face. In surgical face masks, there is an adjustable plastic stripe on the nose clip. Moreover, the elastic straps can be altered with a knot according to the user's face size and adjusting the three folding and edge to cover the mouth. Although cloth masks do not have any adjustable plastic stripes or pleats, these masks provide good protection with full coverage and can be purchased in standard sizes such as small, medium and large. However, face masks are loose-fitting PPEs and do not provide maximum protection. Respirators are the best for ultimate protection from airborne pathogens and particulate matter due to their tight-fitting nature on the face (O'Kelly et al. 2021).

O'Kelly et al. (2021) have assessed the face fit of a few different respirators, surgical and cloth masks. It has been revealed that the proper face fit is difficult to identify visually or manually (O'Kelly et al. 2021). An improper fit reduces overall mask efficiency even though it is made out of nanoparticle or nanofibrous membrane. However, researches confirm that poorly fit N95 respirators and surgical and cloth masks have offered a reduced protection to the wearer (O'Kelly et al. 2021). Nevertheless, studies carried out to analyze the proper face fit of face masks and respirators are lacking. Therefore, it is essential to conduct more research to assess the face fit capabilities of different face mask types.

3.3 Reusability

Millions of face masks have been used by healthcare professionals and industry workers around the world on a daily basis. However, there is a high demand for PPEs such as face masks, especially in a pandemic situation that will rapidly increase the number of disposed masks. Polymer-based materials used in the production of face masks will unintentionally lead to a serious environmental issue related to the degradation of disposed face masks. Furthermore, infected face masks disposed to the environment may potentially increase the spreading of air-borne diseases. Therefore, the possibility of reusing face masks is crucial. Additionally, the extended use and reuse of single-use masks and respirators could cause physical damage to the material. During the washing cycles, squashing, bending and stretching may damage the components of the mask, including the membrane. This will cause potential consequences such as loss of fit, discomfort and irritation to the wearer (Toomey et al. 2021).

An interesting study has been carried out in Hong Kong to assess the reuse of face masks among the adult population. The results have confirmed that around one-third of adults in Hong Kong reuse face masks with the uncertainty of insufficient face masks in the market. The informal storing practices and reuse of used face masks could not be safe, and therefore, continuous research to study the reusability of face masks is needed a great deal (Lee et al. 2021).

Single-use face masks such as surgical face masks are very popular owing to their low cost and easy usage. However, reusable face masks endowed with the quality of enhanced protection can reduce the generation of waste and provide better protection. For instance, nanoparticle-assisted face masks have been introduced to the market, which will provide more than 90% filtering efficiency even after a few washing cycles (Kharaghani et al. 2018, Office 2020, Zhao et al. 2013). Orthogonal nanofiber membrane mask developed by the scientists from KAIST has demonstrated the optimal filtration capacity even after 20 washing cycles (Office 2020). Further, it filters nanoparticles (~ 600 nm) with an 80% efficiency even after undergoing bending for 4000 times (Office 2020). Nanofibrous filter synthesized from polyvinylidene difluoride (PVDF) has depicted constant high filtration efficiency after a simple cleaning with 75% ethanol (Ullah et al. 2020).

Moreover, introducing biodegradable face masks with high filtering capacity is also a successful solution. Cellulose nanofiber membrane developed by a research group from QUT, Australia, has proven to be easy to manufacture, of low cost, highly efficient and disposable after one single use due to its biodegradability (Widdowson 2020).

4. Production on Real Scale and Challenges

The high demand for face masks during the COVID-19 pandemic has depicted a market growth of USD 0.9 billion in 2019 to USD 21.5 billion in 2021 (Markets 2020). Studies have predicted that it will be then reduced to USD 2.7 billion by 2026. Between 2019 and 2021, a Compound Annual Growth Rate (CAGR) of 381.8% has been estimated whereas from 2021 to 2026 it will be an overall CAGR of –33.7% (Markets 2020). The shortage of face masks, especially in the markets of import-dependent countries has been created due to market restraints announced by some governments of face masks or material manufacturing countries. The purpose of these restraints was to fulfill the local demand before supplying the export market. According to the WHO, there should be at least a 40% increase in face mask production to accomplish the global demand (Markets 2020). However, the mandatory laws of many different countries to wear face masks in public and a positive attitude toward wearing face masks boosted by media have directly influenced the demand.

Furthermore, the volatility of raw materials expenditure is one of the major restraints. Polypropylene (PP) is a polymer derived from crude oil purification, which is commonly used in the production of nonwoven face masks (Markets 2020). Nevertheless, due to high seed capital, limited number of companies in the world produce a PP electret melt-blown nonwoven special fabric. This includes expenditures for crude oil and machinery, for instance, hoppers, extruders and melt spinning systems (Markets 2020). Volatility in material prices and machinery costs are major challenges for new manufacturers who try to enter the face mask production market. Additionally, the fluctuating prices of the other raw materials such as metal for nose metal strips also play a major role in face mask market dynamics (Markets 2020). In the end, all these costs will influence the product price in the market.

Although reusable face masks such as nanoparticle-assisted face masks are capable of reducing the number of disposed masks, the initial expenditures on mass-scale production are comparatively higher than those of melt-blown nonwoven masks (Gomollón-Bel 2020). Moreover, a low yield in the synthesis of nanoparticles and fibers, and limited production capacity are a few major challenges scientists should overcome. For example, "4C Air" a nanomaterial production company in the USA (ideal for filtering applications) has mentioned that they have a small production capacity of 2 million masks per month whereas "Bioinicia", another nanomask company, hopes to produce 11 million masks per week (Gomollón-Bel 2020). However, it is important to have a proper balance in the production of nanoparticle-assisted face masks and other masks available in the market to avoid creating supply shortages. Interestingly, emerging economies such as India, China and South Africa provide better opportunities for face mask manufacturers to commercialize their products (Gomollón-Bel 2020). It is very much important to support research and development for the production of face masks since those are widely used in both health care and industrial sectors even without a pandemic situation in the world.

5. Future Perspectives

- To overcome the shortage of face masks and materials in the market, scientists and medical professionals have encouraged the utilization of reusable face masks. As a result, many different types of reusable face masks have been introduced into the market. Therefore, a proper system should be combined to categorize reusable face masks available in the market based on their efficiencies and number of washing cycles.

- According to the estimations, the high demand created for face masks due to the pandemic situation will be slowly reduced by 2026. Therefore, the face mask production market should expand according to a plan to avoid the supply over the demand in a few years. For instance, establishing new face mask factories should be done with a proper plan to regulate the loss of initial production cost due to the future demand decrease.

- Furthermore, the quality of modified face masks should be assured with appropriate clinical trials before releasing to the market. Face masks that are currently available in the market should also include the procedure to ensure the quality and safety of the product for use by customers.

- Moreover, since nanoparticles are ultra-small and easily breathable, an in-depth study should be carried out to assess the possibility of breathing nanoparticles bound in face masks membranes and potential health issues that can be occurred.

- Polymer-based compositions have been used in most face masks membranes. Therefore, the environmental impact that can occur in the future with disposed face masks should be assessed. Used face masks disposed into the environment would cause plastic pollution in soil and aquatic environments. Furthermore, this would lead to the generation of microplastics as a result of the degradation of polymers. Plastic pollution and related ecotoxicity is a serious and irreversible drawback in the utilization of disposable face masks.

References

Abbasinia, M., S. Karimie, M. Haghighat and I. Mohammadfam. 2018. Application of nanomaterials in personal respiratory protection equipment: a literature review. Safety 4(4): 47.

Akduman, C. and E.P.A. Kumbasar. 2018. Nanofibers in face masks and respirators to provide better protection. IOP Conf. Ser Mater Sci. Eng. 460: 1–5

Aziz, N., M. Faraz, M.A. Sherwani, T. Fatma and R. Prasad. 2019. Illuminating the anticancerous efficacy of a new fungal chassis for silver nanoparticle synthesis. Front Chem. 7: 65.

Barhate, R.S., S. Sundarrajan, D. Pliszka and S. Ramakrishna. 2008. Fine chemical processing: The potential of nanofibres in filtration. Filtr. Sep. 45(4): 32–35.

Biswas, J.K., M. Rai, A.P. Ingle, M. Mondal and S. Biswas. 2018. pp. 19–36. *In*: Rai, M. and J.K. Boswas (ed.). Nanomaterials: Ecotoxicity, Safety, and Public Perception. Springer Nature, Switzerland.

Bouwmeester, H., S. Dekkers, M.Y. Noordam, W.I. Hagens, A.S. Bulder, C. de Heer et al. 2009. Review of health safety aspects of nanotechnologies in food production. Regul. Toxicol. Pharmacol. 53(1): 52–62.

Bright, K.R., E.E. Sicairos-Ruelas, P.M. Gundy and C.P. Gerba. 2009. Assessment of the antiviral properties of zeolites containing metal ions. Food Environ. Virol. 1(1): 37–41.

Brown, L., J.M. Wolf, R. Prados-Rosales and A. Casadevall. 2015. Through the wall: extracellular vesicles in Gram-positive bacteria, mycobacteria and fungi. Nat. Rev. Microbiol. 13(10): 620–630.

Chandrashekar, M.R., K.C. Rathish and C.N. Nagesha. 1997. Reservoirs of nosocomial pathogens in neonatal intensive care unit. J. Indian Med. Assoc. 95(3): 72–74, 77.

Chatterjee, S., A. Bandyopadhyay and K. Sarkar. 2011. Effect of iron oxide and gold nanoparticles on bacterial growth leading towards biological application. J. Nanobiotechnol. 9(1): 34.

Chen, L. and J. Liang. 2020. An overview of functional nanoparticles as novel emerging antiviral therapeutic agents. Mat. Sci. Eng. C 112: 110924.

Chua, M.H., W. Cheng, S.S. Goh, J. Kong, B. Li, J.Y.C. Lim et al. 2020. Face Masks in the New COVID-19 Normal: Materials, Testing, and Perspectives. Research, 7286735.

Chuanfang, Y. 2012. Aerosol filtration application using fibrous media—an industrial perspective. Chin. J. Chem. Eng. 20(1): 1–9.

Courtney, C.M., S.M. Goodman, J.A. McDaniel, N.E. Madinger, A. Chatterjee and P. Nagpal. 2016. Photoexcited quantum dots for killing multidrug-resistant bacteria. Nat. Mat. 15(5): 529–534.

Docter, D., D. Westmeier, M. Markiewicz, S. Stolte, S.K. Knauer and R.H. Stauber. 2015. The nanoparticle biomolecule corona: lessons learned—challenge accepted? Chem. Soc. Rev. 44(17): 6094–6121.

Dong, X., W. Liang, M.J. Meziani, Y.-P. Sun and L. Yang. 2020. Carbon dots as potent antimicrobial agents. Theranostics 10(2): 671.

Du, T., J. Liang, N. Dong, J. Lu, Y. Fu, L. Fang et al. 2018. Glutathione-capped Ag_2S nanoclusters inhibit coronavirus proliferation through blockage of viral RNA synthesis and budding. ACS Appl. Mater Interfaces 10(5): 4369–4378.

Dumas, E.-M., V. Ozenne, R.E. Mielke and J.L. Nadeau. 2009. Toxicity of CdTe quantum dots in bacterial strains. IEEE Transac Nanobiosci. 8(1): 58–64.

Espinosa-Cristóbal, L.F., G.A. Martínez-Castañón, R.E. Martínez-Martínez, J.P. Loyola-Rodriguez, N. Patino-Marin, J.F. Reyes-Macias et al. 2009. Antibacterial effect of silver nanoparticles against Streptococcus mutans. Mater Let. 63(29): 2603–2606.

Feliu, N., D. Docter, M. Heine, P. Del Pino, S. Ashraf, J. Kolosnjaj-Tabi et al. 2016. *In vivo* degeneration and the fate of inorganic nanoparticles. Chem. Soc. Rev. 45(9): 2440–2457.

Gajjar, P., B. Pettee, D.W. Britt, W. Huang, W.P. Johnson and A.J. Anderson. 2009. Antimicrobial activities of commercial nanoparticles against an environmental soil microbe, *Pseudomonas putida* KT2440. J. Biol. Eng. 3(1): 1–13.

Garcia-Garcia, E., K. Andrieux, S. Gil, H.R. Kim, T. Le Doan, D. Desmaële et al. 2005. A methodology to study intracellular distribution of nanoparticles in brain endothelial cells. Int. J. Pharm. 298(2): 310–314.

Ghaffari, H., A. Tavakoli, A. Moradi, A. Tabarraei, F. Bokharaei-Salim, M. Zahmatkeshan et al. 2019. Inhibition of H1N1 influenza virus infection by zinc oxide nanoparticles: another emerging application of nanomedicine. J. Biomed. Sci. 26(1): 1–10.

Gomollón-Bel, F. 2020. Nanofibers put a new spin on COVID-19 masks. Chemical and Engineering News.

Gupta, A., R.F. Landis and V.M. Rotello. 2016. Nanoparticle-based antimicrobials: surface functionality is critical. F1000Res. 5: 364.

Hajipour, M.J., K.M. Fromm, A.A. Ashkarran, D.J. de Aberasturi, I.R. de Larramendi, T. Rojo et al. 2012. Antibacterial properties of nanoparticles. Trends Biotechnol. 30(10): 499–511.

Hasan, J., Y. Xu, T. Yarlagadda, M. Schuetz, K. Spann and P.K.D.W. Yarlagadda. 2020. Antiviral and antibacterial nanostructured surfaces with excellent mechanical properties for hospital applications. ACS Biomat. Sci. Eng. 6(6): 3608–3618.

Hashmi, M., S. Ullah and I.S. Kim. 2019. Copper oxide (CuO) loaded polyacrylonitrile (PAN) nanofiber membranes for antimicrobial breath mask applications. Curr. Res. Biotechno. 1: 1–10.

Hiragond, C.B., A.S. Kshirsagar, V.V. Dhapte, T. Khanna, P. Joshi and P.V. More. 2018. Enhanced anti-microbial response of commercial face mask using colloidal silver nanoparticles. Vacuum 156: 475–482.

Horváth, E., L. Rossi, C. Mercier, C. Lehmann, A. Sienkiewicz and L. Forró. 2020. Photocatalytic nanowires-based air filter: towards reusable protective masks. Adv. Func. Mat. 30(40): 2004615.

Ingle, A.P., S. Wagh, J. Biswas, M. Mondal, C.M. Feitosa and M. Rai. 2020. Phyto-fabrication of different nanoparticles and evaluation of their antibacterial and anti-biofilm efficacy. Curr. Nanosci. 16(6): 1002–1015.

Ingle, P.U., J.K. Biswas, M. Mondal, M.K. Rai, P.S. Kumar and A.K. Gade. 2021. Assessment of *in vitro* antimicrobial efficacy of biologically synthesized metal nanoparticles against pathogenic bacteria. Chemosphere 132676.

Kang, S., M. Pinault, L.D. Pfefferle and M. Elimelech. 2007. Single-walled carbon nanotubes exhibit strong antimicrobial activity. Langmuir 23(17): 8670–8673.

Kang, S., M.S. Mauter and M. Elimelech. 2009. Microbial cytotoxicity of carbon-based nanomaterials: implications for river water and wastewater effluent. Env. Sci. Technol. 43(7): 2648–2653.

Khan, A., A.M. El-Toni, S. Alrokayan, M. Alsalhi, M. Alhoshan and A.S. Aldwayyan. 2011. Microwave-assisted synthesis of silver nanoparticles using poly-N-isopropylacrylamide/acrylic acid microgel particles. Colloids Surf, A Physicochem Eng. Asp. 377(1): 356–360.

Kharaghani, D., M.Q. Khan, A. Shahzad, Y. Inoue, T. Yamamoto, S. Rozet et al. 2018. Preparation and *in-vitro* assessment of hierarchal organized antibacterial breath mask based on polyacrylonitrile/silver (PAN/AgNPs) Nanofiber Nanomat. 8(7): 461.

Kumar, A., A.K. Pandey, S.S. Singh, R. Shanker and A. Dhawan. 2011. Cellular uptake and mutagenic potential of metal oxide nanoparticles in bacterial cells. Chemosphere 83(8): 1124–1132.

Kumar, S., M. Karmacharya, S.R. Joshi, O. Gulenko, J. Park, G.-H. Kim et al. 2021. Photoactive antiviral face mask with self-sterilization and reusability. Nano Let. 21(1): 337–343.

Lan, M., J. Wu, W. Liu, W. Zhang, J. Ge, H. Zhang et al. 2012. Copolythiophene-derived colorimetric and fluorometric sensor for visually supersensitive determination of lipopolysaccharide. J. Am. Chem. Soc. 134(15): 6685–6694.

Lee, L.Y.-k., I.C.-w. Chan, O.P.-m. Wong, Y.H.-y. Ng, C.K.-y. Ng, M.H.-w. Chan et al. 2021. Reuse of face masks among adults in Hong Kong during the COVID-19 pandemic. BMC Public Health 21(1): 1267.

Leichman, A.K. 2020. Israeli Antimicrobial Washable Facemasks Enter US Market, Israel21c.

Li, X. and Y. Gong. 2015. Design of polymeric nanofiber gauze mask to prevent inhaling PM2.5 particles from haze pollution. J. Chem. 2015: 460392.

Li, Y., P. Leung, L. Yao, Q.W. Song and E. Newton. 2006. Antimicrobial effect of surgical masks coated with nanoparticles. J. Hosp. Infect. 62(1): 58–63.

Li, Z., J.C. Barnes, A. Bosoy, J.F. Stoddart and J.I. Zink. 2012. Mesoporous silica nanoparticles in biomedical applications. Chem. Soc. Rev. 41(7): 2590–2605.

Marambio-Jones, C. and E.M.V. Hoek. 2010. A review of the antibacterial effects of silver nanomaterials and potential implications for human health and the environment. J. Nanopart. Res. 12(5): 1531–1551.

Markets, M. 2020. Face Mask Market by Nature (Disposable, Reusable), Material Type (PP, PU, Polyester, Cotton), Type (Surgical, Respirator), End-Use (Hospitals & Clinics, Industrial & Institutional, Personal/Individual Protection), and Region - Global Forecast to 2026.

Matuschek, C., F. Moll, H. Fangerau, J.C. Fischer, K. Zänker, M. van Griensven et al. 2020. The history and value of face masks. Eur. J. Med. Res. 25(1): 23–23.

Mazurkova, N.A., Y.E. Spitsyna, N.V. Shikina, Z.R. Ismagilov, S.N. Zagrebel'Nyi and E.I. Ryabchikova. 2010. Interaction of titanium dioxide nanoparticles with influenza virus. Nanotechnologies Rus 5(5): 417–420.

Mollasalehi, H. and R. Yazdanparast. 2013. An improved non-crosslinking gold nanoprobe-NASBA based on 16S rRNA for rapid discriminative bio-sensing of major salmonellosis pathogens. Biosens Bioelectron 47: 231–236.

Morones, J.R., J.L. Elechiguerra, A. Camacho, K. Holt, J.B. Kouri, J.T. Ramírez et al. 2005. The bactericidal effect of silver nanoparticles. Nanotechnol. 16(10): 2346.

O'Kelly, E., A. Arora, S. Pirog, J. Ward and P.J. Clarkson. 2021. Comparing the fit of N95, KN95, surgical, and cloth face masks and assessing the accuracy of fit checking. PLoS One 16(1): e0245688.

Office, P. 2020. Recyclable Nano-Fiber Filtered Face Masks a Boon for Supply Fiasco, Korea Advanced Institute of Science and Technology.

Oh, N. and J.-H. Park. 2014. Endocytosis and exocytosis of nanoparticles in mammalian cells. Int. J. Nanomedicine 9(1): 51.

Pakrashi, S., S. Dalai, D. Sabat, S. Singh, N. Chandrasekaran and A. Mukherjee. 2011. Cytotoxicity of Al_2O_3 nanoparticles at low exposure levels to a freshwater bacterial isolate. Chem. Res. Toxicol. 24(11): 1899–1904.

Panáček, A., L. Kvítek, M. Smékalová, R. Večeřová, M. Kolář, M. Röderová et al. 2018. Bacterial resistance to silver nanoparticles and how to overcome it. Nat. Nanotechnol. 13(1): 65–71.

Park, S., S. Lee, B. Kim, S. Lee, J. Lee, S. Sim et al. 2012. Toxic effects of titanium dioxide nanoparticles on microbial activity and metabolic flux. Biotechnol Bioprocess Eng. 17(2): 276–282.

Pelletier, D.A., A.K. Suresh, G.A. Holton C.K. McKeown, W. Wang, B. Gu et al. 2010. Effects of engineered cerium oxide nanoparticles on bacterial growth and viability. Appl. Environ. Microbiol. 76(24): 7981–7989.

Peters, A., B. Veronesi, L. Calderón-Garcidueñas, P. Gehr, L.C. Chen, M. Geiser et al. 2006. Translocation and potential neurological effects of fine and ultrafine particles a critical update. Part Fibre Toxicol. 3(1): 1–13.

Piret, J. and G. Boivin. 2021. Pandemics throughout history. Front Microbiol. 11: 631736–631736.

Premanathan, M., K. Karthikeyan, K. Jeyasubramanian and G. Manivannan. 2011. Selective toxicity of ZnO nanoparticles toward Gram-positive bacteria and cancer cells by apoptosis through lipid peroxidation. Nanomed Nanotechnol. Biol. Med. 7(2): 184–192.

Purchas, D. and K. Sutherland. 2002. Handbook of Filter Media. Elsevier.

Radniecki, T.S., D.P. Stankus, A. Neigh, J.A. Nason and L. Semprini. 2011. Influence of liberated silver from silver nanoparticles on nitrification inhibition of Nitrosomonas europaea. Chemosphere 85(1): 43–49.

Rai, M. and J.K. Biswas. 2018. Nanomaterials: Ecotoxicity, Safety, and Public Perception. Springer.

Ray, P.C., H. Yu and P.P. Fu. 2009. Toxicity and environmental risks of nanomaterials: challenges and future needs. J. Environ. Sci. Health C 27(1): 1–35.

Riding, M.J., F.L. Martin, J. Trevisan, V. Llabjani, I.I. Patel, K.C. Jones et al. 2012. Concentration-dependent effects of carbon nanoparticles in gram-negative bacteria determined by infrared spectroscopy with multivariate analysis. Environ. Pollut. 163: 226–234.

Risom, L., P. Møller and S. Loft. 2005. Oxidative stress-induced DNA damage by particulate air pollution. Mutat. Res. 592(1-2): 119–137.

Rubbo, S.D. and L.R. Abbott. 1968. Filtration efficiency of surgical masks: a new method of evaluation. ANZ J. Surg. 38(1): 80–83.

Sadiq, I.M., B. Chowdhury, N. Chandrasekaran and A. Mukherjee. 2009. Antimicrobial sensitivity of *Escherichia coli* to alumina nanoparticles. Nanomed. Nanotechnol. Biol. Med. 5(3): 282–286.

Sawosz, E., A. Chwalibog, J. Szeliga, F. Sawosz, M. Grodzik, M. Rupiewicz et al. 2010. Visualization of gold and platinum nanoparticles interacting with *Salmonella enteritidis* and *Listeria monocytogenes*. Int. J. Nanomedicine 5: 631–637.

Sayes, C.M., J.D. Fortner, W. Guo, D. Lyon, A.M. Boyd, K.D. Ausman et al. 2004. The differential cytotoxicity of water-soluble fullerenes. Nano Let. 4(10): 1881–1887.

Shahverdi, A.R., A. Fakhimi, H.R. Shahverdi and S. Minaian. 2007. Synthesis and effect of silver nanoparticles on the antibacterial activity of different antibiotics against *Staphylococcus aureus* and *Escherichia coli*. Nanomed Nanotechnol. Biol. Med. 3(2): 168–171.

Shimizu, M., H. Tainaka, T. Oba, K. Mizuo, M. Umezawa and K. Takeda. 2009. Maternal exposure to nanoparticulate titanium dioxide during the prenatal period alters gene expression related to brain development in the mouse. Part Fibre Toxicol. 6(1): 1–8.

Simon-Deckers, A., S. Loo, M. Mayne-L'hermite, N. Herlin-Boime, N. Menguy, C. Reynaud et al. 2009. Size-, composition- and shape-dependent toxicological impact of metal oxide nanoparticles and carbon nanotubes toward bacteria. Env. Sci. Technol. 43(21): 8423–8429.

Singh, L., H.G. Kruger, G.E.M. Maguire, T. Govender and R. Parboosing. 2017. The role of nanotechnology in the treatment of viral infections. Ther Adv. Infect. Dis. 4(4): 105–131.

Sinha, R., R. Karan, A. Sinha and S.K. Khare. 2011. Interaction and nanotoxic effect of ZnO and Ag nanoparticles on mesophilic and halophilic bacterial cells. Biores. Technol. 102(2): 1516–1520.

Sinha, R. and S.K. Khare. 2013. Molecular basis of nanotoxicity and interaction of microbial cells with nanoparticles. Curr. Biotechnol. 2(1): 64–72.

Skalny, A.V., L. Rink, O.P. Ajsuvakova, M. Aschner, V.A. Gritsenko, S.I. Alekseenko et al. 2020. Zinc and respiratory tract infections: Perspectives for COVID-19. Int. J. Mol. Med. 46(1): 17–26.

Skaria, S.D. and G.C. Smaldone. 2014. Respiratory source control using surgical masks with nanofiber media. Ann. Occup. Hyg. 58(6): 771–781.

Smith, M. and J.C. Smith. 2020. Repurposing therapeutics for COVID-19: Supercomputer-based docking to the SARS-CoV-2 viral spike protein and viral spike protein-human ACE2 interface. ChemRxiv. Preprint.

Sondi, I. and B. Salopek-Sondi. 2004 Silver nanoparticles as antimicrobial agent: a case study on *E. coli* as a model for Gram-negative bacteria. J. Colloid Interface Sci. 275(1): 177–182.

Stauber, R.H., S. Siemer, S. Becker, G.-B. Ding, S. Strieth and S.K. Knauer. 2018. Small meets smaller: effects of nanomaterials on microbial biology, pathology, and ecology. ACS Nano 12(7): 6351–6359.

Stoimenov, P.K., R.L. Klinger, G.L. Marchin and K.J. Klabunde. 2002. Metal oxide nanoparticles as bactericidal agents. Langmuir 18(17): 6679–6686.

Thill, A., O. Zeyons, O. Spalla, F. Chauvat, J. Rose, M. Auffan et al. 2006. Cytotoxicity of CeO_2 nanoparticles for *Escherichia coli*. Physico-chemical insight of the cytotoxicity mechanism. Env. Sci. Technol. 40(19): 6151–6156.

Toomey, E.C., Y. Conway, C. Burton, S. Smith, M. Smalle, X.-H.S. Chan et al. 2021. Extended use or reuse of single-use surgical masks and filtering face-piece respirators during the coronavirus disease 2019 (COVID-19) pandemic: A rapid systematic review. Infect Control Hosp E 42(1): 75–83.

Ullah, S., A. Ullah, J. Lee, Y. Jeong, M. Hashmi, C. Zhu et al. 2020. Reusability comparison of melt-blown vs. Nanofiber face mask filters for use in the coronavirus pandemic. ACS Appl. Nano Mater 3: 7231–7241.

UN. 2022. United Nations Medical Directors'. Risk Mitigation Plan for COVID-19. Recommendations for UN Personnel. Untited Nations. March 2022

Wang, C., L. Wang, Y. Wang, Y. Liang and J. Zhang. 2012. Toxicity effects of four typical nanomaterials on the growth of *Escherichia coli*, *Bacillus subtilis* and *Agrobacterium tumefaciens*. Environ. Earth Sci. 65(6): 1643–1649.

Wang, H., Y. Cai, C. Wang, C. Xu, J. Fang and Y. Yang. 2020. Seeded growth of ZnO nanowires in dye-containing solution: the submerged plant analogy and its application in photodegradation of dye pollutants. Cryst. Eng. Comm. 22(24): 4154–4161.

Wang, N., Y. Yang, S.S. Al-Deyab, M. El-Newehy, J. Yu and B. Ding. 2015. Ultra-light 3D nanofibre-nets binary structured nylon 6–polyacrylonitrile membranes for efficient filtration of fine particulate matter. J. Mat. Chem. A 3(47): 23946–23954.

Westmeier, D., G. Posselt, A. Hahlbrock, S. Bartfeld, C. Vallet, C. Abfalter et al. 2018. Nanoparticle binding attenuates the pathobiology of gastric cancer-associated *Helicobacter pylori*. Nanoscale 10(3): 1453–1463.

Widdowson, N. 2020. New Mask Material can Remove Virus-Size Nanoparticles. Queensland University of Technology.

Wiehe, A., J.M. O'Brien and M.O. Senge. 2019. Trends and targets in antiviral phototherapy. Photochem. Photobiol Sci. 18(11): 2565–2612.

Wu, J., A. Zawistowski, M. Ehrmann, T. Yi and C. Schmuck. 2011. Peptide functionalized polydiacetylene liposomes act as a fluorescent turn-on sensor for bacterial lipopolysaccharide. J. Am. Chem. Soc. 133(25): 9720–9723.

Xia, T., M. Kovochich, J. Brant, M. Hotze, J. Sempf, T. Oberley et al. 2006. Comparison of the abilities of ambient and manufactured nanoparticles to induce cellular toxicity according to an oxidative stress paradigm. Nano Letters 6(8): 1794–1807.

Xiong, S.-W., P.-g. Fu, Q. Zou, L-y. Chen, M.-y. Jiang, P. Zhang et al. 2021. Heat conduction and antibacterial hexagonal boron nitride/polypropylene nanocomposite fibrous membranes for face masks with long-time wearing performance. ACS Appl. Mater Interfaces 13(1): 196–206.

Zhang, C., M. Zhang, Y. Li and D. Shuai. 2019. Visible-light-driven photocatalytic disinfection of human adenovirus by a novel heterostructure of oxygen-doped graphitic carbon nitride and hydrothermal carbonation carbon. Appl. Catal. B 248: 11–21.

Zhang, W., J. Hughes and Y. Chen. 2012. Impacts of hematite nanoparticle exposure on biomechanical, adhesive, and surface electrical properties of *Escherichia coli* cells. Appl. Environ. Microbiol. 78(11): 3905–3915.

Zhao, J., B. Deng, M. Lv, J. Li, Y. Zhang, H. Jiang et al. 2013. Graphene oxide-based antibacterial cotton fabrics. Adv. Healthc Mater 2(9): 1259–1266.

Zhu, N., D. Zhang, W. Wang, X. Li, B. Yang, J. Song et al. 2020. A novel coronavirus from patients with pneumonia in China, 2019. N Engl. J. Med. 382: 727–733

Ziem, B., W. Azab, M.F. Gholami, J.P. Rabe, N. Osterrieder and R. Haag. 2017. Size-dependent inhibition of herpesvirus cellular entry by polyvalent nanoarchitectures. Nanoscale 9(11): 3774–3783.

Index